Probiotics *in* Food Safety *and* Human Health

T0187542

Probiotics *in* Food Safety *and* Human Health

Edited by

Ipek Goktepe
Vijay K. Juneja
Mohamed Ahmedna

CRC Press
Taylor & Francis Group
Boca Raton London New York

CRC Press is an imprint of the
Taylor & Francis Group, an **informa** business

A TAYLOR & FRANCIS BOOK

CRC Press
Taylor & Francis Group
6000 Broken Sound Parkway NW, Suite 300
Boca Raton, FL 33487-2742

First issued in paperback 2019

ISBN-13: 978-1-57444-514-5 (hbk)
ISBN-13: 978-0-367-39199-7 (pbk)
Library of Congress Card Number 2005046938

Library of Congress Cataloging-in-Publication Data

Goktepe, Ipek.
    Probiotics in food safety and human health / Ipek Goktepe, Vijay K. Juneja and Mohamed Ahmedna.
       p. cm.
    Includes bibliographical references and index.
    ISBN 1-57444-514-6 (alk. paper)
    1. Intestines--Microbiology. 2. Food--Microbiology. 3. Microorganisms--Therapeutic use. 4. Functional foods. I. Juneja, Vijay K., 1956- II. Title.

QR171.I6G645 2005
616.3'301--dc22   2005046938

Visit the Taylor & Francis Web site at
http://www.taylorandfrancis.com

and the CRC Press Web site at
http://www.crcpress.com

PREFACE

Our understanding of the functions of intestinal microflora and the use of probiotic microorganisms, a novel concept, to improve human health has made significant strides. Further, as our knowledge of gastrointestinal diseases increases, the use of probiotics offers an innovative approach for new food product development in functional foods for specific diseases. Likewise, the need for better control of microbial contamination in foods has been paramount in recent years. Foods previously thought not to be involved in foodborne illness, or believed to be infrequent sources of foodborne illness, have been associated with outbreaks, or sporadic episodes of illness, which sometimes have been fatal. Our understanding of foodborne pathogens has dramatically increased at an unprecedented rate in the last two decades, and so have the types of microorganisms, previously unknown, or not known to be causes of foodborne illness, which have recently been linked with documented outbreaks of illness. This makes it necessary for scientists, industry, and regulators to reconsider the traditional approach to food preservation for pathogen control in order to enhance food safety.

The variety of medical approaches to tackle gastrointestinal diseases and the challenge in ensuring microbiological safety of our food supply emphasize the need for a comprehensive book on probiotics. These observations, and our involvement through the years in research addressing microbiological safety of foods, led us to the conclusion that such a book is timely. The intent of this book is to present an in-depth characterization and diagnostics of probiotic strains, their mechanism of action in humans, food applications role in the development of new products that contribute to our well-being by guarding against gastrointestinal diseases and regulatory status. Every effort was pronounced to make this book as comprehensive and current as possible. This book is written at a level which presupposes a general background in medical and food microbiology needed to understand the basic mechanisms of probiotic action and the functionality of foods. The unique feature of this book is its thorough coverage of the various topics pertaining to probiotics, thereby creating new opportunities for food and nutrition scientists to develop functional foods with specific health benefits for different subpopulations. Such a thorough coverage of probiotics is expected to be equally useful to healthcare professionals. The material in each chapter is arranged in a logical, systematic, and concise sequence. Each chapter is written in an easy-to-follow format by a nationally and internationally renowned expert or team of experts in their particular fields. At the end of each chapter, citations from the scientific literature by experts in that particular field of probiotics are included.

It is necessary for the food industry and regulatory agencies to have personnel who are knowledgeable of the functional properties of foods aimed at improving human health. Currently, such information is available in a variety of diverse sources that are not always readily available. Accordingly, this book should be of special benefit to individuals who have little or no opportunity for additional classroom training and is a valuable text for those who directly or indirectly are involved in food product development or serve as nutritionists or doing research on food functionality and gastrointestinal disease control, which includes individuals in academic, industrial, and government institutions, including federal, state, and local agencies, food consultants, and food lobbyists.

We are grateful to all of our coauthors for their relentless effort in contributing the chapters. The credit for making this book a reality goes to them. We hope that this book will help in identifying potential new approaches to develop new microbiologically safe functional foods and significantly contribute to decreasing the incidence of bacterial foodborne illness outbreaks and gastrointestinal diseases.

THE EDITORS

IPEK GOKTEPE Ipek Goktepe is an assistant professor of food microbiology and toxicology and teaches principles of toxicology, environmental toxicology, and food biotechnology at North Carolina Agricultural and Technical State University in Greensboro, North Carolina, where she has been a faculty member since 2000. She received her B.S. in fisheries from the University of Istanbul, Turkey in 1993, an M.S. in food science in 1996, a Ph.D. in food science in 1999, and a second M.S. in environmental toxicology in 2000, all from Louisiana State University. She is a member of the Institute of Food Technologists (IFT), the American Society for Microbiology (ASM), the American Association of Family and Consumer Sciences (AAFCS), the Society of Environmental Toxicology and Chemistry (SETAC), Gamma Sigma Delta, Honor Society of Agriculture; The Honor Society of Phi Kappa Phi; Phi Tau Sigma, Honorary Society of Food Science; and Kappa Omicron Nu Honor Society of Family and Consumer Sciences. Dr. Goktepe is the author of over 70 papers and abstracts on various topics in environmental toxicology, food microbiology, and food science. Her research focuses on the use of probiotic bacteria in food safety and human health, development of new antimicrobial and anticarcinogenic compounds from select plants, and toxicological assessment of select pesticides on farm workers. Dr. Goktepe is also co-inventor of two patent applications entitled "A Fiber-Optic Biosensor for Rapid Detection of Pathogens in Poultry Products" and "Special Packaging Technique to Preserve Freshness of Exotic Mushrooms." Dr. Goktepe is the recipient of the 2004 Gamma Sigma Delta Honor Society Excellence in Research Award, the 2005 NCART Outstanding Young Investigator Award, and serves as an associate editor for the *Journal of the Science of Food and Agriculture*.

VIJAY K. JUNEJA Vijay K. Juneja is a supervisory microbiologist and lead scientist in the Microbial Food Safety Research Unit at the Eastern Regional Research Center (ERRC) of the Agricultural Research Service (ARS) branch of the United States Department of Agriculture (USDA) in Wyndmoor, Pennsylvania. Dr. Juneja received his B.V.Sc. and A.H. (D.V.M.) from G. B. Pant University of Agriculture and Technology, India, in 1978 and an M.S. (Animal Science) and Ph.D. degrees in Food Technology and Science from the University of Tennessee in 1988 and 1991, respectively. Soon after receiving his Ph.D., he was appointed as a microbiologist at the ERRC-USDA. Dr. Juneja has developed a nationally and internationally recognized research program on foodborne pathogens, with emphasis on microbiological safety of minimally processed foods and predictive microbiology. He is a coeditor of three books including one entitled, *Control of Foodborne Microorganisms* and serves on the editorial board of the *Journal of Food Protection, Foodborne Pathogens & Disease* and the *International Journal of Food Microbiology*. He also serves as an associate editor for the "Food Microbiology and Safety Section" of the *Journal of Food Science*. Dr. Juneja is recipient of several awards including the ARS, North Atlantic Area (NAA), Early Career Research Scientist of the Year, 1998; ARS–FSIS Cooperative Research Award, 1998; ARS, NAA, Senior Research Scientist of the year, 2002; USDA–ARS Certificate of Merit for Outstanding Performance, 1994, 1996, 1999, 2001, 2002, 2003. Currently, Dr. Juneja is a group leader for a multidisciplinary research project concerning the assurance of microbiological safety of processed foods. He develops strategies for research plans, oversees projects, reports results to user groups, and advises regulators (FDA, FSIS, etc.) on technical matters, i.e., research needs, and emerging issues. His research interests include intervention strategies for control of foodborne pathogens and predictive modeling. Dr. Juneja's research program has been highly productive, generating over 110 publications including 70 peer-reviewed articles, 3 books (coeditor), plus 105 abstracts of presentations at national and international scientific meetings, primarily in the area of food safety and predictive microbiology.

MOHAMED AHMEDNA Mohamed Ahmedna is an associate professor of food science in the Department of Human Environment and Family Sciences at North Carolina Agricultural and Technical State University, Greensboro, North Carolina. Dr. Ahmedna received a B.S. in Fisheries Engineering in 1989 from the Institut Agronormique et Veterinaire Hassan II, Morocco. Later he received an M.S. and a Ph.D. in Food Science in 1995 and 1998, respectively, and an M.S. in Applied Statistics in 1998, all from Louisiana State University, Baton Rouge. Prior to joining North Carolina A&T faculty, he worked as a senior research and development scientist at Technology International Inc., LaPlace, Louisiana. Dr. Ahmedna's research focuses on product development with emphasis on functional foods and the development of value-added products from underutilized agricultural by-products. He is the author of over 80 peer-reviewed scientific publications, abstracts, and proceedings, and has a patent on the development of a biosensor for rapid detection of pathogens in foods. In addition to teaching several graduate courses, including Research Methods in Food Science and Nutrition, Food Preservation, and Food Product Development, Dr. Ahmedna serves as the North Carolina A&T State University campus coordinator for the North Carolina Agromedicine Institute. He is the recipient of numerous awards including the 2001 Gamma Sigma Delta Award of Excellence in Research and the 2002 North Carolina A&T Outstanding Young Investigator Award. He also holds membership in several professional and honorary societies, including Gamma Sigma Delta honor society of Agriculture, Kappa Phi Kappa honor society, Phi Tau Sigma honor society of Food Science, Kappa Omicron Nu honor society for Family and Consumer Sciences, the Institute of Food Technologists, the American Statistical Association, the American

Peanut Research and Education Society, and the North Carolina Agromedicine Institute. Dr. Ahmedna served on several federal review panels, including the U.S. Department of Agriculture (USDA), National Research Initiative (NRI), Small Business Innovative Research (SBIR), U.S. Department of Health and Human Services (USDHHS), and as a technical reviewer for reputable journals such as the *Journal of Food Science*, *Bioresource Technology*, and the *Journal of Food Engineering*.

CONTRIBUTORS

Analía G. Abraham Centro de Investigación y Desarrollo en Criotecnología de Alimentos (CIDCA), CONICET–Facultad de Ciencias Exactas, Universidad Nacional de La Plata, La Plata, Argentina

Farid E. Ahmed Department of Radiation Oncology, Leo W. Jenkins Cancer Center, The Brody School of Medicine, East Carolina University, Greenville, North Carolina, USA

Mohamed Ahmedna Food Science & Nutrition, Department of Human Environment & Family Sciences, North Carolina Agricultural and Technical State University, Greensboro, North Carolina, USA

Robin C. Anderson U.S. Department of Agriculture, Agricultural Research Services, Southern Plains Agricultural Research Center, Food and Feed Safety Research Unit, College Station, Texas, USA

Graciela L. De Antoni Centro de Investigación y Desarrollo en Criotecnología de Alimentos (CIDCA), CONICET-Facultad de Ciencias Exactas, Universidad Nacional de La Plata and Comisión de Investigaciones Científicas de la Provincia de Buenos Aires (CIC PBA), La Plata, Argentina

Juha Apajalahti Danisco Innovation Kantvik, Kantvik, Finland

Todd R. Callaway U.S. Department of Agriculture, Agricultural Research Services, Southern Plains Agricultural Research Center, Food and Feed Safety Research Unit, College Station, Texas, USA

Maria L. Callegari Centro Ricerche Biotecnologiche, Cremona, Italy

Nigel Cook Central Science Laboratory, Sand Hutton, York, UK

Collette Desmond Teagasc, Dairy Products Research Centre, Fermoy, Co. Cork, Ireland

Francisco Diez-Gonzalez Department of Food Science and Nutrition, University of Minnesota, St. Paul, Minnesota, USA

Gary W. Elmer Medicinal Chemistry, University of Washington, Seattle, Washington, USA

Gerald F. Fitzgerald Department of Microbiology, University College, Cork, Ireland

Alberto C. Fossati Cátedra de Inmunología, Facultad de Ciencias Exactas, Universidad Nacional de La Plata, La Plata, Argentina

Udo Friedrich Danisco Innovation Niebüll, Niebüll, Germany

Yoichi Fukushima Nutrition Business Group, Nestlé Japan Ltd., Higashi-Shinagawa, Shinagawa-ku, Tokyo, Japan

Graciela L. Garrote Centro de Investigación y Desarrollo en Criotecnología de Alimentos (CIDCA), CONICET–Facultad de Ciencias Exactas, Universidad Nacional de La Plata, La Plata, Argentina

Kenneth J. Genovese U.S. Department of Agriculture, Agricultural Research Services, Southern Plains Agricultural Research Center, Food and Feed Safety Research Unit, College Station, Texas, USA

Ipek Goktepe Department of Human Environment and Family Sciences, Food and Nutritional Program, North Carolina Agricultural and Technical State University, Greensboro, North Carolina, USA

Herman Goossens Department of Medical Microbiology, University of Antwerp, Wilrijk, Belgium

Egon Bech Hansen Danisco Innovation Copenhagen, Copenhagen, Denmark

Roger B. Harvey U.S. Department of Agriculture, Agricultural Research Services, Southern Plains Agricultural Research Center, Food and Feed Safety Research Unit, College Station, Texas, USA

Martine Heyman INSERM EMI 0212, Faculté Necker, Vaugirard, Paris, France

Kazuhiro Hirayama Department of Veterinary Public Health, The University of Tokyo, Tokyo, Japan

Wilhelm H. Holzapfel Institute of Hygiene and Toxicology, BFEL, Karlsruhe, Germany

Geert Huys   Department of Biochemistry, Physiology and Microbiology, Ghent University, Ghent, Belgium

Hisakazu Iino   Department of Food Science & Nutrition, Faculty of Practical Science, Showa Women's University, Setagaya-ku, Tokyo, Japan

Marlene E. Janes   Assistant Professor, Department of Food Science, Louisiana State University Agricultural Center, Baton Rouge, Louisiana, USA

Anu Lähteenmäki   Functional Foods Forum, University of Turku, Turku, Finland

Liisa Lähteenmäki   VTT Biotechnology, Espoo, VTT, Finland

Liesbeth Masco   Laboratory of Microbial Ecology and Technology, Ghent University, Ghent, Belgium

Tiina Mattila-Sandholm   Valio Ltd., Helsinki, Espoo, Finland

Lynne V. McFarland   Health Services, University of Washington, Seattle, Washington, USA

Sandrine Ménard   INSERM EMI 0212, Faculté Necker, Vaugirard, Paris, France

Lorenzo Morelli   Instituto di Microbiologia, Piacenza, Italy

David J. Nisbet   U.S. Department of Agriculture, Agricultural Research Services, Southern Plains Agricultural Research Center, Food and Feed Safety Research Unit, College Station, Texas, USA

Daniel J. O'Sullivan   Department of Food Science and Nutrition, Center for Microbial and Plant Genomics, University of Minnesotta, St. Paul, Minnesota, USA

Arthur C. Ouwehand   Department of Biochemistry and Food Chemistry & Functional Foods Forum, University of Turku, Turku, Finland

Joseph Rafter   Department of Medical Nutrition, Karolinska Institutet, Novum, Huddinge, Sweden

Paul Ross   Teagasc, Dairy Products Research Centre, Fermoy, Co. Cork, Ireland

Maria Saarela   VTT Biotechnology, VTT, Espoo, Finland

Gerry P. Schamberger Department of Food Science and Nutrition, University of Minnesota, St. Paul, Minnesota, USA

Ralf-Christian Schlothauer Danisco Innovation Niebüll, Niebüll, Germany

Katja Schmid Danisco Innovation Niebüll, Niebüll, Germany

María Serradell Cátedra de Inmunología, Facultad de Ciencias Exactas, Universidad Nacional de La Plata, La Plata, Argentina

Catherine Stanton Teagasc, Dairy Products Research Centre, Fermoy, Co. Cork, Ireland

Christine Staudt Danisco Innovation Niebüll, Niebüll, Germany

Jean Swings Department of Biochemistry, Physiology and Microbiology, Ghent University, Ghent, Belgium

Robin Temmerman Laboratory of Microbial Ecology and Technology, Faculty of Bioscience Enginneering, Ghent University, Ghent, Belgium

Marc Vancanneyt Department of Biochemistry, Physiology and Microbiology, Ghent University, Ghent, Belgium

Vanessa Vankerckhoven Department of Medical Microbiology, University of Antwerp, Wilrijk, Belgium

ACKNOWLEDGMENTS

A text of this undertaking on the broad topic of probiotics would not be possible without the invaluable contributions of the authors who graciously wrote chapters in their fields of specialty. We especially appreciate the contributors' patience during the completion of this text. In addition, such an undertaking would have not been possible without the support provided by our employers, North Carolina Agricultural and Technical State University and the United States Department of Agriculture and Agricultural Research Services. We also thank Anita Lekhwani for initiating and coordinating the entire project. The contributions of the editorial and production staff at Marcel Dekker and CRC Press are gratefully acknowledged for their support, expertise, and hard work. Finally, we owe a thank you to our families for their support, love, and patience during the completion of this book.

CONTENTS

1

Introduction to Prebiotics and Probiotics

Wilhelm H. Holzapfel

CONTENTS

1.1 INTRODUCTION

1.1.1 History

During the second part of the 19th century, early scientific studies on microorganisms have also dealt with their interactions with the human host, albeit primarily from a negative perspective. However, as early as 1885, Escherich (1) described the microbiota and in 1886 early colonization (2) of the infant gastrointestinal tract (GIT) and suggested their benefit for digestion, whereas Döderlein (3) was probably the first scientist to suggest the beneficial association of vaginal bacteria by production of lactic acid from sugars, thereby preventing or inhibiting the growth of pathogenic bacteria. These findings and other information on the early stages of development toward biotherapeutic concepts and the utilization of functional bacteria are summarized in Table 1.1. Bacteria producing lactic acid as the major metabolic product were generally grouped as "lactic acid bacteria" (LAB) even in those early days, and their association with fermented milk products was also recognized. Recent research has underlined the importance of a vital and "healthy" microbial population of the GIT. Particularly, the beneficial association of LAB with the human host, suggested more than 100 years ago on the basis of gut ecological and taxonomic studies by Moro in 1900 (4), Beijerinck (5), and Cahn (6), has been confirmed and extended by increasing research efforts during the last three decades. Metschnikoff (7, 8) in his bestseller *The Prolongation of Life* was probably the first to advocate, or rather postulate, the health benefits of LAB associated with fermented milk products. He suggested the longevity of the Caucasians to be related to the high intake of fermented milk products. Although Metschnikoff viewed gut microbes as detrimental rather than beneficial to human health, he considered substitution of gut microbes by yogurt bacteria to be beneficial. He considered that lactic acid production, resulting from sugar fermentation by LAB, to be particularly beneficial. The bifidobacteria, another group producing lactic acid, phylogenetically distant but commonly accepted to form part of the LAB, were discovered in 1889 and described in the early 1900s by Tissier (9, 10) to be typically associated with the feces especially of breast-fed infants. When compared to formula-fed infants, a lower incidence of intestinal upsets was observed for infants receiving mother's milk. Thereby the assumption was made about the benefical association of bifidobacteria with the human GIT.

1.1.2   Definitions

The expression "probiotic" was probably first defined by Kollath in 1953 (11), when he suggested the term to denote all organic and inorganic food complexes as "probiotics," in contrast to harmful antibiotics, for the purpose of upgrading such food complexes as supplements. In his publication "Anti- und Probiotika," Vergio (12) compared the detrimental effects of antibiotics and other antimicrobial substances with favorable factors ("Probiotika") on the gut microbiology. Lilly and Stillwell (13) proposed probiotics to be "microorganisms promoting the growth of other microorganisms." Although numerous definitions have been proposed since then (see Table 1.1), none has been completely satisfactory because of the need for additional explanations, e.g., with regard to statements such as "beneficial balance," "normal population," or "stabilization of the gut flora." A consensus and somewhat generalized definition as suggested by

TABLE 1.1

Chronology (Arbitrary) and Development of the Concept of Biotherapy and Probiotics

Period	Time	Concept/Approach/Definition	Literature
"Empiric"	<1850	Fermented foods (yogurt) consumed for therapy against diarrhea	
Early developments of microbiology as science	1850–1890	1857: LAB discovered and lactic acid fermentation described by Pasteur	
		1878: Lister isolates LAB ("*Bacterium lactis*") in pure culture from fermented milk	
		Particular micro-organisms beneficial for GI tract	Escherich, 1885 (1)
		The microbiota of the neonate and breast-fed infant	Escherich, 1886 (2)
		Early bacterial colonization of the infant GI tract (by *E. coli*) and relationship to digestion	
		1889: *Bifidobacterium* discovered in feces of breast-fed infants	Tissier, 1900; 1905 (9; 10)
		1890: First "commercial" starter cultures for sour milk and cheese in Copenhagen and Kiel	
Microbiology as basis for scientific approaches	1890–1930	Positive association of lactic acid bacteria in the stabilization of the vagina	Döderlein, 1892 (3)
		Discovery of *Lactobacillus acidophilus*	Moro, 1900 (4)
		"Industrial" lactic acid bacteria	Beijerinck, 1901 (5)
		Rod-shaped bacteria (lactobacilli) of the infant feces	Cahn, 1901 (6)
		Longevity of the Caucasians related to the high intake of fermented milk products. Gut microbes more detrimental but substitution of gut microbes by yogurt bacteria beneficial	Metchnikoff (1907; 1908) (7; 8)
		Prophylactic substitution by non-pathogenic, "physiological" *E. coli* directly after birth	Nissle, 1916 (14)
		"Antagonistic" treatment of chronic intestinal inflammation	Nissle, 1918 (15)
		"Mutaflor" treatment of diarrhea and dysentery	Nissle, 1919 (16)
Development of concepts toward probiotics and biotherapeutics, and their functions	1930–1990	1936: Isolation and early biotherapeutic application of "*Lb. casei*" Shirota	Kollath, 1953 (11)
		1953: First suggestion and definition of the term "probiotic," denoting all organic and inorganic food complexes as probiotics in contrast to harmful antibiotics – for the purpose of upgrading as supplements	
		"Probiotic" first defined: Promotion of body functions and beneficial microorganisms by microbes and their metabolites	Vergio, 1954 (12)
		Prophylactic treatment of acute infections with "physiological" bacteria	Kolb, 1955 (17)

TABLE 1.1

Chronology (Arbitrary) and Development of the Concept of Biotherapy and Probiotics (continued)

Period	Time	Concept/Approach/Definition	Literature
		Characterization of typical lactobacilli and bifidobacteria from different regions of the human GI tract	Lerche and Reuter, 1962 (18); Reuter, 1963 (19); 1965 (20); 1969 (21)
		Probiotic defined as microbiologically produced substances ("factors") which promote growth of other organisms	Lilly and Stillwell, 1965 (13)
		Oral administration of beneficial lactobacilli ("*Lb. acidophilus* Shirota") influences intestinal population of infants	Shirota et al., 1966 (22)
		Role of LAB and their fermentation products in antitumor activity and modification of biological responses	Reddy et al., 1973 (23); Kato et al., 1981 (24); Yokokura et al., 1981 (25)
		Feed supplements for animals – defined as "organisms and substances that have a beneficial effect on the host animal by contributing to its intestinal microbial balance"	Parker, 1974 (26)
		Intestinal population of breast-fed and infants established and similar to those receiving formula-milk	Hoogkamp-Korstanje et al.,1979 (27)
		Modulation of immune response	Schwab, 1977 (28); Conge et al., 1980 (29)
		Definition by Fuller: "live microbial feed supplements which beneficially affect the host animal by improving its intestinal microbial balance"	Fuller, 1989 (30)
Probiotics toward functional strains and understanding of mechanisms	1990– present- day	Improved definition: "mono- or mixed cultures of live micro-organisms which, when applied to animal or man, beneficially affect the host by improving the properties of the indigenous microflora"	Havenaar et al., 1992 (31)
		Definition: "viable microorganisms (bacteria or yeasts) that exhibit a beneficial effect on the health of the host when they are ingested"	Salminen et al., 1998a (32)
		Definition: "living microorganisms, which upon ingestion in certain numbers, exert health benefits beyond inherent basic nutrition"	Guarner and Schaafsma, 1998 (33)
		Consensus definition: "Probiotics are defined, live microorganisms, which when reaching the intestines in sufficient numbers (e.g., administered via food), will exert positive effects"	BgVV, 1999 (34)

the Bundesinstitut für gesundheitlichen Verbraucherschutz und Veterinär-medizin (BgVV; now called BfR) states that probiotics are defined, live microorganisms, which when reaching the intestines in sufficient numbers (e.g., administered via food), will exert positive effects (34). The present-day concept refers to viable microorganisms that promote or support a beneficial balance of the autochthonous microbial population of the GIT. These microorganisms may not necessarily be constant inhabitants of the GIT, but their "...beneficial effect on the general and health status of man and animal" (26, 30) should be ascertained. This is also reflected in the suggestion of Havenaar et al. (31), defining probiotics as "...mono- or mixed cultures of live microorganisms which, when applied to animal or man, beneficially affect the host by improving the properties of the indig-enous microflora." Probiotics are best known by the average consumer with relation to food where they are defined by the EU Expert Group on Functional Foods in Europe (FUFOSE) as "viable preparations in foods or dietary supplements to improve the health of humans and animals" (35). Yet, particular pharmaceutical preparations containing viable microorgan-isms in capsules and which are being used for the restoration of the gas-trointestinal population, e.g., after or during antibiotic treatment, have also been known as "biotherapeutics" for many years.

1.1.3 Administration and Consumption of Probiotics

Viable strains of especially the *Lactobacillus acidophilus* "group" and *Bifido-bacterium bifidum* were introduced into dairy products in Germany during the late 1960s because of their expected adaptation to the intestine and the sensory benefits for producing mildly acidified yogurts. Such products first became known in Germany as mild yogurts or "bio-yogurts", while in the USA, acidophilus milk was better known (36, 37).

As is shown in Figure 1.1, probiotics are available and may be adminis-tered in different forms, comprising foods, mainly in a fermented state, and pharmaceutic products, mainly as capsules or in microencapsulated form. By definition, probiotic strains may even be undefined organisms from fermented foods, which survive the gut passage and may exert pos-itive effects in the GIT. If probiotic microorganisms constitute a defined part of a food, they are defined by FUFOSE as "live consituents of a food which exert positive effects on health" (32, 38). Probiotic foods comprise between 60 and 70% of the total functional food market. A continued increase is observed among the dairy-type probiotic foods, but even in the range of nondairy probiotic food products such as fermented meats and vegetable and fruit juices. Taking into account the wide range of potential (fermentable) substrates and the different conditions under which LAB strains may be challenged for "functional performance," it can be expected that developments toward new food-based probiotics will proceed further in the future.

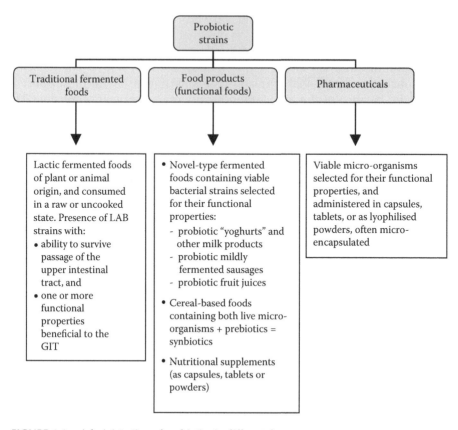

FIGURE 1.1 Administration of probiotics in different forms.

It is postulated that this positive effect is achieved when the proportion of lactobacilli and bifidobacteria in the intestinal population increases, either by increased intake of typical gut bacteria (e.g., as fermented foods or dehydrated preparations), or indirectly as a result of the stimulation of autochthonous gut bacteria belonging to these groups. The lactobacilli and bifidobacteria associated with the GIT are generally considered beneficial for such things as combating disturbances of the mucosa associated immune system and of the established gut population.

A particular feature of probiotic cultures is that they regulate the balance of the gut bacterial population, e.g., by competition for epithelium contact sites and nutrients and also by modulation of the pH value. Other features refer to the support of absorption of nutrients and the synthesis of vitamins such as riboflavin. Further stabilization of the gut microbiota is associated with the synthesis of nutritional physiologically important short-chain fatty acids (SCFAs) by which the gut mucosa is supported. In addition, probiotic cultures are also suggested to stimulate the immune system. These functional aspects are briefly discussed in Section 1.2.

1.2   GUT MICROBIAL ECOLOGY

1.2.1   The Gastrointestinal Tract as Ecosystem

The GIT with its diverse and concentrated microbial population is one of the key organs of the human body; it is in fact an ecosystem of highest complexity that mediates numerous interactions with the chemical (and nutritional) environment. The mucosal surface increase by circular folds, intestinal villi, and microvilli provides a large area for such interactions associated with digestion, adhesion to the mucosal wall, and colonization (39). Compared to about 2 m^2 of skin surface area of the average human body, the gastrointestinal system comprises an area of 150 to 200 m^2 (40), making available the necessary space for digestive interactions and for adhesion and colonization associated with the mucosal wall. The circular folds contribute to about a 3-fold increase, a 7- to 10-fold increase by the epithelium folding (intestinal villi), and a 15- to 40-fold increase by the microvilli in the enterocyte resorptive luminal membrane.

1.2.2   Microbiota of the Human Gastrointestinal Tract

Increasing microbial populations are found throughout the GIT (Table 1.2), ranging from varying numbers of food-associated bacteria in the esophagus, to 10^1 - to 10^3/ml (or g) in the stomach, 10^7/ml in the jejunum (comprising mainly lactobacilli, *Enterobacteriaceae* and streptococci), up to 10^9 CFU/g in the terminal ileum, and ca. 5×10^{11}/g in the distal colon. Many microbes isolated from the duodenum and jejunum are considered to be typical transients, especially considering rapid chymus flow; indigenous colonization is, however, more likely to occur in the lower parts of the ileum. The estimated total population of 10^{14} viable bacteria in the adult human GIT (41) represents about 10 times more than all body tissue cells. The microbial population therefore represents an immense metabolic potential that not only supports the digestion processes but is also interactive in detoxification and toxification processes and, most importantly, comprises the major part of the human immune system. *Bacteroides* and the Gram-positive, anaerobic genera *Eubacterium* and *Bifidobacterium* predominate in the densely populated large intestine. Other groups such as the clostridia, peptostreptococci, "streptococci," and lactobacilli also play an important role, e.g., in the maintenance of a stable gut mucosa and in the generation of SCFAs in a beneficial ratio. The role of the lactobacilli may be more important in the small intestines where they comprise a higher proportion of the total population. In healthy humans, lactobacilli are normally present in the oral cavity (10^3 to 10^4 CFU/g), the ileum (10^3 to 10^7 CFU/g), and colon (10^4 to 10^8 CFU/g) and are the dominant microorganism in the vagina (68-70).

TABLE 1.2

Estimated Numbers (Per ml or g of Intestinal Contents) and Suggested Role (Postulated Effects) of Major Microbial Population Groups in Different Segments of the Gastrointestinal Tract

Microbial Group	Stomach $10^1 – 10^3$ CFU/ml	Duodenum $10^1 – 10^4$ CFU/ml	Jejunum + Ileum $10^5 – 10^8$ CFU/g	Colon $10^9 – 5\times10^{11}$/g	Positive Effects	Negative Effects
Actinomyces spp.			$10^4 – 10^6$		?	
Bacteroides-Prevotella-Porphyromonas Group	Up to 10^2	ca. 10^3	$10^4 – 10^7$	$10^9 – 10^{11}$		+
Bifidobacterium spp.				$10^9 – 10^{10}$	+	
Clostridium spp.			$10^4 – 10^5$	$10^8 – 10^9$	(+)	+
Coprococcus cutactus				$10^7 – 10^8$		
Enterobacteriaceae	Up to 10^2	$10^2 – 10^4$	$10^3 – 10^6$	$10^5 – 10^7$	(+)	(+)
Enterococcus spp.			$10^2 – 10^4$	$10^3 – 10^6$		
Eubacterium spp.				$10^9 – 10^{11}$	+	
Fusobacterium spp.			$10^3 – 10^5$	$10^5 – 10^7$		
Lactobacillus spp.	$10^1 – 10^3$	$10^2 – 10^4$	$10^4 – 10^6$	$10^5 – 10^8$	+	
Megamonas hypermegas				$10^7 – 10^8$		
Megasphaera elsdenii				$10^7 – 10^8$		
Methanobacteria				up to 10^9	(+)	(+)
Peptostreptococcus spp.			$10^2 – 10^6$	$10^8 – 10^9$	(+)	(+)
Proteus spp.				$10^3 – 10^6$		
Pseudomonas spp.				$>10^3$		
Staphylococci				ca. 10^3		
Streptococcus spp.	$10^1 – 10^3$		$10^3 – 10^8$	up to 10^7		
Veillonella spp.			$10^3 – 10^7$	$10^5 – 10^8$		+
Yeasts				ca. 10^3		(+)

Source: Modified according to Holzapfel et al. (39); Tannock (42, 43); Sullivan et al. (44); Holzapfel (45).

Interestingly, *Streptococcus intermedius* and *Haemophilus parahaemolyticus* could not be detected in the jejunum of 18 patients with gastrointestinal diseases, as compared to healthy subjects, whereas lactobacilli were detected more commonly in diseased than healthy subjects (44).

In spite of increased research on gut microbial ecology, still only a relatively small number of the ca. 400 genera and species have been cultivated and studied with regard to their physiology, metabolic interactions, and taxonomy.

Considered as beneficial bacterial groups of major importance to the gut ecosystem, special attention was given to the Gram-positive genera *Lactobacillus, Bifidobacterium,* and, more recently, to *Eubacterium*. All Gram-positive bacteria cluster in two of the formerly recognized 17 eubacterial phyla that also coincide with their DNA base composition (71, 72). Practically all organisms used in probiotic foods or food supplements are representatives of the genera *Lactobacillus, Enterococcus* or *Bifidobacterium*. The genus *Bifidobacterium* shares some phenotypic features with the "typical" LAB, but traditionally and also for practical purposes they are still considered to form part of the LAB. The bifidobacteria exhibit a relatively high guanine plus cytosine (G+C) content of 55 to 67 mol% in the DNA and are phylogenetically distinct from the "true" LAB and form part of the so-called *Actinomycetes* branch. The "true" LAB form part of the so-called *Clostridium* branch which is characterized by a G+C content of <55 mol% in the DNA (70, 71). Based on the comparison of 16S rRNA sequences, *Carnobacterium, Enterococcus, Vagococcus, Aerococcus, Tetragenococcus,* and the newly described genus *Lactosphaera* are more closely related to each other than to any other LAB. *Lactococcus* and *Streptococcus* appear to be relatively closely related genera, whereas the genus *Lactobacillus* is phylogenetically diverse. Comparison of 16S rRNA sequencing data showed the genera *Lactobacillus* and *Pediococcus* to be phylogenetically intermixed as 5 species of *Pediococcus* cluster with 32 homo- and heterofermentative *Lactobacillus* species in the so-called *Lactobacillus casei-Pediococcus* group (73). The 16S rRNA sequence data of pediococci and lactobacilli clearly indicate that the taxa generated on the basis of phenotypic properties, such as cell morphology and fermentation type, do not correspond with the phylogenetic branching. Therefore, a number of species of LAB may have to be reclassified; this may have important consequences for commercial probiotic strains (37).

The early observations and hypotheses in the 20th century pointed toward the beneficial role of the LAB in food fermentations (46), the GIT (4, 6, 7), and the vagina (3) (*vide supra*; Table 1.1). Even so, studies on the types and numbers of LAB of the different regions of the human GIT were rare. Among the first comprehensive studies were those by Reuter and coworkers (18-21, 47). Thanks to their precise and well-documented observations, the three major groups of homofermentative lactobacilli, typical of the intestinal tract of the human host, were characterized in the 1960s and were confirmed by later taxonomic investigations, supported by improved sampling techniques and molecular biological methods comprising:

- The "*Lactobacillus acidophilus* group" involving strains that are presently recognized as *L. acidophilus*, *L. gasseri*, *L. crispatus*, and *L. johnsonii* (discussed in Section 1.3; see also Table 1.3)
- "*Lactobacillus salivarius*"
- The "*Lactobacillus casei* group," comprising strains of *L. paracasei* and *L. rhamnosus*

The heterofermentative lactobacilli were shown to comprise a major phenotypic group (later classified as *L. reuteri*) and, to some extent, also *L. fermentum* and *L. oris* (47).

Apart from the bifidobacteria, the LAB in the gut are mainly represented by the lactobacilli, but in contrast to their domination in the ileum, they only form a minor population in the colon. The lactobacilli, as major LAB representatives, in fact only make up about 1% of the total bacterial population of human feces but may be more numerous in the proximal colon (48). They do not appear to be detectable by conventional culture methods in the feces of all adults; yet, they seem to be consistently present in the colon, albeit in relatively low numbers. This may in part also result from the consumption of fermented food products (42, 49, 50). This has in fact been shown by Dal Bello et al. (51) using "alternative" incubation conditions (30°C, 2% O_2) and confirmed by polymerase chain reaction (PCR) – denaturing gradient gel electrophoresis (DGGE) analyses of resuspended bacterial biomass obtained from agar plates for revealing of the species composition. These workers in fact reported that food-associated LAB, such as *Lactobacillus sakei* and *Leuconostoc mesenteroides*, hitherto not described as intestinal inhabitants, were more easily detected with the alternative incubation condition (see also Table 1.4). Randomly picked colonies grown under the alternative condition showed *L. sakei* as one of the predominant food-associated LAB species, to reach counts of up to 10^6 CFU/g feces.

TABLE 1.3

Features of the Species of the So-Called "Acidophilus" Group of Lactobacilli

Species	Habitat[a]	mol% G+C in the DNA	"Biotypes" acc. to	DNA Homology groups acc. to	
			Lerche and Reuter (1962)	Lauer et al. (1980)	Johnson et al. (1980)
L. acidophilus	HSCP	32–37	I,/II	I a	A-1
L. amylovorus	S/C	40	IV (III)	I b	A-3
L. crispatus	H/P	35–38	III	I c	A-2
L. gallinarum	P	33–36	-	I d	A-4
L. gasseri	H/C	33–35	I	II a	B-1
L. johnsonii	H/S/P	32–38	I, II	II b	B-2

[a] H = humans; S = pigs; C = cattle; P = poultry.

Source: Modified according to Mitsuoka (70); Reuter (47); Holzapfel et al. (39).

TABLE 1.4

LAB Typically Associated with the Human Host

Lactobacilli	Other LAB
Intestinal Bacteria	
Lactobacillus acidophilus "group"	*Bifidobacterium adolescentis**
L. acidophilus senso strictu	*B. angulatum*
L. animalis	*B. bifidum*
*L. brevis**	*B. breve*
L. buchneri	*B. cantenulatum*
L. crispatus	*B. dentium**
L. curvatus	*B. infantis*
*L. deLrueckii **	*B. longum*
L. fermentum	*B. pseudocantenulatum*
*L. gasseri**	
L. johnsonii	*Enterococcus faecalis**
*L. paracasei**	*E. faecium**
*L. plantarum**	
*L. reuteri**	*Leuc. mesenteroides*
*L. rhamnosus**	
L. ruminis	*Pediococcus pentosaceus**
*L. salivarius**	
L. sakei	*Weissella confusa*
Vaginal Bacteria	
*Lactobacillus acidophilus**	*Bifidobacterium bifidum*
L. fermentum	*B. longum*
*L. casei**	*B. infantis*
*L. rhamnosus**	*B. breve*
L. cellobiosus	*B. catenulatum*
*L. plantarum**	*B. dentium*
*L. brevis**	
*L. delbrueckii **	
*L. salivarius**	
*L. jensenii**	
L. vaginalis	
*L. gasseri**	
L. crispatus	

[a] Also found in clinical samples.

Source: From References 42, 47, 50, 51.

Even so, detailed and scientifically well-founded studies on other "beneficial" groups of the human GIT were particularly rare until the last decade of the 20th century. This was mainly due to technical restrictions related to sampling techniques, and detection and cultivation methods. In contrast to the oxygen-tolerant lactobacilli, the study of anaerobic groups such as the bifidobacteria and eubacteria was made possible by the development of anaerobic techniques developed in the early 1970s (52, 53), which were further improved in combination with improved cultivation media. Compared to their domination in infants, bifidobacteria comprise only up to 3% of the total fecal bacteria of humans and up to 10% of the total culturable population (54-56). With increasing age, however, their numbers in feces are reported to decline in adults (43, 57). They comprise up to 91% of the total population in breast-fed babies and up to 75% in formula-fed infants (58). A reduced environment and special media are applied for the selective cultivation of bifidobacteria; yet, such media do not equally support the growth of all *Bifidobacterium* species present in human feces (59). Moreover, the identification of *Bifidobacterium* species by phenotypic characteristics is difficult and unreliable (43, 60). These culture-related and other factors limit the research data and their quality with regard to bifidobacterial and other gastrointestinal populations. Also, it can be expected that the population detected in feces may probably more correspond to that of the distal colon than the proximal region. It is in the distal colon where fermentable but nondigestible carbohydrates (so-called "prebiotics") may play an important role in stimulating the *Bifidobacterium* and *Eubacterium* populations, or perhaps particular species only. With the aid of genetic fingerprinting techniques, it could be shown that particular *Bifidobacterium* and *Lactobacillus* strains are unique to each individual (43). In addition, it was suggested that the composition of these populations remains relatively constant for some individuals and to fluctuate considerably for others (43, 56). Using fluorescent *in situ* hybridization (FISH) with group-specific 16S rRNA-targeted oligonucleotide probes, it was possible to detect variations in bifidobacterial populations in the feces of different age groups. The percentage of bifidobacteria in the feces ranged from 0 to 78.9%, depending on the age group, with large variations within each group (58, 61, 62). Moreover, DGGE banding patterns of human gut bacteria have been found to differ significantly from those of other mammals. Furthermore, 16S rDNA sequences showed three bacterial species, *Ruminococcus obeum*, *Eubacterium halii*, and *Fusobacterium prausnitzii* to be most probably ubiquitous to humans; these groups were therefore suggested to play an important role in the human GIT (63-66).

Matsuki et al. (67) investigated the population structure of the human fecal microorganisms by applying 16S rRNA-gene-targeted group-specific oligonucleotide primers for the *B. fragilis* group, *Bifidobacterium* spp., the *C. coccoides* group, and *Prevotella*, and thereby detected and identified 74% of the predominant bacteria in the feces of six healthy volunteers. The other isolates were identified by 16S ribosomal DNA sequence analysis and consisted of *Collinsella*, the *Clostridium leptum* subgroup, and isolates of other clusters. As shown in Table 1.1, major microbial groups of the human GIT vary in numbers and their distribution among the different regions of the gut. Recent observations suggest the *Bacteroides-Prevotella-Porphyromonas* group, with numbers of up to 10^{11}/g, to dominate the colon population together with the Eubacterium group, and to reach 10- to 100-fold higher numbers than the bifidobacteria (43). As for the *Lactobacillus* species, the *Bacteroides-Prevo-*

tella-Porphyromonas group appears to be present in all regions of the GIT, indicating that not all representatives are strictly anaerobic (Table 1.2).

1.2.3 Role and Functions of the Microorganisms of the Gut

A healthy intestinal epithelium, in association with an established and stable intestinal microbial population, presents a vital barrier against the invasion or uptake of pathogenic microorganisms, antigens, and harmful compounds from the gut lumen, while the intestinal mucosa also efficiently assimilates antigens (36). Specific immune responses are evoked by the specialized antigen transport mechanisms in the villus epithelium and Peyer's patches (74). The positive role of gut microorganisms in human health was largely overlooked for a long time, and the main focus was placed on enteric pathogens and factors leading to gastrointestinal disorders or "dysbiosis" (36). A stable barrier, typical of healthy individuals, ensures host protection and serves as support for normal intestinal function and immunological resistance. The gut-associated lymphoid tissues (GALT) are considered to be the largest "immune organ" in the human body, and its "barriers" serve for intrinsic protection against infective agents. Around 80% (10^{10}) of all immunoglobulin-producing cells are found in the small bowel (75), while the gut microbial population is essential for mucosal immune stimulation and amplification of immunocompetent cells. Numerous physiological functions have been ascribed to the "normal" gut microbial population; some of the major functions are considered the following (36; 39; 76):

* Maintenance and restoration of barrier function
* Stimulation of the immune system
* Maintenance of mucosa nutrition and circulation
* Improvement of bioavailability of nutrtients
* Stimulation of bowel motility and reduction of constipation

1.3 Probiotic Microorganisms

1.3.1 Examples of Probiotic Microorganisms

Probably the longest history of proven health benefits and "safe-use" of probiotic bacteria in food is documented for *L. casei* strain "Shirota" (22) and some strains of the *L. acidophilus* group. Since at least 40 years in Japan and more than 30 years in Germany, LAB cultures of human origin are applied in the manufacture of fermented milk products. Viable strains of especially

"Lactobacillus acidophilus" and *Bifidobacterium bifidum* were introduced in Germany during the late 1960s into dairy products because of their expected adaptation to the intestine and the sensory benefits for producing mildly acidified yogurts (77). In Germany, such products first became known as mild yogurts or "bio-yogurts", whereas in the USA, acidophilus milk was developed. The functional properties and safety of particular strains of *L. casei/paracasei, L. rhamnosus, L. acidophilus,* and *L. johnsonii* have extensively been studied and are well documented (32, 78-80).

Viable probiotic strains with beneficial functional properties are at present found among a wide and diverse number of microbial species and genera. They are supplied in the market either as fermented (mainly "yogurt"-type) food commodities or in lyophilized form, both as food supplements and as pharmaceutical preparations. Most strains currently in use as probiotics in food, nutrition, and in pharmaceutical preparations are members of the LAB (Table 1.5). A number of "nonlactic" strains, e.g., *Bacillus cereus* ("*toyoi*"), *B. clausii, B. pumilis* (146), *Escherichia coli* (Nissle) (16), *Propionibacterium freudenreichii, P. jensenii, P. acidopropionici, P. thoenii* (147), and *Saccharomyces cerevisiae* ("*boulardii*"), are also available in the market mainly as pharmaceutic preparations and some also as animal feed supplements (39, 79) (Table 1.5).

With 65%, the probiotic milk products (mainly "yogurt"-like) represent the largest segment of the functional foods market in Europe, while in Japan they are estimated to comprise about 75% of the foods for specified health uses (FOSHU)) market. Initiated by a national project team under the auspices of the Japan Ministry of Education and Science, specific regulatory measures on functional foods were first initiated in Japan in 1984. This

TABLE 1.5

Microorganisms Reported to Find Application as Probiotics Mainly for Humans

Lactobacillus Species	Bifidobacterium Species	Other LAB	"Non-lactics"[c]
L. acidophilus	*B. adolescentis*	*Ent. faecalis* [a]	*Bacillus cereus*
L. amylovorus	*B. animalis*	*Ent. faecium*	("toyoi")[a,c]
(*L. casei*)	*B. bifidum*	*Sporolactobacillus*	*Escherichia coli*
L. crispatus	*B. breve*	*inulinus* [a]	(Nissle 1917)[c]
L. delbrueckii subsp.	*B. infantis*		*Propionibacterium*
bulgaricus[c]	*B. lactis*[b]		*freudenreichii*[a,c]
L. gallinarum[a]	*B. longum*		*Saccharomyces*
L. gasseri			*cerevisiae*
L. johnsonii			("boulardii")[c]
L. paracasei			
L. plantarum			
L. reuteri			
L. rhamnosus			

[a] Mainly for animals.
[b] Synonym of *B. animalis.*
[c] Mainly in pharmaceutical preparations.

Source: Modified from Holzapfel et al. (39).

triggered the beginning of numerous academic and industrial studies on functional foods in relation to nutrition and evidence in support of functional claims. The Japan Ministry of Health and Welfare thereupon established a specific policy on FOSHU in 1993, by which health claims of some selected functional foods are legally permitted. The developments of functional food science in Japan focused, among others, on minimizing undesirable and maximizing desirable food factors. Three major requirements had to be met for FOSHU approval, viz.:

- Scientific evidence of the efficacy, including clinical testing
- Safety for consumption
- Analytical determination of the effective component

By the end of 1999, 167 items were approved as FOSHU, as compared to 293 by 2002. In April 2001, the Japanese government introduced a new regulatory system (foods with health claims), comprising FOSHU and foods with nutrient function claims (FNFC). Most of the descriptions of foods under the FOSHU system are similar to the category of enhanced function claims of Codex (81-83).

Functional food products primarily contain strains of the "acidophilus group" (mainly *L. acidophilus, L. crispatus,* and *L. johnsonii*), *L. casei/paracasei* and *Bifidobacterium* spp.; enterococci are rarely used in probiotic milk products (22, 32, 77-80). Information on the typical LAB species associated with probiotic milk products in the European market is given in Table 1.6. The problems still encountered with the correct identification of these strains are evident from these data (see also Temmerman et al., Reference 84) and may (among others) be related to the use of unreliable phenotypic methods (compare also Table 1.3 with regard to the acidophilus group). Although phenotypically difficult to assess, the heterogeneity of *L. acidophilus,* one of the most important "probiotic" species, was recognized in the 1960s by Reuter and coworkers (19), who suggested four different "biotypes". DNA-DNA hybridization studies reported in 1980 (20, 21) confirmed this heterogeneity, suggesting the existence of six different homology groups (see Table 1.3). Consequently, only strains belonging to the similarity group and showing a high degree of DNA relatedness with the type strain of *L. acidophilus* remained in this species, while members of the other homology groups were classified as separate species, i.e., *L. amylovorus, L. gallinarum, L. crispatus, L. gasseri,* and *L. johnsonii*. Although they are regarded as separate species, they are closely related and have been suggested to belong to one phylogenetic "group" or branch (37, 39, 47, 72). The exact identification of members of the "*L. acidophilus* group" is an important aid toward indication of the origin and typical host of a species (see Table 1.3).

Identification studies on various mild yogurts and novel-type probiotic yogurt-type dairy products showed the 26 isolated *Lactobacillus* strains to represent *L. acidophilus, L. johnsonii, L. crispatus, L. casei, L. paracasei,* and *L.*

TABLE 1.6

Lactic Acid Bacteria in Commercial Probiotic Dairy Products: Comparison between Claimed Identity and Identification Results

Product Name	Producer / Distributor/ (Country)	Strain Identity Claimed on Product	Confirmed Identity
ABC	Söbbecke (D)	L. acidophilus, L. casei	L. acidophilus, L. paracasei
Actimel	Danone (F)	L. casei Actimel ("Immunitas")	L. paracasei
Andechser Bioaktiv	Bioland (D)	BIOGARDE cultures	L. johnsonii
B'A Fruits	B'A France (F)	Bifidobacterium ("active bifidus")	S. thermophilus
BI'AC	TMA (D)	L. acidophilus, L. casei	L. acidophilus; L. paracasei ssp. paracasei; S. thermophilus
Biogarde plus (naturel)	Almhof (NL)	L. acidophilus, L. casei, Bifidobacterium	L. acidophilus; S. thermophilus
Bio Snac'	Danone (F)	Bifidobacterium, living yogurt cultures	Lc. lactis subsp. lactis
Biotic	Aldi (D)	L. acidophilus LA7	L. acidophilus
Do-filus	Arla (S)	L. acidophilus	L. acidophilus
Fitness Quark	Onken (D)	L. acidophilus, Bifidobacterium	L. johnsonii, S. thermophilus
Fysiq (Mona)	Campina (NL)	L. acidophilus Gilliland, L. casei	L. crispatus, L. paracasei ssp. paracasei
Gaio (Causido®)	MD Foods A/S (DK)	Enterococus faecium, S. thermophilus	Enterococus faecium, S. thermophilus
Gefilus	Valio (FIN)	Lactobacillus GG, living yogurt cultures	L. rhamnosus
Kinderjoghurt mild	J. Bauer KG (D)	L. acidophilus, L. bifidus	L. acidophilus; L. johnsonii, S. thermophilus
Lc1	Nestlé (D)	L. acidophilus LA-1	L. johnsonii
Probiotic LA7-Plus	Bauer (D)	L. acidophilus LA-7	L. acidophilus
Procult Drink	Müller (D)	B. longum, live yogurt cultures	L. acidophilus, S. thermophilus
Natreen Pro 3+	Milchwerke Köln (D)	L. acidophilus LA-H3, L. casei LC-H2	ND
Primo	Zott (D)	BactoLab cultures	L. acidophilus
Symbalance	Toni Lait (CH)	L. acidophilus, L. casei, L. reuteri	L. acidophilus, L. paracasei, L. reuteri
Vifit	Südmilch (D)	L. casei GG	L. rhamnosus, L. acidophilus
Vifit Drink	Mona (NL)	L. casei GG, L. acidophilus, B. bifidum	L. acidophilus, L. rhamnosus
Yakult	Yakult, Europe / (D)	L. casei Shirota	L. paracasei
Yogosan	Lidl (D)	L. casei	L. paracasei (casei)

Note: CH = Switzerland; D = Germany; DK = Denmark; F = France; FIN = Finland; NL = The Netherlands; S = Sweden

Source: Modified and extended data from References 36, 39, 84, 85, 89.

rhamnosus, revealing that some strains had been misclassified (85). Some strains designated as *L. acidophilus* were shown to belong either to *L. johnsonii* or *L. crispatus*. Most strains currently designated as *L. casei* may in fact be members of either *L. paracasei* or *L. rhamnosus* (85, 86); (compare also Collins et al. [87] and Dicks et al. [88]). Viable numbers of lactobacilli in mild and probiotic yogurts varied greatly, whereas a few products contained only low *Lactobacillus* numbers (85). This was followed up by a recent study which, once more, showed the identity given for strains in some products to differ from that found by DNA homology studies (36). An important but still controversial issue regarding the "minimal effective dose" of viable bacteria by which scientifically confirmed beneficial effects may be expected is still under discussion.

In addition to the wide range of probiotic "yogurt"-like dairy products in the market, increased attention is also given to other foods as carriers for probiotic lactobacilli and bifidobacteria (90). Particular interest is focused on different types of cheese with "added functional value" by addition of strains of, e.g., *L. acidophilus*, *L. paracasei,* and *Bifidobacterium* species to different cheese types, including Cheddar, Tallaga, and Ras, and soft cheeses (90, 91). Moreover, the use of *B. bifidum* , *L. acidophilus,* and *L. rhamnosus* GG has also been suggested for ice cream (92, 93), frozen dairy desserts (94), of *L. plantarum* 299v (95) or bifidobacteria (96) for a fermented oatmeal gruel, and of *Bifidobacterium* spp. for fermented sausages and ham (97, 98). Lee (99) suggests many traditional fermented foods to have functional properties, resulting both from the microbial strains involved (mostly LAB) and from other functional components, either originating from the ingredients or formed during fermentation. Several well-known traditional fermented foods may serve as examples, e.g., Korean kimchi (100), sauerkraut (101), and a number of African cereal gruels (e.g., ogi and uji), Nigerian gari, and Asian vegetable foods (99). This hypothesis is also supported by Molin (102), who focused on the role of *Lactobacillus plantarum*, an extremely heterogeneous species.

Numerous probiotic food supplements are available in the market, most commonly as capsules but also as powders and tablets. As for the probiotic milk products, controversies exist also for some products between label claims and sound scientific identifications (84). In addition to the *Lactobacillus* and *Bifidobacterium* species, typical of the probiotic milk products, nonlactics are also being applied, albeit only in a few products, e.g., *Bacillus* IP5832 (identified as *Bacillus cereus*) in "Bactisubtil" (Synthelabo Belgium), *Saccharomyces cerevisiae* in "Bifidus complex" (Biover, Belgium). Furthermore, *Enterococus faecium, L. reuteri,* and even *Pediocooccus acidilactici* have been used in some products (84).

Biotherapeutics for clinical applications are also based on selected probiotic strains, mainly LAB but may also include *Escherichia coli* strains (e.g., the "Nissle" strain), *Saccharomyces cerevisiae (boulardii),* and also a number of *Enterococcus faecium* and *E. fecalis* strains, the latter being marketed under the name of "Symbioflor 1".

1.3.2   Selection of Appropriate Strains

The prevalence of bifidobacteria in the feces of breast-fed infants may have been a major reason for selecting strains of this group for use as probiotics (13). Decisions on the use of *Lactobacillus* strains as probiotics have been determined by a number of favorable factors such as:

- Their association with traditional fermented foods earlier noted by Metchnikoff (7) when he postulated benefits from the consumption of yogurt by the Caucasians), together with the high acceptability of lactic fermented foods
- Their association with the human GIT, together with observations on their beneficial interactions in the gut ecosystem
- The adaptation of many lactobacilli to milk and other food substrates and the relatively long history of technical application of LAB with the use of the first industrial strains dating back to 1890 (see Table 1.1)

The selection of new strains presents a major challenge, both to science and industry. The primary objective is to select microbial strains with one or more proven functional properties. Even when probiotic microorganisms are suggested to promote health and well-being, the challenge remains to define particular end points or biomarkers by which such strains can be characterized and particular claims be sustained — either by *in vivo* or validated *in vitro* tests — even when all the mechanisms involved have not yet been fully elucidated (38). Approaches for selection of an "ideal" strain are therefore still difficult and indeed require considerable resources. Desirable technical features and factors related to health promotion or sustaining health serve as important criteria for strain selection. Five major aspects may be taken into account as key criteria for the selection of an appropriate functional strain (36, 38, 39, 86, 103, 104), viz.:

1. General aspects, e.g., origin, identity, and resistance to mutations
2. Technical aspects (growth properties *in vitro* and during processing, survival and viability during transport and storage)
3. General physiological aspects (resistance against environmental stress and to the antimicrobial factors prevailing in the upper GIT as encountered during the stomach-duodenum passage [pH 2.5, gastric juice, bile acid, pancreatic juice], adhesion potential to intestinal epithelium)
4. Functional aspects and beneficial features (adhesion, colonization potential of the mucosa, competitiveness, specific antimicrobial antagonism against pathogens, stimulation of immune response, selective stimulation of beneficial autochthonous bacteria, restoration of the "normal" population)

5. Safety aspects (no invasive potential, no transferable resistance against therapeutic antibiotics, no virulence factors)

Research during the past two decades focused mainly on functional features of strains selected for inclusion, e.g., in functional foods. Considering the worldwide increase in the consumption of dairy products containing probiotic strains of the bacterial genera *Bifidobacterium* and *Lactobacillus* during this period, relatively little attention has been given to technical and sensory properties of these strains and/or the resulting products (107). For the producer, technical properties related to growth, adaptation, and persistence of some probiotic strains, and also the sensory properties of the resulting products, are still major obstacles toward the large-scale production of functional foods containing probiotic strains. Information on particular production steps and modification of growth conditions are still well-protected industrial secrets for the technical production of some strains. Technical production of especially the bifidobacteria in milk substrate constituted a considerable technical challenge but was at least partly solved by some industries during the 1960s (77). Still, it is known that particularly strains of the "acidophilus" group and also bifidobacteria are not well adapted to the milk substrate and, in addition, do not influence the sensory properties of a product positively. Such strains therefore still constitute special technical challenges (108; Holzapfel et al., unpublished data).

From the viewpoint of regulatory authorities, the safety and nonpathogenicity of a new strain is considered of major importance. Ongoing and partly controversial discussions are particularly directed toward the assessment of new strains without a previous "history of safe use" and the definition of minimal requirements to be met before it can be classified as "safe" or "GRAS". According to Marteau (105), an extremely low potential of four types of side effects may exist for probiotic bacteria, viz., systemic infections, deleterious metabolic activities, excessive immune stimulation in susceptible individuals, and gene transfer. The following approaches for assessing the safety of probiotic and starter strains have been recommended by Salminen et al. (78):

- Characterization of the genus, species, and strain and its origin that will provide an initial indication of the presumed safety in relation to known probiotic and starter strains
- Studies on the intrinsic properties of each specific strain and its potential virulence factors
- Studies on adherence, invasion potential, and the pharmacokinetics of the strain
- Studies into interactions between the strain, intestinal and mucosal microflora, and the host

Lactobacilli and bifidobacteria are extremely rare causes of infection in humans. This, even more so, applies to probiotics based on these organisms; in fact, very few cases of adverse effects have been related to the consumption of probiotics. Even when in rare cases strains of some LAB are isolated from clinical specimens (see also Table 1.4), there is apparently no indication that the general public is at risk from the consumption of lactobacilli or bifido-bacteria used as probiotics or starters (36, 78, 105, 106). In a paper published in 2003 by the Health & Consumer Protection Directorate-General of the European Commission (SANCO Secretariat) entitled "A generic approach to the safety assessment of micro-organisms used in feed/food production," a "qualified presumption of safety" (QPS) system was suggested as an approach for the safety assessment of microorganisms for use in food and feed. This suggests a key decision which takes into account, among other things, (a) experience and a history of safe use and (b) the detection of known strain-specific risk factors.

In addition to the LAB, strains of other microbial groups such as *Bacillus* spp. (146) and *Propionibacterium* spp. (147) have also been reported to show probiotic or functional properties. The safety, particularly of some "probi-otic" strains of *Bacillus cereus,* may be questioned, some of which have been shown to produce Hbl and Nhe enterotoxins (146).

1.3.3   Functional Properties

In spite of research progress in recent years, our understanding of the gut ecosystem is still fragmentary and consequently limits our comprehension of a normal or balanced microbial population. Thus, the impact of a func-tional strain on the composition and function of the intestinal population is still difficult to ascertain (39, 109). Numerous beneficial functions have been suggested for probiotic bacteria (36, 109), e.g.:

- Nutritional benefits:
  - Vitamin production, availability of minerals and trace ele-ments
  - Production of important digestive enzymes (e.g., β-galac-tosidase)
  - Production of β-galactosidase for alleviation of lactose in-tolerance

- Barrier, restoration, antagonistic effects against:
  - Infectious diarrhea (traveller's diarrhea, children's acute vi-ral diarrhea)
  - Antibiotic-associated diarrhea, irradiation-associated di-arrhea

- Cholesterol-lowering effects by:
  - Cholesterol assimilation
  - Modifcation of bile salt hydrolase activities
  - Antioxidative effect

- Stimulation and improvement of the immune system, e.g., by:
  - Strengthening of nonspecific defense against infection
  - Increasing phagocytic acitivity of white blood cells
  - Increasing IgA production
  - Regulating the Th1/Th2 balance; induction of cytokine synthesis

- Enhancement of bowel motility, relief from constipation
- Reduction of inflammatory or allergic reactions, by:
  - Restoration of the homeostasis of the immune system
  - Regulation of cytokine sysnthesis

- Adherence and colonization resistance
- Anticarcinogenic effects in the colon by:
  - Mutagen binding
  - Inactivation of carcinogens or procarcinogens, or prevention of their formation
  - Modulation of metabolic activities of colonic microbes
  - Immune response

- Maintenance of mucosal integrity
- Antioxidative activities (110)

Effects such as lowering of the serum cholesterol level are not fully substantiated yet by placebo-controlled, double-blind, randomized clinical trials. On the other hand, strain-specific effects of probiotic lactic cultures on the human immune system and on diarrhea are well documented, e.g., for counteracting rotavirus or antibiotic-associated diarrhea, by application of strains such as the LGG strain of *L. rhamnosus* and the Shirota strain of *L. casei* (*L. paracasei*) (48, 111, 112). Therapeutic use is also considered successful in cases of lactose intolerance, irritable bowel syndrome, colon cancer, and *Helicobacter pylori* infection (109). Complex underlying mechanisms, such as adhesive and immunomodulating properties of effective strains, are major challenges remaining to be solved by intensified research (36, 80).

Apparently, adhering probiotic strains may transiently colonize the GIT, and thereby cause an increase in IgA levels (113, 114), resulting in the enhancement of serum IgA response to pathogens such as attentuated *Sal-*

monella typhi Ty21a (115). Moreover, many probiotic effects are mediated through immune regulation and especially through balance control of proinflammatory and anti-inflammatory cytokines, thereby suggesting the use of probiotics as innovative tools to alleviate intestinal inflammation, normalize gut mucosal dysfunction, and down-regulate hypersensitivity (116). In the ideal situation, immune stimulation by probiotic strains would be based on transient or longer-term colonization through adhesion and aggregation without invasion (115).

1.4   Prebiotics

Prebiotics are defined as nondigestible but fermentable food ingredients that beneficially affect the host by selectively stimulating the growth and/or activitiy of one or a limited number of bacteria in the colon, and thus improve host health (116). The major prebiotics are resistant dietary carbohydrates, but noncarbohydrates are not excluded from this definition. In theory, Hartemink (118) states that "…any antibiotic that would reduce the number of potentially harmful bacteria and favour health-promoting bacteria or activities, can be considered as a prebiotic". Although these definitions do not highlight any specific bacterial group as such, prebiotics are considered to stimulate selectively bacterial groups such as bifidobacteria, lactobacilli, and eubacteria resident in the colon. These are considered particularly beneficial for the human host. Resistant short-chain carbohydrates (SCCs) are also referred to as nondigestible oligosaccharides (119) or low-digestible carbohydrates (LDCs) (119). These SCC or LDCs provide interesting possibilities for inclusion into conventional food products for their "bifidogenic" effects (36, 118). Several such "candidate prebiotics" are currently under consideration by the industry for human consumption (120). Inulin and fructo-oligosaccharides (FOSs) are considered as typical "bifidogenic factors" and are probably the most commonly used prebiotics in the market (121–123). In addition, Bouhnik et al. (124) have shown that ingestion of 10 g of lactulose per day increases fecal bifidobacterial counts. Other promising prebiotic oligosaccharides under consideration are galacto-oligosacccharides, isomalto-oligosaccharides, soybean oligosaccharides, lactosucrose, and xylo-oligosaccharides (125). A quantitative tool has been developed by Palframan et al. (126) for the comparison of the prebiotic effect of dietary oligosaccharides; the quantitatiove probiotic index (PI) equation may find application in quantifying prebiotic effects *in vitro*.

In some cases, prebiotics such as FOSs are added to probiotic yogurts, the combination of which would thereby result in a "synbiotic" (36). These substances should ideally be well tolerated in the GIT and also reach the cecum where they will be availabe to benefit bacterial groups such as the bifidobacteria and some lactobacilli and eubacteria for fermentation. In some instances, however, dose-related undesirable effects, due to osmotic potential and/or excessive fermentation, may occur, e.g., excessive flatus, bloating, abdominal cramps, and even diarrhea. Although dose-related intolerance symptoms may occur after ingestion of LDCs, the dose of intolerance generally appears to be high, thereby allowing a relatively broad "therapeutic window," i.e., the dose above the minimal effective level (36, 125, 127). Although it is generally established that bifidobacterial numbers increase in the feces of humans upon ingestion of FOSs (123), the average increase is considered small, whereas the "biological significance" for the human population and in specific disease situations appears not to be fully clarified (43, 128, 129). When compared to the observations of Bouhnik et al. (124) with lactulose, it appears that also the type and quantity of probiotic ingested might be decisive. Another factor may, however, be the underestimation of the bifidobacterial population by selective plating techniques, with recovery rates ranging from 17 to 58% (depending on the species), as compared to 85% by culture-independent methods (130).

Still, general agreement seems to exist on a number of beneficial effects of prebiotics, which point to the favorable influence on the small bowel by improved sugar digestion and absorption, glucose and lipid metabolism, and protection against known risk factors of cardiovascular disease. In the actual "target region," the colon, the fermentative production of SCFAs is in fact considered a major beneficial feature related to the primary prevention of colorectal cancer (131). Other confirmed effects from prebiotics are related to the low energy value (<9 kJ/g) resulting from their nondigestibility, to an increase in stool volume, to the modulation of the colonic flora by stimulation of beneficial bacteria (*Bifidobacterium*, *Lactobacillus*, and *Eubacterium* spp.), inhibition of "undesirable" bacteria (*Clostridium* and *Bacteroides*), and colonization resistance against *Clostridium difficile* (132). Some of the postulated effects that have not been finally confirmed refer to the prevention of intestinal infections, the modulation of the immune response, the prevention of colorectal cancer, reduction of the serum cholesterol level, and to improved bioavailability (36). In spite of strong indications on the positive role of LDCs in the maintenance of the human GIT, this issue is not fully clarified and deserves further attention (131).

By definition, a synbiotic refers to a product in which a probiotic and a prebiotic are combined. The postulated synbiotic effect may involve two different "target regions" of the GIT, comprising both the small and the large intestines. Moreover, the growth of a probiotic strain that is able to utilize a prebiotic will be selectively stimulated in the gut. This combination of pre- and probiotics in a single product has been shown to confer benefits beyond those of either on its own. Convincing data showed, e.g., an enhanced reduction in the number of colonic aberrant crypt foci (ACF) (133) and for colon carcinogenesis in rats (134). Also, antibiotic-associated diarrhea could be prevented by the combined application of *Lactobacillus sporogenes* (syn: probably *Bacillus Coaguleus)* and fructo-oligosaccharides in children (135). On the other hand, synbiotic therapy did not result in any improvement of gut barrier function in elective surgical patients (136).

1.5   Conclusions

The establishment of scientifically confirmed evidence of "functional" effects related to pre-, pro-, and synbiotics presents a tremendous challenge to interdisciplinary scientific research. More than a century has passed since the early scientific observations and careful reporting, with increased research efforts and continuous development of new and improved hypotheses in the last two decades. The important role of the intestinal flora in the maintenance of health and in the prevention of disease is well recognized and acknowledged. Disturbance of the delicate balance of the gut microbial ecosystem may lead to dysbiosis and other disorders and thus facilitate establishing a state of disease (39). Particular interest is increasingly focusing on the continuous interaction or "communication" of the gut microbiota with the environment, the central nervous system, the endocrine system, the immune system, and the complex underlying mechanisms (43, 137–140). Based on *in vivo* and *in vitro* studies, Freitas et al. (141) suggest that both the established intestinal bacteria and probiotic strains are able to modulate host–pathogen interactions in the gut. It appears that species-specific modulations of intestinal cell glycosylation may represent a simple, general, and efficient mechanism to adapt the host defense toward pathogens. The strong focus of recent research efforts and observations on the role of the gut microflora and probiotics (functional strains) in immunomodulation is extensively addressed by Fuller and Perdigón (142) and a number of experts in a book on this topic.

Even when Tannock (43) does not see much progress on the "understanding of how probiotics work," he admits that pre- and probiotics have stimulated and generated new interest by the medical profession in the gut microbiota. It is envisaged that probiotic and prebiotic products of the future may be targeted for use in the prevention or alleviation of symptoms of specific diseases, provided that "abnormal microfloras" can be recognized and the safety of probiotics be guaranteed also to immunologically dysfunctional persons (43, 143) and at the same time effect the modulation of the immune response of the immunodeficient host (144).

Another major challenge concerns the development and validation of *in vivo* and *in vitro* test models. Significant progress has been made in recent years in conducting placebo-controlled, double-blind clinical studies, for which important functional effects could be verified. Yet, both for pre- and probiotics, a number of postulated effects still need to be confirmed (36, 39, 43). For prebiotics, studies may particularly be directed toward their influence on blood serum cholesterol values, the role of some dominant but hardly studied bacterial groups in the colon (*Fusobacterium, Eubacterium, Veillonella, Peptostreptococcus*, etc.), and the influence of composition of the colonic microbial population on a "favorable" ratio of SCFAs (36). Microbiologists, particularly, should play a major role in isolating strains and testing mechanisms of action and in "packaging these into reliable products for human use" (145). A particular challenge would be to present probiotic strains in substrates different than milk and to utilize raw food materials more readily available and accepted in specific regions, e.g., maize porridge in African countries (148).

It may be accepted that pre- and probiotics have different "target" regions (36). *In vivo* studies on colonization and interactions of probiotics with the gut mucosa, especially in the small intestines, constitute both a challenging and complex area for investigations. For a "safety and acceptability record" of a probiotic strain, experience and history are still important. In spite of an explosion in recent years of publications dealing with probiotic organisms by clinicians, microbiologists, food scientists, and nutritionists, vital information is still needed as a basis for decision making, e.g., by scientists, industries, and regulatory authorities.

References

1. T Escherich. Die Darmbakterien des Neugeborenen und Säuglings. *Fortschr. Med.*, 3: 515–522, 1885.
2. T Escherich. Die Darmbakterien des Säuglings und ihre Beziehung zur Physiologie der Verdauung. F. Enke, Arbeit an dem pathologischen Institut zu München. Stuttgart 2: 1–180, 1886. (Hadorn, H-B., 1998. Theodor Escherich (1857–1911), Festvortrag. In: *3. Interdisziplinäres Symposium. Darmflora in Symbiose und Pathogenität*. Alfred-Nissle-Gesellschaft e.V. (ed.), D-58089 Hagen.
3. A Döderlein. Das Scheidensekret und seine Bedeutung für das Puerperalfieber (The vaginal transudate and its significance for childbed fever). *Centralbl. Bakteriol.*, 11: 699–700, 1892.
4. E Moro. Über den *Bacillus acidophilus* n. spec. Ein Beitrag zur Kenntnis der normalen Darmbacterien des Säuglings (*Bacillus acidophilus* n. spec. A contribution to the knowledge of the normal intestinal bacteria of infants). *Jahr. Kinderheilkunde*, 52: 38–55, 1900.
5. MW Beijerinck. Sur les ferments de lactique de l'industrie (Lactic acid bacteria of the industry). *Arch. Neerl. Sci. Exact. Nat.*, 6: 212–43, 1901.
6. Dr. Cahn. Über die nach Gram färbbaren Bacillen des Säuglingsstuhles (Bacilli of infant stools stainable according to Gram). *Centralbl. Bakteriol. eI. Abt. Orig.*, 30: 721–726, 1901.
7. E Metschnikoff. *The Prolongation of Life. Optimistic Studies*. William Heinemann, London, 1907.
8. E Metschnikoff. *Prolongation of Life*. Putnam, New York, 1908.
9. H Tissier. *Recherches sur la flore intestinale. Normale et pathologique du nourrisson*. Georges Carre et C. Naud, Paris, 1900.
10. H Tissier. Repartition des microbes dans l'intestin du nourrisson. *Ann. Inst. Pasteur, Paris*, 19: 109–123, 1905.

11. W Kollath. Ernährung und Zahnsystem. Deutsch. Zahnaerzt. Z., 8: 7–16, 1953.
12. F Vergio. Anti- und Probiotika. *Hippokrates*, 4: 116–119, 1954.
13. DM Lilly, RH Stillwell. Probiotics: growth promoting factors produced by microorganisms. *Science*, 147: 747–748, 1965.
14. A Nissle. Über die Grundlagen einer neuen ursächlichen Bekämpfung der pathologischen Darmflora. *Dtsch. Med. Wochenschr.*, 42: 1181–1184, 1916.
15. A Nissle. Die antagonistische Behnadlung chronischer Darmstörungen mit Colibakterien. *Med. Klinik*, Nr. 2: 29–30, 1918.
16. A Nissle. Weiteres über die Mutaflorbehandlung unter besonderer Berücksichtigung der chronischen Ruhr. *Mnch. Med. Wochenschr.*, Nr. 25: 678–681, 1919.
17. H Kolb. Die Behandlung akuter Infekte unter dem Gesichtswinkel der Prophylaxe chronischer Leiden. Über die Behandlung mit physiologischen Bakterien. *Microecol. Ther.*, 1: 15–19, 1955.
18. M Lerche, G Reuter. Das Vorkommen aerob wachsender gram-positiver Stäbchen des Genus *Lactobacillus* Beijerinck im Darminhalt erwachsener Menschen (Occurrence of aerobic Gram-positive rods of the genus *Lactobacillus* Beijerinck in the intestinal contents of adult humans). *Zentralbl. Bakteriol. I. Abt. Orig.*, 185: 446–481, 1962.
19. G Reuter. Vergleichende Untersuchungen über die Bifidus-Flora im Säuglings- und Erwachsenenstuhl. *Zentralbl. Bakteriol. I. Orig. A*, 191: 486–507, 1963.
20. G Reuter. Das Vorkommen von Laktobazillen in Lebensmitteln und ihr Verhalten im menschlichen Intestinaltrakt. *Zentralbl. Bakteriol. I. Orig.*, 197: 468–487, 1965.
21. G Reuter. Zusammensetzung und Anwendung von Bakterienkulturen für therapeutische Zwecke. *Arzneimittelforschung*, 19: 103–109, 1969.
22. M Shirota, K Aso, A Iwabuchi. Study of microflora of human intestine: I. The lateration of the constitution of intestinal flora by oral administration of *L. acidophilus* strain Shirota to healthy infants. *Jpn. J. Bacteriol.*, 21: 274–283, 1966.
23. GV Reddy, KM Shahani, MR Banerjee. Inhibitory effect of yogurt on Ehrlich ascites tumor-cell proliferation. *J. Natl. Cancer Inst.*, 50: 815–817, 1973.
24. I Kato, S Kobayashi, T Yokokura, M Mutai. Anti-tumour activity of *Lactobacillus casei* in mice. *Gann*, 72: 517–523, 1981.
25. T Yokokura, I Kato, M Mutai. Antitumour effect of *Lactobacillus casei* (LC 9018). In: T Mitsuoka, Ed., *Intestinal Flora and Carcinogenesis*, Gakkai-Syuppan Center, Tokyo, 1981, pp. 72–88.
26. RB Parker. Probiotics, the other half of the antibiotic story. *Anim. Nutr. Health*, 29: 4–8, 1974.
27. JAA Hoogkamp-Korstanje, JGEM Lindner, JH Marcelis, H Den Daas-Slagt, NM De Vos. Composition and ecology in the human intestional flora. *Antonie van Leeuwenhoek*, 45, 35–40, 1979.
28. JH Schwab. Modulation of the immune response by bacteria, In: D Schlessinger, Ed., *Microbiology, 1977*, American Society for Microbiology, Washington D.C., 1977, pp. 366–373.
29. GA Conge, P Gouache, JP Desormeau-Bedot, F Loisillier, D Lemonnier. Comparative effects of a diet enriched in live or heated yogurt on the immune system of the mouse. *Reprod. Nutr. Dev.*, 20 (4A), 929–938, 1980.
30. R Fuller. Probiotics in man and animals. *J. Appl. Bacteriol.*, 66: 365–378, 1989.
31. R Havenaar, B Ten Brink, JHJ Huis in't Veld. Selection of strains for probiotic use. In: R Fuller, Ed., *Probiotics: The Scientific Basis*, Chapman & Hall, London, 1992, pp. 209–224.

32. S Salminen, MA Deighton, Y Benno, SL Gorbach. Lactic acid bacteria in health and disease. In: S Salminen, A Von Wright, Eds., *Lactic Acid Bacteria: Microbiology and Functional Aspects*, 2nd ed., Marcel Dekker, New York, 1998, pp. 211–253.
33. F Guarner, GJ Schaafsma. Probiotics. *Int. J. Food Microbiol.*, 39, 237–238, 1998.
34. BgVV (Bundesinstitut für gesundheitlichen Verbraucherschutz und Veterinär-medizin), (1999). Final report of the working group: "Probiotische Mikroogan-ismenkulturen in Lebensmitteln," BgVV, Berlin, October 1999.
35. FUFOSE Working Group. Scientific Concepts of Functional Foods in Europe–Consensus Document. *Br. J. Nutr.*, 81: S1–S27, 1999.
36. WH Holzapfel, U Schillinger. Introduction to pre- and probiotics. *Food Res. Int.*, 35: 109–116, 2002.
37. WH Holzapfel, P Haberer, R Geisen, J Björkroth, U Schillinger. Taxonomy and important features of probiotic microorganisms in food and nutrition. *Am J. Clin. Nutr.*, 73 (Suppl.): 365–373, 2001.
38. S Salminen, C Bouley, M-C Boutron-Ruault, JH Cummings, A Franck, GR Gibson, E Isolauri, M-C Moreau, M Roberfroid, I Rowland. Functional food science and gastrointestional physiology and function. *Br. J. Nutr.* 80, (Suppl. 1): S147–S171, 1998b.
39. WH Holzapfel, P Haberer, J Snel, U Schillinger, JHJ Huis in't Veld. Overview of gut flora and probiotics. *Int. J. Food Microbiol.*, 41: 85–101, 1998.
40. F Waldeck. Funktionen des Magen-Darm-Kanals. In: RF Schmidt, G. Thews, *Physiologie des Menschen*. (24. Aufl.). Springer-Verlag, Berlin, 1990.
41. TD Luckey, MH Floch. Introduction to intestinal microecology. *Am. J. Clin. Nutr.*, 25: 1291–1295, 1972.
42. GW Tannock, K Munro, HJM Harmsen, GW Welling, J Smart, PK Gobal. Anal-yses of the fecal microflora of human subjects consuming a probiotic product containing *Lactobacillus rhamnosus* DR20. *Appl. Environ. Microbiol.*, 66: 2578–2588, 2000.
43. GW Tannock. Probiotics and Prebiotics: where are we going? In: GW Tannock, Ed., *Probiotics and Prebiotics. Where Are We Going?* Caister Academic Press, London, 2002, pp. 1–40.
44. Å Sullivan, H Törnblom, G Lindberg, B Hammarlund, A-C Palmgren, C Ein-arsson, CE Nord. The micro-flora of the small bowel in health and disease. *Anaerobe*, 9: 11–14, 2003.
45. WH Holzapfel. *Lexikon Lebensmittel-Mikrobiologie und -Hygiene*. Behr's Verlag, Hamburg, 2004.
46. S Orla-Jensen. *The Lactic Acid Bacteria*. Fred Host and Son, Copenhagen, 1919.
47. G Reuter. Present and future of probiotics in Germany and in central Europe. *Rev. Biosci. Microfl.*, 16: 43–51, 1997.
48. P Marteau, M De Vrese, CJ Cellier, J Schrezenmeir. Protection from gastrointes-tinal diseases with the use of probiotics. *Am. J. Clin. Nutr.*, 73: 430S–436S, 2001.
49. SM Finegold, VS Sutter, GE Mathisen. Normal indigenous intestinal flora. In: DJ Hentges, Ed., *Human Intestinal Microflora in Health and Disease*. Academic Press, New York, 1983, pp. 3–31.
50. J Walter, C Hertel, GW Tannock, CM Lis, K Munro, WP Hammes. Detection of *Lactobacillus, Pediococcus, Leuconostoc,* and *Weissella* species in human feces by using group-specific PCR primers and denaturing gradient gel electrophoresis. *Appl. Environ. Microbiol.*, 67: 2578–2585, 2001.

51. F Dal Bello, J Walter, WP Hammes, C Hertel. Increased complexity of the species composition of lactic acid bacteria in human feces revealed by alternative incubation condition. *Microb. Ecol.*, 45: 455–463, 2003.

52. LV Holdeman, WE Moore. Roll-tube techniques for anaerobic bacteria. *Am. J. Clin. Nutr.*, 25: 13141–1317, 1972.

53. WEC Moore, LV Holdeman. Human fecal flora: the normal flora of 20 Japanese-Hawaiians. *Appl. Microbiol.*, 27: 961–979, 1974.

54. PS Langendijk, F Schut, GJ Jansen, GC Raangs, GR Kamphuis, HF Wilkinson, GW Welling. Quantitative fluorescence in situ hybridisation of *Bifidobacterium* spp. with genus-specific 16S rRNA-targeted probes and its application in faecal samples. *Appl. Environ. Microbiol.*, 61: 3069–3075, 1995.

55. A Sghir, G Gramet, A Suau, V Rochet, P Pochart, J Dore. Quantification of bacterial groups within human fecal flora by oligonucleotide probe hybridization. *Appl. Environ. Microbiol.*, 66: 2263–2266, 2000.

56. T Requena, J Burton, T Matsuki, K Munro, MA Simon, R Tanaka, K Watanabe, GW Tannock. Identification, detection, and enumeration of *Bifidobacterium* species by PCR targeting the transaldolase gene. *Appl. Environ. Microbiol.*, 68: 2420–2427, 2002.

57. T Mitsuoka. Intestinal flora and aging. *Nutr. Rev*, 50: 438–446, 1992.

58. HJM Harmsen, ACM Wildeboer, GC Raangs, AA Wagendorp, N Klijn, JG Bindels, GW Welling. Analysis of intestinal flora development in breast-fed and formula-fed infants by using molecular identification and detection methods. *J. Pediatr. Gastroenterol. Nutr.*, 30: 61–67, 2000.

59. H Beerens. Detection of bifidobacteria by using propionic acid as a selective agent. *Appl. Environ. Microbiol.*, 57: 2418–2419, 1991.

60. GW Tannock. Identification of lactobacilli and bifidobacteria, In: GW Tannock, Ed., *Probiotics: A Critical Review*. Horizon Scientific Press, Wymondham, U.K., 1999, pp. 45–56.

61. AH Franks, HJ Harmsen, GC Raangs, GJ Jansen, F Schut, GW Welling. Variations of bacterial populations in human feces measured by fluorescent in situ hybridization with group-specific 16S rRNA-targeted oligonucleotide probes. *Appl. Environ. Microbiol.*, 64: 3336–3345, 1998.

62. HJM Harmsen, GR Gibson, P Elfferich, GC Raangs, AC Wildeboer-Veloo, A Argaiz, MB Roberfroid, GW Welling. Comparison of viable cell counts and fluorescence in situ hybridization using specific rRNA-based probes for the quantification of human fecal bacteria. *FEMS Microbiol. Lett.*, 183: 125–129, 2000.

63. EG Zoetendal, AD Akkermans, WM De Vos. Temperature gradient gel electrophoresis analysis of 16S rRNA from human fecal samples reveals stable and host-specific communities of active bacteria. *Appl. Environ. Microbiol.*, 64: 3854–3859, 1998.

64. ADL Akkermans, EG Zoetendal, CF Favier, GHJ Heilig, WM Akkermans-van Vliet, WM de Vos. Temperature and denaturing-gradient gel electrophoresis analysis of 16S rRNA from human faecal samples. *Biosci. Microflora*, 19: 93–98, 2000.

65. EG Zoetendal, K Ben-Amor, AD Akkermans, T Abee, WM de Vos. DNA isolation protocols affect the detection limit of PCR approaches of bacteria in samples from the human gastrointestinal tract. *Syst. Appl. Microbiol.*, 24: 405–410, 2001.

66. EG Zoetendal, A von Wright, T Vilpponen-Salmela, K Ben-Amor, AD Akker-mans, WM de Vos. Mucosa-associated bacteria in the human gastrointestinal tract are uniformly distributed along the colon and differ from the community recovered from feces. *Appl. Environ. Microbiol.*, 68: 3401–3407, 2002.

67. T Matsuki, K Watanabe, J Fujimoto,Y Miyamoto, T Takada, K Matsumoto, H Oyaizu, R Tanaka. Development of 16S rRNA-gene-targeted group-specific primers for the detection and identification of predominant bacteria in human feces. *Appl. Environ. Microbiol.*, 68: 5445–5451, 2002.

68. A Lidbeck, CE Nord. Lactobacilli in relation to human ecology and antimicro-bial therapy (review). *Int. J. Tissue React.*, 13: 115–22, 1991.

69. A Lidbeck, CE Nord. Lactobacilli and the normal human anaerobic microflora (review). *Clin. Infect. Dis.*, 16 (Suppl. 4): S181–187, 1993.

70. T Mitsuoka. The human gastrointestinal tract. In: BJB Wood, Ed., *The Lactic Acid Bacteria, Vol. 1. The Lactic Acid Bacteria in Health and Disease.* Elsevier Applied Science, London, 1992, pp. 69–114.

71. KH Schleifer, W Ludwig. Phylogeny of the genus *Lactobacillus* and related genera. *Syst. Appl. Microbiol.*, 18: 461–467, 1995.

72. KH Schleifer, W Ludwig. Phylogenetic relationships of lactic acid bacteria. In: BJB Wood, WH Holzapfel, Eds., *The Genera of Lactic Acid Bacteria.* Chapman & Hall, London, 1995, pp. 7–18.

73. MD Collins, U Rodrigues, C Ash, M Aguirre, JAE Farrow, A Martinez-Murcia, BA Phillips, AM Williams, S Wallbanks. Phylogenetic analysis of the genus *Lactobacillus* and related lactic acid bacteria as determined by reverse tran-scriptase sequencing of 16S rRNA. *FEMS Microbiol. Lett.*, 77:5–12, 1991.

74. M Heyman, R Ducroc, JF Desjeux, JL Morgat. Horseradish peroxidase transport across adult rabbit jejunum *in vitro. Am. J. Physiol.*, 242: G558–G564, 1982.

75. F Shanahan. The intestinal immune system. In: LR Johnson, J Christensen, E Jacobsen, JH Walsh, Eds., *Physiology of the Gastrointestinal Tract*, 3rd ed., Raven Press, New York, 1994, pp. 643–684.

76. CA Edwards, AM Parrett. Intestinal flora during the first months of life: new perspectives. *Br. J. Nutr.*, 88, (Suppl. 1): S11–S18, 2002.

77. R Schuler-Malyoth, A Ruppert, F Müller. Die Mikroorganismen der Bifidus-gruppe (syn. *Lactobacillus bifidus*). 2. Mitteilung: Die Technologie der Bifiduskul-tur im milchverarbeitenden Betrieb. *Milchwissenschaft*, 23: 554–558, 1968.

78. S Salminen, A von Wright, AC Ouwehand, WH Holzapfel. Safety assessment of starters and probiotics. In: M Adams, R Nout, Eds., *Fermentation and Food Safety.* Aspen Publishers, Gaithersburg, MD, 2000, pp. 239–252.

79. LJ Fooks, GR Gibson. Probiotics as modulators of the gut flora. *Br. J. Nutr.*, 88, (Suppl. 1): S39–S49, 2002.

80. S Salminen, EM Tuomola. Adhesion of some probiotic and dairy *Lactobacillus* strains to Caco-2 cell cultures. *Int. J. Food Microbiol.*, 41: 45–51, 1998.

81. M Groeneveld. *Funktionelle Lebensmittel. 2. Dokumentation zur aktuellen wissen-schaftlichen Diskussion.* ILWI, Bonn, 2002.

82. S Arai. Functional food science in Japan: state of the art. *Biofactors*, 12: 13–16, 2000.

83. T Shimizu. Newly established regulation in Japan: foods with health claims. *Asia Pac. J. Clin. Nutr.*, 11: S94–6, 2002.

84. R Temmerman, B Pot, G Huys, J Swings. Identification and antibiotic suscep-tibility of bacterial isolates from probiotic products. *Int. J. Food Microbiol.*, 81: 1–10, 2003.

85. U Schillinger. Isolation and identification of lactobacilli from novel-type probiotic and mild yoghurts and their stability during refrigerated storage. *Int. J. Food Microbiol.*, 47: 79–87, 1999.

86. G Reuter, G Klein, M Goldberg. Identification of probiotic cultures in food samples. *Food Res. Int.*, 35: 117–124, 2002.

87. MD Collins, BA Phillips, P Zanoni. Deoxyribonucleic acid homology studies of *Lactobacillus casei, Lactobacillus paracasei* sp. nov., subsp. *paracasei* and *tolerans*, and *Lactobacillus rhamnosus* sp. nov., comb. nov. *Int. J. Syst. Bacteriol.*, 39: 105–108, 1989.

88. LMT Dicks, EM Du Plessis, F Dellaglio, E Lauer. Reclassification of *Lactobacillus rhamnosus* ATCC 15820 as *Lactobacillus zeae* nom. rev., designation of ATCC 334 as neotype of *L. casei* subsp. *casei*, and rejection of the name *Lactobacillus paracasei*. *Int. J. Syst. Bacteriol.*, 46: 337–340, 1996.

89. MC Bertolami, ER Farnworth. The Properties of *Enterococcus faecium* and the fermented milk product Gaio®. In: ER Farnworth, Ed., *Handbook Fermented Functional Foods. Functional Foods and Nutraceuticals Series.* CRC Press, Boca Raton, FL, 2003, pp. 59–75.

90. C Stanton, C Desmond, M Coakley, JK Collins, G Fitzgerald RP Ross. Challenges facing development of probiotic-containing functional foods. In: ER Farnworth, Ed., *Handbook Fermented Functional Foods, Functional Foods and Nutraceuticals Series.* CRC Press, Boca Raton, FL, 2003, pp. 27–58.

91. KJ Heller, W Bockelmann, J Schrezenmeir, M deVrese. Cheese and its potential as a probiotic food. In: ER Farnworth, Ed., *Handbook Fermented Functional Foods, Functional Foods and Nutraceuticals Series.* CRC Press, Boca Raton, FL, 2003, pp. 203–226.

92. S Hekmat, DJ McMahon. Survival of *Lactobacillus acidophilus* and *Bifidobacterium bifidum* in ice cream for use as a probiotic food. *J. Dairy Sci.*, 75: 1415–1422, 1992.

93. M Hagen, JA Narvhus. Production of ice cream containing probiotic bacteria. *Milchwissenschaft*, 54: 265–268, 1999.

94. SH Hong, RT Marshall. Natural exopolysaccharides enhance survival of lactic acid bacteria in frozen dairy desserts. *J. Dairy Sci.*, 84: 1367–1374, 2001.

95. G Molin. Probiotics in foods not containing milk or milk constituents, with special reference to *Lactobacillus plantarum* 299v. *Am. J. Clin. Nutr.*, 73: 380S–385S, 2001.

96. R Laine, S Salminen, Y Benno, AC Ouwehand. Performance of bifidobacteria in oat-based media. *Int. J. Food Microbiol.*, 83: 105–109, 2003.

97. S Työppönen, E Petäja, T Mattila-Sandholm. Bioprotectives and probiotics for dry sausages. *Int. J. Food Microbiol.*, 83: 233–244, 2003.

98. WP Hammes, D Haller, MG Gänzle. Fermented meat. In: ER Farnworth, Ed., *Handbook Fermented Functional Foods, Functional Foods and Nutraceuticals Series.* CRC Press, Boca Raton, FL, 2003, pp. 251–276.

99. C-H Lee. Creative fermentation technology for the future. *J. Food Sci.*, 69: CRH31–CRH36, 2004.

100. Y Kwon, Y-K Lee Kim. Korean fermented foods: kimchi and doenjang. In: ER Farnworth, Ed., *Handbook Fermented Functional Foods, Functional Foods and Nutraceuticals Series.* CRC Press, Boca Raton, FL, 2003, pp. 287–304.

101. W Holzapfel, U Schillinger, HJ Buckenhüskes. Sauerkraut. In: ER Farnworth, Ed., *Handbook Fermented Functional Foods, Functional Foods and Nutraceuticals Series.* CRC Press, Boca Raton, FL, 2003, pp. 343–360.

102. G Molin. The Role of *Lactobacillus plantarum* in foods and in human health. In: ER Farnworth, Ed., *Handbook Fermented Functional Foods, Functional Foods and Nutraceuticals Series*. CRC Press, Boca Raton, FL, 2003, pp. 305–342.
103. L Maré, M du Toit. Why humans should swallow live bugs — probiotics. *South African J Epid Infect.*, 17: 60–69, 2002.
104. WP Hammes, C Hertel. Research approaches for pre- and probiotics: challenges and outlook. *Food Res. Int.*, 35: 165–170, 2002.
105. P Marteau. Safety aspects of probiotic products. *Scand. J. Nutr.*, 45: 22–24, 2001.
106. SP Borriello, WP Hammes, WH Holzapfel, P Marteau, J Schrezenmeir, M Vaara, V Valtonen. Safety of probiotics that contain lactobacilli or bifidobacteria. *Clin. Infect. Dis.*, 36: 775–780, 2003.
107. M Saarela, G Mogensen, R Fondén, J Mättö, T Mattila-Sandholm. Probiotic bacteria: safety, functional and technological properties. *J. Biotechnol.*, 84: 197–215, 2000.
108. HM Østlie, MH Helland, JA Narvhus. Growth and metabolism of selected strains of probiotic bacteria in milk. *Int. J. Food Microbiol.*, 87: 17–27, 2003.
109. A Mercenier, S Pavan, B Pot. Probiotics as biotherapeutic agents: present knowledge and future prospects. *Curr. Pharm. Design*, 8: 99–110, 2002.
110. T Kullisaar, M Zilmer, M Mikelsaar, T Vihalemm, H Annuk, C Kairane, A Kilk. Two antioxidative lactobacilli strains as promising probiotics. *Int. J. Food Microbiol.*, 72: 215–224, 2002.
111. S Salminen. Functional dairy foods with *Lactobacillus* strain GG. *Nutr. Rev.*, 54: 99–101, 1996.
112. S Salminen, E Isolauri, E Salminen. Clinical uses of probiotics for stabilising the gut mucosal barrier: successful strains for future challenges. *Antonie Van Leeuwenhoek*, 70: 347–358, 1996.
113. EJ Schiffrin, F Rochat, H Link-Amster, JM Aeschlimann, A Donnet-Hughes. Immunomodulation of human blood cells following the ingestion of lactic acid bacteria. *J. Dairy Sci.*, 78: 491–497, 1995.
114. R Tanaka. The effects of the ingestion of fermented milk with *Lactobacillus casei* Shirota on the gastrointestinal microbial ecology in healthy volunteers. In: AR Leeds, IR Rowland, Eds., *Gut Flora and Health — Past, Present and Future*. Royal Society of Medicine Press, London, 1996, pp. 37–45.
115. JK Collins, G Thornton, GO O'Sullivan. Selection of probiotic strains for human applications. *Int. Dairy J.*, 8: 487–490, 1998.
116. E Isolauri, Y Sütas, P Kankaanpää, H Arvilommi, S Salminen. Probiotics: effects on immunity. *Am. J. Clin. Nutr.*, 73: S444–S450, 2001.
117. G Gibson, MB Roberfroid. Dietary modulation of the human colonic microbiota: introducing the concept of prebiotics. *J. Nutr.*, 125: 1401–1412, 1995.
118. R Hartemink. Prebiotic effects of non-digestible oligo- and polysaccharides. PhD Dissertation, ISBN 90-5808-051-X. Wageningen University, Wageningen, The Netherlands, 1999.
119. JH Cummings, S Christie, TJ Cole. A study of fructo-oligosaccharides in the prevention of traveller's diarrhoea. *Aliment. Pharmacol. Ther.*, 15: 1139–1145, 2001.
120. P Marteau, B Flourié. Tolerance to low-digestible carbohydrates: symptomatology and methods. *Br. J. Nutr.*, 85: (Suppl. 1): 817–821, 2001.
121. Y Bouhnik, B Flourié M Riottot, N Bisetti, MF Gailing, A Guibert. Effects of fructo-oligosaccharides ingestion on fecal bifidobacteria and selected metabolic indexes of colon carcinogenesis in healthy humans. *Nutr. Cancer*, 26: 21–29, 1996.

122. Y Bouhnik, B Flourié, L D'Agay-Abensour, P Pochart, G Gramet, M Durand, JC Rambaud. Administration of transgalacto-oligosaccharides increases faecal bifidobacteria and modifies colonic fermentation metabolism in healthy humans. *J. Nutr.*, 127: 444–448, 1997.

123. Y Bouhnik, Kvahedi, L Achour, A Attar, J Salfati, P Pochart, P Marteau, B Flourié, F Bornet J-C Rambaud. Short-chain fructo-oligosaccharide administration dose-dependently increases fecal bifidobacteria in healthy humans. *J. Nutr.*, 129: 113–116, 1999.

124. Y Bouhnik, A Attar, F A Joly, M Riottot, F Dyard, B Flourié. Lactulose ingestion increases faecal bifidobacterial counts: a randomised double-blind study in healthy humans. *Eur. J. Clin. Nutr.*, 58: 462–466, 2004.

125. RA Rastall, GR Gibson. Prebiotic oligosaccharides: evaluation of biological activities and potential future developments. In: GW Tannock, Ed., *Probiotics and Prebiotics. Where Are We Going?* Caister Academic Press, London, 2002, pp. 107–148.

126. R Palframan, GR Gibson, RA Rastall. Development of a quantitative tool for the comparison of the prebiotic effect of dietary oligosaccharides. *Lett. Appl. Microbiol.*, 37: 281–284, 2003.

127. MD Collins, GR Gibson. Probiotics, prebiotics and synbiotics: dietary approaches for the modulation of microbial ecology. *Am. J. Clin. Nutr.*, 69: 1052–1057, 1999.

128. KM Tuohy, S Kolida, AM Lustenberger, GR Gibson. The prebiotic effects of biscuits containing partially hydrolysed guar gum and fructo-oligosaccharides--a human volunteer study. *Br. J. Nutr.*, 86: 341–348, 2001.

129. KM Tuohy, HM Probert, CW Smejkal, GR Gibson. Using probiotics and prebiotics to improve gut health. (Review). *Drug Discov. Today*, 8: 692–700, 2003.

130. JHA Apajalahti, A Ketunen, PH Nurminen, H Jatila, WE Holben. Selective plating underestimates abundance and shows different recovery of bifidobacterial species from human feces. *Appl. Environ. Microbiol.*, 69: 5731–5735.

131. W Scheppach, H Luehrs, T Menzel. Beneficial health effects of low-digestible carbohydrate consumption. *Br. J. Nutr.*, 85, (Suppl. 1): 823–930, 2001.

132. MJ Hopkins, GT Macfarlane. Nondigestible oligosaccharides enhance bacterial colonization resistance against *Clostridium difficile in vitro. Appl. Environ. Microbiol.*, 69: 1920–1927, 2003.

133. JR Rowland, CJ Rumney, JT Coutts, LC Lievense. Effect of *Bifidobacterium longum* and insulin on gut bacterial metabolism and carcinogen-induced aberrant crypt foci in rats. *Carcinogenesis*, 19: 281–285, 1998.

134. DD Gallaher, J Khil. The effect of synbiotics on colon carcinogenesis in rats. *J. Nutr.*, 129 (Suppl.): 1483S–1487S, 1999.

135. M La Rosa, G Bottaro, N Gulino, F Gambuzza, F Di Forti, G Ini, E Tornambe. Prevention of antibiotic-associated diarrhea with *Lactobacillus sporogenes* and fructo-oligosaccharides in children. A multicentric double-blind vs placebo study (Italian). *Minerva Pediatr.*, 55: 447–452, 2003.

136. AD Anderson, CE McNaught, PK Jain, J MacFie. Randomised clinical trial of symbiotic therapy in elective surgical patients. *Gut*, 53: 241–245, 2004.

137. F Shanahan. A gut reaction: lymphoepithelial communication in the intestine. *Science*, 275: 1897–1898, 1997.

138. Y Umesaki, Y Okada, A Imaoka, H Setoyama, S Matsumoto. Interactions between epithelial cells and bacteria, normal and pathogenic. *Science*, 276: 964–965, 1997.

139. J Wang, M Whetsell, JR Klein. Local hormone networks and intestinal T cell homeostasis. *Science*, 275: 1937–1939, 1997.
140. S Blum, EJ Schiffrin. Intestinal microflora and homeostasis of the mucosal immune response: implications for probiotics? In: GW Tannock, Ed., *Probiotics and Prebiotics. Where Are We Going?* Caister Academic Press, London, 2002, pp. 311–330.
141. M Freitas, E Tavan, C Cayuela, L Diop, C Sapin, G Trugnan. Host-pathogens cross-talk. Indigenous bacteria and probiotics also play the game. *Biol. Cell*, 95: 503–506, 2003.
142. R Fuller, G Perdigón, Eds., *Probiotics 3. Immunomodulation by the Gut Microflora and Probiotics*. Kluwer, Dordrecht, 2000.
143. RD Wagner, T Warner, L Roberts, J Farmer, E Balish. Colonization of congenitally immunodeficient mice with probiotic bacteria. *Infect. Immun.*, 65: 3345–3351, 1997.
144. G Perdigón, G Oliver. Modulation of the immune response of the immunosuppressed host by probiotics. In: R Fuller, G Perdigón, Eds., *Probiotics 3. Immunomodulation by the Gut Microflora and Probiotics*. Kluwer, Dordrecht, 2000, pp. 148–175.
145. C Potera. Probiotics gaining recognition. *ASM News*, 65: 739–740, 1999.
146. Le H Duc, HA Tong, TM Barbosa, AO Henriques, SM Cutting. Characterization of Bacillus probiotics available for human use. *Appl. Environ. Microbiol.*, 70: 2161–2171, 2004.
147. Y Huang, MC Adams. *In vitro* assessment of the upper gastrointestinal tolerance of potential probiotic dairy propionibacteria. *Int. J. Food Microbiol.*, 91: 253–260, 2004.
148. MH Helland, T Wicklung, JA Narvhus. Growth and metabolism of selected strains of probiotic bacteria in maize porridge with added malted barley. *Int. J. Food Microbial*, 91: 305–313, 2004.

2

Development of Probiotic Food Ingredients

Katja Schmid, Ralf-Christian Schlothauer, Udo Friedrich,
Christine Staudt, Juha Apajalahti, and Egon Bech Hansen

CONTENTS

2.1 Introduction

2.1.1 History of Probiotics

The idea that some bacteria contained in our food may have beneficial effects
is much older than the term probiotic. At the beginning of the 20th century,
the Russian Nobel Prize laureate Elie Metchnikoff associated the observed

longevity of Bulgarian peasants with their high consumption of live microbes in fermented milk products, as he reported in his book *The Prolongation of Life* (1). In 1930, the Japanese scientist Minoru Shirota isolated a lactic acid bacterium from the feces of a healthy infant. Five years later, one of the first fermented milk drinks thought to support intestinal health was produced with the strain he developed and was named "Yakult." The concept of probiotics was already successful in Asia for many years when the first probiotic fermented milk products were eventually introduced in Europe in the 1980s. Today, probiotic food products containing bifidobacteria and/or lactobacilli are consumed by millions of people worldwide.

2.1.2   Definition of Probiotics

The term probiotic ("for life") was originally proposed in 1965 by Lilley and Stillwell (2) as an antonym to the term antibiotic and was thought to be used for microbial substances, which promote the growth of other microorganisms. Some years later the term probiotic was used in the context of animal feeds by Parker (3) ("Organisms and substances which contribute to intestinal microbial balance") and Fuller (4) ("A live microbial feed supplement which beneficially affects the host animal by improving its intestinal microbial balance"). A more recent definition intended for human nutrition has been proposed in 1998 by Salminen et al. (5): "a live microbial food ingredient that is beneficial to health". Gibson and Roberfroid found it necessary also to define a term for ingredients stabilizing a beneficial flora and defined the term prebiotic: "A nondigestible food ingredient that beneficially affects the host by selectively stimulating the growth and/or activity of one or a limited number of bacteria in the colon, that can improve the host health" (6). Taken together, probiotics and prebiotics define a direct and an indirect route to establish a beneficial intestinal flora.

2.2   Intestinal Flora — Important "Organ" in Human Nutrition and Health

2.2.1   Role of Intestinal Microbial Flora in Digestion and Release of Nutrients

The human intestinal flora is composed of about 10^{14} bacterial cells (7), which belong to hundreds of different species — most of them are probably still unknown. The metabolic activity of the flora is as diverse as the flora itself. The bacteria can ferment various substances that the host is not able to digest himself (e.g., prebiotics). The bacteria and their fermentation products influence the anatomy, physiology, and immunology of the host (8). Beneficial metabolic activities of the intestinal bacteria include the synthesis of B and K vitamins, production of short-chain fatty acids (SCFAs), which serve as an energy source for host tissues (e.g., butyric acid is a preferred energy source for intestinal epithelium), and conversion of dietary carcinogens to inactive compounds and the conversion of prodrugs to active drugs (9).

2.2.2 Role of Beneficial Flora in Exclusion of Pathogens

The indigenous intestinal microflora provides a colonization resistance against potential pathogenic microorganisms. In part, this colonization resistance is based on occupation of available niches (competitive inhibition of binding sites) and autogenic regulation factors (e.g., synthesis of fatty acids, hydrogen peroxide, and bacteriocins). An additional aspect might be the non-specific activation of the immune system (10). The importance of the colonization resistance becomes obvious when a broad-spectrum antibiotic therapy disturbs the bacterial balance and decreases the colonization resistance, which may lead to an overgrowth of enteropathogenic bacteria such as *Clostridium difficile* (9).

2.2.3 Disorders Associated with Disturbed Intestinal Flora

The intestinal microflora plays a fundamental role in the perpetuation of the host's normal intestinal functions (11). The colonization of the intestine with a balanced microflora is, however, also of importance for the correct development of the immune system (12). Exposure to commensal bacteria is essential for fine-tuning of T-cell receptor function and mucosal cytokine profiles (13). The intestinal microflora and the intestinal immune system influence each other and together have an impact on the host, also beyond the intestine (12). Some diseases involve changes in the composition of the normal microflora. These changes may be a consequence or the cause of an enteric disease (12). In the case of infectious diarrhea, it is clear that a pathogenic microorganism causes the disease (12). However, in many other intestinal diseases the question of cause or consequence cannot yet be answered. There is growing evidence that, depending on the genetic background of the host, an abnormal interaction between the gut flora and the local immune response may lead to inflammatory bowel disease (IBD) (9). Alterations in the balance and composition of the microflora (e.g., decreases in lactobacilli and bifidobacteria) have been reported in both disease models in experimental animals and human patients (11). Gut bacteria also seem to play a role in allergies. Changes in the environment, including reduced microbial exposure and altered food consumption, are likely to explain the increase in the prevalence of allergic disease during the past decades (14). A different microbial exposure and changed eating habits could influence the indigenous microflora. Kalliomaki et al. (15)

found differences in the neonatal gut microflora between infants who did or did not develop atopy. Infants with allergic disorders were found to have increased levels of clostridia and reduced levels of bifidobacteria (14). This indicates the importance of the indigenous intestinal bacteria for the development of human immunity to a nonatopic mode (15). Irritable bowel syndrome (IBS) is another intestinal disorder that has been suggested to be accompanied by an abnormal fecal microflora with high numbers of facultative anaerobes and low numbers of lactobacilli and bifidobacteria. However, the cause of IBS is still unknown (16). Endogenous microflora may also be involved in the onset of colon cancer (17). One proposed mechanism is that intestinal bacteria release carcinogens from conjugates formed in the liver. As the gut microflora affects the immune system, it is conceivable that a change in its composition may exert influence on host health also beyond the intestinal tract.

2.3   Analysis of the Effect of Probiotics

2.3.1   *In Vitro* Systems and Assays

2.3.1.1   *Selection of Candidate Strains*

The first task in the development of new probiotic strains is the isolation and selection of appropriate candidates from a pool of biodiversity. This can be done either by looking into existing culture collections (18), by isolation from traditional fermented culture products (19), or by isolation from healthy animals or humans (20–24). Once a number of potential candidates exists, which can be cultured at least under laboratory conditions, the second task is to identify the best subselection of probiotic candidates by applying simple *in vitro* assays that mimic, for example, the digestive tract or the intestinal epithelium.

2.3.1.1.1   *Acid and Bile Tolerance*

One of the most obvious tests to rank probiotic candidates for further screening is their tolerance to conditions in the stomach and the upper gut. Therefore, *in vitro* assays have been developed for studying acid and bile tolerance of bacteria (25–27). In the acid test, candidate strains are exposed to low pH in a buffer solution or medium for a period of time, and then the viability of the bacteria is determined. Some research groups have tried to make the assay more realistic by taking into account the food matrix. There were significantly higher cell counts detectable if the acid exposure was not done in a buffer system but in a simulated food matrix, e.g., infant formula model (Apajalahti et al., unpublished results) or using an even more heterogeneous food matrix like fermented meat or fat from a sausage (Elsser Dieter, Danisco, Germany, personal communication). An improved acid tolerance assay measures the ability of the strains to still acidify substrate solution after the treatment (Apajalahti et al., unpublished results). Bile tolerance can be investigated by incubating the bacteria for 24 hours in a milk–yeast medium containing different concentrations of bile extract and checking the cell count and pH at the beginning and after 24 hours of incubation.

2.3.1.1.2 In Vitro *Adherence Models*

For certain health benefits, it is desirable that probiotic bacteria at least transiently colonize the intestine. Adhesion of the microorganisms to the intestinal mucosa is thought to be a precondition for colonization. The adhesive properties of bacteria can be tested either with monolayers of intestinal tissue culture cells or with human intestinal mucus (28). Different types of differentiated human intestinal cell lines such as HT-29, Caco-2, and HT29-MTX can be used (20, 29). Figure 2.1 shows electron microscopic photographs of bifidobacteria adhering to Caco-2 cells. Most research groups used radiolabeled cells to verify the binding capacity of the strains (29, 30). Also, direct microscopy using Gram-stained bacteria can be employed to assess the binding capacity. With the same cell culture model, the antagonistic activity of probiotic bacteria against pathogenic strains at the mucosal interface can also be studied (29).

2.3.1.1.3 In Vitro *Pathogen Challenge Test*

A third *in vitro* test criterion for probiotic candidate strains may be an *in vitro* challenge against pathogenic bacteria. A fast first screening is usually done with an agar spot test. A selection of probiotic bacteria is spotted on agar plates and overlaid with agar containing certain pathogens. The zone of inhibition around the spots of the probiotic strain is used as a quantitative measure of the inhibition capacity (31, 32). It should be noted, however, that acid production is contributing significantly to the inhibition observed in this test.

2.3.1.1.4 *Immunoassays*

In vitro cultures of intestinal or systemic immune cells can be applied in order to investigate the influence of probiotic bacteria on immunocompetent cells. Usually immune cells from animals or humans (e.g., macrophages, dendritic cells, natural killer cells, other lymphocyte subsets) are incubated with dead or live bacteria, and cytokine release, activation of cytoplasmic transcription factors, expression of activation antigens, or proliferation of immunocompetent cells is observed (33–36). These immunoassays can also be done with intestinal immune cells isolated from patients suffering from, e.g., Crohn's disease (37).

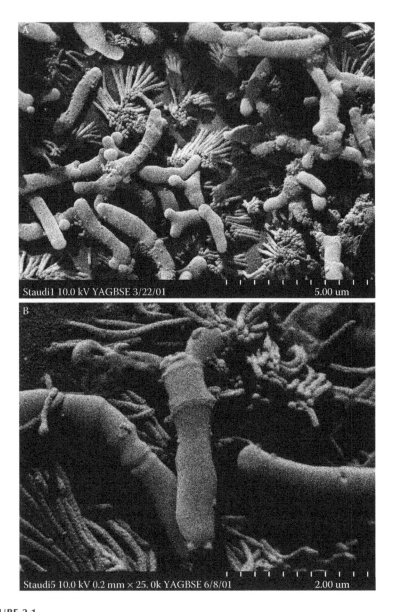

FIGURE 2.1

Electron microscopic photographs showing the interaction between bifidobacteria and the colon adenocarcinoma cell line Caco-2, which was used after complete differentiation. (A) and (B) show the adhesion of breast-fed infant isolate *Bifidobacterium bifidum* S16 to the microvilli of the eukaryotic cells in independent experiments. The whole length of the scales in the right corner of the photographs corresponds to 5 μm (A) and 2 μm (B) (20).

2.3.1.1.5 Safety Tests

Fermented foods have a long-standing history in our daily diet. Up to one quarter of our food intake consists of fermented foodstuffs (38, 39). Since probiotics are designed to have a beneficial effect on the health of the consumer, the safety considerations need to go beyond the history of safe use of lactic acid bacteria. Several safety biomarkers have been discussed in the literature (40–42). Straub et al. (43) gave a detailed analysis of several lactic acid bacteria (LAB) strains that have been screened for their potential to form biogenic amines (BAs). Interestingly, some species known to be probiotic (e.g., *Lactobacillus reuteri*) produce significant amounts of BAs. The safety check of probiotic candidates may involve screening for potential virulence factors (44) and enzyme activities involved in the formation of putatively genotoxic metabolites, including β-glucuronidase, nitroreductase, and azoreductase (40). A further safety assay for potential probiotic candidate strains make use of an *in vitro* mucin degradation model. In this model, any damage or disturbance of the mucin layer is considered to compromise the host's mucosal defense function. Indeed, it has been shown that pathogenic strains of *Escherichia coli* and *Salmonella enterica* do grow on intestinal mucus (45). It should be noted, however, that bacteria utilizing luminal mucin, which has peeled off from the epithelium, might be totally harmless (46, 47).

The spread of food-borne pathogens carrying multiple antibiotic resistances is currently a serious food safety problem. It is consequently very important to assure that industrially used bacterial cultures do not directly or indirectly contribute to increase the antibiotic resistance problem. The possible exchange of antibiotic resistance markers between pathogens and food microorganisms has been raised as a possible food safety issue (48). The work of several research groups evaluating the actual risk of probiotic cultures participating in the spread of antibiotic resistance has been reviewed by Salminen et al. (49). In order to eliminate even a theoretical risk of the spread of antibiotic resistance through the cultures used in food, it has now become common practice to check strains for the sensitivity to commonly used antibiotics (31, 50). LAB are intrinsically resistant to many antibiotics. Nevertheless, there would only be an actual risk for the consumer in case of a transmissible resistance. In most cases the resistance will not be transmissible, and this type of intrinsic resistance is not to be considered a safety problem.

2.3.1.2 Gut Model Reactors

Several reactor systems have been developed to simulate the gastrointestinal tract. They are used for studying the effects of pro- and prebiotics on the composition and metabolism of the microflora. Most models consist of several sequential vessels that represent different parts of the intestinal tract. Macfarlane and colleagues developed a three-stage continuous culture system modeling the proximal, transverse, and distal colon. The model system was validated against gut contents from human sudden death victims (51, 52). The basic medium in which the products are tested is an artificial ileostomy fluid, which mimics the composition of digested contents in the host colon. Another design is the Simulator of the Human Intestinal Microbial Ecosystem (SHIME). This system consists of six vessels simulating the following parts of the gastrointestinal tract: stomach, duodenum, small intestine, ascending colon, transverse colon, and descending colon (8, 53, 54). TIM-1 and TIM-2 are The Netherlands Organisation for Applied Scientific Research's (TNO) gastrointestinal models. TIM-1 simulates the stomach and small intestine, whereas TIM-2 is the colon model. Both systems consist of a number of glass units with flexible walls inside. The flexible walls allow peristaltic movements by pumping water into the space between the flexible walls and the glass jacket. The system has been validated for several applications (55).

Apajalahti and Siikanen developed a miniature multistage (2 to 8), computer-controlled simulator (Figure 2.2), which can be adjusted to mimic fermentation in the distal intestine of various host species (unpublished). The basic medium in this small-scale simulator is authentic extract of ileal digested contents from pigs or other animals, which have been fed a human diet or other diet relevant for the specific study. This complex undefined medium is used to ensure that bacteria are provided with all, also undiscovered, nutrients, vitamins, and cofactors present in the authentic intestinal environment. Multiple channels in the simulator allow simultaneous comparison of several test products under identical conditions by using identical inoculum.

2.3.1.3   Analytical Procedures for Analysis of Complex Microbial Ecosystems

As a consequence of their low morphological diversity, microorganisms cannot be identified under the microscope. Therefore, chemotaxonomic markers, growth requirements, biochemical properties, and genetic characteristics are used to differentiate and enumerate these organisms in the system under investigation. The detection and enumeration of microorganisms can be performed using cultivation-dependent and -independent techniques.

2.3.1.3.1   Cultivation-Dependent Techniques

Conventional plating procedures are still the most widely used enumeration methodologies in various fields of microbiology. There are different strategies for differential detection and enumeration based on cultivation. The general principle of selective plating is that selective media should specifically support the growth of the limited number of bacteria, e.g., species of a single genus. No medium or culture conditions support the growth of all bacteria. Nevertheless, bacterial counts obtained using nutrient-rich media are often called total bacterial counts. For *Lactobacillus acidophilus* and *Bifidobacterium* species, several differential plating media have been described (56). For probiotic bifidobacteria as feed additives, a plating method has recently

FIGURE 2.2
The multistage, computer-controlled laboratory simulator shown mimics bacterial fermentations in the gastrointestinal tract. This EnteroMix simulator was developed at the Danisco Innovation Centre, Kantvik, Finland.

been validated as an official enumeration control method (57). For a detailed summary of cultivation methodologies and media for probiotic *Lactobacillus* and *Bifidobacterium* species, the interested reader is referred to the review of Charteris et al. (56). Cultivation-dependent techniques have also been used in various clinical and intervention studies of probiotic bacteria (58–61). Unfortunately, there are no standard bacterial enumeration methods, which would have been used in these studies, and, therefore, they are seldom comparable. Indeed, it has been shown that even the most commonly used selective media for bifidobacteria recover these bacteria at different efficiencies. All selective media are likely to underestimate the abundance of bifidobacteria and, furthermore, different media select for different species of bifidobacteria (62). The requirement for anaerobiosis of the strictly anaerobic microorganisms of the gastrointestinal tract can complicate sample handling and preparation for cultivation (63). Moreover, it has been known for a long time that plate counting is inadequate to quantify all microorganisms in environmental samples. By calculating the percentages of viable vs. total cell counts, it was estimated that only 0.001 to 15% of the cells occurring in various ecosystems can be cultivated using standard techniques (64). This so-called "great plate count anomaly" (65) can either be explained by (i) the presence of known microorganisms being in a nonculturable state, or for which the chosen cultivation methods were not suitable; or (ii) by the presence of unknown species which cannot be cultured using the applied techniques (64). It is also possible that a substantial fraction of bacteria die or become seriously injured during sample handling. For studies of complex microbial communities, these drawbacks of cultivation-dependent techniques should be taken into account. Cultivation-independent techniques have greatly improved our understanding of complex microbial systems over the past years. Especially the full-cycle rRNA approach, consisting of sequence retrieval and subsequent probing (64), has improved our understanding of microbial diversity.

2.3.1.3.2 Determination of Bacterial Viability

The viability of bacteria may be crucial for the application of probiotic microorganisms. As opposed to the determination of colony-forming units more recently developed techniques make use of cultivation-independent measures of viability. Cell viability can be inferred from enzymatic activities such as esterase conversion of carboxyfluorescein diacetate (cFDA) (66); the reduction of tetrazolium salts such as CTC (67, 68); or dyes such as carboxyfluorescein, propidium iodide, TOTO-1, SYTO 9, and oxonol that have been used to determine viability indicators such as membrane integrity and membrane potential (62, 69–73). Bunthof et al. (66) combined these measures with determination of culturability and showed that cFDA labels the culturable subpopulation, whereas TOTO-1 labels the nonculturable one. These authors also showed that cFDA and TOTO-1 were accurate indicators of culturability for cultures that had been exposed to deconjugated bile salts or to acid. A number of detection tools have been used for these assays including fluorescence microscopy, confocal laser scanning microscopy, and flow cytometry (66, 69–75). For a comparison of these techniques, see the review by Lipski et al. (76). Flow cytometry is of particular interest in the probiotic and starter-culture industry due to the high number of cells that can be analyzed in a given time. Thus, flow-cytometry-based viability tests for probiotic bacteria are likely to replace time-consuming cultivation-dependent techniques in the future.

2.3.1.3.3 Percentage of Guanine+Cytosine Profiling

Percentage of guanine+cytosine (G+C) profiling separates the DNA of the component populations of the bacterial community based on their characteristic G+C content through differential density, which is imposed by the AT-dependent DNA-binding dye bisbenzimidazole (77). Determination of the percentage of G+C content represented by each gradient fraction can be accomplished by regression analysis of data obtained from gradients containing standard DNA samples of known percentage of G+C content (78). This method is little used since it is somewhat laborious. However, it is one of the few methods depicting the total bacterial community independently of any previous knowledge of the component bacteria or their DNA sequences. It is worth noting that most molecular methods are dependent on previous knowledge of DNA sequences for primer and probe design or rely on the presence of widely universal sequence regions. We have previously shown that percentage of G+C profiling depicts the total bacterial community of the gastrointestinal tract in a highly reproducible manner (78, 79). Thus, G+C profiling is a useful approach to study the structure of the entire gastrointestinal microbial community independent of cultivation bias (Figure 2.3). It can also be used in combination with other methods such as 16S rDNA sequencing and denaturing gradient gel electrophoresis (DGGE) and thereby significantly improve the resolution power of these methods. This approach has been successfully applied to studying the effects of inulin on the microbial community of the mouse cecum (80) and among other examples the total bifidobacteria community in human feces has been determined (62).

2.3.1.3.4 PCR-Based Identification and Detection Techniques

Polymerase-chain reaction (PCR) has caused a quantum leap in biology since it was introduced in the late 1980s (81). Also in the field of bacterial detection and identification, PCR-based techniques have become the methods of choice in most microbiological laboratories around the world. In several studies concerning the gastrointestinal microbial system, specific PCR-based detection or identification assays have been developed (82–87). In most studies bacteria are cultivated first, and the isolates are then identified using specific PCR primers. Conventional PCR is a purely qualitative tool, i.e., cell densities or DNA copy numbers cannot be inferred from the amount of PCR product produced. Therefore, if PCR is applied to DNA extracted from fecal samples,

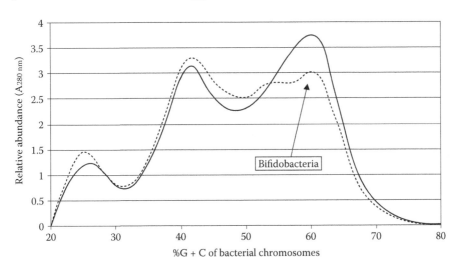

FIGURE 2.3
Percentage Guanine+Cytosine profiling of fecal bacterial communities. The method fractionates bacterial chromosomes according to their characteristic percentage G+C. This allows culture-independent profiling of total bacterial communities. The figure shows the structure of the fecal bacterial community of a human volunteer before (dotted line) and after (solid line) ingestion of a probiotic bacterium *Bifidobacterium* sp. 420.

no information can be obtained concerning the quantitative presence of a bacterial phylotype of interest. Therefore, other methodologies such as PCR-DGGE, which has been used to study the effect of probiotic instillation (88), do not allow quantitative interpretations. A further improvement of PCR has been achieved by monitoring the PCR process in real time (89), i.e., by quantifying the amount of PCR product during the process using DNA binding fluorescent dyes or fluorescent probes. This technique is also referred to as quantitative PCR (Q-PCR). Although Q-PCR has been a particularly active field of research in the past few years (90–102), to our knowledge there has been only one report concerning the quantification of potentially probiotic bacteria (84). Considering the flexibility of Q-PCR in terms of the variety of potential applications such as genetic identification and transcription profiling as well as its potential for high throughput analyses, we expect it to be a widely used method in future probiotic research.

2.3.1.3.5 *Whole-Cell Fluorescent* In Situ *Hybridization*

In contrast to Q-PCR or hybridization techniques that are based on extracted nucleic acids, whole-cell hybridization is applied to morphologically intact cells. Thus, information about the cell concentration can be obtained and, if applied *in situ*, the spatial distribution of specific microorganisms within their microenvironment can be studied. Oligonucleotide probes used for whole-cell hybridization are usually labeled using fluorescent markers such as the cyanine dyes CY3 or CY5, which can be detected by flow cytometry, epifluorescence microscopy, and confocal laser scanning microscopy. Typical microscopic images obtained with CY3-labeled probes are shown in Figure 2.4. If epifluorescence microscopy or confocal laser scanning microscopy are applied, the method is usually referred to as fluorescence *in situ* hybridization (FISH), although the sample microstructure is frequently not maintained to allow realistic *in situ* studies. Prior to whole-cell hybridization, cells need to be permeabilized. Generally, this can be achieved by fixation of the sample with paraformaldehyde (PFA) or ethanol. In some cases, additional permeabilization steps such as treatment with lysozyme (103), mutanolysin (104), lipase (105), or HCl (106) may improve the permeability of cell walls for oligonucleotide probes. To obtain a total bacterial detection rate, the domain-specific probe EUB338 (107) is usually applied to PFA-fixed samples, and percentages are calculated relative to the counts obtained by the nucleic acid stain 4',6-diamidino-2-phenylindole (DAPI) (Figure 2.4). There have been several reports concerning the application of FISH in studies of the gastrointestinal microbial system (58, 108–112). FISH is a particularly useful method to study the composition and structure of complex microbial communities because the fixation largely maintains the *in situ* situation. However, counting procedures are usually very laborious if procedures for computer-aided image analysis are not applied. Moreover, if FISH is applied in combination with retrieval of rRNA and rDNA sequences, yet uncultured bacterial species can be detected (64). This approach allows insights into

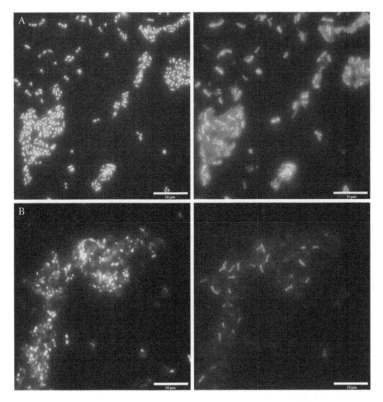

FIGURE 2.4
Microscopic photographs showing fluorescence conferred by DAPI-stained (left panel) and probe-labeled cells (right panel). In (A), cells of a pure culture of probiotic *Bifidobacterium lactis* 420 were hybridized with a specific 16S rRNA-targeting CY3-labeled probe using two helper oligonucleotides. Note that virtually all cells confer both DAPI and CY3 fluorescence. In (B), a fecal sample of a healthy subject was hybridized using the probe Bif164, being complementary to the primer Bif164 published previously (83). The scale bars represent 10 μm.

relationships among microorganisms or with gastrointestinal cell lines that are not accessible using conventional cultivation-dependent techniques. The high degrees of conservation of the ribosomal RNAs usually limit the taxonomic resolution of FISH to the species or subspecies level.

2.3.2 Animal Trials

Animal trials can be very helpful in identifying potential health benefits of pre- and probiotics in a safer, faster, and cheaper way than with human trials. In addition, they offer an opportunity to make investigations that would not be acceptable in human studies (e.g., *Salmonella* challenge studies). Apart from all the advantages of animal trials, the results cannot always be directly transferred to humans. The safety of new probiotic strains can be tested *in vitro* or in animal studies. BALB/c mice may be used in order to study adverse effects, oral toxicity, and translocation potential of new probiotic strains (113–115). Challenge experiments with, for example, *Salmonella* have also been performed in BALB/c mice. In these studies, mice were fed with probiotic products or placebo and received a single oral dose of *Salmonella*. Differences in survival and immunological parameters between probiotic- and placebo-fed mice can be observed (116-118). In general, most health benefits of probiotics that can be observed in humans can also be investigated in animals, e.g., stimulation of the immune system in mice (119) and protection against rotavirus diarrhea in suckling rats (120). In some studies, germ-free animals, monoassociated or human flora-associated animals are used (121-123). Furthermore, there are animal models available for several human diseases that allow studying the effect of pre- and probiotics on these diseases. The transferability of the findings on human patients is a critical point also for these model systems. Ovariectomized rats serve as an osteoporosis model for investigations on calcium absorption and bone density (124). Interleukin-10 gene-deficient mice develop a Crohn's disease-like chronic colitis; hence, they are used as a model for IBD (125). Congenitally immunodeficient mice may help to assess the safety of probiotics for immunodeficient humans including neonates (126). In addition to several animal cancer models for different types of cancer exist (127–129), other examples are rat and mouse models for hypertension (130), allergies (131), and arthritis (132).

2.3.3   Human Trials

Human trials (clinical or dietary intervention studies) are essential for proving health benefits of probiotic strains. Different designs have been applied such as pre- and postintervention designs and placebo-controlled designs, parallel and crossover designs, and case-control studies. The best evidence is probably coming from double-blind, randomized, placebo-controlled trials. In addition to the study design, the methods applied for analysis of various parameters are also of great importance. Epidemiologic evidence relating probiotics or probiotic-containing foods and disease incidence would be valuable but is hardly available. However, these studies would be difficult to control as fundamental parameters such as specific strain and dose would be unknown for most probiotic-containing food products (133). In human studies certain markers are usually applied in order to observe the effect of pre- and probiotics or functional foods in general on the human body. Markers can be classified into three categories: [a] markers that relate to the exposure to the food component under study, [b] markers that relate to the target function or biological response, and [c] markers that relate to an appropriate intermediate endpoint (134). It was suggested that markers of type [b] might lead to enhanced function claims, whereas markers of type [c] might allow reduced risk of disease claims (134). In the case of probiotics, the detection of a certain probiotic strain in feces could be classified as a type [a] marker, an increase in natural killer cell activity and phagocytosis as type [b] (enhanced functioning of the immune system), and a reduced incidence of respiratory symptoms (e.g., cough, sore throat, runny nose) as type [c] (reduced risk of respiratory tract infections).

2.4 Health Benefits of Probiotics

2.4.1 Health Benefits of Probiotics Established in Human Studies

Originally, probiotics were thought to balance disturbances of the gut microflora and thereby prevent or correct gastrointestinal-related dysfunctions. However, some health benefits, e.g., immune modulation, may be achieved even with dead bacteria (135). Many health benefits of probiotic bacteria have been shown in human studies. However, the mechanism of action behind most of these effects remains to be ascertained.

2.4.1.1 *Diarrhea*

The most common cause of acute diarrhea in childhood is rotavirus (5). Several probiotic strains — especially *Lactobacillus rhamnosus* GG — have been shown to prevent or alleviate infantile diarrhea (5, 136–138). It is also well-established that some probiotic strains can both prevent and shorten antibiotic-associated disorders (137–139). However, the evidence for the effects of probiotics on traveler's diarrhea remains low, because few studies have been conducted and they showed contradictory results (137, 138).

2.4.1.2 *Stimulation of the Immune System*

Many human studies have been performed to investigate the effects of probiotic cultures on the immune system. Some studies focused on the intestinal immune system, others on the systemic immunity. These studies reveal that probiotic bacteria are able to enhance both innate and acquired immunity by increasing natural killer cell activity and phagocytosis, changing cytokine profiles, and increasing levels of immunoglobulins (140–142). Two probiotic strains have been developed with a particular focus on their enhancing effects on immune responses: HOWARU™ Bifido (*Bifidobacterium lactis* HN019) and HOWARU™ Rhamnosus (*Lactobacillus rhamnosus* HN001) (113). Both strains have been demonstrated in several studies to enhance natural immune function in healthy people (143–149).

2.4.1.3 Inflammatory Bowel Disease

There is growing evidence that probiotics have a potential therapeutic benefit for patients suffering from IBD. Controlled clinical studies have shown that probiotics are efficacious in the maintenance of remission of pouchitis, prophylaxis of pouchitis after the formation of an ileoanal reservoir, maintenance of remission of ulcerative colitis, and treatment of Crohn's disease. The probiotics that have been used in these controlled clinical trials are two single strains (*Escherichia coli* Nissle 1917; *Saccharomyces boulardii*) and a product called VSL#3 that consists of a mixture of four strains of lactobacilli, three strains of bifidobacteria, and one strain of *Streptococcus salivarius* subsp. *thermophilus*, as has been reviewed lately by Hart et al. and Marteau et al. (11, 138).

2.4.1.4 Irritable Bowel Syndrome

The level of evidence that probiotics may alleviate the symptoms of subjects with IBS is low to date. Varying results have been obtained in the trials that have been conducted so far. Most trials have concentrated on the reduction or cure of symptoms. However, it is possible that the future role of probiotics may be in prevention rather than cure of IBS (16, 138).

2.4.1.5 Lactose Intolerance

Bacterial cultures — yogurt starter cultures as well as some probiotic cultures — are known to improve the lactose digestion in lactose maldigestors. The concentration of the lactose-cleaving enzyme β-galactosidase is too low in subjects suffering from lactose intolerance. Bacteria in fermented or unfermented food products release their β-galactosidase in the small intestine, where it supports lactose digestion. However, probiotic bacteria seem to promote lactose digestion in the small intestine less efficiently than do conventional yogurt cultures, but they may alleviate clinical symptoms arising from the undigested lactose (133, 138, 150).

2.4.1.6 Allergies

Pelto et al. (151) found that *Lactobacillus rhamnosus* GG confers an immunostimulatory effect in healthy adults, whereas the same strain downregulates the immunoinflammatory response in milk-hypersensitive subjects when challenged with milk. Moreover, probiotics have been applied successfully in the management of atopic eczema in infants (152). Furthermore, *Lactobacillus rhamnosus* GG was shown to be effective in the prevention of early atopic disease in children at high risk. The *Lactobacillus rhamnosus* GG product was given prenatally to mothers and postnatally for 6 months to the mothers or to their infants directly. The frequency of atopic eczema in the probiotic group was half that of the placebo group at the age of 2 years. The preventive effect was reconfirmed at the age of 4 years (153, 154).

2.4.1.7 Cancer

A few epidemiological studies indicate an association between a lower incidence of colorectal cancer and consumption of fermented dairy products containing lactobacilli or bifidobacteria. However, there is no direct experimental evidence that probiotics reduce the risk of colon cancer in man, but there is some indirect evidence based on several markers applied in human studies (e.g., fecal enzyme activities, fecal mutagenicity and genotoxicity, immunological markers) (5, 17, 140). The effect of *Lactobacillus casei* strain Shirota on recurrence of superficial bladder cancer was studied by Aso et al. (155, 156). The 50% recurrence-free intervals after tumor resection were significantly higher (1.8 times) for the probiotic group compared with the placebo group. A case-control study conducted in Japan with 180 cases and 445 controls revealed that habitual intake of lactic acid bacteria reduces the risk of bladder cancer (157).

2.4.1.8 Respiratory Tract Infections

The evidence for a potential positive effect of probiotic bacteria on respiratory tract infections has been hitherto very low (158). A probiotic yogurt drink containing *Lactobacillus rhamnosus* GG, *Bifidobacterium* species 420, and *Lactobacillus acidophilus* 145 was shown to reduce significantly the occurrence of potentially pathogenic bacteria in the nose compared with a control yogurt (159). Hatakka et al. (160) conducted a long-term study with 571 Finnish children attending day care centers. They found a slight reduction in the incidence of respiratory infections and antibiotic treatments after 7 weeks' consumption of a milk containing *Lactobacillus rhamnosus* GG compared with a control milk.

2.4.1.9 Constipation

Some studies have been carried out on the effects of lactic acid bacteria on constipation and intestinal motility (161). A reduced severity of constipation and an improved bowel movement frequency and stool consistency have been observed in constipated but otherwise healthy people after consumption of a fermented milk drink containing *Lactobacillus casei* strain Shirota (162). Administration of *Bifidobacterium longum* BB536 to constipated women resulted in a significantly increased defecation frequency and stool softness (163). A positive influence of *Bifidobacterium longum* BB536 on the "regularity" was also reported for elderly people (21).

2.4.1.10 Urogenital Tract Infections

Apart from the intestine, the urogenital tract is a promising field of application for probiotic bacteria. A case-control study with 139 females with acute urinary tract infection and 185 controls revealed that consumption of fermented milk products containing probiotic bacteria was associated with a decreased risk of recurrence of urinary tract infection (164). To date there is just a small number of human studies showing positive effects of probiotics in urinary tract infections (158). Nevertheless, these studies suggest that probiotic preparations given orally or intravaginally may provide a therapeutic source of lactobacilli to help control urogenital infections in women (133, 165).

2.4.1.11 Helicobacter pylori *Infection*

Colonization of the stomach mucosa with *Helicobacter pylori* has been associated with gastritis, stomach carcinoma, gastric ulcer, and lymphomas. Several probiotic strains have been shown to inhibit *Helicobacter pylori in vitro*. Human studies confirmed this inhibitory effect on *Helicobacter pylori*, which seems to be independent of the viability of the bacteria (137, 138, 140, 158).

2.4.1.12 High Cholesterol

Many human studies have evaluated the effects of culture-containing dairy products or probiotic bacteria on cholesterol levels with equivocal results (140). Some examples are given below. A fermented milk containing *Enterococcus faecium* and *Streptococcus thermophilus* was reported to produce a small but significant decrease in total and LDL-cholesterol in patients with primary hypercholesterolemia. However, some subjects did not respond to the product and even showed a cholesterol increment (166). Richelsen et al. (167) investigated the effect of a long-term (6 months) consumption of the same fermented milk product. In normocholesterolemic subjects, the fermented milk resulted in a rapid reduction of LDL-cholesterol, but after 6 months the effect was similar to the placebo milk. In another long-term study (6 months), a yogurt containing *Lactobacillus acidophilus* 145, *Bifidobacterium longum* 913, and 1% oligofructose did not have a significant effect on total cholesterol and LDL-cholestrol in normo- and hypercholesterolemic women. But as the HDL-cholesterol concentration increased significantly, the ratio of LDL to HDL cholesterol decreased significantly (168).

2.5 Technology of Probiotics

2.5.1 Application of Probiotic Cultures in Food Products

Probiotic bacteria are applied in many different products worldwide. In addition to food products, probiotic cultures are also used in pharmaceuticals and animal feed. Most definitions of probiotics are based on live bacteria that confer a health benefit for the consumer. Thus, it is considered as important that probiotic products contain an effective dose of living cells during their whole shelf life. However, for some health benefits, viability of the microorganisms does not seem to be essential. Nonviable bacteria are, for example, applied in some pharmaceuticals and food supplements. The selective enumeration of probiotic species in a fermented product is sometimes impossible due to the product's background flora. Also, the choice of media may have a major impact on the viable cell counts.

2.5.1.1 Dairy Products

Probiotic bacteria have been applied in fermented dairy products for many years. In some cases fermented milk products are monocultures of probiotic bacteria, but usually support cultures are applied to speed up the acidification process and provide the desired texture and flavor. Many lactobacilli and bifidobacteria survive in fermented milk products for 4 to 8 weeks. There are several parameters that may influence the growth and survival of the probiotics, e.g., the starter culture, fermentation temperature, pH, sugar content, presence of oxygen, packaging material, fruit preparations, and other ingredients. Therefore, the survival of a probiotic culture should be reconfirmed in the final product formulation. Probiotics may also be applied to unfermented milk products such as milk-based sweet or acidified drinks and ice cream.

2.5.1.2 Other Food Products

The applicability of probiotics in food products depends in general on factors like water activity, processing and storage temperature, shelf life, oxygen content, pH, mechanical stress, salt content, and content of other harmful or essential ingredients. For many products, excess water activity is a critical parameter that increases the death rate of bacteria. Products with an unfavorable water activity are, for example, cereals, chocolate, marmalade, honey, and toffees. These products are too "dry" for applying live bacteria and too "wet" for the application of freeze-dried bacteria. Freeze-dried bacteria could be applied in these products if the bacteria could be protected from moisture, as small amounts of moisture can be very detrimental to the dried culture. In addition to dairy products, fruit juices have been shown to be suitable carriers for probiotics. The limiting factor for many of the probiotic strains is the low pH of the juices. There is growing interest in applying probiotics to fermented meat products. Lactic acid bacteria have been used for the fermentation of meat products for many years, and today some strains are also utilized as protective cultures. Probiotics might be an instrument to change the perception of meat products toward a healthier image. This might, however, also be a hurdle in the marketing of probiotic sausages. Freeze-dried probiotic bacteria are applied to infant nutrition powders and powdered milk drinks. In these products, the water activity is very low, which is essential for the stability of freeze-dried bacteria.

2.5.1.3 *Food Supplements and Over-the-Counter Products*

Most probiotic food supplements and over-the-counter (OTC) products are available as powders, tablets, and capsules. As these products also contain dried bacteria, the water activity in the final product must be very low. Another critical parameter for tablets is the pressure applied in tableting and the heat that is produced. An enteric coating can be applied on tablets and capsules in order to protect the bacteria from the acidic environment in the stomach and improve their survival rate.

2.6 Conclusions

A very short conclusion to this chapter could be that probiotic food ingredients must be safe and have a well-documented functionality. When taken to the extreme, this statement describes the requirements set for all food ingredients. What makes probiotics different from most other ingredients is that the functionality — improving the health of the consumer — is not seen in the food product itself, but after consumption. The documentation of such functionality requires a substantial amount of science.

During the next few years we will see new probiotic products being developed through the use of genomics. The use of genomics on microorganisms has the potential to reveal the complete metabolic potential of each of the bacterial strains. Several probiotic strains have already been completely sequenced (169, 170). The genome of *Bifidobacterium longum* revealed a large number of genes potentially coding for enzymes in the metabolism of prebiotic carbohydrates (169). This opens the possibility to construct new combinations of pre- and probiotics. Also, the application of human genomics is likely to show a large impact on the innovation in this area. With the availability of the sequence of the entire human genome (171, 172), it is now possible to design DNA chips for the analysis of the regulation of particular genes or even to analyze simultaneously the regulation of all human genes. The combined use of genomics on the microorganisms as well as the human host is likely to result in the design of probiotics with increased health-improving effects.

In addition, the safety aspect of probiotics is somewhat different from most other food ingredients. The production on an industrial scale of highly concentrated live bacterial cultures without undesirable contamination by harmful microorganisms requires competencies possessed only by the dedicated food culture manufacturers and a few scientifically based food companies. Legislation on probiotics and microbial food cultures in general differs among major markets. In the European Union, the regulatory requirements are going to be harmonized and therefore subject to change. Hopefully, Europe in this area will avoid the usual tendency to overregulate. Future innovations in probiotic food ingredients will be determined by a balance between the rapidly expanding scientific achievements and the regulatory framework imposed on the area. The scientific achievements open new possibilities, but they also open our eyes for previously unknown risks. Regulatory measures are justifiable if a real safety problem is being discovered. If, however, the problem is still hypothetical, regulation will probably be harmful, as innovation in the area will be delayed by costly approval procedures. This regulatory issue is also relevant for other food ingredients in the area of food safety.

Danisco is a company producing a large range of food ingredients including, among others, products for food safety, cultures, probiotics, and prebiotics. The company is actively conducting research to support the development of innovative products for food safety and healthy food.

References

1. E Metchnikoff. *The Prolongation of Life: Optimistic Studies.* Heinemann, London, 1907.
2. DM Lilley, RH Stillwell. Probiotics: growth promoting factors produced by microorganisms. *Science*, 147: 747–748, 1965.
3. RB Parker. Probiotics, the other half of the antibiotic story. *Anim. Nutr. Health*, 29: 4–8, 1974.
4. R Fuller. Probiotics in man and animals. *J. Appl. Bacteriol.*, 66(5): 365–378, 1989.
5. S Salminen, C Bouley, MC Boutron-Ruault, JH Cummings, A Franck, GR Gibson. Functional food science and gastrointestinal physiology and function. *Br. J. Nutr.*, 80 (Suppl 1): S147–S171, 1998.
6. GR Gibson, MB Roberfroid. Dietary modulation of the human colonic microbiota: introducing the concept of prebiotics. *J. Nutr.*, 125(6): 1401–1412, 1995.
7. T Mitsuoka. The human gastrointestinal tract. In: BJB Wood, Ed., *The Lactic Acid Bacteria, Vol 1: The Lactic Acid Bacteria in Health and Disease.* Elsevier Applied Science, London, UK, 1992, pp. 69–114.
8. M Gmeiner, W Kneifel, KD Kulbe, R Wouters, P De Boever, L Nollet. Influence of a synbiotic mixture consisting of *Lactobacillus acidophilus* 74-2 and a fructooligosaccharide preparation on the microbial ecology sustained in a simulation of the human intestinal microbial ecosystem (SHIME reactor). *Appl. Microbiol. Biotechnol.*, 53(2): 219–223, 2000.
9. F Shanahan. Probiotics and inflammatory bowel disease: from fads and fantasy to facts and future. *Br. J. Nutr.*, 88:S5–S9, 2002.
10. R Havenaar, I Huis. Probiotics: a general view. In: Wood BJB, Ed., *The Lactic Acid Bacteria, Vol 1: The Lactic Acid Bacteria in Health and Disease.* JHS Huis Int. Veld. Elsevier Applied Science, London, UK, 1992, pp. 151–170.
11. AL Hart, AJ Stagg, MA Kamm. Use of probiotics in the treatment of inflammatory bowel disease. *J. Clin. Gastroenterol.*, 36(2): 111–119, 2003.
12. A Ouwehand, E Isolauri, S Salminen. The role of the intestinal microflora for the development of the immune system in early childhood. *Eur. J. Nutr.*, 41 (Suppl 1): I32–I37, 2002.

13. GA Rook, JL Stanford. Give us this day our daily germs. *Immunol. Today*, 19(3): 113–116, 1998.
14. K Laiho, A Ouwehand, S Salminen, E Isolauri. Inventing probiotic functional foods for patients with allergic disease. *Ann. Allergy Asthma Immunol.*, 89(6 Suppl 1): 75–82, 2002.
15. M Kalliomaki, P Kirjavainen, E Eerola, P Kero, S Salminen, E Isolauri. Distinct patterns of neonatal gut microflora in infants in whom atopy was and was not developing. *J. Allergy Clin. Immunol.*, 107(1): 129–134, 2001.
16. JA Madden, JO Hunter. A review of the role of the gut microflora in irritable bowel syndrome and the effects of probiotics. *Br. J. Nutr.*, 88 (Suppl 1): S67–S72, 2002.
17. J Rafter. Lactic acid bacteria and cancer: mechanistic perspective. *Br. J. Nutr.*, 88: S89–S94, 2002.
18. J Prasad, H Gill, J Smart, PK Gopal. Selection and characterisation of *Lactobacillus* and *Bifidobacterium* strains for use as probiotics. *Int. Dairy J.*, 8: 993–1002, 1998.
19. VM Kimaryo, GA Massawe, NA Olasupo, WH Holzapfel. The use of a starter culture in the fermentation of cassava for the production of "kivunde", a traditional Tanzanian food product. *Int. J. Food Microbiol.*, 56(2–3): 179–190, 2000.
20. C Staudt. Wechselwirkungen von Bifidobakterien mit Darmepithelzellen und mit extrazellulären Matrix- und Plasmaproteinen. Department of Microbiology and Biotechnology, University of Ulm, Ulm, 2002.
21. M Seki, M Igarashi, Y Fukuda, S Shimamura, T Kawashima, K Ogasa. The effect of Bifidobacterium cultured milk on the "regularity" among an aged group. *J. Jpn. Soc. Nutr. Food Sci.*, 34: 379–387, 1978.
22. M Saxelin, Valio-Limited. Lactobacillus GG and its health effects. LGG Summatim 1999 Valron, Helsinki, Finland.
23. G Gardiner, RP Ross, JK Collins, G Fitzgerald, C Stanton. Development of a probiotic cheddar cheese containing human-derived *Lactobacillus paracasei* strains. *Appl. Environ. Microbiol.*, 64(6): 2192–2199, 1998.
24. D Haller, H Colbus, MG Ganzle, P Scherenbacher, C Bode, WP Hammes. Metabolic and functional properties of lactic acid bacteria in the gastro-intestinal ecosystem: a comparative *in vitro* study between bacteria of intestinal and fermented food origin. *Syst. Appl. Microbiol.*, 24(2): 218–226, 2001.
25. C Dunne, L O'Mahony, L Murphy, G Thornton, D Morrissey, S O'Halloran, et al. *In vitro* selection criteria for probiotic bacteria of human origin: correlation with *in vivo* findings. *Am. J. Clin. Nutr.*, 73(Suppl 2): 386S–392S, 2001.
26. LS Chou, B Weimer. Isolation and characterization of acid- and bile-tolerant isolates from strains of *Lactobacillus acidophilus*. *J. Dairy Sci.*, 82(1): 23–31, 1999.
27. MA Ehrmann, P Kurzak, J Bauer, RF Vogel. Characterization of lactobacilli towards their use as probiotic adjuncts in poultry. *J. Appl. Microbiol.*, 92(5): 966–975, 2002.
28. AC Ouwehand, PV Kirjavainen, MM Gronlund, E Isolauri, SJ Salminen. Adhesion of probiotic microorganisms to intestinal mucus. *Int. Dairy J.*, 9: 630, 1999.
29. PK Gopal, J Prasad, J Smart, HS Gill. *In vitro* adherence properties of *Lactobacillus rhamnosus* DR20 and *Bifidobacterium lactis* DR10 strains and their antagonistic activity against an enterotoxigenic *Escherichia coli*. *Int. J. Food Microbiol.*, 67(3): 207–216, 2001.

30. EM Tuomola, SJ Salminen. Adhesion of some probiotic and dairy *Lactobacillus* strains to Caco-2 cell cultures. *Int. J. Food Microbiol.*, 41(1): 45–51, 1998.
31. MM Brashears, D Jaroni, J Trimble. Isolation, selection, and characterization of lactic acid bacteria for a competitive exclusion product to reduce shedding of *Escherichia coli* O157:H7 in cattle. *J. Food Prot.*, 66(3): 355–363, 2003.
32. MF Fernandez, S Boris, C Barbes. Probiotic properties of human lactobacilli strains to be used in the gastrointestinal tract. *J. Appl. Microbiol.*, 94(3): 449–455, 2003.
33. HR Christensen, H Frokiaer, JJ Pestka. *Lactobacilli* differentially modulate expression of cytokines and maturation surface markers in murine dendritic cells. *J. Immunol.*, 168(1): 171–178, 2002.
34. KY Ng, MW Griffiths. Enhancement of macrophage cytokine release by cell-free fractions of fermented milk. *Milchwissenschaft*, 57(2): 66–70, 2002.
35. M Miettinen, A Lehtonen, I Julkunen, S Matikainen. *Lactobacilli* and *Streptococci* activate NF-kappa B and STAT signaling pathways in human macrophages. *J. Immunol.*, 164(7): 3733–3740, 2000.
36. D Haller, S Blum, C Bode, WP Hammes, EJ Schiffrin. Activation of human peripheral blood mononuclear cells by nonpathogenic bacteria *in vitro*: evidence of NK cells as primary targets. *Infect. Immun.*, 68(2): 752–759, 2000.
37. N Borruel, M Carol, F Casellas, M Antolin, F de Lara, E Espin, et al. Increased mucosal tumour necrosis factor alpha production in Crohn's disease can be downregulated ex vivo by probiotic bacteria. *Gut*, 51(5): 659–664, 2002.
38. WP Hammes, PS Tichaczek. The potential of lactic acid bacteria for the production of safe and wholesome food. *Z. Lebensm. Unters. Forsch.*, 198(3): 193–201, 1994.
39. SP Borriello, WP Hammes, W Holzapfel, P Marteau, J Schrezenmeir, M Vaara, et al. Safety of probiotics that contain *lactobacilli* or *bifidobacteria*. *Clin. Infect. Dis.*, 36(6): 775–780, 2003.
40. KJ Heller. Evaluating the safety of probiotic cultures. *Dtsch. Milchwirtsch.*, 51: 820–823, 2000.
41. DC Donohue, S Salminen. Safety of probiotic bacteria. *Asia Pac. J. Clin. Nutr.*, 5: 25–28, 1996.
42. DC Donohue, S Salminen, P Marteau. Safety of probiotic bacteria. In: S Salminen, A von Wright, Eds., *Lactic Acid Bacteria: Microbiology and Functional Aspects*. Marcel Dekker, New York, 1998, pp. 369–383.
43. BW Straub, M Kicherer, SM Schilcher, WP Hammes. The formation of biogenic amines by fermentation organisms. *Z. Lebensm. Unters. Forsch.*, 201(1): 79–82, 1995.
44. CM Franz, WH Holzapfel, ME Stiles. *Enterococci* at the crossroads of food safety. *Int. J. Food Microbiol.*, 47(1–2): 1–24, 1999.
45. S Edelman, S Leskela, E Ron, J Apajalahti, TK Korhonen. *In vitro* adhesion of an avian pathogenic *Escherichia coli* O78 strain to surfaces of the chicken intestinal tract and to ileal mucus. *Vet. Microbiol.*, 91(1): 41–56, 2003.
46. JS Zhou, PK Gopal, HS Gill. Potential probiotic lactic acid bacteria *Lactobacillus rhamnosus* (HN001), *Lactobacillus acidophilus* (HN017) and *Bifidobacterium lactis* (HN019) do not degrade gastric mucin *in vitro*. *Int. J. Food Microbiol.*, 63(1–2): 81–90, 2001.
47. AC Ouwehand, H Lagstrom, T Suomalainen, S Salminen. Effect of probiotics on constipation, fecal azoreductase activity and fecal mucin content in the elderly. *Ann. Nutr. Metab.*, 46(3–4): 159–162, 2002.

48. M Teuber, L Meile, F Schwarz. Acquired antibiotic resistance in lactic acid bacteria from food. *Antonie Van Leeuwenhoek*, 76(1–4): 115–137, 1999.

49. S Salminen, A von Wright, L Morelli, P Marteau, D Brassart, WM de Vos. Demonstration of safety of probiotics — a review. *Int. J. Food Microbiol.*, 44(1–2): 93–106, 1998.

50. M Danielsen, A Wind. Susceptibility of *Lactobacillus* spp. to antimicrobial agents. *Int J Food Microbiol* 82(1): 1–11, 2003.

51. GT Macfarlane, S Macfarlane, GR Gibson. Validation of a three-stage compound continuous culture system for investigating the effect of retention time on the ecology and metabolism of bacteria in the human colon. *Microb. Ecol.*, 35(2): 180–187, 1998.

52. GR Gibson, R Fuller. Aspects of *in vitro* and *in vivo* research approaches directed toward identifying probiotics and prebiotics for human use. *J. Nutr.*, 130(2 Suppl): 391S–395S, 2000.

53. AT Meddah, A Yazourh, I Desmet, B Risbourg, W Verstraete, MB Romond. The regulatory effects of whey retentate from bifidobacteria fermented milk on the microbiota of the Simulator of the Human Intestinal Microbial Ecosystem (SHIME). *J. Appl. Microbiol.*, 91(6): 1110–1117, 2001.

54. P De Boever, B Deplancke, W Verstraete. Fermentation by gut microbiota cultured in a simulator of the human intestinal microbial ecosystem is improved by supplementing a soygerm powder. *J. Nutr.*, 130(10): 2599–2606, 2000.

55. MJ van der Werf, K Venema. Bifidobacteria: genetic modification and the study of their role in the colon. *J. Agric. Food Chem.*, 49(1): 378–383, 2001.

56. WP Charteris, PM Kelly, L Morelli, JK Collins. Selective detection, enumeration and identification of potentially probiotic *Lactobacillus* and *Bifidobacterium* species in mixed bacterial populations. *Int. J. Food Microbiol.*, 35(1): 1–27, 1997.

57. RG Leuschner, J Bew, P Simpson, PR Ross, C Stanton. A collaborative study of a method for the enumeration of probiotic bifidobacteria in animal feed. *Int. J. Food Microbiol.*, 83(2): 161–170, 2003.

58. GW Tannock, K Munro, HJ Harmsen, GW Welling, J Smart, PK Gopal. Analysis of the fecal microflora of human subjects consuming a probiotic product containing *Lactobacillus rhamnosus* DR20. *Appl. Environ. Microbiol.*, 66(6): 2578–2588, 2000.

59. M Gmeiner, W Kneifel, KD Kulbe, R Wouters, P De Boever, L Nollet, et al. Influence of a synbiotic mixture consisting of *Lactobacillus acidophilus* 74-2 and a fructooligosaccharide preparation on the microbial ecology sustained in a simulation of the human intestinal microbial ecosystem (SHIME reactor). *Appl. Microbiol. Biotechnol.*, 53(2): 219–223, 2000.

60. N Yuki, K Watanabe, A Mike, Y Tagami, R Tanaka, M Ohwaki, et al. Survival of a probiotic, *Lactobacillus casei* strain Shirota, in the gastrointestinal tract: selective isolation from faeces and identification using monoclonal antibodies. *Int. J. Food Microbiol.*, 48(1): 51–57, 1999.

61. CN Jacobsen, V Rosenfeldt Nielsen, AE Hayford, PL Moller, KF Michaelsen, A Paerregaard, et al. Screening of probiotic activities of forty-seven strains of *Lactobacillus* spp. by *in vitro* techniques and evaluation of the colonization ability of five selected strains in humans. *Appl. Environ. Microbiol.*, 65(11): 4949–4956, 1999.

62. JHA Apajalahti, A Kettunen, H Nurminen, H Jatila, WE Holben. Selective plating underestimates abundance and shows differential recovery of bifidobacterial species from human feces. *Appl. Environ. Microbiol.*, 69: 5731–5735, 2003.
63. SC Ricke, SD Pillai. Conventional and molecular methods for understanding probiotic bacteria functionality in gastrointestinal tracts. *Crit. Rev. Microbiol.*, 25(1): 19–38, 1999.
64. RI Amann, W Ludwig, KH Schleifer. Phylogenetic identification and *in situ* detection of individual microbial cells without cultivation. *Microbiol. Rev.*, 59(1): 143–169, 1995.
65. JT Staley, A Konopka. Measurement of *in situ* activities of nonphotosynthetic microorganisms in aquatic and terrestrial habitats. *Annu. Rev. Microbiol.*, 39: 321–346, 1985.
66. CJ Bunthof, K Bloemen, P Breeuwer, FM Rombouts, T Abee. Flow cytometric assessment of viability of lactic acid bacteria. *Appl. Environ. Microbiol.*, 67(5): 2326–2335, 2001.
67. V Creach, AC Baudoux, G Bertru, BL Rouzic. Direct estimate of active bacteria: CTC use and limitations. *J. Microbiol. Methods*, 52(1): 19–28, 2003.
68. VK Bhupathiraju, M Hernandez, D Landfear, L Alvarez-Cohen. Application of a tetrazolium dye as an indicator of viability in anaerobic bacteria. *J. Microbiol. Methods*, 37(3): 231–243, 1999.
69. KB Amor, P Breeuwer, P Verbaarschot, FM Rombouts, AD Akkermans, WM de Vos, et al. Multiparametric flow cytometry and cell sorting for the assessment of viable, injured, and dead bifidobacterium cells during bile salt stress. *Appl. Environ. Microbiol.*, 68(11): 5209–5216, 2002.
70. CJ Bunthof, S van den Braak, P Breeuwer, FM Rombouts, T Abee. Rapid fluorescence assessment of the viability of stressed *Lactococcus lactis*. *Appl. Environ. Microbiol.*, 65(8): 3681–3689, 1999.
71. CJ Bunthof, S van Schalkwijk, W Meijer, T Abee, J Hugenholtz. Fluorescent method for monitoring cheese starter permeabilization and lysis. *Appl. Environ. Microbiol.*, 67(9): 4264–4271, 2001.
72. MA Auty, GE Gardiner, SJ McBrearty, EO O'Sullivan, DM Mulvihill, JK Collins, et al. Direct *in situ* viability assessment of bacteria in probiotic dairy products using viability staining in conjunction with confocal scanning laser microscopy. *Appl. Environ. Microbiol.*, 67(1): 420–425, 2001.
73. KB Rechinger, H Siegumfeldt. Rapid assessment of cell viability of *Lactobacillus delbrueckii* subsp. *bulgaricus* by measurement of intracellular pH in individual cells using fluorescence ratio imaging microscopy. *Int. J. Food Microbiol.*, 75(1–2): 53–60, 2002.
74. CJ Bunthof, T Abee. Development of a flow cytometric method to analyze subpopulations of bacteria in probiotic products and dairy starters. *Appl. Environ. Microbiol.*, 68(6): 2934–2942, 2002.
75. CJ Bunthof, S van den Braak, P Breeuwer, FM Rombouts, T Abee. Fluorescence assessment of *Lactococcus lactis* viability. *Int. J. Food Microbiol.*, 55(1–3): 291–294, 2000.
76. A Lipski, U Friedrich, K Altendorf. Application of rRNA-targeted oligonucleotide probes in biotechnology. *Appl. Microbiol. Biotechnol.*, 56(1–2): 40–57, 2001.
77. WE Holben, D Harris. DNA-based monitoring of total bacterial community structure in environmental samples. *Mol. Ecol.*, 4(5): 627–631, 1995.

78. JH Apajalahti, A Kettunen, MR Bedford, WE Holben. Percent G+C profiling accurately reveals diet-related differences in the gastrointestinal microbial community of broiler chickens. *Appl. Environ. Microbiol.*, 67(12): 5656–5667, 2001.

79. JH Apajalahti, LK Sarkilahti, BR Maki, JP Heikkinen, PH Nurminen, WE Holben. Effective recovery of bacterial DNA and percent-guanine-plus-cytosine-based analysis of community structure in the gastrointestinal tract of broiler chickens. *Appl. Environ. Microbiol.*, 64(10): 4084–4088, 1998.

80. JH Apajalahti, H Kettunen, A Kettunen, WE Holben, PH Nurminen, N Rautonen, et al. Culture-independent microbial community analysis reveals that inulin in the diet primarily affects previously unknown bacteria in the mouse cecum. *Appl. Environ. Microbiol.*, 68(10): 4986–4995, 2002.

81. RK Saiki, DH Gelfand, S Stoffel, SJ Scharf, R Higuchi, GT Horn, et al. Primer-directed enzymatic amplification of DNA with a thermostable DNA polymerase. *Science*, 239: 487–491, 1988.

82. M Ventura, R Reniero, R Zink. Specific identification and targeted characterization of *Bifidobacterium lactis* from different environmental isolates by a combined multiplex-PCR approach. *Appl. Environ. Microbiol.*, 67(6): 2760–2765, 2001.

83. RG Kok, A de Waal, F Schut, GW Welling, G Weenk, KJ HellingwerfJ. Specific detection and analysis of a probiotic *Bifidobacterium* strain in infant feces. *Appl. Environ. Microbiol.*, 62(10): 3668–3672, 1996.

84. T Requena, J Burton, T Matsuki, K Munro, MA Simon, R Tanaka, et al. Identification, detection, and enumeration of human *Bifidobacterium* species by PCR targeting the transaldolase gene. *Appl. Environ. Microbiol.*, 68(5): 2420–2427, 2002.

85. A Tilsala-Timisjarvi, T Alatossava. Development of oligonucleotide primers from the 16S-23S rRNA intergenic sequences for identifying different dairy and probiotic lactic acid bacteria by PCR. *Int. J. Food Microbiol.*, 35(1): 49–56, 1997.

86. T Matsuki, K Watanabe, R Tanaka, M Fukuda, H Oyaizu. Distribution of bifidobacterial species in human intestinal microflora examined with 16S rRNA-gene-targeted species-specific primers. *Appl. Environ. Microbiol.*, 65(10): 4506–4512, 1999.

87. Y Song, N Kato, C Liu, Y Matsumiya, H Kato, K Watanabe. Rapid identification of 11 human intestinal *Lactobacillus* species by multiplex PCR assays using group- and species-specific primers derived from the 16S-23S rRNA intergenic spacer region and its flanking 23S rRNA. *FEMS Microbiol. Lett.*, 187(2): 167–173, 2000.

88. JP Burton, PA Cadieux, G Reid. Improved understanding of the bacterial vaginal microbiota of women before and after probiotic instillation. *Appl. Environ. Microbiol.*, 69(1): 97–101, 2003.

89. R Higuchi, C Fockler, G Dollinger, R Watson. Kinetic PCR analysis: real-time monitoring of DNA amplification reactions. *Biotechnology (NY)*, 11(9): 1026–1030, 1993.

90. CF Amar, PH Dear, J McLauchlin. Detection and genotyping by real-time PCR/RFLP analyses of *Giardia duodenalis* from human faeces. *J. Med. Microbiol.*, 52(8): 681–683, 2003.

91. MA Verboon-Maciolek, M Nijhuis, AM van Loon, N van Maarssenveen, H van Wieringen, MA Pekelharing-Berghuis, et al. Diagnosis of enterovirus infection in the first 2 months of life by real-time polymerase chain reaction. *Clin. Infect. Dis.*, 37(1): 1–6, 2003.

92. J Menotti, B Cassinat, R Porcher, C Sarfati, F Derouin, JM Molina. Development of a real-time polymerase-chain-reaction assay for quantitative detection of *Enterocytozoon bieneusi* DNA in stool specimens from immunocompromised patients with intestinal microsporidiosis. *J. Infect. Dis.*, 187(9): 1469–1474, 2003

93. A Heim, C Ebnet, G Harste, P Pring-Akerblom. Rapid and quantitative detection of human adenovirus DNA by real-time PCR. *J. Med. Virol.*, 70(2): 228–239, 2003.

94. SD Belanger, M Boissinot, N ClairouxN, FJ Picard, MG Bergeron. Rapid detection of *Clostridium difficile* in feces by real-time PCR. *J. Clin. Microbiol.*, 41(2): 730–734, 2003.

95. J Blessmann, H Buss, PA Nu, BT Dinh, QT Ngo, AL Van, et al. Real-time PCR for detection and differentiation of *Entamoeba histolytica* and *Entamoeba dispar* in fecal samples. *J. Clin. Microbiol.*, 40(12): 4413–4417, 2002.

96. DM Wolk, SK Schneider, NL Wengenack, LM Sloan, JE Rosenblatt. Real-time PCR method for detection of *Encephalitozoon intestinalis* from stool specimens. *J. Clin. Microbiol.*, 40(11): 3922–3928, 2002.

97. T Matsuki, K Watanabe, J Fujimoto, Y Miyamoto, T Takada, K Matsumoto, et al. Development of 16S rRNA-gene-targeted group-specific primers for the detection and identification of predominant bacteria in human feces. *Appl. Environ. Microbiol.*, 68(11): 5445–5451, 2002.

98. PB Kurowski, JL Traub-Dargatz, PS Morley, CR Gentry-Weeks. Detection of *Salmonella* spp in fecal specimens by use of real-time polymerase chain reaction assay. *Am. J. Vet. Res.*, 63(9): 1265–1268, 2002.

99. SD Belanger, M Boissinot, C Menard, FJ Picard, MG Bergeron. Rapid detection of Shiga toxin-producing bacteria in feces by multiplex PCR with molecular beacons on the smart cycler. *J. Clin. Microbiol.*, 40(4): 1436–1440, 2002.

100. T Watkins-Riedel, M Woegerbauer, D Hollemann, P Hufnagl. Rapid diagnosis of enterovirus infections by real-time PCR on the LightCycler using the TaqMan format. *Diagn. Microbiol. Infect. Dis.*, 42(2): 99–105, 2002.

101. JM Logan, KJ Edwards, NA Saunders, J Stanley. Rapid identification of *Campylobacter* spp. by melting peak analysis of biprobes in real-time PCR. *J. Clin. Microbiol.*, 39(6): 2227–223, 2001.

102. T Bellin, M Pulz, A Matussek, HG Hempen, F Gunzer. Rapid detection of enterohemorrhagic *Escherichia coli* by real-time PCR with fluorescent hybridization probes. *J. Clin. Microbiol.*, 39(1): 370–374, 2001.

103. C Beimfohr, A Krause, R Amann, W Ludwig, KH Schleifer. *In situ* identification of *Lactococci, Enterococci* and *Streptococci. Syst. Appl. Microbiol.*, 16: 450–456, 1993.

104. M Schuppler, M Wagner, G Schön, UB Göbel. *In situ* identification of nocardioform actinomycetes in activated sludge using fluorescent rRNA-targeted oligonucleotide probes. *Microbiology*, 144(Part 1): 249–259, 1998.

105. RJ Davenport, TP Curtis, M Goodfellow, FM Stainsby, M Bingley. Quantitative use of fluorescent in situ hybridization to examine relationships between mycolic acid-containing actinomycetes and foaming in activated sludge plants. *Appl. Environ. Microbiol.*, 66(3): 1158–1166, 2000.

106. SJ Macnaughton, AG O'Donnell, TM Embley. Permeabilization of mycolic-acid-containing actinomycetes for in situ hybridization with fluorescently labelled oligonucleotide probes. *Microbiology*, 140(Part 10): 2859–2865, 1994.

107. RI Amann, BJ Binder, RJ Olson, SW Chisholm, R Devereux, DA Stahl. Combination of 16S rRNA-targeted oligonucleotide probes with flow cytometry for analyzing mixed microbial populations. *Appl. Environ. Microbiol.*, 56(6): 1919–1925, 1990.

108. AH Franks, HJ Harmsen, GC Raangs, GJ Jansen, F Schut, GW Welling. Variations of bacterial populations in human feces measured by fluorescent *in situ* hybridization with group-specific 16S rRNA-targeted oligonucleotide probes. *Appl. Environ. Microbiol.*, 64(9): 3336–3345, 1998.

109. PS Langendijk, F Schut, GJ Jansen, GC Raangs, GR Kamphuis, MH Wilkinson, et al. Quantitative fluorescence *in situ* hybridization of *Bifidobacterium* spp. with genus-specific 16S rRNA-targeted probes and its application in fecal samples. *Appl. Environ. Microbiol.*, 61(8): 3069–3075, 1995.

110. HJ Harmsen, GR Gibson, P Elfferich, GC Raangs, AC Wildeboer-Veloo, A Argaiz, et al. Comparison of viable cell counts and fluorescence *in situ* hybridization using specific rRNA-based probes for the quantification of human fecal bacteria. *FEMS Microbiol. Lett.*, 183(1): 125–129, 2000.

111. HJ Harmsen, AC Wildeboer-Veloo, GC Raangs, AA Wagendorp, N Klijn, JG Bindels, et al. Analysis of intestinal flora development in breast-fed and formula-fed infants by using molecular identification and detection methods. *J. Pediatr. Gastroenterol. Nutr.*, 30(1): 61–67, 2000.

112. A Sghir, G Gramet, A Suau, V Rochet, P Pochart, J Dore. Quantification of bacterial groups within human fecal flora by oligonucleotide probe hybridization. *Appl. Environ. Microbiol.*, 66(5): 2263–2266, 2000.

113. JS Zhou, Q Shu, KJ Rutherfurd, J Prasad, MJ Birtles, PK Gopal, et al. Safety assessment of potential probiotic lactic acid bacterial strains *Lactobacillus rhamnosus* HN001, *Lb. acidophilus* HN017, and *Bifidobacterium lactis* HN019 in BALB/c mice. *Int. J. Food Microbiol.*, 56(1): 87–96, 2000.

114. JS Zhou, Q Shu, KJ Rutherfurd, J Prasad, PK Gopal, HS Gill. Acute oral toxicity and bacterial translocation studies on potentially probiotic strains of lactic acid bacteria. *Food Chem. Toxicol.*, 38(2–3): 153–161, 2000.

115. Q Shu, JS Zhou, KJ Rutherfurd, MJ Birtles, J Prasad, PK Gopal, et al. Probiotic lactic acid bacteria (*Lactobacillus acidophilus* HN017, *Lactobacillus rhamnosus* HN001 and *Bifidobacterium lactis* HN019) have no adverse effects on the health of mice. *Int. Dairy J.*, 9: 831–836, 1999.

116. HS Gill, Q Shu, H Lin, KJ Rutherfurd, ML Cross. Protection against translocating *Salmonella typhimurium* infection in mice by feeding the immuno-enhancing probiotic *Lactobacillus rhamnosus* strain HN001. *Med. Microbiol. Immunol., (Berlin)*, 190(3):97–104, 2001.

117. Q Shu, H Lin, KJ Rutherfurd, SG Fenwick, J Prasad, PK Gopal, et al. Dietary *Bifidobacterium lactis* (HN019) enhances resistance to oral *Salmonella typhimurium* infection in mice. *Microbiol. Immunol.*, 44(4): 213–222, 2000.

118. M Paubert-Braquet, Gan Xiao-Hu, C Gaudichon, N Hedef, A Serikoff, C Bouley, et al. Enhancement of host resistance against *Salmonella typhimurium* in mice fed a diet supplemented with yogurt or milks fermented with various *Lactobacillus casei* strains. *Int. J. Immunother.*, 11(4): 153–161, 1995.

119. HS Gill, KJ Rutherfurd, J Prasad, PK Gopal. Enhancement of natural and acquired immunity by *Lactobacillus rhamnosus* (HN001), *Lactobacillus acidophilus* (HN017) and *Bifidobacterium lactis* (HN019). *Br. J. Nutr.*, 83(2): 167–176, 2000.

120. C Guerin-Danan, JC Meslin, A Chambard, A Charpilienne, P Relano, C Bouley, et al. Food supplementation with milk fermented by *Lactobacillus casei* DN-114 001 protects suckling rats from rotavirus-associated diarrhea. *J. Nutr.*, 131(1): 111–117, 2001.

121. Z Djouzi, C Andrieux, MC Degivry, C Bouley, O Szylit. The association of yogurt starters with *Lactobacillus casei* DN 114.001 in fermented milk alters the composition and metabolism of intestinal microflora in germ-free rats and in human flora-associated rats. *J. Nutr.*, 127(11): 2260–2266, 1997.

122. S Yamazaki, K Machii, S Tsuyuki, H Momose, T Kawashima, K Ueda. Immunological responses to monoassociated *Bifidobacterium longum* and their relation to prevention of bacterial invasion. *Immunology*, 56(1): 43–50, 1985.

123. R Oozeer, N Goupil-Feuillerat, CA Alpert, GM van de Guchte, J Anba, J Mengaud. *Lactobacillus casei* is able to survive and initiate protein synthesis during its transit in the digestive tract of human flora-associated mice. *Appl. Environ. Microbiol.*, 68(7): 3570–3574, 2002.

124. M Igarashi, Y Iiyama, R Kato, M Tomita, N Asami, I. Ezawa. Effect of *Bifidobacterium longum* and lactulose on the strength of bone in ovariectomized osteoporosis model rats. *Bifidus*, 7: 139–147, 1994.

125. KL Madsen. Inflammatory bowel disease: lessons from the IL-10 gene-deficient mouse. *Clin. Invest. Med.*, 24(5): 250–257, 2001.

126. RD Wagner, T Warner, L Roberts, J Farmer, E Balish. Colonization of congenitally immunodeficient mice with probiotic bacteria. *Infect. Immun.*, 65(8): 3345–3351, 1997.

127. T Mizutani, T Mitsuoka. Inhibitory effect of some intestinal bacteria on liver tumorigenesis in gnotobiotic C3H/He male mice. *Cancer Lett.*, 11(2): 89–95, 1980.

128. M Asano, E Karasawa, T Takayama. Antitumor activity of *Lactobacillus casei* (LC 9018) against experimental mouse bladder tumor (MBT-2). *J. Urol.*, 136(3): 719–721, 1986.

129. M Fukui, T Fujino, K Tsutsui, T Maruyama, H Yoshimura, T Shinohara, et al. The tumor-preventing effect of a mixture of several lactic acid bacteria on 1,2-dimethylhydrazine-induced colon carcinogenesis in mice. *Oncol. Rep.*, 8(5): 1073–1078, 2001.

130. A Fuglsang, D Nilsson, NC Nyborg. Cardiovascular effects of fermented milk containing angiotensin-converting enzyme inhibitors evaluated in permanently catheterized, spontaneously hypertensive rats. *Appl. Environ. Microbiol.*, 68(7): 3566–3569, 2002.

131. K Shida, R Takahashi, E Iwadate, K Takamizawa, H Yasui, T Sato, et al. *Lactobacillus casei* strain Shirota suppresses serum immunoglobulin E and immunoglobulin G1 responses and systemic anaphylaxis in a food allergy model. *Clin. Exp. Allergy*, 32(4): 563–570, 2002.

132. I Kato, K Endo-Tanaka, T Yokokura. Suppressive effects of the oral administration of *Lactobacillus casei* on type II collagen-induced arthritis in DBA/1 mice. *Life Sci.*, 63(8): 635–644, 1998.

133. ME Sanders. Considerations for use of probiotic bacteria to modulate human health. *J. Nutr.*, 130(2S Suppl): 384S–390S, 2000.

134. AT Diplock, PJ Aggett, M Ashwell, F Bornet, EB Fern, MB Roberfroid. Scientific concepts of functional foods in Europe: consensus document. *Br. J. Nutr.*, 81(Suppl. 1): S1–S27, 1999.

135. K Laiho, U Hopp, AC Ouwehand, S Salminen, E Isolauri. Probiotics: on-going research on atopic individuals. *Br. J. Nutr.*, 88: S19–S27, 2002.

136. H Szajewskah, JZ Mrukowicz. Probiotics in the treatment and prevention of acute infectious diarrhea in infants and children: a systematic review of published randomized, double-blind, placebo-controlled trials. *J. Pediatr. Gastroenterol. Nutr.*, 33 Suppl 2: S17–S25, 2001.

137. LJ Fooks, GR Gibson. Probiotics as modulators of the gut flora. *Br. J. Nutr.*, 88: S39–S49, 2002.

138. P Marteau, P Seksik, R Jian. Probiotics and intestinal health effects: a clinical perspective. *Br. J. Nutr.*, 88: S51–S57, 2002.

139. JF Colombel, A Cortot, C Neut, C Romond. Yoghurt with *Bifidobacterium longum* reduces erythromycin-induced gastrointestinal effects. *Lancet*, 2(8549): 43, 1987.

140. ME Sanders. Probiotics. *Food Technol.*, 53(11): 67–77, 1999.

141. HS Gill. Stimulation of the immune system by lactic cultures. *Int. Dairy J.*, 8: 535–544, 1998.

142. AE Wold. Immune effects of probiotics. *Scand. J. Nutr.*, 45: 76–85, 2001.

143. K Arunachalam, HS Gill, RK Chandra. Enhancement of natural immune function by dietary consumption of *Bifidobacterium lactis* (HN019). *Eur. J. Clin. Nutr.*, 54(3): 263–267, 2000.

144. BL Chiang, YH Sheih, LH Wang, CK Liao, HS Gill. Enhancing immunity by dietary consumption of a probiotic lactic acid bacterium (*Bifidobacterium lactis* HN019): optimization and definition of cellular immune responses. *Eur. J. Clin. Nutr.*, 54(11): 849–855, 2000.

145. HS Gill, KJ Rutherfurd, ML Cross, PK Gopal. Enhancement of immunity in the elderly by dietary supplementation with the probiotic *Bifidobacterium lactis* HN019. *Am. J. Clin. Nutr.*, 74(6): 833–839, 2001.

146. HS Gill, KJ Rutherfurd, ML Cross. Dietary probiotic supplementation enhances natural killer cell activity in the elderly: an investigation of age-related immunological changes. *J. Clin. Immunol.*, 21(4): 264–271, 2001.

147. HS Gill, KJ Rutherfurd. Probiotic supplementation to enhance natural immunity in the elderly: effects of a newly characterized immunostimulatory strain *Lactobacillus rhamnosus* HN001 (DR20) on leucocyte phagocytosis. *Nutr. Res.*, 21: 183–189, 2001.

148. YH Sheih, BL Chiang, LH Wang, CK Liao, HS Gill. Systemic immunity-enhancing effects in healthy subjects following dietary consumption of the lactic acid bacterium *Lactobacillus rhamnosus* HN001. *J. Am. Coll. Nutr.*, 20(2 Suppl): 149–156, 2001.

149. HS Gill, ML Cross, KJ Rutherfurd, PK Gopal. Dietary probiotic supplementation to enhance cellular immunity in the elderly. *Br. J. Biomed. Sci.*, 58(2): 94–96, 2001.

150. M de Vrese, A Stegelmann, B Richter, S Fenselau, C Laue, J Schrezenmeir. Probiotics — compensation for lactase insufficiency. *Am. J. Clin. Nutr.*, 73(2 Suppl): 421S–429S, 2001.

151. L Pelto, E Isolauri, EM Lilius, J Nuutila, S Salminen. Probiotic bacteria downregulate the milk-induced inflammatory response in milk-hypersensitive subjects but have an immunostimulatory effect in healthy subjects. *Clin. Exp. Allergy*, 28(12): 1474–1479, 1998.

152. E. Isolauri, T Arvola, Y Sutas, E Moilanen, S Salminen. Probiotics in the management of atopic eczema. *Clin. Exp. Allergy*, 30(11): 1604–1610, 2000.

153. M Kalliomaki, S Salminen, H Arvilommi, P Kero, P Koskinen, E Isolauri. Probiotics in primary prevention of atopic disease: a randomised placebo-controlled trial. *Lancet*, 357(9262): 1076–1079, 2001.

154. M Kalliomaki, S Salminen, T Poussa, H Arvilommi, E Isolauri. Probiotics and prevention of atopic disease: 4-year follow-up of a randomised placebo-controlled trial. *Lancet*, 361(9372): 1869–1871, 2003.

155. Y Aso, H Akaza, T Kotake, T Tsukamoto, K Imai, S Naito. Preventive effect of a *Lactobacillus casei* preparation on the recurrence of superficial bladder cancer in a double-blind trial. The BLP Study Group. *Eur. Urol.*, 27(2): 104–109, 1995.

156. Y Aso, H Akazan. Prophylactic effect of a *Lactobacillus casei* preparation on the recurrence of superficial bladder cancer. BLP Study Group. *Urol. Int.*, 49(3): 125–129, 1992.

157. Y Ohashi, S Nakai, T Tsukamoto, N Masumori, H Akaza, N Miyanaga, et al. Habitual intake of lactic acid bacteria and risk reduction of bladder cancer. *Urol. Int.*, 68(4): 273–280, 2002.

158. M de Vrese, J Schrezenmeir. Probiotics and non-intestinal infectious conditions. *Br. J. Nutr.*, 88: S59–S66, 2002.

159. U Gluck, JO Gebbers. Ingested probiotics reduce nasal colonization with pathogenic bacteria (*Staphylococcus aureus, Streptococcus pneumoniae*, and beta-hemolytic streptococci). *Am. J. Clin. Nutr.*, 77(2): 517–520, 2003.

160. K Hatakka, E Savilahti, A Ponka, JH Meurman, T Poussa, L Nase, et al. Effect of long term consumption of probiotic milk on infections in children attending day care centres: double blind, randomised trial. *Br. Med. J.*, 322(7298): 1327, 2001.

161. AS Naidu, WR Bidlack, RA Clemens. Probiotic spectra of lactic acid bacteria (LAB). *Crit. Rev. Food Sci. Nutr.*, 39(1): 13–126, 1999.

162. C Koebnick, I Wagner, K Ising, U Stern. Die Wirkung eines probiotischen Getränks auf gastrointestinale Symptome und das Befinden von Patienten mit chronischer Obstipation. *Ernaehr. Umsch.*, 48: 392–396, 2001.

163. T Ogata, T Nakamura, K Anjitsu, T Yaeshima, S Takahashi, Y Fukuwatari, et al. Effect of *Bifidobacterium longum* BB536 administration on the intestinal environment, defecation frequency and fecal characteristics of human volunteers. *Biosci. Microflora*, 16: 53–58, 1997.

164. T Kontiokari, J Laitinen, L Jarv, T Pokka, K Sundqvist, M Uhari. Dietary factors protecting women from urinary tract infection. *Am. J. Clin. Nutr.*, 77(3): 600–604, 2003.

165. G Reid. Probiotic agents to protect the urogenital tract against infection. *Am. J. Clin. Nutr.*, 73(2 Suppl): 437S–443S, 2001.

166. MC Bertolami, AA Faludi, M Batlouni. Evaluation of the effects of a new fermented milk product (Gaio) on primary hypercholesterolemia. *Eur. J. Clin. Nutr.*, 53(2): 97–101, 1999.

167. B Richelsen, K Kristensen, SB Pedersen. Long-term (6 months) effect of a new fermented milk product on the level of plasma lipoproteins — a placebo-controlled and double blind study. *Eur. J. Clin. Nutr.*, 50(12): 811–815, 1996.

168. G Kiessling, J Schneider, G Jahreis. Long-term consumption of fermented dairy products over 6 months increases HDL cholesterol. *Eur. J. Clin. Nutr.*, 56(9): 843–849, 2002.

169. MA Schell, M Karmirantzou, B Snel, D Vilanova, B Berger, G Pessi, et al. The genome sequence of *Bifidobacterium longum* reflects its adaptation to the human gastrointestinal tract. *Proc. Natl. Acad. Sci. U. S. A.*, 99(22): 14422–14427, 2002.

170. DOE JOINT GENOME INSTITUTE, http://www.jgi.doe.gov.
171. JC Venter, MD Adams, EW Myers, PW Li , RJ Mural, GG Sutton, et al. The sequence of the human genome. *Science*, 291(5507): 1304–1351, 2001.
172. ES Lander, LM Linton, B Birren, C Nusbaum , MC Zody, J Baldwin, et al. Initial sequencing and analysis of the human genome. *Nature*, 409(6822): 860–921, 2001.

3

Taxonomy and Biology of Probiotics

Lorenzo Morelli and Maria L. Callegari

CONTENTS

3.1 Introduction

The problem of the identification and naming of probiotic organisms arose together with the first suggestion that viable microrganisms could be an aid to human well-being.

Metchnikoff, the scientist credited as being the first to characterize a specific *Lactobacillus* strain able to survive in the human intestinal tract, described this bacterium, isolated from Bulgarian yogurts by a Swiss scientist and then studied at the Pasteur Institute, as "the most active bacillus in causing souring of milk." According to the original source, Metchnikoff named it *"Bacillus bulgaricus"*. However, this name is misleading, and we cannot identify this bacterium with the species named, in modern times, *L. bulgaricus* or *L. delbrueckii* subsp. *bulgaricus* and used for yogurt production. The original description of Metchnikoff deals with a bacterium able to accumulate 25 g/l of lactic acid when grown in milk, an amount of lactic acid not typical of the "bulgaricus" strains and reached only by *L. helveticus* strains (Kandler and Weiss, 1986). To strengthen this doubt, it is to be noted that in international culture collections there is one strain identified as "one of the original Metchnikoff strains," and it was deposited as *L. jugurti* (ATCC cat. N.521), a species nowadays recognized as a biotype of *L. helveticus* (Dellaglio et al., 1973; Kandler and Weiss, 1986).

Misclassification of probiotic bacteria has occurred also in recent times. For example, *Lactobacillus* GG, one of the most successful probiotic organisms (Gorbach, 2002) was originally identified as belonging to the *L. acidophilus* species (Gorbach and Goldin, 1989).

This is surprising, as this species is unable to ferment five carbon sugars and produce DL-lactic acid as a final product of glucose fermentation. On the contrary, *Lactobacillus* GG is able to ferment pentoses and produces L(+)-lactic acid, and this is the normal behavior of strains belonging to *L. rhamnosus* as *Lactobacillus* GG; nowadays *Lactobacillus* GG is correctly identified as a lactose-negative biotype of the species *Lactobacillus rhamnosus*.

What is more surprising is the misclassification of bacteria not only at the species but even at the genus level. A case history could be represented by probiotic products claimed to contain the so-called *Lactobacillus sporogenes*. This species was proposed in 1932 but was immediately recognized as a misclassification, and it was officially rejected in 1939 (!) in *Bergey's Manual*. This rejection was confirmed in the 5th (p. 377), 6th (p. 762), and 8th (p. 577) editions of *Bergey's Manual* (*Bergey's Manual of Determinative Bacteriology*, 5th (1939), 6th (1948), and 8th (1974) editions, William & Wilkins, Baltimore). Clearly, the name "*Lactobacillus sporogenes*" has no scientific validity, but it is still in use in several labels of probiotic products (Sanders et al., 2003).

In addition, recent surveys on the identification of spore-forming probiotic bacteria have shown that in a relevant number of cases, nomenclature of bacteria used for labeling is incorrect or out of date (Hamilton-Miller et al., 1999).

Ambiguous nomenclature has plagued not only probiotic bacteria but also yeast. An example is *Saccharomyces boulardii,* a well-studied yeast used as a probiotic for humans and animals and supported by a robust scientific literature. This probiotic has been used in several clinical trials and has been included as a positive example of the action of probiotics in several meta-analyses (Cremonini et al., 2002; D'Souza et al., 2002), but its allotment to a specific taxonomic group has been achieved only recently by means of molecular biology techniques, which suggest that this organism could be included in *S. cerevisiae* (Mitterdorfer et al., 2002; Hennequin et al., 2001; Dujon, 2001)

It is also of note that several products that are on the market and claim to contain "bifid" bacteria, report on their labels the presence of bacteria named according to very old nomenclature, no longer accepted by the scientific community. Several papers have been published on this matter, and they point out the use of old names such as "*Lactobacterium bifidum*" or "*Lactobacillus bifidum*," ignoring the existence of a genus called "*Bifidobacterium*". On the other hand, the same authors note the presence of lactobacilli misclassified as *L. acidophilus* but belonging to one of the six species obtained in 1980 by DNA-DNA analysis of strains phenotypically identified as *L. acidophilus*.

What is reported above shows that difficulties in selecting probiotic bacteria is an old but still existing problem that has to be solved in order to assure consumers about the real composition of probiotic products. Tools are available for the purpose, and they are neither expensive nor too sophisticated. It is to be noted that the expert consultation set up jointly by the Food and Agriculture Organization and the World Health Organization (see www.fao.org and www.who.int) has strongly advised identifying probiotic bacteria by means of molecular biology techniques.

In the same documents it was suggested that probiotic bacteria should be deposited in an international culture collection, in order to facilitate the assessment and further identification of the active ingredients of probiotic products.

3.2 Identification and Classification

In order to introduce this section, it could be worthwhile to provide some basic definitions (Benner et al., 2001) taken from the latest edition of *Bergey's Manual of Systematic Bacteriology.*

- **Classification** is the arranging of organisms into taxonomic groups (taxa) on the basis of similarities or relationships.
- **Nomenclature** is the assignment of names to the taxonomic groups according to rules.
- **Identification** is the process of determining that a new isolate belongs to one of the established, named taxa.

Taxonomic ranks are levels of the hierarchical organization used in bacterial taxonomy. As regards bacteria, the ranking levels are reported in Table 3.1.

Species is the basic group for systematic classification of prokaryotes, and it can be defined as a collection of strains that have in common many features and that differ in a significant way from all other species. The word "strain" means the product of a succession of cultures derived from an initial single colony. It is a bacterial culture composed of cells derived from a single initial cell. For each species, one strain is designated as the type of strain, and all other strains that are sufficiently similar to it are included with it in the same species. However, the concept of species is still under debate and revision (Rossello-Mora and Amann, 2001).

TABLE 3.1

Taxonomic Ranking Used for Bacterial Classification

Ranks	Example
Domain	Bacteria
Class	*Actinobacteria* Stackebrandt et al., 1997
Order	*Bifidobacteriales* Stackebrandt et al., 1997
Family	*Bifidobacteriaceae* Stackebrandt et al., 1997
Genus	*Bifidobacterium* Orla-Jensen, 1924
Species (one example)	*Bifidobacterium bifidum* (Tissier, 1900) Orla-Jensen 1924, species.

According to Benner et al. (2001), a subspecies is "based on minor but consistent phenotypic variations within the species or on genetically determined clusters of strains within the species"; another definition of subspecies was provided by Wayne et al. (1987): "Subspecies designations can be used for genetically close organisms that diverge in phenotype."

As further subdivisions, bacteria are sometimes grouped according to special characters. These groups, however, have no official standing in official nomenclature.

It is relevant to note that, as is pointed out in the chapter devoted to bacterial classification of the *Bergey's Manual*, that "There is no 'official' classification of bacteria. (This is in contrast to bacterial nomenclature, where each taxon has one and only one valid name, according to international agreed-upon rules)."

The International Committee on Systematics of Prokaryotes (formerly the International Committee on Systematic Bacteriology) is a branch of the International Union of Microbiological Societies and is in charge of updating the International Code of Nomenclature of Prokaryotes (formerly the International Code of Nomenclature of Bacteria), which contains principles, rules, and recommendations in order to name bacteria correctly.

Since 1975, this code has required a new step for bacterial nomenclature, i.e., the valid publication of names of bacteria. It means that publication of the name and description of a new taxon in a recognized scientific printed publication is required in order to include this new taxon in the Approved Lists of Bacterial Names, which is the inventory of the accepted nomenclature for bacteria. The publications must be in accord with requirements set up in the International Code of Nomenclature of Prokaryotes.

The publication of the first Approved Lists of Bacterial Names (Johnson et al., 1980), which contained all the bacterial names having standing in nomenclature as of January 1, 1980, was a starting point in the renewal of bacterial nomenclature. The names validly published before January 1, 1980 but not present in these lists are no longer valid but are available for revival individually if all requirements for doing so are met.

Updates of "Approved Lists of Bacterial Names" are regularly published in the *International Journal of Systematic and Evolutionary Microbiology* (formerly the *International Journal of Systematic Bacteriology*). The lists are also available via the Internet (http://www.bacterio.cict.fr/). The use of these lists is of relevance not only for scientific purposes but also for industrial uses.

Another keystone of the bacterial taxonomy that also has practical implications is the designation of the taxon type, which is a requirement peculiar to systematic bacteriology and extremely relevant not only for taxonomists but also for applied microbiology.

The type of each taxon must be designated by the author at the time the name of the taxon is validly published in the appropriate scientific journal (see the example reported in Table 3.1).

This rule was set because it is impossible to fully describe a microorganism, and it is necessary to have a reference organism available to the scientific community for purposes of comparisons, studies, etc. This availability is guaranteed in an International Culture Collection.

It seems worthwhile to point out also that patenting procedures do require the deposit of the biological material that is the object of patent, i.e., the bacterial strain. This quite sensitive matter is regulated by rules listed in the Budapest Treaty on the International Recognition of the Deposit of Microorganisms for the Purposes of Patent Procedure. To quote from the introduction to the treaty: "To guarantee full disclosure of an invention involving microorganisms, the Budapest Treaty provides for the deposit of (micro) biological material with a recognised culture collection, known as an 'International Depositary Authority (IDA)'. An IDA must make the material publicly available at the appropriate point in the patenting procedure."

Deposit in the International Culture Collection of the strains used as probiotic supplement to animal feed is also requested by the European Union (EU). Since 1993, the EU Directive No. 93/113 states, under Article 7, that the following is mandatory: "B. For microorganisms and their preparations: (a) the identifications of the strain(s) according to a recognized international code of nomenclature and the deposit number of the strain(s); (b) the number of colony-forming units (CFU/g)." It is highly relevant to note that this directive requires the use of the international code of bacterial nomenclature and the deposit of the strains.

We then arrive at the core of our considerations about taxonomy and the probiotic use of bacteria. Bacterial taxonomy is based on the characterization of a single strain, the prototype of the taxon, and the following efforts are based on comparative analysis between the prototype and the organism under study.

There will be no bacterial taxonomy without deposited strain types, but also industrial applications require the identification and public availability of strains. The question arises about the tools that are available to identify the strains and to perform the comparison among them. Increasing knowledge of DNA properties and the development of more and more sophisticated techniques of molecular biology have made a considerable contribution to species identification.

The first method based on nucleic acid applied for the description of taxa was the determination of the guanine and cytosine (G+C) content of the DNA. This method is considered a classical way to distinguish between strains genetically different that are phenotypically similar. It is one of the genomic characteristics recommended for considering a strain as belonging to a species. Variation values higher than 5% indicate heterogeneity within a species. It is important to underline that the G+C contents do not take into consideration the linear nucleotide sequences of the DNA analyzed, and for this reason this parameter can only be used negatively. In other words, this method can be used for separating groups, but a similar value of G+C does not indicate a close relationship between two strains.

In contrast, a parameter that allows the delimitation of species borders is the determination of whole genome similarity since a strong correlation has been found between this value and phenotypic similarity. For this reason DNA hybridization became a standard technique in bacterial taxonomy. A characteristic property of DNA molecules, exploited in this technique, is its ability for reassociation or hybridization. The denatured DNA molecules are able to reassociate, under appropriate experimental conditions, to reform native duplex structure. DNAs from different microorganisms can reassociate depending on the degree of similarity of their sequences. It is currently recommended that a strain be considered as belonging to a given species if there is a value of 70% or higher degree of similarity. Then the percentage of homology or similarity is the quantification of relatedness of microorganisms. This represents a very important practical advantage because it allows a sharper definition, when compared with the phenotypic methods, of groups of microorganisms. This method allowed the solving of some identification problems as for example the confused situation of *L. acidophilus* species. Historically, probiotic roles have been attributed primarily to this species that were extremely heterogeneous on the basis of phenotypic characterization. In 1980 (Johnson et al. 1980), DNA-DNA hybridization studies suggested that the *L. acidophilus* species was composed of six different species. Strain ATCC4356 with a higher degree of similarity with *L. acidophilus* remained in the original species, whereas the other homology groups are now considered separate species.

Techniques based on ribosomal RNA or rRNA gene analysis brought about a complete revolution in bacterial taxonomy. In fact, the development of molecular-genetic methods led to identifying in ribosomal RNA molecules very good candidates for phylogenetic studies because the 16S, 5S, and 23S genes are highly conserved molecules that are present in all microorganisms. These genes contain conserved region related to their structural and functional obligation; these regions are very important for phylogenetic investigations. Highly variable regions are also present in these molecules, and their sequences can be used for differentiation between closely related species. The amplification of the 16S rRNA gene and sequencing can be performed using universal primers designed on the basis of the conserved region sequences, making this identification method suitable for a wide range of organisms. A very practical advantage is that DNA can be extracted directly from a single colony-forming unit. These practical considerations allow understanding of the widespread use of this technique in microbiology laboratories.

Comparison of sequences of rRNA genes is currently considered to be one of the most powerful and accurate methods for species identification. In recent years, sequencing of 16S, 5S, and 23S rRNA genes has led to a rapid expansion of large international sequence databases (Maidak et al., 1999) such as the Ribosomal Database Project (RDP). The huge number of sequences available allows the creation of phylogenetic trees or dendrograms, and in some cases the taxa created on the basis of phenotypic properties do not correspond to the phylogenetic branching. As a consequence, for certain species their classification has been reconsidered.

It is commonly accepted that 16S rDNA sequencing and more generally methods based on the rRNA genes are a good alternative to many classical methods, but the last guideline gives precedence to the phenotype over the genotype. The complete sequence of the 16S rRNA gene is not suitable as a unique parameter for species identification, but additional investigations using other methods are needed for a reliable taxonomy description of bacterial isolates. In fact, it has been demonstrated that the resolving power of 16S rRNA sequences is limited when closely related organisms are examined (Stackebrandt and Goebel, 1994). Thus, the decision if a new isolate belongs or not to a given species has to take into consideration results provided with other standard methods. At present, in accordance with *Bergey's Manual*, a species includes strains with genomic similarities of 70% or greater and a 97% of 16S rRNA sequence similarity.

Sequences of the variable regions of rRNA genes can also be used as a target for highly specific probes or species-specific primers, and for these reasons a series of rapid identification methods has been developed.

DNA hybridization probes have found a large spectrum of applications. Probes have been used for bacteria identification and for studies of the structure and dynamics of complex ecosystems.

Probes commonly used are 16S and 23S rRNA target nucleotides, DNA probes, and probes based on key enzyme sequences. Pot et al. (1993) used a 23S rRNA targeted oligonucleotide for identification of *L. acidophilus*, *L. gasseri*, and *L. johnsonii*. Results reported by the authors show that this procedure allowed the correct identification of strains belonging to the respective species. This method has had a certain success but has been increasingly substituted by other methods principally involving the polymerase chain reaction (PCR).

A species-specific PCR method was developed by Drake et al (1996), and it was possible to identify strains belonging to the following species: *L. casei* subsp. *casei*, *L. casei* subsp. *rhamnosus*, *L. acidophilus*, *L. delbrueckii* subsp. *bulgaricus*, and *L. delbrueckii* subsp. *lactis*. The taxonomic position of the *L. casei* group has been a much-debated question. *Bergey's Manual of Systematic Bacteriology* lists four subspecies of *L. casei* (Kandler and Weiss, 1986) whereas Collins et al. (1989) reclassified strains of this group into three species on the basis of DNA-DNA homology. Ten years later, Mori et al. (1997) described the sequence signature of 16S rDNA that discriminates among these species. The exploitation of these sequences allowed Ward and Timmins. (1999) to identify these species by a species-specific PCR analysis.

In order to develop an accurate and convenient method for the characterization of bifidobacteria in the intestinal microflora, Matsuki et al. (1998, 1999) set up a species-specific PCR for all known species of bifidobacteria that inhabit the human intestinal tract. Authors compared this identification method with the traditional culture methods and concluded that the species-specific PCR technique was able to detect a wider range of species than the other method. However, *B. longum* and *B. suis* were not distinguishable from each other by this technique.

An alternative way to exploit the phylogenetic information of rDNA molecules is a method called restriction fragment-length polymorphism (RFLP). For this technique universal primers of conserved regions are used to amplify parts of 16S including or not the ribosomal spacer region. Products of amplification reaction are digested by endonucleases, and patterns obtained are examined by electrophoresis agarose gel. These patterns are usually species-specific.

Chen et al. (2000) used this technique in order to demonstrate a polymorphism in the spacer region between the 23S and 5S rRNA genes of bacteria belonging to the *L. casei* group. Grouping obtained with RFLP analysis was in agreement with the reclassification proposed by Collins et al. (1989).

Another method for species-specific identification that has been developed on the basis of 16S rRNA gene sequences is the amplified ribosomal DNA restriction analysis (ARDRA). This procedure provides the amplification of the 16S rRNA gene followed by a restriction analysis. The ARDRA technique was used to identify different species of *Lactobacillus* isolated from human feces and vagina (Ventura et al., 2000). The method was set up using strain types and then was applied to fresh isolates. Strains were first grouped by physiological tests, and then 16S mrDNA samples were obtained by PCR amplification. A set of four restriction enzymes was able to differentiate 14 species of *Lactobacillus*. This procedure was validated using 16S rDNA sequencing. The same method has also been used to identify bifidobacteria (Ventura et al., 2000).

This procedure allowed the identification of 15 species of bifidobacteria; however, either the ARDRA technique or the species-specific PCR primers described by Matsuki et al. (1999) were unable to distinguish *B. longum* from *B. suis*. The authors suggested that these two species should be taxonomically combined on the basis of DNA-DNA homology value (from 75 to 78%) and 16S rDNA similarity (>90%). In contrast with the results obtained by species-specific PCR, the ARDRA technique was able to discriminate *B. catenulatum* from *B. pseudocatenulatum*. ARDRA seems to be a reliable and reproducible molecular identification method either for lactobacilli or bifidobacteria. In ecology studies when it is important to evaluate the predominant species, it becomes very important to dispose of rapid and accurate techniques of identification because of the high number of colonies that it is necessary to analyze.

Industrial interest in the use of specific bacterial strains such as food additives in dairy products is rapidly growing. This development has led to the requirement for accurate quality control of probiotic products and hence methods for specific identification of probiotic strains. Such accurate identification procedures could also serve in accurately monitoring the development of populations of specific probiotic strains during the passage through the gastrointestinal tract.

Techniques for plasmid isolation and characterization are routinely used in several laboratories; they are easily performed and reproducible. It should also be pointed out that several properties that are relevant for defining probiotic properties of one strain (i.e., the ability to produce antibacterial compounds such as bacteriocins) are generally plasmid encoded. To monitor the plasmid profile of a probiotic strain, it is then useful not only for strain identification, but also for checking its genetic stability. This typing technique can obviously be applied only for bacteria containing plasmids, and this represents an important limitation and reduces the proportion of strains typeable by this method. For this reason, and the instability of these extrachromosomal elements, this method has been substituted by techniques based on chromosomal DNA that have become more popular.

Genomic methods are normally PCR methods such as randomly amplified polymorphic DNA (RAPD), amplified fragment-length polymorphism (AFLP), arbitrary primer PCR (AP-PCR), repetitive extragenic palindromic-PCR (Rep-PCR), or various restriction analysis such as restriction endonuclease analysis (REA), pulsed field gel electrophoresis (PFGE), or ribotyping. AFLP is a combination of PCR and restriction analysis.

Chromosomal DNA restriction endonuclease analysis (REA) was the first of the chromosomal DNA-based typing schemes. In this method genomic DNA is cut with frequent cutting enzymes and then is examined on an agarose gel. Patterns obtained after electrophoresis are highly reproducible because of the high specificity of endonucleases and the stability of the chromosomal DNA. The differences of banding patterns between strains are due to the differences in DNA base composition of the organisms examined. The most important inconvenience of this technique is the very complex banding patterns obtained using the common agarose gel migrations because numerous bands are normally obtained with the high-frequency cutting enzymes.

A method that overcomes the limitations of REA is the PFGE technique. This method is considered by some authors to be the "gold standard" for strain identification. PFGE is a special methodology to avoid shearing the DNA into random fragments. Genomic DNA is prepared by embedding intact microorganisms in agarose plugs. After cell lysis, plugs of isolated DNA are digested *in situ* with a low-frequency cutting enzyme. Electrophoresis involves periodic orientation changes of the electric field. This migration allows a separation of fragments of very high molecular weight. The combination of migration condition and use of low-frequency cutting enzymes produces restriction patterns composed of a relatively low number of fragments, making their visual examination easy. PFGE has been described as a reproducible and highly discriminatory method used for typing of important probiotic bacteria such as *L. casei* (Ferrero et al., 1996; Tynkkynen et al., 1999), bifidobacteria (McCartney et al., 1996), and *L. rhamnosus* (Tynkkynen et al., 1999).

An interesting aspect has been pointed out by Tynkkynen et al. (1999). These authors applied PFGE, RAPD, and ribotyping on 24 strains belonging to the *L. casei* group in order to compare the discriminatory power of these methods. PFGE was able to discriminate 17 genotypes for 24 strains examined against 15 and 12 genotypes individuated by ribotyping and the RAPD method, respectively. Also, McCartney et al. (1996) obtained the same results for bifidobacteria and lactobacilli isolated from human fecal samples comparing PFGE and ribotyping.

Another way for typing probiotic bacteria is the ribotyping method. In this case genomic DNA is digested by restriction endonucleases and transferred to a membrane for a subsequent hybridization analysis. Probes used are 5S, 23S, and 16S rRNA genes. After probing, each fragment of bacteria containing rRNA genes is detected creating a fingerprint composed of 1 to 15 bands. These profiles can be easily compared among strains. Reproducibility and high-discriminatory power are important advantages of this genotyping method.

Ribotyping presents a very important practical advantage in that, because of the similarity of ribosomal genes, a universal probe can be used. This method is used for typing, but also as a species identification method (Zhong et al., 1998). PFGE and ribotyping are considered two very highly discriminatory and reliable methods, but obviously they present some disadvantages. Both methods are time-consuming and a little more tedious than other methods. Moreover, the number of strains that it is possible to examine using these techniques is quite limited compared with PCR-based methods. Another aspect that is important to evaluate is that PFGE is expensive and requires considerable experimental experience.

The introduction of PCR methodology into microbiology laboratories led to the development of a series of PCR-based methods for strain typing. RAPD is based on a PCR reaction in which short arbitrary sequences are used as primers. The amplification products are constituted by random-sized DNA fragments. RAPD is a very simple, rapid technique. Moreover, selection of primers used in the PCR reaction does not depend on a known species or strain-specific sequence, and for this reason it can be a universally applicable technique. For these properties the RAPD technique has been considered a suitable method for routine use, and as a consequence of that, it has been widely used in microbiology laboratories. Unfortunately, the reproducibility of RAPD is occasionally poor because small changes in annealing conditions can produce changes in pattern profiles. It is important to point out that the PCR reaction in the RAPD technique is performed using a single short primer, making this method more sensitive to changing PCR conditions. Factors that can influence annealing conditions have been well-documented (Meunier and Grimont, 1993; Penner et al., 1993). RAPD profiles have been used for typing of *Bifidobacterium* (Roy et al., 1996., Du Plessis and Dicks, 1995) and for strains belonging to the *L. acidophilus* group. RAPD has also been used as a typing method for *L. rhamnosus* and *L. casei* strains (Tynkkynen et al., 1999) in comparison with PFGE and ribotyping. As already mentioned in the description of the PFGE technique, RAPD was the least discriminatory among the studied methods.

Attempts to overcome RAPD limitations led to the development of methods able to generate characteristic DNA fingerprint patterns using primers that recognized conserved regions of chromosomal DNA. One of them is the AP-PCR technique. PCR reactions are performed using a specific primer targeting a highly conserved region within the 16S rRNA gene. A variation of this method is a procedure designed as triplicate arbitrary primed PCR (TAP-PCR). In this case three different annealing temperatures are used simultaneously in a triplicate reaction. The advantage of this variation in the procedure is the easier identification on agarose gel of bands that are sensitive to small changes in temperatures. This technique has been successfully used for the typing of isolates from major genera of lactic acid bacteria and bifidobacteria (Cusik and O'Sullivan, 2000).

Repetitive extragenic palindromic (Rep) elements and enterobacterial repetitive intergenic consensus (ERIC) sequences are conserved regions dispersed on genomic DNA. These regions are the targeting of primers used in amplification reactions in Rep-PCR and in ERIC-PCR techniques, depending on the type of regions used for typing. Using these methods it is possible to obtain profiles due to the amplification of inter-Rep and inter-ERIC distances. Profiles obtained are typical of bacterial species or sometimes are strain specific.

A combination of PCR and restriction enzyme analysis yielded the AFLP technique. Genomic DNA is digested with two different restriction endonucleases that yield restriction fragments with two different types of sticky ends. Adapters are ligated to these ends in order to prepare the template for the PCR reaction. The two different primers used in the reaction mixture contain the same sequences of the adapters.

AFLP is a very highly discriminatory and reliable method, but like PFGE and ribotyping is a time-consuming technique and requires strict standardization conditions.

Recently, AFLP, PFGE, Rep-PCR, TAP-PCR, and ERIC-PCR techniques have been compared in order to evaluate their discriminatory power for the identification of *Lactobacillus johnsonii* strains. Results obtained suggested that PCR fingerprinting methods like ERIC and Rep-PCR are highly reliable and rapid methods for characterization of *L. johnsonii* at their strain level.

Taxonomy for probiotic bacteria is essential for quality assurance as required by consumers and state law. It is obvious that the identification of species may indicate the safety of a strain and its industrial applicability for use in probiotic products. In spite of all the molecular tools now available for taxonomic purposes, it is clear that in the absence of a gold standard the so-called polyphasic approach seems to be the only way to solve bacterial identification problems (Vandamme et al., 1996). A classification that takes into consideration all information on all possible known aspects of a particular microorganism in fact covers the essential part of its genome.

Identification of probiotic bacteria and/or enteric microorganisms could be of interest even when applied *"in vivo"*. The following two techniques are detailed just as an example of the potential of the molecular approach to the ecology of the gut.

Recent advances in rRNA-based molecular techniques make it possible to identify microorganisms among complex bacterial populations without prior cultivation (Aman et al., 1995). The detection of all members within natural communities can be achieved by amplification of 16S rRNA genes from samples. These PCR products can be cloned and sequenced, but this approach is time-consuming and very laborious. This problem has been overcome by the development of denaturing gradient gel electrophoresis (DGGE). This technique, in fact, allows the separation of 16S rDNA fragments of the same length. Primers used for the amplification reaction are designed on the basis of conserved regions of 16S rRNA gene but additionally consisting of a GC clamp of 40 nucleotide GC.

Fragments are separated by a polyacrylamide gel with a denaturant gradient. Separation of the mixture of fragments is based on the mobility of partially melted DNA molecules that are the same size but have a different base composition. The DGGE technique presents, however, some disadvantages. The most important is the short fragment available for sequencing, making the correct bacteria identification difficult. The second problem arises because DNA extracted from natural samples may not be representative of the total community because of a difference in cell wall composition that can make a difference in the extraction yield. The DGGE technique has been used in several studies of intestinal communities (Zoetendal et al., 2001; 2002), for bifidobacteria communities (Satokari et al., 2001) or for monitoring changes in fecal bacterial populations after the ingestion of exogenous *Lactobacillus* strain (Simpson et al., 2000; Tannock et al., 2000).

The second technique used for the study of complex bacterial communities is fluorescence *in situ* hybridization (FISH) using specific oligonucleotides designed on the basis of 16S rRNA sequence as probe. This method has been described as a very useful tool for bacterial identification within a complex matrix. The application of FISH to human fecal samples in order to detect bifidobacteria demonstrates that this method could be as accurate as the traditional enumeration techniques (Langendijk et al., 1995). On the contrary, it has been reported that the number of lactobacilli detected by FISH was higher than plate counts. Authors suggested that this difference could be due to the detection by FISH of nonviable cells. However, for accurate and fast identification, the FISH method is probably more appropriate, because it is based on molecular markers (Harmsen et al., 2000).

3.3   The Physiology of Probiotic Bacteria

Whereas dairy starter bacteria are selected for their ability to rapidly acidify milk or produce desirable organoleptic characters in dairy products, probiotic bacteria are selected for their potential to promote a health benefit after ingestion. It is generally agreed that probiotics should be viable in the gut in order to exert their beneficial effects, with the exception of the alleviation of lactose intolerance, which is linked to an enzymatic action.

Mandatory physiological traits for maintaining viable ingested bacteria are resistance to gastric conditions and to the action bile salts. Pronounced differences have been observed in the survival rate of probiotic strains, even among individuals belonging to the same species, passing through the stomach and the upper intestinal tract. Survival rates have been estimated at 20 to 40% for properly selected strains (Bezkorovainy, 2001), whereas others have a survival rate less than 0.1%, the main obstacles to survival being gastric acidity and the action of bile salts.

Human gastric juice has been used (Conway et al., 1987; Goldin et al., 1992), but *in vitro* tests simulating a gastric environment have been more widely applied (Charteris et al., 1998a) in order to predict the behavior of the strains in *in vivo* conditions. These assays have been successful in selecting new probiotic organisms, which performed well when tested in humans. *In vitro* assays (Hood and Zottola, 1988; Charteris et al., 1998b) show that enteric lactobacilli could tolerate the exposure to pH 2 for several minutes, while higher pH slightly affects their counts, and pH 1 is destructive for all tested lactobacilli.

It is to be noted, however, that results obtained by means of these assays are to be critically considered; data recently published by a Danish group (Jacobsen et al., 1999) showed that strains of lactobacilli which have a documented ability to survive and reproduce in the human gut scored poorly when challenged *in vitro* for 4 hours at pH 2.5.

These differences could be explained by the conditions applied, probably too harsh, but they also point out that little is known about the real physiological conditions of bacteria during the gastric passage. Are they metabolizing cells, or does the acidic pH keep them in a dormant state? Do they use the metabolic pool of ATP, if present, or are they able to ferment and to produce new ATP? What is clearly established is that the resistance to gastric conditions is strain dependent, as all published studies report a different scoring among strains belonging to one species.

The ability to survive the action of bile salts is an absolute requirement of probiotic bacteria, and it is generally included among the criteria used to select potentially probiotic strains. Gilliland et al. (1984) were the first to suggest the importance of assessing bile tolerance to select lactobacilli for probiotic use.

Several authors used this assay to assess the bile resistance of potential or already commercialized probiotic lactobacilli. Results obtained in all these papers clearly showed, also for this trait, that the amount of delay detected in the growth curve of lactobacilli challenged with oxgall was strain- and not species-dependent. The mechanisms underlining the resistance are at the moment under investigation.

Genes involved in hydrolysis of bile salts have been cloned and sequenced (Elkins et al., 2001), but it should be pointed out that the deconjugation and resistance are unrelated activities (Moser and Savage, 2001). Cholic acid was found to be highly deleterious for viability of lactobacilli (De Boever and Verstraete, 1999) A novel interaction between lactobacilli and bile has been recently described (Kurdi et al., 2000); cholic acid was found to accumulate in *Lactobacillus* cells by means of a transmembrane proton gradient.

An additional assay generally used for probiotic selection and related to the physiology of these bacteria is the measurement of the ability of the potential probiotic strain to adhere to surfaces mimicking the intestinal epithelia tissue, such as cell cultures, mucus, etc. (Blum et al., 1999; Ouwehand et al., 2001) as it is generally agreed that a strain needs to adhere to intestinal surfaces to persist in a moving environment such as the gut. It is to be noted, however, that some authors have suggested that a potential risk could arise from adhering probiotic bacteria (Apostolou et al., 2001).

Adhesion assays have provided a lot of information on the strain to variation of adhesion properties, but they do not seem to provide clear-cut results, as reported differences among adhering and nonadhering strains of lactobacilli are small variations (i.e., 1 vs. 1.5 bacterium per Caco-2 cell) that are unlikely to be able to predict significant differences in the *in vivo* situation (Morelli, 2000).

Furthermore, in a survey undertaken in the frame of a European Union-funded project, it turned out that the outcome of adhesion assays are deeply influenced (Blum and Reniero, 2000) by the following factors:

- pH
- Presence of calcium ions
- Dose of lactobacilli used
- Presence of supernatants
- Growth phase in which bacteria are collected

Moreover, only strains scoring high in these *in vitro* tests have been used for confirmation assays *in vivo*. It could be of interest to validate these tests, to assess also the strains with the lowest *in vitro* scores for the ability to survive and persist in the gut under *in vivo* conditions.

We can observe that all the above-cited ambiguity about *in vitro* assays reflects the lack of knowledge about the physiological properties that are responsible for acid and bile resistance, whereas
ce data are available for the molecular mechanisms involved in adhesion.

Surface layer proteins have been shown to have a role in determining adhesion to surfaces (Kos et al., 2003; Matsumura et al., 1999, Schneitz et al., 1993). Cloning of a surface layer from an adhering *Lactobacillus brevis* to a nonadhering *Lactococcus lactis* has transferred to the recipient strain the adhesion properties, providing good evidence of the involvement of these surface structures in adhesion processes (Avall-Jaaskelainen et al., 2003). A detailed study of the surface layer domains involved in adherences to a number of surfaces has been published (Antikainen et al., 2002).

It was recently shown, by means of atomic force microscopy, that the presence and absence of surface layer protein has a major impact in the surface topography of the cell wall. Moreover, this analysis showed adhesion peaks caused by the presence of polysaccharides on the cell surface of bacteria lacking the surface layer proteins. This is a new field of investigation, strictly related to the physiological composition of the cell surfaces, and worthy of further development (Schaer-Zammaretti and Ubbink, 2003).

The ability to bind mucus has been related to the presence of specific proteins. In *Lactobacillus fermentum* 104R a 29-kDa surface protein (then smaller than surface layer proteins) has been related to the mucus-binding property of this strain (Rojas et al., 2002), while in *L. reuteri* 1063, the gene encoding for a 358-kDa protein was cloned and sequenced. A clear link was established between this protein and the observed ability of this strain to bind to mucus (Roos and Jonsson, 2002).

A third mechanism that establishes a link between a physiological trait and adhesion is the aggregation phenotype. Some strains are able to grow as clumps or sand-like particles; these strains have been isolated from human gut and vagina, but also from animals (Kos et al., 2003; Cesena et al., 2001; Roos et al., 1999).

Aggregating strains are generally hydrophobic and score well when assayed for their ability to survive and persist in the intestinal tract (Cesena et al., 2001).

A model system has also been established in which a spontaneous mutant, lacking the aggregation phenotype, has also lacked the ability to reproduce well in the human gut; this result could establish a link between aggregation and such a probiotic property (Cesena et al., 2001).

An *in vitro* model has also been used to evaluate some probiotic activities possibly exerted by select strains in the intestinal tract. Adhesion could have a prohealth effect by blocking the attachment sites of pathogenic bacteria and viruses, and thus probiotic bacteria might contribute to the prevention of infection by pathogens.

The nonmucus-secreting enterocyte-like intestine Caco-2 cell line displays typical features of intestinal cells, and they are well-known and widely used model systems for evaluating probiotic bacteria. By using this cell line as an *in vitro* model, it was shown that *L. acidophilus* and *L. rhamnosus* strains adhere in relatively high numbers of organisms and are able to prevent attachment of pathogenic microorganisms such as *Salmonella typhimurium*, *Yersinia enterocolitica*, and enteropathogenic *E. coli* (Coconnier et al., 1993; Bernet et al., 1994). The value of these observations can, however, be questioned. First of all, only a threefold difference between "good" and "poor" adherent was detected (Lehto and Salminen, 1997), which microbiologically is not highly significant. In addition, the growth conditions may vitally influence the expression of bacterial surface structures. The cell surface, for example, of an overnight culture is quite different from that of bacteria which have just passed the stomach and small intestine and have been stressed by gastric acid, bile, and pancreatic juice.

In recent work, a dramatic difference in the morphology of bifidobacteria grown in laboratory media in the presence or absence of mucus has been shown (Figure 3.1). These results are in accordance with those already reported for lactobacilli (Jonsson et al., 2001). The addition of mucin to the growth medium of different strains of *Lactobacillus reuteri* improved the mucus-binding activity of most strains that initially showed very poor binding when grown in MRS broth (De Man et al., 1960). At the molecular level, it was shown that proteolytic treatment of bacteria grown in the presence of mucin eliminated the adhesion, suggesting that mucin induces the production of cell surface proteins that possess mucus-binding properties.

As an example of the relevance of the physiology of probiotic bacteria for their beneficial roles, even in healthy subjects, it is possible to cite what has been named "colonization resistance" (Vollaard and Clasener, 1994), meaning the mechanism physiologically used by bacteria already present in the gut to maintain their presence in this environment and to avoid colonization of the same intestinal sites by freshly ingested microorganisms, including pathogens. Nowadays this concept is applied to healthy individuals in order to reduce the risk of infections. Most of the so-called probiotic products on the market today are based on the assumption that healthy individuals are also better protected by a food supplement of bacteria selected to improve colonization resistance.

However, the overall mechanism of action of probiotic bacteria toward the other bacteria inhabiting the gut is not well understood. It is thought that feeding probiotics could alter the intestinal population as determined by fecal enumeration. However, in individuals where the number of already present enteric bacteria is low (e.g., after antibiotic therapy), a probiotic supplement can increase the number of fecal counts of lactic bacteria. It seems difficult to increase the presence of the total population of lactic acid bacteria in healthy subjects. Some studies suggest that a replacement effect takes place; in other words, while the total lactic counts

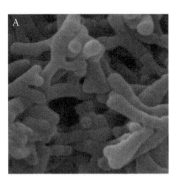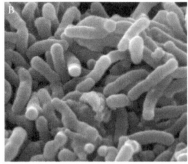

FIGURE 3.1
Different morphologies of *Bifidobacterium longum* M4 strain grown in the presence or absence of mucus. (A) Cells grown in TPY medium with mucin added (Scardovi, 1986); (B) cells grown without mucin. Courtesy of T. Vignali; Istituto di Microbiologia UCSC; graduation thesis, July 2003.

remain stable, the fed strain(s) replace some of the previously present lactic strains.

A large number of well-conducted trials (Cremonini et al., 2002; D'Souza et al., 2002) have shown *in vivo* efficacy against rotavirus and antibiotic-associated diarrhea. Although the efficacy has been demonstrated in a quite acceptable way, the mechanism underlying this efficacy is little understood.

It could be suggested that two potential ways of physiological action could be envisaged for probiotic bacteria: (1) adhesion to epithelia and (2) interaction with the gut-associated lymphoid tissue (GALT). Adhering probiotic bacteria seem to out-compete pathogens in the colonization process, and in this way they protect the host from infections. These observations, however, are based mainly on *in vitro* experiments, and molecules possibly involved are not described.

Interactions of probiotic bacteria with GALT were discovered after several workers noticed *in vivo* the ability of some selected strains to regulate some immunological functions. Investigations in this field have been widened, and now a complex picture of the immune-regulating properties of strains of lactic bacteria is available. A careful selection and characterization of the strains have to be done in order to avoid the use of strains able to elicit a proinflammatory response from the host. This is a quite new and relevant point to be raised from the safety point of view.

On the other hand, an exciting field of research and application has been opened by the studies which have shown that interaction of probiotic bacteria with the host immune functions could be used to reduce problems of allergy in children. Although the mechanisms are not fully understood, the clinical observations seem more than promising.

A further area of probiotic efficacy is related to the action of these bacteria on nutrients. The role of lactic bacteria to favor lactose digestion in lactose-intolerant subjects has been fully demonstrated. The mechanism of action seems to be linked to the release of β-galactosidase from bacterial cells during the transit through the small intestine. This evidence has been obtained for the yogurt starter cultures, but the mechanism could be similar also for enteric lactic acid bacteria. However, it is to be noted that lactic bacteria utilizing lactose via a phosphoenol pyruvate (PEP)-dependent phosphotransferase system (PTS) are less efficient in lactose digestion during gut transit.

Additional effects of probiotic bacteria on ingested food, such as the binding of mutagenic substances, have also been shown, but the evidence is slight. The bacterial strain is the core of the probiotic action; it is essential, in the future, to have a deeper knowledge of each strain and its physiology during its life in the intestinal tract to have a more science-based application of probiotic supplementation of foods.

References

Amann RI, Ludwig W, Schleifer KH. Phylogenetic identification and in situ detection of individual microbial cells without cultivation. *Microbiol. Rev.*, 1995; 59: 143–169.

Antikainen J, Anton L, Sillanpaa J, Korhonen TK. Domains in the S-layer protein CbsA of *Lactobacillus crispatus* involved in adherence to collagens, laminin and lipoteichoic acids and in self-assembly. *Mol. Microbiol.*, 2002; 46: 381–394.

Apostolou E, Kirjavainen PV, Saxelin M, Rautelin H, Valtonen V, Salminen SJ, Ouwehand AC. Good adhesion properties of probiotics: a potential risk for bacteremia? *FEMS Immunol. Med. Microbiol.*, 2001; 31: 35–39.

Avall-Jaaskelainen S, Lindholm A, Palva A. Surface display of the receptor-binding region of the *Lactobacillus brevis* S-layer protein in *Lactococcus lactis* provides nonadhesive lactococci with the ability to adhere to intestinal epithelial cells. *Appl. Environ. Microbiol.*, 2003; 69: 2230–2236.

Benner DJ, Staley JT, Kieg NR. Classification of procaryotic organisms and the concept of bacterial speciation In: *Bergey's Manual of Systematic Bacteriology, Vol 1*. Boone DR, Castenholz RW, Garrity GM, Eds., Williams & Wilkins, Baltimore, 2001, pp. 27–31.

Bernet MF, Brassart D, Neeser JR, Servin AL. *Lactobacillus acidophilus* LA 1 binds to cultured human intestinal cell lines and inhibits cell attachment and cell invasion by enterovirulent bacteria. *Gut*, 1994; 35: 483–489.

Bezkorovainy A. Probiotics: determinants of survival and growth in the gut. *Am. J. Clin. Nutr.*, 2001; 73l: 399S–405S.

Blum S, Reniero R, Schiffrin EJ, Crittenden R, Mattila-Sandholm T, Salminen S, von Whright A, Ouwehand AC, Saarela M, Saxelin M, Collins K, Morelli L. Adhesion studies for probiotics: need for validation and refinement. *Trends Food Sci. Technol.*, 1999; 10, 405–410.

Blum S, Reniero R. Industrial panel statements: adhesion of selected *Lactobacillus* strains to enterocyte-like Caco-2 cells *in vitro*: a critical evaluation of the reliability of *in vitro* adhesion assays. Proceedings of the 4th Workshop of the PROBEMO- FAIR CT 96-1028 project "Functional Foods for EU health in 2000." Rovaniem, Finland, February 25–28 VTT-Helsinki (published).

Cesena C, Morelli L, Alander M, Siljander T, Tuomola E, Salminen S, Mattila-Sandholm T, Vilpponen-Salmela T, von Wright A. *Lactobacillus crispatus* and its nonaggregating mutant in human colonization trials. *J. Dairy Sci.*, 2001; 84: 1001–1010.

Charteris WP., Kelly PM., Morelli L., Collins JK. Ingredient selection criteria for probiotic micro-organisms in functional dairy foods. *Int. J. Dairy Technol.*, 1998a; 51: 123–136.

Charteris WP, Kelly PM, Morelli L, Collins JK. Development and application of an *in vitro* methodology to determine the transit tolerance of potentially probiotic *Lactobacillus* and *Bifidobacterium* species in the upper human gastrointestinal tract. *J. Appl. Microbiol.*, 1998b; 84: 759–768.

Chen H, Lim CK, Lee YK, Chan YN. Comparative analysis of the genes encoding 23S-5S rRNA intergenic spacer regions of *Lactobacillus casei*-related strains. *Int. J. Syst. Evol. Microbiol.*, 2000; 50 471–478.

Coconnier MH, Bernet MF, Kerneis S, Chauviere G, Fourniat J, Servin AL. Inhibition of adhesion of enteroinvasive pathogens to human intestinal Caco-2 cells by *Lactobacillus acidophilus* strain LB decreases bacterial invasion. *FEMS Microbiol. Lett.*, 1993; 110:299–305.

Collins MD, Phillips BA, Zanoni P. Deoxyribonucleic acid homology studies of *Lactobacillus casei, Lactobacillus paracasei* sp. nov. subsp. *paracasei* and subsp. *tolerans* and *Lactobacillus rhamnosus* sp. nov. comb. nov. *Int. J. Syst. Bacteriol.*, 1989; 39: 105–108.

Conway PL, Gorbach SL, Goldin BR. Survival of lactic acid bacteria in the human stomach and adhesion to intestinal cells. *J. Dairy Sci.*, 1987; 70: 1–12.

Cremonini F, Di Caro S, Nista EC, Bartolozzi F, Capelli G, Gasbarrini G, Gasbarrini A. Meta-analysis: the effect of probiotic administration on antibiotic-associated diarrhoea. *Aliment. Pharmacol. Ther.*, 2002; 16: 1461–1467.

Cusick SM, O'Sullivan DJ. Use of a single, triplicate arbitrarily primed-PCR procedure for molecular fingerprinting of lactic acid bacteria. *Appl. Environ. Microbiol.*, 2000; 66: 2227–2231.

De Boever P, Wouters R, Verschaeve L, Berckmans P, Schoeters G, Verstraete W. Protective effect of the bile salt hydrolase-active *L. reuteri* against bile salt cytotoxicity. *Appl. Microbiol. Biotechnol.*, 1999; 53: 709–714.

De Man JC, Rogosa M, Sharpe ME. A medium for the cultivation of lactobacilli. *J. Appl. Bacteriol.*, 1960; 23: 130–135.

Dellaglio F, Bottazzi V, Trovatelli LD. Deoxyribonucleic acid homology and base composition in some thermophilic lactobacilli. *J. Gen. Microbiol.*, 1973; 74: 289–297.

Drake M, Small CL, Spence KD, Swanson BG. Rapid detection and identification of *Lactobacillus* spp. in dairy products by using the polymerase chain reaction. *J. Food Prot.*, 1996; 59: 1031–1036.

D'Souza AL, Rajkumar C, Cooke J, Bulpitt CJ. Probiotics in the prevention of antibiotic associated diarrhoea: meta-analysis. *Br. Med. J.*, 2002; 324(7350): 1361.

Dujon B. Microsatellite typing as a new tool for identification of *Saccharomyces cerevisiae* strains. *J. Clin. Microbiol.*, 2001; 39: 551–559.

Du Plessis EM, Dicks LM. Evaluation of random amplified polymorphic DNA (RAPD)-PCR as a method to differentiate *Lactobacillus acidophilus, Lactobacillus crispatus, Lactobacillus amylovorus, Lactobacillus gallinarum, Lactobacillus gasseri,* and *Lactobacillus johnsonii. Curr. Microbiol.*, 1995; 31: 114–118.

Elkins CA, Savage DC. Identification of genes encoding conjugated bile salt hydrolase and transport in *Lactobacillus johnsonii* 100–100. *J. Bacteriol.*, 1998; 180: 4344–4349.

Ferrero M, Cesena C, Morelli L, Scolari G, Vescovo M. Molecular characterization of *Lactobacillus casei* strains. *FEMS Microbiol. Lett.*, 1996; 140: 215–219.

Gilliland SE, Staley TE, Bush LJ. Importance of bile tolerance of *Lactobacillus acidophilus* used as Dietary Adjunct. *J. Dairy Sci.*, 1984; 67: 3045–3051.

Goldin BR, Gorbach SL, Saxelin M, Barakat S, Gualtieri L, Salminen S. Survival of *Lactobacillus* species (strain GG) in human gastrointestinal tract. *Dig. Dis. Sci.*, 1992; 173: 121–128.

Gorbach SL, Goldin BR. Lactobacillus strains and methods of selection. United States Patent No. 4,839,281. 1989.

Gorbach SL. Probiotics in the third millennium. *Dig. Liver Dis.*, 2002; 34: S2–S7.

Hamilton-Miller JM, Shah S, Winkler JT. Public health issues arising from microbiological and labelling quality of foods and supplements containing probiotic microorganisms. *Public Health Nutr.*, 1999; 2: 223–229.

Harmsen HJ, Gibson GR, Elfferich P, Raangs GC, Wildeboer-Veloo AC, Argaiz A, Roberfroid MB, Welling GW. Comparison of viable cell counts and fluorescence in situ hybridization using specific rRNA-based probes for the quantification of human fecal bacteria. *FEMS Microbiol. Lett.*, 2000; 183: 125–129.

Hennequin C, Thierry A, Richard GF, Lecointre G, Nguyen HV, Gaillardin C, Dujon B. Microsatellite typing as a new tool for identification of *Saccharomyces cerevisiae* strains. *J. Clin. Microbiol.*, 2001; 39: 551–559.

Hood SK, Zottola EA. Effect of low pH on the ability of *Lactobacillus acidophilus* to survive and adhere to human intestinal cells. *J. Food. Sci.*, 1988; 53: 1514–1516.

Jacobsen CN, Rosenfeldt Nielsen V, Hayford AE, Moller PL, Michaelsen KF, Paerregaard A, Sandstrom B, Tvede M, Jakobsen M. Screeing of probiotic activities of forty-seven strains of *Lactobacillus* spp. by *in vitro* techniques and evaluation of the colonization ability of five selected strains in humans. *Appl. Environ. Microbiol.*, 1999; 65: 4949–4956

Johnson JL, Phelps CF, Cummins CS, London J, Gasser F. Taxonomy of the *Lactobacillus acidophilus* group. *Int. J. Syst. Bacteriol.*, 1980; 30: 53–68 and 225–420.

Jonsson H, Strom E, Roos S. Addition of mucin to the growth medium triggers mucus-binding activity in different strains of *Lactobacillus reuteri in vitro*. *FEMS Microbiol. Lett.*, 2001 16; 204:19–22.

Kandler O, Weiss N. 1986. Genus *Lactobacillus* Beijerinck 1901. In Sneath PHA., Mair NS, Sharpe ME, Holt JG., Eds., *Bergey's Manual of Determinative Bacteriology*. Vol. 2. Williams & Wilkins, Baltimore, pp. 1209–1234.

Kos B, Suskovic J, Vukovic S, Simpraga M, Frece J, Matosic S. Adhesion and aggregation ability of probiotic strain *Lactobacillus acidophilus* M92. *J. Appl. Microbiol.*, 2003; 94: 981–987.

Kurdi P, van Veen HW, Tanaka H, Mierau I, Konings WN, Tannock GW, Tomita F, Yokota A. Cholic acid is accumulated spontaneously, driven by membrane delta pH, in many lactobacilli. *J. Bacteriol.*, 2000; 182: 6525–6528.

Langendijk PS, Schut F, Jansen GJ, Raangs GC, Kamphuis GR, Wilkinson MH, Welling GW. Quantitative fluorescence in situ hybridization of *Bifidobacterium* spp. with genus-specific 16S rRNA-targeted probes and its application in fecal samples. *Appl. Environ. Microbiol.*, 1995; 61: 3069–3075.

Lehto EM, Salminen SJ. Inhibition of *Salmonella typhimurium* adhesion to Caco-2 cell cultures by *Lactobacillus* strain GG spent culture supernate: only a pH effect? *FEMS Immunol. Med. Microbiol.*, 1997; 18: 125–132.

Maidak BL, Cole JR, Parker CT Jr, Garrity GM, Larsen N, Li B, Lilburn TG, McCaughey MJ, Olsen GJ, Overbeek R, Pramanik S, Schmidt TM, Tiedje JM, Woese CR. A new version of the RDP (Ribosomal Database Project). *Nucleic Acids Res.*, 1999; 27: 171–173.

Matsuki T, Watanabe K, Tanaka R, Oyaizu H. Rapid identification of human intestinal bifidobacteria by 16S rRNA-targeted species- and group-specific primers. *FEMS Microbiol. Lett.*, 1998; 167: 113–121.

Matsuki T, Watanabe K, Tanaka R, Fukuda M, Oyaizu H. Distribution of bifidobacterial species in human intestinal microflora examined with 16S rRNA-gene-targeted species-specific primers. *Appl. Environ. Microbiol.*, 1999; 65: 4506–4512.

Matsumura A, Saito T, Arakuni M, Kitazawa H, Kawai Y, Itoh T. New binding assay and preparative trial of cell-surface lectin from *Lactobacillus acidophilus* group lactic acid bacteria. *J. Dairy Sci.*, 1999; 82: 2525–2529.

McCartney AL, Wenzhi W, Tannock GW. Molecular analysis of the composition of the bifidobacterial and lactobacillus microflora of humans. *Appl. Environ. Microbiol.*, 1996; 62: 4608–4613.

Meunier JR, Grimont PA. Factors affecting reproducibility of random amplified polymorphic DNA fingerprinting. *Res. Microbiol.*, 1993; 144: 373–379.

Mitterdorfer G, Mayer HK, Kneifel W, Viernstein H. Clustering of *Saccharomyces boulardii* strains within the species *S. cerevisiae* using molecular typing techniques. *J. Appl. Microbiol.*, 2002; 93: 521–530.

Morelli L. *In vitro* selection of probiotic lactobacilli: a critical appraisal. *Curr. Issues Intest. Microbiol.*, 2000; 1: 59–67.

Mori K, Yamazaki K, Ishiyama T, Katsumata M, Kobayashi K, Kawai Y, Inoue N, Shinano H. Comparative sequence analyses of the genes coding for 16S rRNA of *Lactobacillus casei*-related taxa. *Int. J. Syst. Bacteriol.*, 1997; 47: 54–57.

Moser SA, Savage DC. Bile salt hydrolase activity and resistance to toxicity of conjugated bile salts are unrelated properties in lactobacilli. *Appl. Environ. Microbiol.*, 2001; 67: 3476–3480.

Orla-Jensen. La classification de bactéries lactiques. *Lait*, 1924; 4: 468–474.

Ouwehand AC, Tuomola EM, Tolkko S, Salminen S. Assessment of adhesion properties of novel probiotic strains to human intestinal mucus. *Int. J. Food Microbiol.*, 2001; 64: 119–126.

Penner GA, Bush A, Wise R, Kim W, Domier L, Kasha K, Laroche A, Scoles G, Molnar SJ, Fedak G. Reproducibility of random amplified polymorphic DNA (RAPD) analysis among laboratories. *PCR Methods Appl.*, 1993; 2: 341–345.

Pot B, Hertel C, Ludwig W, Descheemaeker P, Kersters K, Schleifer KH. Identification and classification of *Lactobacillus acidophilus*, *L. gasseri*, and *L. johnsonii* strains by SDS-PAGE and rRNA-targeted oligonucleotide probe hybridization. *J. Gen. Microbiol.* 1993; 139: 513–517.

Rojas M, Ascencio F, Conway PL. Purification and characterization of a surface protein from *Lactobacillus fermentum* 104R that binds to porcine small intestinal mucus and gastric mucin. *Appl. Environ. Microbiol.*, 2002; 68: 2330–2336.

Roos S, Jonsson H. A high-molecular-mass cell-surface protein from *Lactobacillus reuteri* 1063 adheres to mucus components. *Microbiology*, 2002; 148: 433–442.

Roos S, Lindgren S, Jonsson H. Autoaggregation of *Lactobacillus reuteri* is mediated by a putative DEAD-box helicase. *Mol. Microbiol.*, 1999; 32: 427–436.

Rossello-Mora R, Amann R. The species concept for prokaryotes. *FEMS Microbiol. Rev.*, 2001; 25: 39–67.

Roy D, Ward P, Champagne G. Differentiation of bifidobacteria by use of pulsed-field gel electrophoresis and polymerase chain reaction. *Int. J. Food Microbiol.*, 1996;29:11–29.

Sanders ME, Morelli L., Tompkins TA. Sporeformers as human probiotics: *Bacillus*, *Spolactobacillus*, and *Brevibacillus*. *Compr. Rev. Food Sci. Food Safety*, 2003; 2: 101–110.

Satokari RM, Vaughan EE, Akkermans AD, Saarela M, de Vos WM. Bifidobacterial diversity in human feces detected by genus-specific PCR and denaturing gradient gel electrophoresis. *Appl. Environ. Microbiol.*, 2001; 67: 504–513.

Scardovi V. Genus *Bifidobacterium*. In: *Bergey's Manual of Systematic Bacteriology*, Vol. 2 (ed. Sneath, P.H.A., Mair, N.S., Sharpe, M.E. and Holt, J.G.), p. 1418. New York: Williams & Wilkins.

Schaer-Zammaretti P, Ubbink J. Imaging of lactic acid bacteria with AFM elasticity and adhesion maps and their relations hip to biological and structural data. *Ultramicroscopy*, 2003; 97: 199–208.

Schneitz C, Nuotio L, Lounatma K. Adhesion of *Lactobacillus acidophilus* to avian intestinal epithelial cells mediated by the crystalline bacterial cell surface layer (S-layer). *J. Appl. Bacteriol.*, 1993; 74: 290–294.

Simpson JM, McCracken VJ, Gaskins HR, Mackie RI. Denaturing gradient gel electrophoresis analysis of 16S ribosomal DNA amplicons to monitor changes in fecal bacterial populations of weaning pigs after introduction of *Lactobacillus reuteri* strain MM53. *Appl. Environ. Microbiol.*, 2000; 66: 4705–4714.

Stackebrandt E, Goebel BM. Taxonomic note: a place for DNA-DNA reassociation and 16S rRNA sequence analysis in the present species definition in bacteriology. 1994; 846–849.

Stackebrandt E, Rainey FA, Ward-Rainey N. Proposal for a new hierarchic classification system, *Actinobacteria* classis nov. *Int. J. Syst. Bacteriol.*, 1997; 47: 479–491.

Tannock GW, Munro K, Harmsen HJ, Welling GW, Smart J, Gopal PK. Analysis of the fecal microflora of human subjects consuming a probiotic product containing *Lactobacillus rhamnosus* DR20. *Appl. Environ. Microbiol.*, 2000; 66: 2578–2588.

Tissier MH, (1900) Reserches sur la flore intestinale normale et pathologique du nourisson. Thesis. University of Paris.

Tynkkynen S, Satokari R, Saarela M, Mattila-Sandholm T, Saxelin M. Comparison of ribotyping, randomly amplified polymorphic DNA analysis, and pulsed-field gel electrophoresis in typing of *Lactobacillus rhamnosus* and *L. casei* strains. *Appl. Environ. Microbiol.*, 1999; 65: 3908–3914.

Vandamme P, Pot B, Gillis M, de Vos P, Kersters K, Swings J. Polyphasic taxonomy, a consensus approach to bacterial systematics. *Microbiol. Rev.*, 1996; 60: 407–438.

Ventura M, Casas IA, Morelli L, Callegari ML. Rapid amplified ribosomal DNA restriction analysis (ARDRA) identification of *Lactobacillus* spp. isolated from fecal and vaginal samples. *Syst. Appl. Microbiol.*, 2000; 23: 504–509.

Ventura M, Zink R. Specific identification and molecular typing analysis of *Lactobacillus johnsonii* by using PCR-based methods and pulsed-field gel electrophoresis. *FEMS Microbiol. Lett.*, 2000; 217: 141–154.

Vollaard EJ, Clasener HA. Colonization resistance. *Antimicrob. Agents Chemother.*, 1994; 38: 409–414.

Ward LJ, Timmins MJ. Differentiation of *Lactobacillus casei, Lactobacillus paracasei* and *Lactobacillus rhamnosus* by polymerase chain reaction. *Lett. Appl. Microbiol.*, 1999; 29: 90–92.

Wayne LG, Brenner DJ, Colwell RR, Grimont PAD, Kandler O, Krichevsky MI, Moore LH, Moore WEC, Murray RGE, Stackebrandt E, Starr MP, Truper HG. Report of the ad hoc Committee on Reconciliation of Approaches to Bacterial Systematics. *Int. J. Syst. Bacteriol.*, 1987; 37, 463–464.

Zhong W, Millsap K, Bialkowska-Hobrzanska H, Reid G. Differentiation of *Lactobacillus* species by molecular typing. *Appl. Environ. Microbiol.*, 1998; 64: 2418–2423.

Zoetendal EG, Ben-Amor K, Akkermans AD, Abee T, de Vos WM. DNA isolation protocols affect the detection limit of PCR approaches of bacteria in samples from the human gastrointestinal tract. *Syst. Appl. Microbiol.*, 2001; 24: 405–410.

Zoetendal EG, von Wright A, Vilpponen-Salmela T, Ben-Amor K, Akkermans AD, de Vos WM. Mucosa-associated bacteria in the human gastrointestinal tract are uniformly distributed along the colon and differ from the community recovered from feces. *Appl. Environ. Microbiol.*, 2002; 68: 3401–3407.

4

Primary Sources of Probiotic Cultures

Daniel J. O'Sullivan

CONTENTS

4.1   Introduction

Probiotic research has exploded at nearly an exponential rate over the last 15 years, since the first workable definition for probiotics was proposed. This definition, proposed by Fuller (1), was "a live microbial feed supplement which beneficially affects the host animal by improving its intestinal microbial balance." Although this definition is still workable in many instances today, an expanding application list for probiotics has resulted in many new variations of the definition being proposed. These variations take into account such applications as benefits to the host outside its microbial balance and applications other than feed, such as topical applications for probiotics. This recent explosion in probiotic research would lead one to think that probiotics are a relatively new concept. However, the concept has been around for 100 years, from the studies of lactobacilli in soured milks by Elie Metchnikoff (2) and the treatment of infant diarrhea with bifidobacteria by Tissier (3). The resurgence is primarily due to a better understanding of the intestinal microorganisms and their effect on intestinal health.

Maintaining or improving human health is not the only target market for probiotics. Farm animals are also a big market as studies have implicated probiotics in many positive attributes such as reduced scouring in calves (4), improved weight gain and feed efficiency for beef cattle (5), improved milk yield in dairy cattle (6), increased weight gain in chickens (7), control of postweaning diarrhea in pigs (8), and reduced *salmonella* in chickens (9). Currently, antibiotics are included in many animal feeds for many of these benefits. With the increasing public resistance against this practice and increasing knowledge of animal probiotics, this market would seem to have a very bright future.

Although the human probiotic market is quite significant in Europe and Asia, particularly in Japan, it only started receiving attention by the U.S. food industry recently (10). The growing U.S. interest in this market is primarily because of the increasing number of health-conscious consumers in the U.S. and the potential health benefits attributed to probiotics. The potential benefits include increased resistance to gastrointestinal (GI) tract infections, alleviation of constipation, reestablishment of a healthy intestinal flora following antibiotic or chemotherapy treatments, stimulation of the immune system, reduction of serum cholesterol, prophylactic for intestinal cancers, and alleviation of the symptoms of lactose intolerance (reviewed in References 10 and 11). The lack of a consensus scientific agreement on the extent of these benefits attributed to specific probiotic strains has slowed regulatory acceptance of any specific functional claims to date. However, with mounting evidence for specific health claims, these may soon be marketable, thus accelerating this market.

4.2   Delivery of Probiotics

In order for probiotics to have their desired effect, they need to be delivered safely to their targeted site of action. The traditional target sites for probiotics are the upper and lower gut, and this chapter will concentrate to a large part on probiotics targeting these areas. However, it is worth summarizing some of the other probiotic targets that contributed to the expansion of the probiotic concept. The oral cavity is generally the most populated body area outside of the large intestine with microbial populations as high as 10^9 CFU/ g. A probiotic product consisting of *Streptococcus salivarius* in a product called

BLIS K12 Throat Guard is marketed in New Zealand to control *Streptococcus pyogenes* (strep throat) infections in children (12). Another target is the vagina for the control of vaginosis, and lactobacilli have been shown to be promising probiotic candidates for this purpose (reviewed in Reference 13). Finally, some ongoing research is looking at the potential of certain probiotic bacteria to prevent the occurrence of inflammation in wounds (14).

Probiotics targeting the intestine clearly encounter the greatest hurdles in order to be delivered to their targeted site. The biggest hurdles are the acid conditions of the stomach and the bile in the duodenum. Currently, probiotic strains that can overcome mere hurdles are obtained by screening for strains that have adapted to these stresses. Although this strategy does achieve the goal of delivering the probiotic safely to its target site, it may in some cases compromise the probiotic bacterium as the adaptation, or attenuation may put it in a competitive disadvantage when it arrives at its targeted niche. The effect of attenuation on probiotic functioning will be discussed in more detail in the following sections. This acid adaptation is especially problematic for probiotics targeting the large intestine as this is normally a buffered, largely neutral pH environment. A more logical approach is to choose a probiotic bacterium based on its *in vivo* characteristics and develop technologies to protect it *in vitro* and during transit through the stomach and duodenum. Studies have been done using microencapsulation technologies to protect probiotic bacteria during processing, storage, and transit through simulated gastric conditions. Some of these studies concentrated on bifidobacteria, which are the primary probiotic bacteria targeting the large intestine. Encapsulation using calcium alginate or kappa-carrageenan have shown good promise at protecting cells during processing and storage (15, 16). However, microencapsulation using cellulose acetate phthalate protected bifidobacteria quite impressively during spray drying and during prolonged exposure to gastric simulated conditions (17). Further developments in technologies such as these will enable food processors to concentrate more on probiotic traits for efficacy *in vivo*, rather than on traits for surviving processing, storage, and transit of the probiotic through the stomach and duodenum.

In the U.S., probiotics are available to consumers in some dairy food products and also in dietary supplements. The food industry is strictly regulated by the Food and Drug Administration (FDA), which limits marketing and label statements concerning the products. Currently, the FDA has not approved any specific health statements concerning probiotics, which greatly hampers the marketing of probiotic foods by the food industry. Nevertheless, consumers are becoming increasingly more aware of these healthful cultures, and their listing on labels appeals to a growing niche market. Dairy products have a long history of association with cultures, which helped probiotics find a home primarily in the dairy industry. It also helps that probiotic bacteria generally demonstrate better survival characteristics in dairy products compared to other foods. Examples of dairy foods that contain probiotic cultures in the U.S. are yogurts, milks, and fermented milks such as kefir. These products generally add a *Bifidobacterium* and a *Lactobacillus acidophilus*, identified as "bifidus" and "acidophilus" on the labels. Sometimes additional probiotic cultures are added such as *L. casei* and *L. reuteri*, which are listed in Stonyfield yogurt. However, other than listing the cultures on the labels, statements pertaining to their benefits are either absent or extremely guarded. Dietary supplements, on the other hand, are largely unregulated, although the FDA in instances does control them when the consumer is either at risk or when grossly unfair practices are used. This allows more liberal statements on probiotic dietary supplements, which are generally supplied as freeze-dried cultures in capsules. However, the general consumer is very cautious with dietary supplements that do not have a long history of safe and effective use, as there have been numerous highly publicized cases of supplements causing severe harm, including death, to individuals. Although the overall dietary supplement market is growing in the U.S., probiotics occupy <1% of this market.

The probiotic market is significantly greater in many other countries, particularly within the European Union (EU) and Asia, caused primarily by a more liberal approach to marketing statements on food labels. Perhaps the most successful type of probiotic product in these markets is a small, sweetened dairy drink with very high numbers (>10^{10}) of viable *Lactobacillus casei*, which is marketed to give the immune system a boost when taken on a daily basis. Examples of these products are Yakult and Danone's Actimel™ (Figure 4.1). It is noteworthy that Danone did some test marketing of Actimel in the U.S. as a dietary supplement in Colorado, Arizona, and Florida. Although the product did acquire a niche market, its growth was hampered by its labeling as a dietary supplement and its small container size! However, after several years of testing, it is now marketed in the U.S. under the brand name DanActive™.

FIGURE 4.1
Probiotic drinks containing *Lactobacillus casei* marketed as immune system boosters when consumed daily. Apple is included for a size perspective.

4.3 Where Do Probiotic Bacteria Come from?

Selection of strains for probiotic use should always follow two general principles: safety of the organism, and possessing desirable characteristics for its intended use. Intestinal probiotics are dominated by members of *Bifidobacterium* and *Lactobacillus*, as these two genera have a long history of safe use and have GRAS (generally regarded as safe) status. They are also very suited to augmenting the intestine, as *Bifidobacterium* is a major inhabitant of the large intestine and *Lactobacillus* is a major inhabitant of the small intestine. The origin of the strains used in probiotics can be either freshly isolated from a human or animal host or from a culture collection. Culture collection strains have generally been extensively cultivated in fermentation systems and thus have likely attenuated to suit the *in vitro* environment. However, as stated by Havenaar et al. (18), the choice of where to get a probiotic strain depends on the specific purpose of the probiotic. For example, if only transient activity of the probiotic is needed, such as for lactose digestion, then it is not necessary for the probiotic to have characteristics that would enable it to colonize the host. Most probiotic effects in the GI system would be enhanced if the probiotic would be able to compete with the indigenous flora. This requires a more careful selection of strains. Although all the criteria for this purpose are not currently known for any intestinal organism, there is one general consensus pertaining to a criterion that is important. That consensus is that the probiotic should originate from the same animal species that it is intended to target (19, 20). The rationale is that the intestinal environments in different animal species are sufficiently different such that the most competitive bacteria in each host species have evolved specific traits for survival in that host (21). Human probiotics, therefore, should originate from a human source if the objective is to effectively modulate the microbial populations at their target sites.

It is noteworthy that when commercial probiotics are fed to human subjects during controlled feeding studies, the probiotic can be detected in high numbers in the feces during the feeding period, but rapidly disappears following cessation of feeding (22). The rate of decrease of the probiotic following feeding is generally less than 1 week (Figure 4.2). Some probiotics such as *Lactobacillus rhamnosus* GG can in some cases persist longer than a week postfeeding (23). The lack of detection of the probiotic postfeeding indicates that it is not able to compete very well with the endogenous strains of that species. This is to be expected because the endogenous strains are adapted to their environment, whereas the probiotic strain may not have all the traits necessary to compete with it. Unfortunately, all of the necessary traits are not yet known, but significant progress has been made, and this is discussed in a following section.

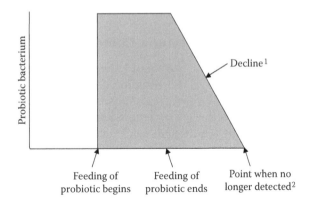

FIGURE 4.2
A typical detection pattern for probiotic bacteria from feces during human feeding trials. (1) The rate of decline in numbers detected following feeding is dependent on the particular probiotic; (2) depends on the probiotic, but is generally <1 week.

4.4   Species Used as Probiotic Cultures

Probiotics for human probiotics are dominated by different species and strains of the genera *Lactobacillus* and *Bifidobacterium* (Table 4.1). A primary reason for this is that both these genera are dominant inhabitants of their respective niches in the intestine, *Lactobacillus* in the small intestine and *Bifidobacterium* in the large intestine. Of the lactobacilli, *L. acidophilus* is by far the most widely used probiotic as it has a long history of research and use. Many of the other *Lactobacillus* species listed in Table 4.1 are closely related to *L. acidophilus*. This long history dates back to the early 1900s when significant research on intestinal flora followed Metchnikoff's work with a lactic acid-producing bacterium, then called *Bacillus bulgaricus*, found in a yogurt product. As *L. acidophilus* is a related bacterium and one of the predominant organisms in the intestinal tract of breast-fed babies, it quickly took the place of *L. bulgaricus* as the probiotic of choice in the U.S. (24). It, therefore, has almost 100 years of use in human diets. Another very popular *Lactobacillus* probiotic is *L. casei*, as it is thought to significantly enhance the immune system by inducing IL-12 and IFNγ expression and activates NK cell tumoricidal activity, as well as stimulating the production of secretory IgA (25, 26). This has been the foundation for the successful probiotic beverages shown in Figure 4.1. It is thought the high immune response is because they colonize the small intestine in the vicinity of the Peyer's patches, which are very immunogenic. Many other lactobacilli, as well as some bifidobacteria, also have been shown to stimulate an enhanced immune response (25).

Of the bifidobacteria, *B. longum* is particularly dominant in human intestines and is one of the few species to regularly harbor plasmids (27). It is a

TABLE 4.1

Microorganisms Used in Human Probiotics[a]

Species Generally Present in the Human GI Tract	Popular Strains and Locations
Lactobacillus acidophilus	NCFM (USA)
	R0010 (Canada)
	LA-1 (Europe)
	SBT-2062 (Japan)
L. casei	Imunitass (Europe)
	Shirota (Japan)
L. rhamnosus	GG (Europe)
	R0052 (Canada)
L. paracasei	CRL 431 (Europe)
L. salivarius	UCC118 (Europe)
L. johnsonii	LA1 (Europe)
L. reuteri	
L. crispatus	
L. fermentum	
L. gasseri	
Bifidobacterium longum	BB536 (Japan)
B. infantis	
B. breve	Yakult (Japan)
B. bifidum	
B. adolescentis	
Enterococcus faecium	Fargo 688 (Europe)
Pediococcus pentosaceus	
P. acidilactici	
E. coli	
Species not generally present in the human GI tract	
Bifidobacterium animalis	
B. lactis	Bb12 (Europe)
	DR10 (New Zealand)
Bacillus subtilis	
Saccharomyces boulardii	

[a] Does not include cultures used for functional purposes in foods but which may have probiotic activities such as *Streptococcus thermophilus, Lactobacillus delbrueckii* subsp. *bulgaricus, L. plantarum, Propionibacterium acidipropionici,* and *Lactococcus lactis.*

very popular "touted" *Bifidobacterium* species in commercial human probiotics (10). More significantly, it is the most frequently reported species associated with many of the purported health benefits of bifidobacteria. These include diarrhea prevention in antibiotic-treated patients (28), cholesterol reduction (29), alleviation of lactose intolerance symptoms (30), immune stimulation (31), and cancer prevention (32). This species is the primary one associated with cancer prevention, and it is reported to protect against many different carcinogens, including methyl quinolines (32), heterocyclic amines (33), nitrosamines (34), and azomethane (35). *B. infantis,* which is the closest genetic relative to *B. longum,* was generally reported in the early literature to be solely an inhabitant of the child intestine. However, it is now regularly found in adults as well (36). The fact that it disappears from the intestines of elderly people makes it a very intriguing probiotic indeed.

Bifidobacterium lactis is a very commonly used probiotic, and it is intriguing that it is not a normal human inhabitant. It was first isolated in 1997 from fermented milk by Meile et al. (37) and was noted to have a higher tolerance to oxygen than other bifidobacteria. Although it was genetically very close to *B. animalis*, it was deemed to have changed sufficiently during adaptation to fermentation environments to warrant a new species. The new name was especially attractive to the food industry, as it is a lot more palatable to list lactis on a label rather than animalis. However, a number of comparative studies between these two species suggest that *B. lactis* should still be named *B. animalis* (38, 39). The changes that occurred in *B. lactis* during its adaptation to fermentation conditions make it a very resilient strain that can remain viable during processing and storage longer than other bifidobacteria. These practical reasons contribute to its popularity. These adaptations, however, would not give it a competitive edge in the intestine as the most competitive strains lose unwanted traits in a natural environment. Although this would limit the full potential of *B. lactis*, it still has the potential for many positives during its transient passage through the intestine.

It is also intriguing that *Bacillus subtilis* is used in many human probiotics. Although it is a safe bacterium, with GRAS status, the genus *Bacillus* in general is not usually an inhabitant of human intestines. It is a lot to ask of a bacterium to modulate an environment that is completely foreign to it. However, its ability to form endospores enables it to survive passage through the stomach and duodenum, which is the primary attraction for using this bacterium. One concern is the purity of the spore preparations, as one study of two popular commercial *B. subtilis* probiotic preparations sold in Europe and Asia found, neither contained *B. subtilis*. One contained a related species to *B. subtilis*, whereas the other contained the unrelated species, *B. alcalophilus* (40).

4.5   What Is Known about Selection Criteria for Probiotic Cultures

When selecting a candidate bacterium for a probiotic, a lot of decisions have to be made beyond the origin of the bacterium and the genus and species. Once a species is chosen, it is then necessary to find a strain of that species with all the traits necessary for an optimum efficacious effect. If there was no knowledge of any trait that was important for a probiotic, it would essentially be impossible to select a strain that had any effect. A comparable analogy would be that if an alien from outer space approached his subordinate and asked to please select five *Homo sapiens* from Earth for a basketball team. Unless the subordinate had any knowledge of what traits a basketball player should have, what are the odds of his picking a Kobe, Shaq, or Michael

Jordan randomly from 6 billion strains of *Homo sapiens*. Fortunately, some knowledge is available of desirable traits for probiotic bacteria, and although it is far from complete, it can help to get strains with desirable traits for specific purposes. The following is a brief summary of the traits that are currently considered important when selecting probiotic strains.

4.5.1 Tolerance to Acid and Bile

Unless a probiotic is going to survive transit through the stomach and duodenun, it is going to be of little or no benefit, unless the purpose was just to deliver β-galactosidase to the small intestine. The high-acid conditions of the stomach require that the organism should have a high tolerance to acid or is protected using encapsulation technology. Acid resistance is frequently measured by evaluating its ability to survive pH 3 or lower for 3 hours, an average passage time through the stomach. Although this is a good indicator, more in-depth studies using simulated transit conditions can also be performed (41). Whereas many probiotics do not have their stress defenses constitutively expressed, and thus would be killed with this treatment, they can often be primed by exposing them to mild stresses that can induce stress defenses to enable them to withstand greater stresses. Therefore, potential isolates for probiotic cultures need not necessarily be directly resistant to pH 3, as long as they tolerate it after prior priming at a higher pH. This may easily be accomplished when yogurt is used as the delivery vehicle for the probiotic, as the mild acid conditions may be sufficient to effectively prime many isolates. Similarly, isolates need sufficient tolerance to bile to enable safe passage through the duodenum to their site of action. This is generally measured by simply plating out isolates on media containing bile salts. This process, however, largely measures direct resistance to bile rather than just tolerance. *In vivo*, a probiotic culture's stress response will already be strongly induced following passage through the stomach. It is known that exposure to one stress can induce a response that protects cells against multiple stresses (42). As the stress response is already induced at that stage, it may be capable of surviving the bile in the duodenum. This is pertinent as many candidate isolates may be overlooked if they do not display direct resistance to bile, when in reality the ability to induce sufficient tolerance is all that is required.

4.5.2 β-Galactosidase Activity

A significant proportion of the population has some degree of intolerance to lactose. This is especially true of certain racial groups, such as African American and Hispanic. People with northern European heritage have the lowest instance of lactose intolerance. The lactic acid bacteria are excellent digesters of lactose due to production of β-galactosidase or phospho-β-galactosidase. This feature of a probiotic organism is pertinent when the culture is intended to reduce the symptoms of lactose maldigestion. Isolates display large differences in the amounts of this enzyme they produce. It should also be noted that most wild-type isolates have very low constitutive levels of this enzyme but have much higher levels following growth in lactose as the sole carbon source. The induced β-galactosidase level is, therefore, the pertinent feature for a probiotic organism intended to aid subjects with lactose digestion. Enzyme activities of potential candidates can also be readily increased up to approximately threefold following a straightforward classical mutagenesis approach (43). As this is a food-grade classical approach, it does not cross the realm into the currently used definition for a genetically modified food.

4.5.3   Adherence to Intestinal Cells

It is intriguing that a lot of emphasis is placed on the ability of probiotic bacteria to adhere to the available human cell lines when there is still no definitive evidence to date that they actually do adhere to cells *in vivo*. Electron micrographs depicting bacteria adhering to intestinal cells can be misleading as the methodology used for this technique dehydrates the mucin layer. Bacteria in the mucin layer would, therefore, appear to be attached to the epithelial cells on electron micrographs. However, ecological evidence from other environmental habitats would suggest that to survive and compete successfully in a natural ecosystem, such as the human intestine that is in constant flux, a bacterium needs to be able to attach to the available attachment sites in the intestine. Studies have shown that probiotic bacteria can attach to mucin, so it is feasible that this may be sufficient to sustain a species (44). Measuring the ability of isolates to adhere to intestinal cells is not very easy given the diversity in human intestines and diversity of cells within. Currently, adherence is measured primarily using two *in vitro* cell lines, Caco-2 and HT-29. As these essentially represent a single cell from the intestines of two individuals, it is not a thorough test on the true adherence abilities of isolates. Positive attachment to these cell lines, however, can be viewed as a good indicator of their potential to attach. However, too much emphasis should not be placed on cell attachment until it is proven that attachment to intestinal cells rather than to mucin is important for colonization.

4.5.4   Bacteriocin Production

Ecological studies have shown that aggressive colonizers can inhibit their competitors using a number of mechanisms. One such mechanism is the ability to produce bacteriocins, which are proteinaceous antimicrobial compounds. Lactobacilli most likely use this approach, as many of them have been shown to produce bacteriocins, including many members of the prevalent probiotic species, *L. acidophilus* (45), *L. casei* (46), *L. reuteri* (47), *L. rhamnosus* (48), *L. gasseri* (49), *L. salivarius* (50), and *L. johnsonii* (51). It is therefore an important trait to consider when selecting strains of lactobacilli for probiotic purposes. Another aspect is the spectrum of antimicrobial activity of the bacteriocins, and this varies greatly among the characterized bacteriocins (52). It is considered more desirable to have a broader spectrum of activity to provide the host better protection against GI infections. Only one species of *Bifidobacterium*, *B. bifidum*, has been shown to produce a bacteriocin (53), suggesting that members of this genera may rely on other competitive features for aggressively colonizing the large intestine.

4.5.5 Competition for Organic Nutritients

Carbon sources available in the intestine can be quite variable. Therefore, strains with a varied metabolic capability would have an advantage over strains with a limited metabolic capability. The ability to metabolize nutrients that are not metabolized by the host would be very advantageous for a species. One example is oligosaccharides, which are not metabolized by humans and, therefore, are available for microbial growth. Very few bacteria can metabolize oligosaccharides, and because bifidobacteria and some lactobacilli are some of the few that can, this can give them a significant advantage when these substrates are present. The area of prebiotics has arisen from a derivation of this concept, that is, feeding people oligosaccharides such as inulin or fructooligosaccharide (FOS) to give the resident bifidobacteria and lactobacilli a competitive advantage in their respective environments. Screening for strains that can efficiently utilize certain prebiotics will therefore enable the probiotic to be ingested with the prebiotic, thus providing a competitive advantage for the organism when both arrive in the nutrient-limiting large intestine. This concept of consuming a probiotic with a prebiotic is referred to as synbiotics (54).

4.5.6 Competition for Iron

Iron is an essential element for essentially all living cells; *Lactobacillus plantarum* and *Borrelia burgdorferi* are notable exceptions (55, 56). This metal is a required cofactor for a variety of basic biochemical mechanisms although it can be toxic at elevated amounts because of its ability to generate free radicals. Not surprisingly, iron uptake and storage are carefully regulated by cells. The solubility and bioavailability of iron is extremely low in neutral pH environments. As iron is acid soluble, it is not a limiting factor for growth in low pH environments, such as the small intestine. However, it is considered a major competitive factor among microbes in neutral pH environments. Certain microorganisms respond to iron limitation by secreting siderophores, low-molecular-weight compounds produced to solubilize, bind, and transport enviromental iron to their cells (57). Dominant colonizers of an environment generally have better scavenging systems and can inhibit the growth of other competing organisms by depriving them of iron (58, 59). As the large intestine is largely neutral pH, competition for iron has to be an important feature for surviving there. Bifidobacteria are superior competitors in the large intestine. It was proposed in this laboratory that bifidobacteria compete successfully in their environment against other bacteria, such as *Escherichia coli*, by depriving them of iron for growth. Previous studies on iron uptake by bifidobacteria *in vitro* suggested they did not secrete iron-binding compounds (60). However, these studies were performed in batch cultures where the production of organic acids would reduce the pH and make iron more bioavailable. Also, Mevissen-Verhage et al. (61) showed that infants fed cows milk fortified with iron had higher counts of *E. coli* and less bifidobacteria in their feces than infants fed unfortified milk. This study strongly suggested that bifidobacteria use iron scavenging to compete against *E. coli* in the large intestine. This hypothesis was tested in my laboratory using a bank of 29 *Bifidobacterium* isolates. Eight of 29 strains were shown to produce a compound during buffered neutral pH conditions that prevented the growth of indicator bacteria. Interestingly, these eight were newly isolated strains, whereas all of the strains from culture collections did not exhibit inhibition, suggesting this trait may be attenuated upon prolonged culture *in vitro*. For further studies, one of the positive isolates was chosen a *B. longum* isolate, which we had previously isolated and characterized (62). To test the range of inhibition, *Lactococcus lactis*, *Clostridium difficile*, *C. perfringens* and *E. coli* were also tested with this *B. longum* isolate, and all were inhibited by the production of a compound. It should be noted that the inhibition is purely static, as addition of iron to the inhibitory zone, and reincubation of the plate, enable the indicator to grow (reviewed in Reference 63).

4.6 How Will the Genomic Era Provide the Knowledge Base for a Scientific Selection Process for Probiotic Cultures?

All areas of biological science are currently being revolutionized by the genomics era. The first microbial genome to be published was *Haemophilus influenzae* in 1995, and in the ensuing 8 years another 127 have been published and many more completed but not published. In addition, there are currently more than 400 ongoing microbial sequencing projects. The use of this sequence information will provide enormous knowledge about the functioning of the respective bacteria.

There are currently six complete genomes for probiotic cultures deciphered (Table 4.2). Three of these were sequenced as part of a consortium that was developed in the U.S. consisting of a group of U.S. researchers interested in the molecular functioning of lactic acid bacteria cultures. This Lactic Acid Bacteria Genome Consortium (LABGC), which consists of 10 scientists, formed a collaboration with the Joint Genome Institute (JGI) funded by the Department of Energy to sequence 11 genomes of commercially relevant

TABLE 4.2

Probiotic Bacteria Whose Genomes Are Completely Sequenced[a]

Bacterium	Strain	Genome Size	Coordinator	Institution
Bifidobacterium longum	NCC2705	2.3	Schell et al. (64)	Nestlé, Switzerland
B. longum[b]	DJO10A	2.3	DJ O'Sullivan	University of Minnesota
Lactobacillus acidophilus	NCFM	2.0	TR Klaenhammer R Cano	North Carolina State University Cal-Poly Technical University
L. gasseri[b]	ATCC33323	1.8	TR Klaenhammer	North Carolina State University
L. casei[b]	ATCC334	2.3	J Broadbent	Utah State University
L. johnsonii	NCC533	2.0	D Pridmore	Nestlé, Switzerland

[a] Does not include cultures used for functional purposes in foods, but which may have probiotic activities (*Lactococcus lactis*, *L. lactis* subsp. *cremoris*, *Streptococcus thermophilus*, *Lactobacillus delbrueckii* subsp. *bulgaricus*, *L. plantarum*). Also does not include the ongoing projects that may be completed prior to publication (*Bifidobacterium breve* and *Lactobacillus rhamnosus*).

[b] Sequenced as part of a collaboration between the Joint Genome Institute and the Lactic Acid Bacteria Genome Consortium; sequence will be made public during 2004.

cultures, including three probiotic cultures. Unlike many genome projects with private funding, the genomes sequenced via the JGI/LABGC will be deposited in public databases upon completion. Currently, publications are being prepared, and the sequence is set to be released upon publication. This information will enable researchers to begin to uncover all the relevant traits of probiotic cultures that are necessary for efficacy.

Selection of probiotic cultures is severely hampered by an incomplete knowledge of what features of the cultures are important for survival during processing and storage, and for survival and efficacy in the human intestine. Knowledge of what genes are expressed in the different environments will reveal the features of the cultures, which are relevant to their survival in dairy products and performance in the intestine. Microarray technology holds the key to uncovering all these traits. This technology involves probes for every gene in an organism being placed on a glass slide to form a "gene chip" of the organism. By hybridizing the gene chips with total RNA from cultures growing in different environments, it is possible to uncover what genes are being expressed. By applying this technology to probiotic cultures, all the traits of the cultures that are relevant for their survival *in vitro* and for competition *in vivo* can be uncovered. This information will greatly improve the selection process for probiotic cultures.

4.7 Conclusions

The field of probiotics has progressed a lot since its inception over 100 years ago. Perhaps the biggest hurdle yet to overcome is understanding mechanisms of probiotic effects and how the cultures compete naturally in the intestine. Although some information is available, the scientific picture of this is still very fuzzy. Uncovering these answers will provide a clearer scientific basis for culture selection for specific probiotic effects. This will greatly improve the marketing of probiotics and also provide cultures with optimum efficacy for their intended probiotic effects. The genomics era appears to hold the key to this information, and the coming years should see a flood of information concerning bacterial traits involved in probiotic functioning. This will enhance the scientific credibility of this field and enable cultures to be obtained that are optimized for all the traits necessary for their probiotic purpose.

References

1. R Fuller. Probiotics in man and animals. *J. Appl. Bacteriol.*, 66: 365–378, 1989.
2. E Metchnikoff. *The Prolongation of Life*. G.P. Putnam's Sons, New York, 1908.
3. H Tissier. Traitement des infections intestinales par la méthode de la flore bactérienne de l'intestin. *Crit. Rev. Soc. Biol.*, 60: 359–361, 1906.
4. K Beeman. The effect of *Lactobacillus* spp. on convalescing calves. *Agri-practice*, 6: 8–10, 1985.
5. I Parra, and A DeCostanzo. Influence of yeast culture supplementation during the initial 23 d on feed, or throughout a 58-d feeding period on performance of yearling bulls. *J. Anim. Sci.*, 70(Suppl 1): 287, 1992.
6. JT Huber. Probiotics in cattle. In: R Fuller, Ed., *Probiotics 2: Applications and Practical Aspects*. Chapman & Hall, London, 1997, pp. 162–186.
7. PTN Lan, LT Binh, and Y Benno. Impact of two probiotic *Lactobacillus* strains feeding on fecal lactobacilli and weight gains in chicken. *J. Gen. Appl. Microbiol.*, 49: 29–36, 2003.
8. SC Kyriakis, VK Tsiloyiannis, J Vlemmas, K Sarris, AC Tsinas, C Alexopoulos, and L Jansegers. The effect of probiotic LSP 122 on the control of post-weaning diarrhoea syndrome of piglets. *Res. Vet. Sci.*, 67: 223–228, 1999.
9. CA Fritts, JH Kersey, MA Motl, EC Kroger, F Yan, J Si, Q Jiang, MM Campos, AL Waldroup, and PW Waldroup. *Bacillus subtilis* C-3102 (Calsporin) improves live performance and microbiological status of broiler chickens. *J. Appl. Poultry Res.*, 9: 149–155, 2000.
10. ME Sanders. Probiotics. *Food Technol.*, 53: 67–77, 1999.
11. DJ O'Sullivan and MJ Kullen. Tracking of probiotic bifidobacteria in the intestine. *Int. Dairy J.*, 8: 513–525, 1998.
12. JR Tagg and KP Dierksen. Bacterial replacement therapy: adapting 'germ warfare' to infection prevention. *Trends Biotechnol.*, 21: 217–223, 2003.

13. G Reid. The role of cranberry and probiotics in intestinal and urogenital tract health. *Crit. Rev. Food Sci. Nutr.*, 42: 293–300, 2002.

14. VN Klimenko, AS Tugushev, AV Zakharchuk, and AV Klimenko. Application of probiotics in the treatment of patients with nonhealing purulent-inflammatory wounds. *Klin. Khir.*, 11–12: 33–34, 2002.

15. A Talwalkar and K Kailasapathy. Effect of microencapsulation on oxygen toxicity in probiotic bacteria. *Aust. J. Dairy Technol.*, 58: 36–39, 2003.

16. K Adhikari, A Mustapha, and IU Grun. Survival and metabolic activity of microencapsulated *Bifidobacterium longum* in stirred yogurt. *J. Food Sci.*, 68: 275–280, 2003.

17. CS Favaro-Trindale and CRF Grosso. Microencapsulation of *L. acidophilus* (La-05) and *B. lactis* (Bb-12) and evaluation of their survival at the pH values of the stomach and in bile. *J. Microencap.*, 19: 485–494, 2002.

18. R. Havenaar, B Ten Brink, and JHJ Huis in 't Veld. Selection of strains for probiotic use. In: R Fuller, Ed., *Probiotics: The Scientific Basis*. Chapman & Hall, London, 1992, pp. 209–224.

19. T Mattila-Sandholm. Demonstration project FAIR CT96-1028. In: AM Kauppila, and T Mattila-Sandholm, Eds., Novel Methods For Probiotic Research: 2nd Workshop Demonstration of the Nutritional Functionality of Probiotic Foods FAIR CT96-1028. Technical Research Center of Finland (VTT), 1997, pp. 11–17.

20. C Dunne, L Murphy, S Flynn, L O'Mahony, S O'Halloran, M Feeney, D Morrissey, G Thornton, G Fitzgerald, C Daly, B Kiely, EMM Quigley, GC O'Sullivan, F Shanahan, and JK Collins. Probiotics: from myth to reality. Demonstration of functionality in animal models of disease and in human clinical trials. *Antonie van Leeuwenhoek*, 76: 279–292, 1999.

21. R. Freter. Factors affecting the microecology of the gut. In: R Fuller, Ed., *Probiotics: The Scientific Basis*. Chapman & Hall, London, 1992, pp. 111–144.

22. RD Berg. Probiotics, prebiotics or 'conbiotics'. *Trends Microbiol.*, 6: 89–92, 1998.

23. M Alander, R Satokari, R Korpela, M Saxelin, T Vilpponen-Salmela, T Mattila-Sandholm, and A von Wright. Persistence of colonization of human colonic mucosa by a probiotic strain, *Lactobacillus rhamnosus* GG, after oral consumption. *Appl. Environ. Microbiol.*, 65: 351–354, 1999.

24. WD Frost and H Hankinson. *Lactobacillus acidophilus*: An Annotated Bibliography to 1931. Davis-Green Corporation, Milton, Wisconsin, 1931.

25. ML Cross. Microbes versus microbes: immune signals generated by probiotic lactobacilli and their role in protection against microbial pathogens. *FEMS Immunol. Med. Microbiol.*, 34: 245–253, 2002.

26. R. Herich and M Levkut. Lactic acid bacteria, probiotics and immune system. *Vet. Med.* 47: 169–180, 2002.

27. V Scardovi. Genus *Bifidobacterium* Orla-Jensen 1924, 472AL. In: PH Sneath, NS Mair, ME Sharpe, and JG Holt, Eds., *Bergey's Manual of Systematic Bacteriology, Vol. 2*. Williams & Wilkins, Baltimore, 1986, pp. 1418–1434.

28. F Black, K Einarsson, A Lidbeck, K Orrhage, and CE Nord. Effect of lactic acid producing bacteria on the human intestinal microflora during ampicillin treatment. *Scand. J. Infect. Dis.*, 23: 247–254, 1991.

29. PC Dambekodi and SE Gilliland. Incorporation of cholesterol into the cellular membrane of *Bifidobacterium longum*. *J. Dairy Sci.*, 81: 1818–1824, 1998.

30. TA Jiang, A Mustapha, and D A Savaiano. Improvement of lactose digestion in humans by ingestion of unfermented milk containing *Bifidobacterium longum*. *J. Dairy Sci.*, 79: 750–757, 1996.

31. T Takahashi, E Nakagawa, T Nara, T Yajima, and T Kuwata. Effects of orally ingested *Bifidobacterium longum* on the mucosal IgA response of mice to dietary antigens. *Biosci. Biotechnol. Biochem.*, 62: 10–15, 1998.

32. BS Reddy and A Rivenson. Inhibitory effect of *Bifidobacterium longum* on colon, mammary, and liver carcinogenesis induced by 2–amino–3–methylimidazo[4,5–f]quinoline, a food mutagen. *Cancer Res.*, 53: 3914–3918, 1993.

33. O Sreekumar and A Hosono. The antimutagenic of a properties of a polysaccharide produced by *Bifidobacterium longum* and its cultured milk against some heterocyclic amines. *Can. J. Microbiol.*, 44: 1029–1036, 1998.

34. JP Grill, J Crociani, and J Ballongue. Effect of bifidobacteria on nitrites and nitrosamines. *Lett. Appl. Microbiol.*, 20: 328–330, 1995.

35. J Singh, A Rivenson, M Tomita, S Shimamura, N Ishibashi, and BS. Reddy. *Bifidobacterium longum*, a lactic acid-producing intestinal bacterium inhibits colon cancer and modulates the intermediate biomarkers of colon carcinogenesis. *Carcinogenesis*, 18: 833–841, 1997.

36. F He, AC Ouwehand, E Isolauri, M Hosoda, Y Benno, and S Salminen. Differences in composition and mucosal adhesion of bifidobacteria isolated from healthy adults and healthy seniors. *Curr. Microbiol.*, 43: 351–354, 2001.

37. L Meile, W Ludwig, U Rueger, C Gut, P Kaufmann, G Dasen, S Wenger, and M Teuber. *Bifidobacterium lactis* sp. *nov*, a moderately oxygen tolerant species isolated from fermented milk. *Sys. Appl. Microbiol.*, 20: 57–64, 1997.

38. YM Cai, M Matsumoto, and Y Benno. *Bifidobacterium lactis* Meile et al. 1997 is a subjective synonym of *Bifidobacterium animalis* (Mitsuoka 1969; Scardovi and Trovatelli 1974). *Microbiol. Immunol.*, 44: 815–820, 2000.

39. M Ventura and R Zink. Rapid identification, differentiation, and proposed new taxonomic classification of *Bifidobacterium lactis. Appl. Environ., Microbiol.*, 68: 6429–6434, 2002.

40. DH Green, PR Wakeley, A Page, A Barnes, L Baccigalupi, E Ricca, and SM Cutting. Characterization of two *Bacillus* probiotics. *Appl. Environ. Microbiol.*, 65: 4288–4291, 1999.

41. WP Charteris, PM Kelly, L Morelli, and JK Collins. Development and application of an *in vitro* methodology to determine the transit tolerance of potentially probiotic *Lactobacillus* and *Bifidobacterium* species in the upper human gastrointestinal tract. *J. Appl. Microbiol.*, 84: 759–768, 1998.

42. P Duwat, B Cesselin, S Sourice, and A Gruss. *Lactococcus lactis*, a bacterial model for stress responses and survival. *Int. J. Food Microbiol.*, 55: 83–86, 2000.

43. SA Ibrahim and DJ O'Sullivan. Use of chemical mutagenesis for the isolation of food grade β-galactosidase overproducing mutants of bifidobacteria, lactobacilli and *Streptococcus salivarius* ssp. *thermophilus. J. Dairy Sci.*, 83: 923–930, 2000.

44. M Matsumoto, H Tani, H Ono, H Ohishi, and Y Benno. Adhesive property of *Bifidobacterium lactis* LKM512 and predominant bacteria of intestinal microflora to human intestinal mucin. *Curr. Microbiol.*, 44: 212–215, 2002.

45. M Yamato, K Ozaki, and F Ota. Partial purification and characterization of the bacteriocin produced by *Lactobacillus acidophilus* YIT 0154. *Microbiol. Res.*, 158: 169–172, 2003.

46. J Palacios, G Vignolo, ME Farias, APD Holgado, G Oliver, and F Sesma. Purification and amino acid sequence of lactocin 705, a bacteriocin produced by *Lactobacillus casei* CRL 705. *Microbiol. Res.*, 154: 199–204, 1999.

47. T Kabuki, T Saito, Y Kawai, J Uemura, and T Itoh. Production, purification and characterization of reutericin 6, a bacteriocin with lytic activity produced by *lactobacillus reuteri* LA6. *Int. J. Food Microbiol.*, 34: 145–156, 1997.

48. M Bernardeau, JP Vernoux, and M Gueguen. Probiotic properties of two *Lactobacillus* strains *in vitro*. *Milchwissenschaft*, 56: 663–667, 2001.

49. T Toba, E Yoshioka, and T Itoh. Potential of *Lactobacillus gasseri* isolated from infant feces to produce bacteriocins. *Lett. Appl. Microbiol.*, 12: 228–231, 1991.

50. K Arihara, S Ogihara, T Mukai, M Itoh, and Y Kondo. Salivacin 140, a novel bacteriocin from *Lactobacillus salivarius* subsp *salicinius* T140 active against pathogenic bacteria. *Lett. Appl. Microbiol.*, 22: 420–424, 1996.

51. T Abee, TR Klaenhammer, and L Letellier. Kinetic studies of the action of lactacin F, a bacteriocin produced by *Lactobacillus johnsonii* that forms poration complexes in the cytoplasmic membrane. *Appl. Environ. Microbiol.*, 60: 1006–1013, 1994.

52. RW Jack, JR Tagg, and B Ray. Bacteriocins of Gram-positive bacteria. *Microbiol. Rev.*, 59: 171–200, 1995.

53. Z Yildirim, DK Winters, and MG Johnson. Purification, amino acid sequence and mode of action of bifidocin B produced by *Bifidobacterium bifidum* NCFB 1454. *J. Appl. Microbiol.*, 86: 45–54, 1999.

54. GR Gibson and MB Roberfroid. Dietary modulation of the human colonic microbiota — introducing the concept of prebiotics. *J. Nutr.*, 125: 1401–1412, 1995.

55. F Archibald. *Lactobacillus plantarum*, an organism not requiring iron. *FEMS Microbiol. Lett.*, 19: 29–32, 1983.

56. JE Posey and FC Gherardini. Lack of a role for iron in the Lyme disease pathogen. *Science*, 288: 1651–1653, 2000.

57. JB Neilands, K Konopka, B Schwyn, M Coy, RT Francis, BH Paw, and A Bagg. Comparative biochemistry of microbial iron assimilation. In: G Winkelmann, D van der Helm, and JB Neilands, Eds., *Iron Transport in Microbes, Plants and Animals*. VCH Publishers, Weinheim, Germany, pp. 3–33, 1987.

58. DJ O'Sullivan. Cloning, organization and regulation of genes involved in iron metabolism in fluorescent *Pseudomonas* spp. with biocontrol potential. PhD dissertation, National University of Ireland, Cork, 1990.

59. DJ O'Sullivan and F O'Gara. Traits of fluorescent *Pseudomonas* spp. involved in the suppression of plant root pathogens. *Microbiol. Rev.*, 56: 662–676, 1992.

60. A Bezkorovainy. Iron transport and utilization by bifidobacteria. In: A Bezkorovainy and R Miller-Catchpole, Eds., *Biochemistry and Physiology of Bifidobacteria*. CRC Press, Boca Raton, Florida, pp. 147–176, 1989.

61. EA Mevissen-Verhage, JH Marcelis, WC Harmsen-Van Amerongen, NM de Vos, and J Verhoef. Effect of iron on neonatal gut flora during the first three months of life. *Eur. J. Clin. Microbiol.*, 4: 273–278, 1985.

62. MJ Kullen, LJ Brady, and DJ O'Sullivan. Evaluation of using a short region of the *recA* gene for rapid intrageneric characterization of dominant bifidobacteria in the human large intestine. *FEMS Microbiol. Lett.*, 154: 377–383, 1997.

63. DJ O'Sullivan. Screening of intestinal microflora for effective probiotic bacteria. *J. Agric. Food Chem.*, 49: 1751–1760, 2001.

64. MA Schell, M Karmirantzou, B Snel, D Vilanova, B Berger, G Pessi, MC Zwahlen, F Desiere, P Bork, M Delley, RD Pridmore, and F Arigoni. The genome sequence of *Bifidobacterium longum* reflects its adaptation to the human gastrointestinal tract. *Proc. Natl. Acad. Sci. U. S. A.*, 99: 14422–14427, 2002.

5

Properties of Evidence-Based Probiotics for Human Health

Lynne V. McFarland and Gary W. Elmer

CONTENTS

5.1   Introduction

Probiotics in the form of fermented dairy products have long been part of attempts to maintain good health in many parts of the world. Until recent years, much of the scientific support for this use has been based on empirical observations. Early uses and studies have employed dairy strains of lactobacilli, which may not be optimal for survival and activity in the environment of the human gastrointestinal or vaginal tracts. Now, however, probiotic lactobacilli, bifidobacteria, and other lactic acid-producing bacteria and a probiotic yeast have been identified that have desirable pharmacological activities and pharmacokinetic properties *in vivo*. Using these microbes as "living drugs" a reasonably large number of controlled clinical trials have been conducted, and the findings are validating the use of probiotics to prevent and treat a variety of diseases. Desirable attributes for probiotic microbial strains would include the following:

1. A stable and well-described microbe
2. Absolute lack of toxicity and pathogenicity
3. The ability to survive and perhaps multiply in the location of desired action but not to persist permanently
4. An ability to associate and adhere to the target tissue to effect a desirable host response
5. For treatment of infections, production of antipathogenic activities including stimulation of a host immune response
6. Efficacy proven in well-designed placebo-controlled clinical trials
7. Ease of large-scale commercial production and distribution

Few probiotics currently meet all of these criteria, although the lack of careful studies precludes a complete assessment. Below, selected important biological properties of probiotic strains of microorganisms are discussed with an emphasis on those microbes that have been tested and found to have efficacy in humans.

5.2 Desirable Properties for Therapeutic Probiotics

There are numerous studies that attempt to define the parameters needed to characterize a potential probiotic candidate, but the consensus agrees on the properties shown in Figure 5.1. The research needed to support a new probiotic strain is not trivial. Dunne et al. (1) screened over 1500 strains before finding a single strain that had probiotic potential. Lee et al. studied 109 lactic acid bacteria isolated from healthy infants and found eight potential probiotic strains (2). McLean and Rosenstein found only 4 of 60 vaginal isolates of lactobacilli strains were inhibitory toward bacterial vaginosis pathogens (3). The pathway to development involves the initial isolation and characterization of the probiotic strain, testing using *in vitro* assays, experiments with animal models, safety and dosing ranging in healthy human volunteers and finally determination of the efficacy in patients with the disease indication (4).

5.2.1 Manufacturing Properties

Probiotics are products that share similar requirements with other sellable manufactured goods on the market. Therefore, the process of producing and manufacturing probiotics should have standardized protocols and quality control procedures. In order for the product to be successful in the marketplace, probiotics need to return a profit. Fortunately, the production of most probiotics is straightforward and relatively inexpensive, at least compared with the manufacturing of medications. The production of microorganisms is based on a long history of fermentation processes and does not rely upon advanced technology.

5.2.1.1   *Strain Identification*

The probiotic strain should have distinct characteristics that allow the taxonomic definition and identification of the probiotic strain from the other members of its species (5–8). Current methods rely on molecular techniques (polymerase chain reaction-based or other genotyping methods), carbohydrate fermentation patterns, biotyping, or DNA and RNA homology grouping (9). Strain identification is important for both quality control and differences in clinical efficacy for the differing strains of probiotics (10).

5.2.1.2   *Potency Properties*

The type of food product that will deliver the probiotic should allow the probiotic to survive in high numbers and have its characteristics compatible with the probiotic. Heller described the various manufacturing steps for various probiotic foods including milk, cheese, yogurts, kefir, and cottage

Manufacturing properties:

Easy to produce
inexpensive to grow and package
competitive marketing edge
lack of antibiotic-resistant plasmids
stability of dosage form (shelf-life)

Within target host:

Resistant to amylase   —————————————

Stimulates immune response   ——————————

Acid resistant (survives stomach)   —————————

Bile acid tolerant   ——————————————

Survival in target organ: —————————————

      persistent in complex microbial ecology
      replication within target organ
      attachment to mucosal surfaces
      unaffected by antibiotics or medications
      production of specific pathogen inhibitors
      (bacteriocins, proteases, H_2O_2)

Safety:
      Lack of systemic translocation
      Lack of side-effects

FIGURE 5.1
Desirable properties of probiotics.

cheese and found that all of these types of foods allowed for the survival of *Lactobacillus* strains (11). If the probiotic is to be added to foodstuffs, the behavior and recovery should be tested using the final food product. It is important that different lots of manufactured probiotics are tested for viability, as studies have found lot-to-lot variability. Clements et al. found lot-to-lot variation in Lactinex® products, with one lot having absolutely no clinical efficacy (12).

5.2.1.3 Stability Properties

Probiotics, whether sold as dried cultures or added to foodstuffs, need to have a sufficiently long shelf life to reach the consumer at high concentrations and in a living state. Lyophilized or freeze-dried preparations of probiotics have the advantage of the longest shelf life and have no need for refrigeration. Dried culture preparations of *L. acidophilus* strain NCFM were found to be viable for up to 8 months (6). Stability of the same strain in yogurts or other foods is also excellent, but the carrier food usually degrades before the probiotic loses significant viability.

5.2.1.4 Formulation and Dose Concerns

There are no standard regulations for the minimum dose of probiotics on the market. The evidence from clinical trials suggests that doses over 10^8 to 10^{10} colony-forming units per day (CFU/d) are necessary before a therapeutic effect is observed (13). Most oral doses to deliver this intake of viable microbes have ranged from 1 to 3 g/d for infants and 1 to 15 g/d in adults (13).

5.2.1.5 Marketing Properties

If the probiotic has a distinct or unique property, such as the production of a substance that targets a specific pathogen or is formulated in a manner that allows more probiotic to reach its target organ, then this would gave the probiotic a competitive edge when it comes to marketing. These properties will be discussed later in this chapter, but include the production of hydrogen peroxide by *Lactobacillus* species (6, 12), production of bacteriocins (13), other inhibitory substances by *L. acidophilus* LA1 (8) or the antimicrobial agent reuterin by *L. reuteri* (14), or toxin-specific proteases against *Clostridium difficile* toxins by *Saccharomyces boulardii* (15). As probiotics are often given to patients with acute disease, another advantage for a potential probiotic is the ability to survive in the presence of antibiotics or concurrent medications. Such probiotics as *Saccharomyces* species have the advantage of not being susceptible to antibiotics (other than antifungal agents).

5.2.1.6 Commercial-Scale Production

Many investigational drugs are not able to make the leap successfully from efficacy based on small, well-controlled production in a laboratory to the mass production needed for a successful marketable product. Promising probiotics from smaller trials may not be amenable to the rigorous stresses that are involved in large-scale manufacturing procedures. Probiotics require a high degree of stability, ability to withstand drying and exposure to air, and survival in the formulation process and packaging lines.

5.2.1.7   Quality Control

As with other products on the market, manufacturers need to have established quality control procedures and methods to test them. Yeung et al. reported seven studies that found discrepancies between probiotic product labeling and independent laboratory analysis of the species contained in the product (7). Hughes and Hillier tested 16 probiotic products and found only 4 contained the *Lactobacillus* species listed on the label, and 69% had contaminants (16). Zhong and Millsap tested 64 *Lactobacillus* strains from various sources and found that 6 contained *L. plantarum*, which was not indicated on the product label (17).

5.2.2   Endogenous Properties

5.2.2.1   Transit to the Target Organ or Surface

In order for a bacterial or yeast probiotic to be effective, it must first survive to reach its target organ. As most probiotics are given orally and are focused on diseases of the gastrointestinal tract, the probiotic must survive through the mouth, stomach, duodenum, and within the colonic environment. Probiotic strains are tested for amylase (to survive passage through the mouth), acid resistance (to survive passage through the stomach), and bile acid tolerance (to survive through the duodenum and intestine). If a strain is highly acid resistant and bile acid tolerant, the strain may be investigated further (1).

5.2.2.2   Persistence within the Target

The gastrointestinal tract may contain over 500 species of bacteria and yeasts, which when disrupted by antibiotics or other processes may produce inflammation or disease symptoms as a result. As most probiotics target the intestinal tract, the probiotic must survive in the complex microbial milieu of the colon. Microbes that inhabit the intestinal tract normally resist the advance of new microbes using multifactorial mechanisms globally termed "colonization resistance" (18). The normal microbial flora presents a formidable barrier to new organisms that involves the physical barrier effect of the existing biofilm, production of bacteriocins, spatial crowding, or changes in pH (production of acidic byproducts). Therefore, the probiotic must be able to circumvent these mechanisms and become established in the intestinal tract. Investigation into potential probiotics is aided if studies are done that document interactions with the normal microbial flora and the potential probiotic (18, 19). Brigidi et al. administered the oral probiotic VSL 3 (a mixture of eight strains) to ten patients with irritable bowel syndrome or functional diarrhea and successfully recovered two of the strains in the VSL 3 mixture from the patients' stools (20). Dunne et al. tested 80 healthy volunteers with a yogurt that had the *Lactobacillus salivarius* UCC118 strain added and found that <10% still carried the probiotic strain 3 weeks after cessation of its administration (1). Fujiwara et al. documented the survival of *Bifidobacterium longum* SBT2928 in the human intestine for 30 days after ingestion (21). Production of antimicrobial substances that inhibit pathogens is also useful (1, 2, 8, 12, 20).

5.2.2.3 Adherence to Mucosal Surfaces

Survival in the target organ is aided by adherence to intestinal epithelial cells or other mucosal surfaces of its target organ (8, 22, 23). Most probiotics have been shown to have adherence to either selected cell lines *in vitro* or within animal models (12, 19, 22).

5.2.2.4 Reproduction within the Target

The ability to colonize the intestinal tract and reproduce may also be a beneficial trait (21).

5.2.2.5 Metabolic Pathway Interference

Another potential property is the ability of the probiotic to influence metabolic activities, such as cholesterol assimilation, production of carcinogens, lactase activity, or the production of hydrogen peroxide (6, 14, 19).

5.2.2.6 Stimulation of the Immune System

As many of the diseases treated or prevented with probiotics involve the immune system, an advantageous property would be the stimulation of the immune response. This property has been found associated with several probiotics (24–26) although absent in others (27).

5.2.2.7 Safety

The probiotic should be safe to give not only to healthy people as a preventive agent, but also to ill patients if the probiotic is to be used as a therapeutic agent. Special concerns should be considered when giving a living organism to a patient who may be severely compromised by illness, takes concurrent medications, or has a disturbed immune system. The probiotic candidate should survive, but not persist indefinitely. Persistence of the strain may affect the balance of the normal microecology or lead to translocation if the immune status of the host is compromised. The probiotic strain should be tested for its pathogenic potential using *in vitro* models and animal models, and in healthy human volunteers.

The safety profile of the probiotics tested in randomized, controlled trials has shown a high level of tolerance and a low incidence of side effects (28, 29). In rare instances, case reports of toxicity associated with probiotic use have been reported. Complications associated with *Lactobacillus* strains have included a fatal case of *L. rhamnosus* septicemia after *Clostridium difficile* infection (30), liver abscess due to *L. rhamnosis* (31), a case of *L. casei* pneumonia and sepsis in a patient with AIDS (32), and an aortic graft infection by *L. casei* (33). Cases of catheter-associated fungemia have been reported associated with the use of *Saccharomyces boulardii* (19, 34, 35) but all cases of fungemia resolved with treatment. There is concern with the use of *Enterococcus* strains as probiotics as this genus develops multiple antibiotic resistance, including resistance to vancomycin (28, 36). Overall, use of probiotics in patients who are not immunocompromised has not been associated with serious side effects. Caution should be exercised when the patient is immunosuppressed, has a central catheter, or is extremely ill. Large controlled studies on the incidence of side effects for most probiotics are lacking.

In summary, there are some valid guidelines to suggest avenues of investigation for the identification of potential probiotic strains, but there is no reliable *in vitro* set of predictors of *in vivo* efficacy. The identification of these probiotics must be based on a variety of *in vitro* and *in vivo* models, with further investigations using randomized, placebo-controlled clinical trials of patients with a specific disease indication.

5.3 Properties of Evidenced-Based Probiotics

5.3.1 By Microbial Species

5.3.1.1 *Lactobacillus Species in General*

Lactobacilli have long been the most prominent of probiotic microorganisms because of their association with popular fermented dairy products. Consumption of these products, e.g., yogurt, has long been associated with good health. Lactobacilli are Gram-positive rods and are part of the large group of lactic acid-producing bacteria. Lactobacilli are important in the food industry, being involved in pickling and cheese and yogurt making. Human strains of lactobacilli are part of the normal flora of the mouth, lower small intestine, colon, and vagina in some but not all people. While they may be present, they are not the dominant species in the intestinal tract. Although they are anaerobic and get their energy from fermentative metabolism, they can survive in the presence of oxygen because they have peroxidase activity to inactivate hydrogen peroxide. The end product of the fermentation of carbohydrates via pyruvic acid is lactic acid, and a distinguishing characteristic of this genus is the ability to survive at low pH. This ability to produce lactic acid gives lactobacilli a competitive niche in environments rich in nutrients and may explain, in part, their probiotic action. They are rarely pathogenic.

5.3.1.2 Lactobacillus rhamnosus

Lactobacillus rhamnosus GG is the most studied lactobacilli-based probiotic. This strain of *Lactobacillus* was isolated from the feces of a healthy human by Gorbach and Goldin (hence the GG) as reported by Silva et al. (13). It has been variously designated *Lactobacillus casei* subspecies *rhamnosus* strain GG or *Lactobacillus rhamnosus* GG. It was deposited with the American Type Culture Collection as *Lactobacillus acidophilus* and assigned number ATCC 53103. Patents have been issued for isolation and use (U.S. Patent 4,839,281, U.S. Patent 5,032,399). *L. rhamnosus* GG is stable to bile and an acidic pH and was described to be adherent to intestinal mucosal cells *in vitro* (37). Silva et al. (13) described a microsin from this strain that had relatively broad-spectrum antibacterial activity as measured by an agar diffusion assay. Although the strain has antagonistic activity against *Salmonella typhimurium* in germ-free mice (38), this finding does not prove *in vivo* production of the antimicrobial substance. It is unknown whether this substance is produced *in vivo*.

An *ex vivo* study (37) demonstrated that *L. rhamnosus* GG survived in gastric juice for 4 h at pHs 7, 5, and 3 but not at pH 1. Fecal samples were analyzed from healthy adult volunteers receiving *L. rhamnosus* GG in the form of fermented milk or whey for 28 d. *L. rhamnosus* GG was recovered in all samples tested. The observation (37) that *L. rhamnosus* GG could be isolated 7 d after stopping ingestion of the fermented whey product indicated that this probiotic has the capacity to persist and colonize. Furthermore, a human study (39) quantitating *L. rhamnosus* GG in fecal and colon biopsy samples revealed that the probiotic was found in all biopsy and fecal samples 14 d after cessation of treatment, but had mostly disappeared at 28 d. There were more positive biopsy samples than fecal samples, indicating that *L. rhamnosus* GG has the ability to adhere and perhaps divide associated with the colonic mucosa. Similarly, *L. rhamnosus* GG could be recovered from fecal samples in over half of premature infants at 3 weeks after administration (40).

Feeding *L. rhamnosus* GG to healthy volunteers for 4 weeks decreased fecal β-glucuronidase specific activity, whereas feeding *Streptococcus thermophilus* or *Lactobacillus bulgaricus* did not (37). Thus, the probiotic seemed to have modulated intestinal metabolic activity. The interest in β-glucuronidase relates to the activity of this enzyme in deconjugating and "activating" potentially harmful chemicals in the gut, but the long-term effects of this modulation are unknown. A study in rats showed that daily feeding of *L. rhamnosus* GG (41) before and during dimethylhydrazine injections decreased the incidence and number of colon tumors, compared to no probiotic feeding (41). Another interesting finding was that *L. rhamnosus* GG enhanced the mucus-binding ability of *Bifidobacterium lactis* (42), indicating that one probiotic might influence the adhesion of another. Relative to the potential application of probiotics in oral health, *L. rhamnosus* GG was found to inhibit *Streptococcus sobrinus in vitro* (43) and to colonize the oral cavity (8 of 9 subjects positive at week 2 after dosing cessation) (44). A 9-month trial of *L. rhamnosus* GG-supplemented milk vs. regular milk in children (n = 451) showed a trend for both a decrease in *Streptococcus mutans* and for less dental caries in the supplemented group (45). *L. rhamnosus* GG, unlike other lactobacilli, does not ferment lactose and is not considered cariogenic. The finding of benefit in decreasing indexes of caries risk from short-term application of *L. rhamnosus* GG in children should be further explored.

L. rhamnosus DR20 has been shown to enhance immunity in animals (46) and in man (47, 48) and is of interest as a probiotic. In one of the most comprehensive studies of probiotic disposition in and effects on the human intestinal tract, Tannock and coworkers (49) studied the composition of the fecal microflora of 10 healthy volunteers after 6 months of ingestion of *L. rhamnosus* DR20. Using a variety of sophisticated genomic and analytical techniques, they were able to quantitate changes in the overall composition of the microbial flora and to specifically follow elimination of the probiotic strain. *L. rhamnosus* DR20 induced increases in total lactobacilli and enterococci, but consistent differences in other microbial populations were not found or were activities of azoreductase or β-glucuronidase consistently affected. *L. rhamnosis* DR20 was detected at $>10^2$ CFU/g in all stool samples in 6 of 10 subjects, but inconsistently or at very low levels in the others. Upon cessation, DR20 did not persist beyond 1 month in 9 of 10 subjects. The average stool concentration (from subjects that excreted DR20) was 3×10^5 CFU/g with an input of 1.6×10^9/d. The authors estimated that the colon contents contained about 10^7 CFU. The percentage recovery of the dose was not calculated, however. Thus, this exhaustive study showed that this probiotic transiently changed *Lactobacillus* and enterococcal levels in the feces, but not other metabolic activities or microbial groups studied. The DR20 strain was not consistently detected in all subjects during dosing and rapidly disappeared after dosing stopped.

5.3.1.3   Lactobacillus reuteri

Another lactic acid-producing probiotic is *L. reuteri*. Strains of this microbe are widespread in nature and can be isolated from a variety of food products, from animals, and from the human gastrointestinal tract. *In vitro, L. reuteri* produces from glycerol, 3-hydroxypropionaldehyde, which has relatively broad-spectrum antimicrobial activity (50). Reutericin 6, a bacteriocin produced by *L. reuteri* strain LA 6, has also been identified (51). It is not clear whether these antimicrobial substances are produced in the human intestine in concentrations high enough to directly inhibit pathogens; nevertheless, *L. reuteri* strains have received considerable commercial attention as a probiotic for both human and animal uses. *L. reuteri* strain 1063 encodes a protein involved in mucosal adhesion in pigs and chickens and is autoaggregative (14, 52).

Jacobsen et al. (53) did an interesting evaluation of 47 *Lactobacillus* strains with respect to properties that might be important for probiotic activity. Among the tested strains, only *L. rhamnosus* GG, *L. rhamnosus* 19070-2, and

L. reuteri DSM 12246 were relatively resistant to acid and bile, were highly adherent to Caco-2 cells in culture, showed antimicrobial activity against enteric pathogens, and could be isolated from most stool samples from 12 volunteers after ingestion. Wolf et al. gave 1×10^{11} *L. reuteri* to 15 volunteers for 21 d (54). A similar group received placebo. Adverse effects were not different between placebo and treatment, and analyses of blood and urine were within expected ranges for both groups. Fecal samples from treated volunteers contained significantly higher concentrations of *L. reuteri* compared to placebo for days 7, 14, 21, and 28, but not at day 77. Thus, some colonization was evident for at least 7 d after cessation of dosing, but not at 30 d. The *L. reuteri*, even at this relatively high dose, was well tolerated. A later study (55) done in 39 HIV-infected adults also showed *L. reuteri* to be well tolerated; however, fecal concentrations appeared to be lower than observed in the earlier study in healthy volunteers. Casas and Dobrogosz (14) have comprehensively reviewed the probiotic activities of *L. reuteri*.

5.3.1.4 Bifidobacterium *species*

Bifidobacterium species are anaerobic lactic-and acetic acid-producing Gram-positive rods or branched rods. They are present in the normal flora but are the predominant member of the intestinal flora of the breast-fed infant. Upon weaning, levels decline. *Bifidobacterium* sp. grow well in milk, and there has been long interest in bifidobacteria-containing probiotics in the form of fermented dairy products. One of the few human pharmacokinetic studies on bacterial probiotics was conducted by Marteau et al. (56) in a rather heroic human study quantitating the transit of both *Lactobacillus acidophilus* and a *Bifidobacterium* species in the upper gastrointestinal tract. Six volunteers took a single dose of milk containing the bacteria or took sterile milk in random order. Ileal fluid was aspirated continuously for 8 h. Total recovery in the ileum of the dose of *L. acidophilus* and *Bifidobacterium* sp. was 1.5 and 37.5%. In a subsequent study using an antibiotic-resistant variant of *Bifidobacterium* that could be distinguished from resident bifidobacteria, the total transit and recovery of this special strain was measured in 8 volunteers receiving a fermented milk product for 8 d. The shape of the fecal concentration curve was similar to *Bacillus stearothermophilus* spores used as a marker. The *Bifidobacterium* was no longer detected 8 d after cessation of dosing. The recovered dose was about 30%, i.e., similar to that recovered in the ileum in the earlier study. It was concluded that *Bifidobacterium* sp. does not colonize the colon. It can also be concluded that much of the destruction of the probiotic took place in the upper intestinal tract. Marteau et al. (57) found that *Bifidobacterium animalis* shortened the colonic transit time in healthy women. Radioopaque pellets were used to measure transit after *B. animalis* or control fermented milk. The mechanism of the effect was not explained by an effect on fecal weight or bacterial mass or modifications of bile salts. Interestingly, the control milk (fermented with yogurt cultures) did not influence transit times. Several *Bifidobacterium* sp. have been shown to adhere and to displace selected pathogens from intestinal cells *in vitro* (58, 59). Recently high fecal recovery (>10% of dose) was obtained in human volunteers receiving a *Bifidobacterium breve* fermented soy milk product (60). Thus, although there is no firm evidence of proliferation of administered *Bifidobacterium* species, these probiotic bacteria have good ability to survive passage through the gastrointestinal tract and may favorably affect intestinal functions.

5.3.1.5   Lactobacillus acidophilus

The presence of *L. acidophilus* is widespread in commercially available probiotic products. It is also found in fermented dairy products and may be part of the intestinal and vaginal microflora. It is clear that the important biological properties of adherence and stability in the gastrointestinal tract are strain specific. Strains optimal for fermenting milk may not be optimal to generate favorable effects in the lower bowel or the vaginal tract. Few direct comparisons of *Lactobacillus* strains are published, so that it is important to be explicit about which strain is being investigated in publications. Selected *L. acidophilus* strains (e.g., LB, LA1, BG2FO4, LCFM) express adhesive factors that foster adhesion to human intestinal cells *in vitro* (6, 61–63). While adherence *in vivo* may occur, this has not been well-demonstrated, and exogenously administered lactobacilli eventually drop below detectable concentrations in stool samples. An inhibitory effect of heat-killed *L. acidophilus* LB organisms against the human intestinal Caco-2 cell adhesion and cell invasion by a large variety of diarrheagenic bacteria was found (61). Furthermore, *L. acidophilus* LB culture supernatant inhibited the ability of *Salmonella enterica* to enter and proliferate in Caco-2 cells. *L. acidophilus* LCFM, a commonly used dairy strain, survives passage through the human gastrointestinal tract, but did not colonize during a 2-week consumption in healthy volunteers (64), but pharmacokinetic details are lacking. It was concluded that daily consumption would be needed to maintain high intestinal levels of this probiotic strain. Several *L. acidophilus* strains have been shown to produce antimicrobial substances *in vitro* (6, 61, 65), but production *in vivo* of concentrations high enough for a direct inhibition of pathogen growth has not been demonstrated. Conconnier et al. (66) have shown decreased *Salmonella typhimurium* concentrations after 4 and 7 d of treatment with a fivefold concentrate of *L. acidophilus* LB supernatant given orally to mice, but this does not prove that observations of inhibition of pathogens *in vivo* are due to an antimicrobial secreted by an ingested *L. acidophilus* LB. However, it seems that the LB strain of *L. acidophilus* encompasses a number of desirable probiotic properties.

5.3.1.6   Lactobacillus casei

Similar to *L. acidophilus*, the probiotic properties of *L. casei* are strain specific. *L. casei* strain Shirota has received much commercial attention. It has been shown to have antagonistic activities against *Escherichia coli* in a mouse model for urinary tract infection (67), against *Listeria monocytogenes* in rats (68), to reduce influenza virus titers in aged mice (69), and to reduce *Helicobacter pylori* in humans (decreased urea breath test) (70). Spanhaak et al. (71) gave *L. casei* strain Shirota to healthy humans in a placebo-controlled trial and demonstrated an increase in fecal bifidobacteria and a decrease in beta-glucuronidase and beta-glucosidase activities, but no effect on immune parameters compared to placebo. Thus, this strain has the capability to modify the composition and metabolic activities of the human intestinal flora. Using a selective medium and monoclonal antibodies, Yuki et al. quantitated fecal recovery of *L. casei* strain Shirota in humans (72). With an input of about 10^{10} /d for 3 d, about 10^7 /g of feces was recovered at day 4. While survival was demonstrated, unfortunately, pharmacokinetic details and a mass balance of dose and recovery were not investigated. Survival of *L. casei* strain. Lcr35 was also studied in healthy volunteers using a specific 16S ribosomal probe. The presence of relatively high levels of Lcr35-like endogenous lactobacilli precluded a precise view of the probiotic strain. Nevertheless, Lcr35-like lactobacilli increased during probiotic ingestion and decreased after cessation.

5.3.1.7 Other Lactobacilli

Jacobsen et al. (53) evaluated 47 *Lactobacillus* strains with respect to properties that might be important for probiotic activity. Among the tested strains, only *L. rhamnosus* GG, *L. rhamnosus* 19070-2, and *L. reuteri* DSM 12246 were relatively resistant to acid and bile, were highly adherent to Caco-2 cells in culture, and showed antimicrobial activity *in vitro* against enteric pathogens. Among the strains failing one or more of these parameters were strains of *L. plantarum, L. fermentum, L. johnsonii (L. acidophilus* LA1), *L. crispatus, L. paracasei, L. acidophilus, L. delbrueckii,* and *L. helveticus.* The authors point to the importance of *in vitro* methods to predict survival in the human intestinal tract, but even highly adherent strains did not persist after stopping dosing. A streptomycin–rifampicin-resistant mutant of *L. gasseri* given to volunteers for 7 d decreased fecal populations of *Staphylococcus* and fecal *p*-cresol (73). Unlike reported studies with other lactobacilli, there was evidence of persistence of *L. gasseri*. Levels at day 90 after cessation of dosing were about 10^4 CFU/g feces in four of eight subjects. This is presumptive evidence of proliferation. Another strain of *L. gasseri* has shown activity in clarithromycin-resistant *Helicobacter pylori* infection (74). Probiotic strains of *L. gasseri* are worthy of further evaluation in large-scale clinical trials.

Reid and coworkers have studied the applications of lactobacilli in urogenital infections (75–77), particularly *L. rhamnosus* GR-1 and *L. fermentum* RC-14 (78–79). In a recent randomized clinical trial in 64 healthy women given a mixture of the two lactobacilli or placebo orally for 2 months (78), the probiotic bacteria favorably affected (compared to placebo) the vaginal flora with respect to the presence of Gram-negative (Bacterial vaginosis appearance) or the predominance of a *Lactobacillus*-dominated flora. Reductions in yeasts and coliforms were also noted in the lactobacilli-treated group. When the two *Lactobacillus* strains were administered vaginally, they were shown to persist in the vaginal tract for up to 19 days (79) and also to colonize the vaginal tract following oral administration. Lactobacilli strains GR-1 and RC-14 were shown to be bile tolerant and to survive passage through the intestinal tract (80). Interestingly, oral *Lactobacillus* GG did not have an effect on the vaginal flora (81), indicating, once again, that strain selection is of paramount importance and that intestinal probiotics may not be the most appropriate for infections at other sites. The findings of Reid and coworkers show that appropriate strains of probiotic lactobacilli have the potential to provide an alternative to routine antimicrobial therapy for vaginal infections.

Another lactic acid-producing commercially available probiotic that has been studied is *Enterococcus faecium* SF68. Although this strain is nontoxic, there has been concern that the probiotic might pick up and share vancomycin-resistant genes in the human gut. Lund and Edlund have used an *in vitro* mating assay to point out that *E. faecium* is a possible recipient of the vanA gene (82). On the other hand, these investigators showed that oral administration of the probiotic strain of *E. faecium* did not correlate with the presence of vancomycin-resistant enterococci, which were found only sporadically in fecal samples (83). The probiotic SF68 strain was able to be detected at the end of dosing but not in samples taken 3 weeks later (83). In vancomycin-treated subjects, no *E. faecium* SF68 could be detected in any samples. Thus, *E. faecium* SF68 seems safe for probiotic use based on current evidence, but given the many other well-studied and effective probiotics that are available, one may question the wisdom of widespread use of a probiotic of *E. faecium*.

5.3.1.8   Saccharomyces boulardii

Another well-studied, commercially available probiotic is *S. boulardii*. McFarland and Bernasconi (18) have reviewed the properties and therapeutic applications of this unusual yeast. Unlike the lactobacilli, *S. boulardii* is not found as a component of the gastrointestinal or vaginal tracts. However, it does have a growth optimum of 37°C and survives passage to the feces in animals and in humans (84). The yeast does not strongly adhere to the intestinal mucosa and is eliminated within 24 to 72 h if not readministered. However, *S. boulardii* was able to colonize germ-free mice and prevent establishment of *Candida albicans* (85). Continuous feeding was protective against *Clostridium difficile* pathology in germ-free mice, but no colonization was evident when a normal microflora was present (86). The elimination pharmacokinetics of *S. boulardii* has been extensively studied in laboratory animals, and these studies have been reviewed by Martin et al. (87). Here we will focus on the biological properties discerned from studies in humans.

As was observed in laboratory animals, *S. boulardii* does not colonize in humans and must be given daily to achieve therapeutic levels. Maximal fecal concentrations were found to be achieved 36 to 60 h after a single dose and were below limits of detection 2 to 5 d later (84, 88). Daily administration of 1 g (3 × 10¹⁰ CFU) resulted in fecal steady-state levels of about 10⁸ CFU/g by day 3, and these concentrations were maintained as long as the yeast was given daily (84, 88). Upon cessation of dosing, *S. boulardii* was eliminated with a half-life of about 6 h. In volunteer subjects, intersubject variability in *S. boulardii* stool levels was found to be large, but intrasubject variability small with repeat dosing (88). Fecal steady-state levels were found to be dose dependent within the dose range of 0.2 to 3 g of lyophilized product taken daily.

S. boulardii is not affected by antibacterial antibiotics directly, yet antibiotic administration to human volunteers, probably by modulating the competing microbial flora, increased steady-state fecal concentrations 2.4-fold (88). Even with antibiotic treatment, the yeast did not persist after stopping dosing. Nystatin treatment, however, reduced steady-state fecal levels to below detection (89). On the other hand, the highly bioavailable antifungal fluconazole had no effect on *S. boulardii* fecal levels provided the doses of yeast and antifungal were separated by 3 h. In rats, the fecal recovery of *S. boulardii* was increased fourfold by high-fiber diets (Elmer et al., unpublished data) and high-fiber diets were shown to modulate onset of *Clostridium difficile* disease in a hamster model (90). In healthy volunteers, *S. boulardii* steady-state levels were positively related to average daily stool weight, suggesting that high-fiber diets might increase intestinal concentrations of the yeast (88). Fiber intake influences on probiotic kinetics and therapeutic action remain an unexplored area of research.

S. boulardii exhibits other biological properties indicative of a desirable probiotic. Interference in adhesion of *Entamoeba histolytica* trophozoites to human erythrocytes has been demonstrated *in vitro* (91) and *S. boulardii* administration decreased mortality in *E. histolytica*-infected rats (92). Berg et al. showed decreased translocation of *Candida albicans* to mesenteric lymph nodes and spleens of immunocompromised mice (93) with *S. boulardii* treatment compared to controls.

Studies have shown that *S. boulardii* has the ability to decrease the action of microbial toxins. In isolated intestinal loops, *S. boulardii* decreased fluid volume secreted in response to cholera toxin (94), and the cAMP increased in response to addition of the toxin to cultured intestinal epithelial cells (95). The cAMP response was elicited by dead yeast or by *S. boulardii*-conditioned culture medium and appeared to be due to a secreted protein.

S. boulardii is the only probiotic shown in controlled clinical trials to decrease the recurrence rate of *C. difficile* disease in humans (96–97), and the mechanism of this effect has been under intense investigation. Pothoulakis et al. (15, 98) described a 54-kDa protease secreted by the yeast that inhibited binding of labeled *C. difficile* toxin A to ileal brush-border membranes, reduced toxin A induced fluid secretion in rat ileal loops, and partially digested toxin A and its receptor. An antiserum raised against the yeast protease blocked these effects and blocked the toxin A and toxin B inhibition of protein synthesis in cultured human colonic cells and created drops in transepithelial resistance (99). Previous studies using the hamster model for *C. difficile* disease indicated that the effect of *S. boulardii* treatment was not in eliminating *C. difficile*, but in decreasing *C. difficile* toxins (100). Together, these data suggest that *S. boulardii* protease plays a large role in the beneficial therapeutic effects of this yeast in *C. difficile* disease.

However, *S. boulardii* enhances host immunological responses, and these and other biological properties need to be also considered as potential mechanisms of action. Buts et al. showed an increase in secretory IgA in the small intestines of rats treated with *S. boulardii* (25) and described a metalloprotease secreted by the yeast *in vivo* (101). An increased total IgA antibody and specific IgA directed against *C. difficile* toxin A has been demonstrated in *S. boulardii*-treated mice (102). In a human study by Caetano et al., *S. boulardii* activated complement and the reticuloendothelial system (103).

Buts et al. (104) and Jahn et al. (105) have studied the response of the intestinal mucosa to exposure to *S. boulardii*. *S. boulardii* was demonstrated to have trophic effects resulting in increases in disaccharidases and alkaline phosphatase activities in the intestinal mucosa without changes in intestinal morphology or invasion into subepithelial layers. The enhanced expression of luminal enzyme activities was suggested as a beneficial effect to improve carbohydrate digestion impaired by diarrhea. Buts et al. have provided evidence that these trophic effects are mediated by release of spermidine and spermine upon catabolism of yeast cells during intestinal transit (106, 107). Thus, the beneficial probiotic activities of *S. boulardii* are most likely due to multifaceted biological properties including immune stimulation, trophic effects on intestinal cells, digestion of toxins and toxin receptors, and antagonistic activities against pathogens.

5.4   Evidence-Based Probiotics

The interest in probiotics has increased due to the heightened flexibility in medical care choices and alternatives to the reliance on antibiotic therapy. As shown in Figure 5.2, the strength of evidence-based proof for health claims for probiotics has been inconsistent (108, 109). Numerous health claims have been based on case reports; small case series; or small, open trials and not blinded, placebo-controlled randomized trials. However, several meta-analyses have been reported for a limited number of health claims (110–115). A complete review of the efficacy of probiotics is beyond the scope of this chapter. Here we provide examples of human studies supporting the use of probiotics for the prevention and treatment of diarrhea.

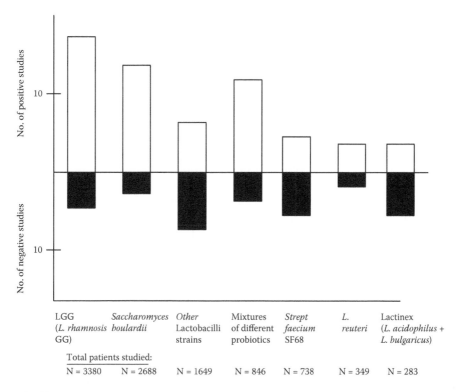

FIGURE 5.2
Frequency of positive and negative findings in controlled trials testing probiotics for human disease.

5.4.1 Antibiotic-Associated Diarrhea

A prevalent and vexing problem is antibiotic-associated diarrhea (AAD). In the hospital setting, antibiotic-associated diarrhea may prolong hospital stay and put undo demands on the time of hospital staff. In hospitalized patients receiving antibiotics, the prevalence of AAD ranges from 5 to 39%, depending upon the antibiotic (116). In the outpatient setting, diarrhea as an adverse event of antimicrobial therapy may decrease patient compliance with treatment and may precipitate a need to switch to a different, less appropriate, more expensive antimicrobial. Prolonged patient exposure to antibiotics increases pathogen resistance. Although the etiologies of antibiotic-associated diarrhea have not been thoroughly established, antimicrobial perturbation of the normal intestinal microflora is at the heart of the problem.

Numerous controlled trials have shown that probiotics can be efficacious in preventing AAD. Various lactobacilli (*L. acidophilus, L. bulgaricus, L. rhamnosus*), *Bifidobacterium longum, Enterococcus faecium,* and the yeast *Saccharomyces boulardii* have been tested in randomized, placebo-controlled clinical trials (117, 118). A meta-analysis of nine such trials (112) revealed an odds ratio of 0.37

(0.26 to 0.53) in favor of probiotic treatment over placebo. Appropriate probiotics should be routinely recommended to prevent this most common adverse effect of antimicrobial therapy.

5.4.2 *Clostridium difficile* Disease

A rare but more serious and even life-threatening adverse event of antimicrobial therapy is overgrowth of *Clostridium difficile*. The resulting elaboration of *C. difficile* toxins A and B is followed by an intense and characteristic ileal and colonic inflammation termed pseudomembranous colitis. Standard treatment is a course of metronidazole. More difficult cases require oral vancomycin therapy. A conceptual problem with treatment is that an antimicrobial agent is used to treat an antimicrobially induced disease! Indeed, once therapy is stopped, relapse is common (96, 97, 119–121). Probiotics have great potential as adjunct agents to standard metronidazole or vancomycin therapy. The approach would be to administer the probiotic during and for some time after antimicrobial *C. difficile* treatment in order to help "normalize" the gut microbial flora and to establish colonization resistance against *C. difficile* regrowth. Few probiotics have been tested to prevent recurrences of *C. difficile*-associated colitis in controlled clinical trials. There is limited evidence for success with *L. rhamnosus* (119, 120), but only *S. boulardii* has been tested in placebo-controlled trials. Encouragingly, this yeast treatment cut recurrences by more than 50% (97, 122). This is an example of a probiotic being successfully used to prevent recurrences of a very serious infection. It is not known whether *S. boulardii* or other probiotics can prevent an initial *C. difficile* infection. A study testing probiotic efficacy in prophylaxis of *C. difficile* disease would be difficult to conduct because pseudomembranous colitis is a serious but rare event following antimicrobial therapy.

5.4.3 Traveler's Diarrhea

Tourists traveling to areas of the world where the risk of acquiring traveler's diarrhea is high are usually given a prescription for an antimicrobial for use if diarrhea develops. Probiotics have the potential to decrease antibiotic use (and hence antibiotic resistance) when taken prophylactically to prevent traveler's diarrhea. Both *L. rhamnosus* and *S. boulardii* have been shown to have modest efficacy (123–125), but more study is needed, and other probiotics should be tested. The market potential for an effective probiotic for traveler's diarrhea is high.

5.4.4 Pediatric Diarrhea

In developing countries, childhood diarrhea is a leading cause of childhood morbidity and mortality. The etiology involves bacteria and parasites. Improved sanitation and nutrition is key to decreasing the extent of the problem, but routine use of an inexpensive probiotic preparation could have value. *L. rhamnosis* has been studied for the prevention of diarrhea in under-nourished Peruvian children (126). In this difficult setting, a modest reduction in episodes was observed compared to placebo. The effect was largely confined to nonbreast-fed, 18- to 29-month-old children (4.77 episodes per child per year in the treated group vs. 6.32 in the placebo group, p = 0.006). In contrast, rotavirus is the most common cause of acute childhood diarrhea in developed countries. In a hospital setting in the U.S., a mixture of *Bifidobacterium bifidum* and *Streptococcus thermophilus* as a supplement to a formula was beneficial in preventing diarrhea in infants (127). A similar study using *L. rhamnosus* reduced the risk (6.7% in treated vs. 33.3% in placebo, p = 0.002) of nosocomial diarrhea in hospitalized infants in Poland (128). Thus, probiotic use shows promise as a benign prophylactic measure to reduce the problem of diarrhea acquisition in hospitalized children.

5.4.5 Treatment of Acute Diarrhea

Oral rehydration is the cornerstone for treatment of acute childhood diarrhea, but several studies show that added probiotics can help speed recovery. Probiotic preparations containing one or more lactobacilli (*L. rhamnosus, L. reuteri, L. acidophilus, L. bulgaricus*), *S. thermophilus, B. infantis,* or *S. boulardii* have been shown to reduce days of diarrhea in studies conducted in seven different countries (129–134) and in a multicentered European trial (135). A recent meta-analysis on studies testing probiotics for acute diarrhea in children was reported by Huang et al. (113). Analysis of 18 eligible studies suggested that probiotic use decreased illness by about 1 day. Similar conclusions were reached in two additional systematic reviews (114–115). The meta-analysis by Van Niel et al. (115) was restricted to studies involving lactobacilli therapy. Summary point estimates showed a reduction of diarrhea of 0.7 d and a reduction in stool frequency on day 2 of 1.6 stools per day. As expected, there was considerable heterogeneity in the data sets; nevertheless, probiotics, particularly lactobacilli, were concluded to be a useful treatment for pediatric diarrhea.

There is less information on the efficacy of probiotics to treat acute diarrhea in adults. The diverse pathogens involved in adult diarrhea contribute to high intersubject response variability. Nevertheless, probiotics hold promise for treatment of some adult diarrheas (130).

5.5 Conclusions

Probiotics have shown an ability to prevent and to treat diarrhea, and have a high safety record. Probiotic use also has the potential to decrease antibiotic use and hence help in the effort to decrease antibiotic resistance. However, many health care providers (in the U.S. especially) remain skeptical about the value of routine probiotic use, and many commercially popular probiotics have not been studied in controlled clinical trials. In order for probiotics to be more widely accepted, it is incumbent on the industry to sponsor, either individually or collectively, further controlled studies to validate therapeutic claims. Studies are also needed to identify the mechanism(s) by which probiotics act beneficially and to better understand the pharmacokinetics of elimination. With this knowledge, the next approach should be to optimize probiotic action by appropriate strain selection and dosing regimens. For the future, genetic engineering has the potential of helping to develop highly effective probiotic microbes targeted for specific diseases. Then these "living drugs" will find more routine and effective use in therapy.

References

1. C Dunne, L O'Mahony, L Murphy, G Thornton, D Morrissey, S O'Halloran, M Feeney, S Flynn, G Fitzgerald, C Daly, B Kiely, GC O'Sullivan, F Shanahan, and JK Collins. *In vitro* selection criteria for probiotic bacteria of human origin: correlation with *in vivo* findings. *Am. J. Clin. Nutr.*, 73: 386S–392S, 2001.
2. YJ Lee, WK Yu, and TR Heo. Identification and screening for antimicrobial activity against *Clostridium difficile* of *Bifidobacterium* and *Lactobacillus* species isolated from healthy infant feces. *Int. J. Antimicrob. Agents*, 21: 340–346, 2003.
3. NW Mclean and IJ Rosenstein. Characterization and selection of a *Lactobacillus* species to re-colonize the vagina of women with recurrent bacterial vaginosis. *J. Med. Microbiol.*, 49: 543–552, 2000.
4. L V McFarland. Quality control and regulatory issues for biotherapeutic agents, in *Biotherapeutic Agents and Infectious Diseases* (GW Elmer, LV McFarland, and CM Surawicz, Eds.), Humana Press, Totowa, NJ, 1999, pp. 159–193.
5. WH Holzapfel, P Haberer, R Geisen, J Bjorkroth, and U Schillinger. Taxonomy and important features of probiotic microorganisms in food and nutrition. *Am. J. Clin. Nutr.*, 73: 365S–373S, 2001.
6. ME Sanders and TR Klaenhammer. The scientific basis of *Lactobacillus acidophilus* NCFM functionality as a probiotic. *J. Dairy Sci.*, 84: 319–331, 2001.
7. PSM Yeung, ME Sanders, CL Kitts, R Cano, and PS Tong. Species-specific identification of commercial probiotic strains. *J. Dairy Sci.*, 85: 1039–1051, 2002.
8. MF Bernet-Camard, V Lievin, D Brassart, JR Neeser, AL Servin, and S Hudault. The human *Lactobacillus acidophilus* strain LA1 secretes a nonbacteriocin antibacterial substance(s) active *in vitro* and *in vivo*. *Appl. Environ. Microbiol.*, 63: 2747–2753, 1997.

9. KJ Heller. Probiotic bacteria in fermented foods: product characteristics and starter organisms. *Am. J. Clin. Nutr., 73:* 374S–379S, 2001.
10. ML Clements, MM Levine, PA Ristaino, VE Daya, and TP Hughes. Exogenous lactobacilli fed to man — their fate and ability to prevent diarrheal disease. *Prog. Food Nutr. Sci., 7:* 29–37, 1983.
11. G Reid, ME Sanders, HR Gaskins, GR Gibson, A Mercenier, R Rastall, M Roberfroid, I Rowland, C Cherbut, and TR Klaenhammer. New scientific paradigms for probiotics and prebiotics. *J. Clin. Gastroenterol., 37:* 105–118, 2003.
12. P Mastromarino, P Brigidi, S Macchia, L Maggi, F Pirovano, V Trinchieri, U Conte, and D Matteuzzi. Characterization and selection of vaginal *Lactobacillus* strains for the preparation of vaginal tablets. *J. Appl. Microbiol., 93:* 884–893, 2002.
13. M Silva, NV Jacobus, C Deneke, and SL Gorbach. Antimicrobial substance from a human *Lactobacillus* strain. *Antimicrobiol. Agents Chemother., 31:* 1231–1233, 1987.
14. I Casas and W Dobrogosz. Validation of the probiotic concept: *Lactobacillus reuteri* confers broad-spectrum protection against disease in humans and animals. *Microb. Ecol. Health Dis., 12:* 247–285, 2000.
15. I Castagliuolo, JM Lamont, ST Nikulasson, and C Pothoulakis. *Saccharomyces boulardii* protease inhibits *Clostridium difficile* toxin an effects in the rat ileum. *Infect. Immun., 64:* 5225–5232, 1996.
16. VL Hughes and SL Hillier. Microbiologic characteristics of *Lactobacillus* products used for colonization of the vagina. *Obstet. Gynecol., 75:* 244–248, 1990.
17. W Zhong, K Millsap, H Bialkowska-Hobrzanska, and G Reid. Differentiation of *Lactobacillus* species by molecular typing. *Appl. Environ. Microbiol., 64:* 2418–2423, 1998.
18. LV McFarland. Normal flora: diversity and functions. *Microb. Ecol. Health Dis., 12:* 193–207, 2000.
19. LV McFarland and P Bernasconi. *Saccharomyces boulardii* — A review of an innovative biotherapeutic agent. *Microb. Ecol. Health Disease, 6:* 157–171, 1993.
20. P Brigidi, B Vitali, E Swennen, G Bazzocchi, and D Matteuzzi. Effects of probiotic administration upon the composition and enzymatic activity of human fecal microbiota in patients with irritable bowel syndrome or functional diarrhea. *Res. Microbiol., 152:* 735–741, 2001.
21. S Fujiwara, Y Seto, A Kimura, and H Hashiba. Intestinal transit of an orally administered streptomycin–rifampicin-resistant variant of *Bifidobacterium longum* SBT2928: its long-term survival and effect on the intestinal microflora and metabolism. *J. Appl. Microbiol., 90:* 43–52, 2001.
22. C Forestier, C De Champs, C Vatoux, and B Joly. Probiotic activities of *Lactobacillus casei rhamnosus: in vitro* adherence to intestinal cells and antimicrobial properties. *Res. Microbiol., 152:* 167–173, 2001.
23. PV Kirjavainen, AC Ouwehand, E Isolauri, and SJ Salminen. The ability of probiotic bacteria to bind to human intestinal mucus. *FEMS Microbiol. Lett., 167:* 185–189, 1998.
24. J McCarthy, L O'Mahony, L O'Callaghan, B Sheil, EE Vaughan, N Fitzsimons, J Fitzgibbon, GC O'Sullivan, B Kiely, JK Collins, and F Shanahan. Double blind, placebo controlled trial of two probiotic strains in interleukin 10 knockout mice and mechanistic link with cytokine balance. *Gut, 52:* 975–980, 2003.

25. JP Buts, P Bernasconi, JP Vaerman, and C Dive. Stimulation of secretors IgA and secretors component of immunoglobulins in small intestine of rats treated with *Saccharomyces boulardii. Dig. Dis. Sci., 35:* 251–256, 1990.

26. E Isolauri, J Joensuu, H Suomalainen, M Luomala, and T Vesikari. Improved immunogenicity of oral D x rrv reassortant rotavirus vaccine by *Lactobacillus casei* GG. *Vaccine, 13:* 310–312, 1995.

27. G Perdigon, M Locascio, M Medici, APD Holgado, and G Oliver. Interaction of bifidobacteria with the gut and their influence in the immune function. *Biocell, 27:* 1–9, 2003.

28. N Ishibashi and S Yamazaki. Probiotics and safety. *Am. J. Clin. Nutr., 73:* 465S–470S, 2001.

29. S Salminen, C Bouley, MC Boutron-Ruault, JH Cummings, A Franck, GR Gibson, E Isolauri, MC Moreau, M Roberfroid, and I Rowland. Functional food science and gastrointestinal physiology and function. *Br. J. Nutr., 80:* S147–S171, 1998.

30. G MacGregor, AJ Smith, B Thakker, and J Kinsella. Yogurt biotherapy: contraindicated in immunosuppressed patients? *Postgrad. Med. J., 78:* 366–367, 2002.

31. M Rautio, H Jousimies-Somer, H Kauma, I Pietarinen, M Saxelin, S Tynkkynen, and M Koskela. Liver abscess due to a *Lactobacillus rhamnosus* strain indistinguishable from *L rhamnosus* strain GG. *Clin. Infect. Dis., 28:* 1159–1160, 1999.

32. PG Rogasi, S Vigano, P Pecile, and F Leoncini. *Lactobacillus casei* pneumonia and sepsis in a patient with AIDS. Case report and review of the literature. *Ann. Ital. Med. Int. 13:* 180–182, 1998.

33. Y Schoon, B Schuurman, AG Buiting, SE Kranendonk, and SJ Graafsma. Aortic graft infection by *Lactobacillus casei*: a case report. *Neth. J. Med., 52:* 71–74, 1998.

34. M Pletinex, J Legain, and Y Vandenplas. Fungemia with *Saccharomyces boulardii* in a 1-year-old girl with protracted diarrhea. *J. Ped. Gastroenterol. Nutr., 21:* 113–115, 1995.

35. P Zunic, J Lacotte, M Pegoix, G Buteux, G Leroy, B Mosquet, and M Moulin. Fungemy associated with *Saccharomyces boulardii* — about one case. *Therapie, 46:* 498, 1991.

36. JA Vanderhoof, DB Whitney, DL Antonson, TL Hanner, JV Lupo, and RJ Young. *Lactobacillus* GG in the prevention of antibiotic-associated diarrhea in children. *J. Ped., 135:* 564–568, 1999.

37. BR Goldin, SL Gorbach, M Saxelin, S Barakat, L Gualtieri, and S Salminen. Survival of *Lactobacillus* species (strain GG) in human gastrointestinal tract. *Dig. Dis. Sci., 37:* 121–128, 1992.

38. S Hudault, V Lievin, MF BernetCamard, and AL Servin. Antagonistic activity exerted *in vitro* and *in vivo* by *Lactobacillus casei* (strain GG) against *Salmonella typhimurium* C5 infection. *Appl. Environ. Microbiol., 63:* 513–518, 1997.

39. M Alander, R Satokari, R Korpela, M Saxelin, T Vilpponen-Salmela, T Mattila-Sandholm, and A von Wright. Persistence of colonization of human colonic mucosa by a probiotic strain, *Lactobacillus rhamnosus* GG, after oral consumption. *Appl. Environ. Microbiol., 65:* 351–354, 1999.

40. MR Millar, C Bacon, SL Smith, V Walker, and MA Hall. Enteral feeding of premature infants with *Lactobacillus* GG. *Arch. Dis. Childhood, 69:* 483–487, 1993.

41. R Goldin, LJ Gualtieri, and RP Moore. The effect of *Lactobacillus* GG on the initiation and promotion of DMH-induced intestinal tumors in the rat. *Nutr. Cancer Int. J., 25:* 197–204, 1996.

42. AC Ouwehand, E Isolauri, PV Kirjavainen, S Tolkko, and SJ Salminen. The mucus binding of *Bifidobacterium lactis* Bb12 is enhanced in the presence of *Lactobacillus* GG and *Lact. delbrueckii* subsp *bulgaricus. Lett. Appl. Microbiol., 30:* 10–13, 2000.

43. JH Meurman, H Antila, A Korhonen, and S Salminen. Effect of *Lactobacillus rhamnosus* strain GG (ATCC-53103) on the growth of *Streptococcus sobrinus in vitro. Eur. J. Oral Sci., 103:* 253–258, 1995.

44. JH Meurman, H Antila, and S Salminen. Recovery of *Lactobacillus* strain GG (ATCC-53103) from saliva of healthy volunteers after consumption of yogurt prepared with the bacterium. *Microb. Ecol. Health Dis., 7:* 295–298, 1994.

45. L Nase, K Hatakka, E Savilahti, M Saxelin, A Ponka, T Poussa, R Korpela, and JH Meurman. Effect of long-term consumption of a probiotic bacterium, *Lactobacillus rhamnosus* GG, in milk on dental caries and caries risk in children. *Caries Res., 35:* 412–420, 2001.

46. Q Shu and HS Gill. Immune protection mediated by the probiotic *Lactobacillus rhamnosus* HN001 (DR20 (TM)) against Escherichia coli O157: H7 infection in mice. *FEMS Immunol. Med. Microbiol., 34:* 59–64, 2002.

47. YH Sheih, BL Chiang, LH Wang, CK Liao, and HS Gill. Systemic immunity-enhancing effects in healthy subjects following dietary consumption of the lactic acid bacterium *Lactobacillus rhamnosus* HN001. *J. Am. Coll. Nutr., 20:* 149–156, 2001.

48. HS Gill, KJ Rutherfurd, ML Cross, and PK Gopal. Enhancement of immunity in the elderly by dietary supplementation with the probiotic *Bifidobacterium lactis* HN019. *Am. J. Clin. Nutr., 74:* 833–839, 2001.

49. GW Tannock, K Munro, HJM Harmsen, GW Welling, J Smart, and PK Gopal. Analysis of the fecal microflora of human subjects consuming a probiotic product containing *Lactobacillus rhamnosus* DR20. *Appl. Environ. Microbiol., 66:* 2578–2588, 2001.

50. TL Talarico, IA Casas, TC Chung, and WJ Dobrogosz. Production and isolation of Reuterin, a growth inhibitor produced by *Lactobacillus reuteri. Antimicrob. Agents Chemother., 32:* 1854–1858, 1988.

51. T Toba, SK Samant, E Yoshioka, and T Itoh. Reutericin-6, a new bacteriocin produced by *Lactobacillus reuteri* La-6. *Lett. Appl. Microbiol., 13:* 281–286, 1991.

52. T Wadstrom, K Andersson, M Sydow, L Axelsson, S Lindgren, and B Gullmar. Surface properties of Lactobacilli isolated from the small intestine of pigs. *J. Appl. Bacteriol., 62:* 513–520,1987.

53. CN Jacobsen, VR Nielsen, AE Hayford, PL Moller, KF Michaelsen, A Paerregaard, B Sandstrom, M Tvede, and M Jakobsen. Screening of probiotic activities of forty-seven strains of *Lactobacillus* spp. by *in vitro* techniques and evaluation of the colonization ability of five selected strains in humans. *Appl. Environ. Microbiol., 65:* 4949–4956, 1999.

54. BW Wolf, KA Garleb, DG Ataya, and IA Casas. Safety and tolerance of *Lactobacillus reuteri* in healthy adult male subjects. *Microb. Ecol. Health Dis., 8:* 41–50, 1995.

55. BW Wolf, KB Wheeler, DG Ataya, and KA Garleb. Safety and tolerance of *Lactobacillus reuteri* supplementation to a population infected with the human immunodeficiency virus. *Food Chem. Toxicol., 36:* 1085–1094, 1998.

56. P Marteau, P Pochart, Y Bouhnik, S Zidi, I Goderel, and JC Rambaud. Survival of *Lactobacillus acidophilus* and *Bifidobacterium* sp ingested in a fermented milk in the small intestine — a rational basis for the use of probiotics in man. *Gastroenterol. Clin. Biol., 16:* 25–28, 1992.

57. P Marteau, E Cuillerier, S Meance, MF Gerhardt, A Myara, M Bouvier, C Bouley, F Tondu, G Bommelaer, and JC Grimaud. Bifidobacterium animalis strain DN-173 010 shortens the colonic transit time in healthy women: a double-blind, randomized, controlled study. *Aliment. Pharmacol. Ther., 16:* 587–593, 2002.

58. PK Gopal, J Prasad, J Smart, and HS Gill. *In vitro* adherence properties of *Lactobacillus rhamnosus* DR20 and *Bifidobacterium lactis* DR10 strains and their antagonistic activity against an enterotoxigenic *Escherichia coli. Int. J. Food Microbiol., 67:* 207–216, 2001.

59. M Matsumoto, H Tani, H Ono, H Ohishi, and Y Benno. Adhesive property of *Bifidobacterium lactis* LKM512 and predominant bacteria of intestinal microflora to human intestinal mucin. *Curr. Microbiol., 44:* 212–215, 2002.

60. Y Shimakawa, S Matsubara, N Yuki, M Ikeda, and F Ishikawa. Evaluation of *Bifidobacterium breve* strain Yakult-fermented soymilk as a probiotic food. *Int. J. Food Microbiol., 81:* 131–136, 2003.

61. MH Coconnier, MF Bernet, S Kerneis, G Chauviere, J Fourniat, and AL Servin. Inhibition of adhesion of enteroinvasive pathogens to human intestinal Caco-2 cells by *Lactobacillus acidophilus* strain Lb decreases bacterial invasion. *FEMS Microbiol. Lett., 110:* 299–305, 1993.

62. MH Coconnier, TR Klaenhammer, S Kerneis, MF Bernet, and AL Servin. Protein-mediated adhesion of *Lactobacillus acidophilus* Bg2Fo4 on human enterocyte and mucus-secreting cell-lines in culture. *Appl. Environ. Microbiol., 58:* 2034–2039, 1992.

63. MF Bernet, D Brassart, JR Neeser, and AL Servin. *Lactobacillus acidophilus* La-1 binds to cultured human intestinal cell lines and inhibits cell attachment and cell invasion by enterovirulent bacteria. *Gut, 35:* 483–489, 1994.

64. J Sui, S Leighton, F Busta, and L Brady. 16S ribosomal DNA analysis of the faecal lactobacilli composition of human subjects consuming a probiotic strain *Lactobacillus acidophilus* NCFM (R). *J. Appl. Microbiol., 93:* 907–912, 2002.

65. AA Aroutcheva, JA Simoes, and S Faro. Antimicrobial protein produced by vaginal *Lactobacillus acidophilus* that inhibits *Gardnerella vaginalis. Infect. Dis. Obstet. Gynecol., 9:* 33–39, 2001.

66. MH Coconnier, V Lievin, MF Bernet-Camard, S Hudault, and AL Servin. Antibacterial effect of the adhering human *Lactobacillus acidophilus* strain LB. *Antimicrob. Agents Chemother., 41:* 1046–1052, 1997.

67. T Asahara, K Nomoto, M Watanuki, and T Yokokura. Antimicrobial activity of intraurethrally administered probiotic *Lactobacillus casei* in a murine model of *Escherichia coli* urinary tract infection. *Antimicrob. Agents Chemother., 45:* 1751–1760, 2001.

68. R de Waard, J Garssen, GCAM Bokken, and JG Vos. Antagonistic activity of *Lactobacillus casei* strain Shirota against gastrointestinal *Listeria monocytogenes* infection in rats. *Int. J. Food Microbiol., 73:* 93–100, 2002.

69. T Hori, J Kiyoshima, K Shida, and H Yasui. Augmentation of cellular immunity and reduction of influenza virus titer in aged mice fed *Lactobacillus casei* strain Shirota. *Clin. Diagn. Lab. Immunol., 9:* 105–108, 2002.

70. MJ Park, JS Kim, HC Jung, IS Song, and JJ Lee. The suppressive effect of a fermented milk containing lactobacilli on *Helicobacter pylori* in human gastric mucosa. *Gastroenterology, 120:* A590, 2001.

71. S Spanhaak, R Havenaar, and G Schaafsma. The effect of consumption of milk fermented by *Lactobacillus casei* strain Shirota on the intestinal microflora and immune parameters in humans. *Eur. J. Clin. Nutr., 52:* 899–907, 1998.

72. N Yuki, K Watanabe, A Mike, Y Tagami, R Tanaka, M Ohwaki, and M Morotomi. Survival of a probiotic, *Lactobacillus casei* strain Shirota, in the gastrointestinal tract: selective isolation from feces and identification using monoclonal antibodies. *Int. J. Food Microbiol., 48:* 51–57, 1999.

73. S Fujiwara, Y Seto, A Kimura, and H Hashiba. Establishment of orally-administered *Lactobacillus gasseri* SBT2055SR in the gastrointestinal tract of humans and its influence on intestinal microflora and metabolism. *J. Appl. Microbiol., 90:* 343–352, 2001.

74. A Ushiyama, K Tanaka, Y Aiba, T Shiba, A Takagi, T Mine, and Y Koga. *Lactobacillus gasseri* OLL2716 as a probiotic in clarithromycin-resistant *Helicobacter pylori* infection, *J. Gastroenterol. Hepatol., 18,* 986–991, 2003.

75. G Reid. Safety of *Lactobacillus* strains as probiotic agents. *Clin. Infect. Dis., 35:* 349–350, 2002.

76. G Reid and AW Bruce. Selection of *Lactobacillus* strains for urogenital probiotic applications. *J. Infect. Dis., 183:* S77–S80, 2001.

77. G Reid. Probiotic agents to protect the urogenital tract against infection. *Am. J. Clin. Nutr., 73:* 437S–443S, 2001.

78. G Reid, D Charbonneau, J Erb, B Kochanowski, D Beuerman, R Poehner, and AW Bruce. Oral use of *Lactobacillus rhamnosus* GR-1 and *L. fermentum* RC-14 significantly alters vaginal flora: randomized, placebo-controlled trial in 64 healthy women. *FEMS Immunol. Med. Microbiol., 35:* 131–134, 2003.

79. GE Gardiner, C Heinemann, AW Bruce, D Beuerman, and G Reid. Persistence of *Lactobacillus fermentum* RC-14 and *Lactobacillus rhamnosus* GR-1 but not *L rhamnosus* GG in the human vagina as demonstrated by randomly amplified polymorphic DNA. *Clin. Diagn. Lab. Immunol., 9:* 92–96, 2002.

80. GE Gardiner, C Heinemann, ML Baroja, AW Bruce, D Beuerman, J Madrenas, and G Reid. Oral administration of the probiotic combination *Lactobacillus rhamnosus* GR-1 and *L. fermentum* RC-14 for human intestinal applications. *Int. Dairy J., 12:* 191–196, 2002.

81. G Reid, D Beuerman, C Heinemann, and AW Bruce. Probiotic *Lactobacillus* dose required to restore and maintain a normal vaginal flora. *FEMS Immunol. Med. Microbiol., 32:* 37–41, 2001.

82. B Lund and C Edlund. Probiotic *Enterococcus faecium* strain is a possible recipient of the van A gene cluster. *Clin. Infect. Dis., 32:* 1384–1385, 2001.

83. B Lund, C Edlund, L Barkholt, CE Nord, M Tvede, and RL Poulsen. Impact on human intestinal microflora of an *Enterococcus faecium* probiotic and vancomycin. *Scand. J. Infect. Dis., 32:* 627–632, 2000.

84. H Blehaut, J Massot, GW Elmer, and RH Levy. Disposition kinetics of *Saccharomyces boulardii* in man and rat. *Biopharm. Drug Disposition, 10:* 353–364, 1989.

85. R Ducluzeau and M Bensaada. Comparative effect of a single or continuous administration of *Saccharomyces boulardii* on the establishment of various strains of *Candida* in the digestive tract of gnotobiotic mice. *Ann. Microbiol., B133:* 491–501, 1982.

86. GW Elmer and G Corthier. Modulation of *Clostridium difficile* induced mortality as a function of the dose and the viability of the *Saccharomyces boulardii* used as a preventative agent in gnotobiotic mice. *Can. J. Microbiol., 37:* 315–317, 1991.

87. SW Martin, AC Heatherington, and GW Elmer. Pharmacokinetics of biotherapeutic agents, in *Biotherapeutic Agents and Infectious Diseases* (Elmer GW, McFarland LV, and Surawicz CM, Eds.), pp. 47–84, Humana Press, Totowa, NJ, 1999.

88. SM Klein, GW Elmer, LV McFarland, CM Surawicz, and RH Levy. Recovery and elimination of the biotherapeutic agent, *Saccharomyces boulardii*, in healthy human volunteers. *Pharm. Res., 10:* 1615–1619, 1993.

89. GW Elmer, KA Moyer, R Vega, CM Surawicz, AC Collier, TM Hooton, and LV McFarland. Evaluation of *Saccharomyces boulardii* for patients with HIV-related chronic diarrhea and in healthy volunteers receiving antifungals. *Microecol. Ther., 25:* 157–171, 1995.

90. KL Horner, RC LeBoeuf, LV McFarland, and GW Elmer. Dietary fiber affects the onset of *Clostridium difficile* disease in hamsters. *Nutr. Res., 20:* 1103–1112, 2000.

91. MC Rigothier, J Maccario, and P Gayral. Inhibitory activity of *Saccharomyces* yeasts on the adhesion of *Entamoeba histolytica* trophozoites to human erythrocytes *in vitro*. *Parasitol. Res., 80:* 10–15, 1994.

92. MC Rigothier, J Maccario, PN Vuong, and P Gayral. Effects of *Saccharomyces boulardii* yeast on trophozoites of *Entamoeba histolytica in vitro* and in cecal amebiasis in young rats. *Ann. Parasitol. Hum. Compar., 65:* 51–60, 1990.

93. R Berg, P Bernasconi, D Fowler, and M Gautreaux. Inhibition of *Candida albicans* translocation from the gastrointestinal tract of mice by oral administration of *Saccharomyces boulardii*. *J. Infect. Dis., 168:* 1314–1318, 1993.

94. N Vidon, B Huchet, and JC Rambaud. Effect of *S boulardii* on water and sodium secretions induced by cholera-toxin. *Gastroenterol. Clin. Biol., 10:* 13–16, 1986.

95. D Czerucka, I Roux, and P Rampal. *Saccharomyces boulardii* inhibits secretagogue-mediated adenosine 3',5'-cyclic-monophosphate induction in intestinal cells. *Gastroenterology, 106:* 65–72, 1994.

96. CM Surawicz, LV McFarland, G Elmer, and J Chinn. Treatment of recurrent *Clostridium difficile* colitis with vancomycin and *Saccharomyces boulardii*. *Am. J. Gastroenterol., 84:* 1285–1287, 1989.

97. CM Surawicz, LV McFarland, RN Greenberg, M Rubin, R Fekety, ME Mulligan, RJ Garcia, S Brandmarker, K Bowen, D Borjal, and GW Elmer. The search for a better treatment for recurrent *Clostridium difficile* disease: use of high-dose vancomycin combined with *Saccharomyces boulardii*. *Clin. Infect. Dis., 31:* 1012–1017, 2000.

98. C Pothoulakis, CP Kelly, MA Joshi, N Gao, CJ Okeane, I Castagliuolo, and JT Lamont. *Saccharomyces boulardii* inhibits *Clostridium difficile* toxin-a binding and enterotoxicity in rat ileum. *Gastroenterology, 104:* 1108–1115, 1993.

99. I Castagliuolo, MF Riegler, L Valenick, JT Lamont, and C Pothoulakis. *Saccharomyces boulardii* protease inhibits the effects of *Clostridium difficile* toxins A and B in human colonic mucosa. *Infect. Immun., 67:* 302–307, 1999.

100. GW Elmer and LV McFarland. Suppression by *Saccharomyces boulardii* of toxigenic *Clostridium difficile* overgrowth after vancomycin treatment in hamsters. *Antimicrob. Agents Chemother., 31:* 129–131, 1987.

101. JP Buts, N DeKeyser, C Stilmant, E Sokal, and S Marandi. *Saccharomyces boulardii* enhances N-terminal peptide hydrolysis in suckling rat small intestine by endoluminal release of a zinc-binding metalloprotease. *Pediatr. Res., 51:* 528–534, 2002.

102. A Qamar, S Aboudola, M Warny, P Michetti, C Pothoulakis, JT Lamont, and CP Kelly. *Saccharomyces boulardii* stimulates intestinal immunoglobulin A immune response to *Clostridium difficile* toxin A in mice. *Infect. Immun., 69:* 2762–2765, 2001.

103. JAM Caetano, MT Parames, MJ Babo, A Santos, AB Ferreira, AA Freitas, MRC Coelho, and AM Mateus. Immunopharmacological effects of *Saccharomyces boulardii* in healthy human volunteers. *Int. J. Immunopharmacol., 8:* 245–247, 1986.

104. JP Buts, P Bernasconi, MP Vancraynest, P Maldague, and R Demeyer. Response of human and rat small intestinal mucosa to oral administration of *Saccharomyces boulardii. Pediatr. Res., 20:* 192–196, 1986.

105. HU Jahn, R Ullrich, T Schneider, RM Liehr, HL Schieferdecker, H Holst, and M Zeitz. Immunological and trophical effects of *Saccharomyces boulardii* on the small intestine in healthy human volunteers. *Digestion, 57:* 95–104, 1996.

106. JP Buts, N Dekeyser, and L Deraedemaeker. *Saccharomyces boulardii* enhances rat intestinal enzyme expression by endoluminal release of polyamines. *Pediatr. Res., 36:* 522–527, 1994.

107. JP Buts, N Dekeyser, J Kolanowski, E Sokal, and F Vanhoof. Maturation of villus and crypt cell functions in rat small intestine — role of dietary polyamines. *Dig. Dis. Sci., 38:* 1091–1098, 2003.

108. LV McFarland. A review of the evidence of health claims for biotherapeutic agents. *Microb. Ecol. Health Dis., 12:* 65–76, 2000.

109. P Marteau, P Seksik, and R Jian. Probiotics and intestinal health effects: a clinical perspective. *Br. J. Nutr., 88:* S51–S57, 2002.

110. L Agerholm-Larsen, ML Bell, GK Grunwald, and A Astrup. The effect of a probiotic milk product on plasma cholesterol: a meta-analysis of short-term intervention studies. *Eur. J. Clin. Nutr., 54:* 856–860, 2000.

111. F Cremonini, S Di Caro, EC Nista, F Bartolozzi, G Capelli, G Gasbarrini, and A Gasbarrini. Meta-analysis: the effect of probiotic administration on antibiotic-associated diarrhoea. *Aliment. Pharmacol. Ther., 16:* 1461–1467, 2002.

112. AL D'Souza, C Rajkumar, J Cooke, and CJ Bulpitt. Probiotics in prevention of antibiotic associated diarrhoea: meta-analysis. *Br. Med. J., 324:* 1361–1364, 2002.

113. JS Huang, A Bousvaros, JW Lee, A Diaz, and EJ Davidson. Efficacy of probiotic use in acute diarrhea in children — A meta-analysis. *Dig. Dis. Sci., 47:* 2625–2634, 2002.

114. H Szajewska and JZ Mrukowicz. Probiotics in the treatment and prevention of acute infectious diarrhea in infants and children: a systematic review of published randomized, double-blind, placebo-controlled trials. *J. Pediatr. Gastroenterol. Nutr., 33:* S17–S25, 2001.

115. CW Van Niel, C Feudtner, MM Garrison, and DA Christakis. Lactobacillus therapy for acute infectious diarrhea in children: a meta-analysis. *Pediatrics, 109:* 678–684, 2002.

116. LV McFarland. Epidemiology, risk factors and treatments for antibiotic-associated diarrhea. *Dig. Dis., 16:* 292–307, 1998.

117. LV McFarland, CM Surawicz, RN Greenberg, GW Elmer, KA Moyer, SA Melcher, KE Bowen, and JL Cox. Prevention of beta-lactam-associated diarrhea by *Saccharomyces boulardii* compared with placebo. *Am. J. Gastroenterol., 90:* 439–448, 1995.

118. CM Surawicz, GW Elmer, P Speelman, LV McFarland, J Chinn, and G Vanbelle. Prevention of antibiotic associated diarrhea by *Saccharomyces boulardii* — a prospective study. *Gastroenterology, 96:* 981–988, 1989.

119. SL Gorbach, TW Chang, and B Goldin. Successful treatment of relapsing *Clostridium difficile* colitis with *Lactobacillus* GG. *Lancet, 2:* 1519, 1987.

120. JA Biller, AJ Katz, AF Flores, TM Buie, and SL Gorbach. Treatment of recurrent *Clostridium difficile* colitis with *Lactobacillus* GG. *J. Pediatr. Gastroenterol. Nutr., 21:* 224–226, 1995.

121. LV McFarland, CM Surawicz, M Rubin, R Fekety, GW Elmer, and RN Greenberg. Recurrent *Clostridium difficile* disease: epidemiology and clinical characteristics. *Inf. Control Hosp. Epidemiol., 20:* 43–50, 1999.

122. LV McFarland, CM Surawicz, RN Greenberg, R Fekety, GW Elmer, KA Moyer, SA Melcher, KE Bowen, JL Cox, Z Noorani, G Harrington, M Rubin, and D Greenwald. Randomized placebo-controlled trial of *Saccharomyces boulardii* in combination with standard antibiotics for *Clostridium difficile* disease. *J. Am. Med. Assoc., 271:* 1913–1918, 1994.

123. PJ Oksanen, S Salminen, M Saxelin, P Hamalainen, A Ihantolavormisto, L Muurasniemiisoviita, S Nikkari, T Oksanen, I Porsti, E Salminen, S Siitonen, H Stuckey, A Toppila, and H Vapaatalo. Prevention of travelers diarrhea by *Lactobacillus* GG. *Ann. Med., 22:* 53–56, 1990.

124. E Hilton, P Kolakowski, C Singer, and M Smith. Efficacy of *Lactobacillus* GG as a diarrheal preventive in travelers. *J. Travel. Med., 4:* 41–43, 1997.

125. H Kollaritsch, H Holst, P Grobara, and G Wiedermann. Prevention of traveler's diarrhea with *Saccharomyces boulardii*. Results of a placebo controlled double-blind study. *Fortschr. Med., 111:* 152–156, 1993.

126. RA Oberhelman, EH Gilman, P Sheen, DN Taylor, RE Black, L Cabrera, AG Lescano, R Meza, and G Madico. A placebo-controlled trial of *Lactobacillus* GG to prevent diarrhea in undernourished Peruvian children. *J. Pediatr., 134:* 15–20, 1999.

127. JM Saavedra, NA Bauman, I Oung, JA Perman, and RH Yolken. Feeding of *Bifidobacterium bifidum* and *Streptococcus thermophilus* to infants in-hospital for prevention of diarrhea and shedding of rotavirus. *Lancet, 344:* 1046–1049, 1994.

128. H Szajewska, M Kotowska, JZ Mrukowicz, M Armanska, and W Mikolajczyk. Efficacy of *Lactobacillus* GG in prevention of nosocomial diarrhea in infants. *J. Pediatr., 138:* 361–365, 2001.

129. GW Elmer, CM Surawicz, and LV McFarland. Biotherapeutic agents — A neglected modality for the treatment and prevention of selected intestinal and vaginal infections. *J. Am. Med. Assoc., 275:* 870–876, 1996.

130. GW Elmer. Probiotics: "Living drugs". *Am. J. Health Syst. Pharm., 58:* 1101–1109, 2001.

131. MC Lee, LH Lin, KL Hung, and HY Wu. Oral bacterial therapy promotes recovery from acute diarrhea in children. *Acta Paediatr. Taiwan, 42:* 301–305, 2001.

132. G Boudraa, M Benbouabdellah, W Hachelaf, M Boisset, JF Desjeux, and M Touhami. Effect of feeding yogurt versus milk in children with acute diarrhea and carbohydrate malabsorption. *J. Pediatr. Gastroenterol. Nutr., 33:* 307–313, 2001.

133. V Rosenfeldt, KF Michaelsen, M Jakobsen, CN Larsen, PL Moller, M Tvede, H Weyrehter, NH Valerius, and A Paerregaard. Effect of probiotic *Lactobacillus* strains on acute diarrhea in a cohort of nonhospitalized children attending day-care centers. *Pediatr. Infect. Dis. J., 21:* 417–419, 2002.

134. V Rosenfeldt, KF Michaelsen, M Jakobsen, CN Larsen, PL Moller, P Pedersen, M Tvede, H Weyrehter, NH Valerius, and A Paerregaard. Effect of probiotic *Lactobacillus* strains in young children hospitalized with acute diarrhea. *Pediatr. Infect. Dis. J., 21:* 411–416, 2002.

135. S Guandalini, L Pensabene, MA Zikri, JA Dias, LG Casali, H Hoekstra, S Kolacek, K Massar, D Micetic-Turk, A Papadopoulou, JS de Sousa, B Sandhu, H Szajewska, and Z Weizman. *Lactobacillus* GG administered in oral rehydration solution to children with acute diarrhea: a multicenter European trial. *J. Pediatr. Gastroenterol. Nutr., 30:* 54–60, 2000.

6

Genetic Engineering of Probiotic Bacteria

Collette Desmond, Paul Ross, Gerald F. Fitzgerald, and Catherine Stanton

CONTENTS

6.1 Introduction

Probiotics are described as "live microorganisms which when consumed in adequate numbers confer a health benefit on the host" (1). Probiotic foods are part of the functional foods sector, i.e., those foods or food ingredients that exert a beneficial effect on host health and/or reduce the risk of chronic disease beyond basic nutritional requirements (2). There is an ever increasing and more convincing body of evidence to support the clinical efficacy of these live bacteria as part of the human diet (3). Some of the best-documented and most clearly reported evidence of the health benefits of specific probiotics concerns the treatment of viral diarrheal disorders (4), prevention of atopic disease (5), and prevention of acute pouchitis (6). In many cases, however, the effects attributed to the ingestion of probiotics remain scientifically unsubstantiated (7), a situation which could be resolved by better defining particular probiotic strains and an understanding of the mechanisms that underlie survival and functionality in the host gastrointestinal tract. Thus, when considering the challenges and outlook for the future of probiotic functional foods, more distinct and consistent probiotic strains may be developed by design rather than screening. In this respect, the era of genomics could make this rationale simpler and more readily achievable, due particularly to genetic engineering technology which allows selected genes to be exchanged from one microorganism to another, even among different species.

Genome sequence information for intestinal microorganisms is increasing rapidly, so there are new potential opportunities to apply this knowledge to the construction of designer strains which are engineered for better performance. Three possible reasons to genetically engineer probiotic strains include (i) to accentuate the health-promoting properties or "functionality" of probiotic bacteria for example, to introduce entirely novel traits in well-characterized probiotic strains; (ii) to improve the technological robustness of probiotic bacteria, enhancing yield and performance during fermentation, food processing, and shelf-life; and (iii) to develop new applications in engineering nonprobiotic strains to promote health and/or produce therapeutic molecules such as antigens for the development of vaccine vectors.

In this review, we will concentrate on genetic technologies that have been developed over the last 30 years for lactic acid bacteria (LAB) and that can be, and are being, applied to probiotic strains. Technologies such as food-grade cloning and transformation techniques, chromosomal mutations, functional genomics, and proteomics are some of the more outstanding techniques employed in genetic engineering (Figure 6.1). We will also examine current sequencing projects that are completed or near completion, that provide a sound understanding of the probiotic genome and how this information may be applied to engineer probiotic strains of the future. A better understanding of the genetics and physiology of probiotic lactobacilli and bifidobacteria in particular will allow for better strain use and selection in the development of probiotic foods. Enterococci are also amenable to genetic manipulation, but concerns over their safety (8) may in the future limit their potential for human therapeutic purposes. Finally, the potential benefits of these genetically modified bacteria to society and the possible risks associated with their use will be discussed in relation to human health.

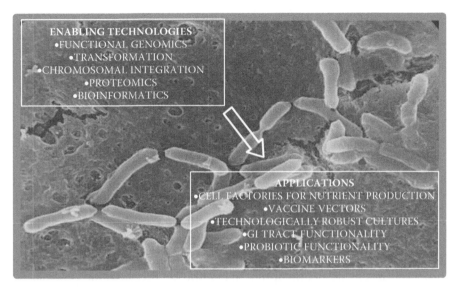

FIGURE 6.1
Technologies involved in the genetic modification of lactobacilli and bifidobacteria, and some of the potential applications that result.

6.2   Development of Improved Genetic Tools

LAB are widely used in the food industry and have been used for millennia in food fermentations. In the last three decades, these microorganisms have been made amenable to genetic manipulation through concentrated efforts of a number of groups worldwide. In particular, *Lactococcus lactis* IL1403, which was the first LAB whose genome was sequenced (9), has been used as a template for genetic manipulations. The information acquired from working with *L. lactis* and the technologies developed for genetic engineering of the strain have in many cases been "loosely applied" to the development of similar technologies for lactobacilli. Lactococci and lactobacilli are similar in that they both have high adenine and guanine (AT) content, and many strains can readily be transformed by electroporation. In contrast, bifidobacteria have a high guanine and cytosine (GC) content, and thus prove more challenging when developing electroporation techniques. The genetic systems used to introduce novel properties in bacteria used in industrial food fermentations must be food grade. Essentially, this means that they should be considered as safe as the host in which they are introduced, well-characterized, and stable as well as flexible (10). For selection purposes, many of the successful vectors developed for LAB to date carry antibiotic resistance genes. Though it is an essential feature of food-grade systems that they should not contain transferable antibiotic resistance markers, it is important

TABLE 6.1

Transformation Efficiencies of Various *Lactobacillus* and
Bifidobacterium Species

Strain	Vector	Frequency	Ref.
L. casei	pSA3	8.5×10^4	(12)
L. acidophilus	pGK12	3.3×10^5	(116)
L. helveticus	pLHR	1.3×10^4	(117)
L. curvatus	pNZ12	2.4×10^5	(15)
L. fermentum	pNZ17	4.8×10^4	(14)
L. casei	pNZ17	3.7×10^6	(14)
L. delbrueckii	pLEM415	10^3 to 10^4	(13, 118)
B. breve	pRM2	1.3×10^4	(20)
B. longum	pDG7	7×10^4	(20)
B. infantis	pNC7	1.2×10^5	(21)

here to consider these genetic systems that are not food grade purely for
proof of concept when attempting to manipulate the probiotic genome.

6.2.1   Transformation

An essential element for the controlled improvement of bacteria and investigation into the genetic linkage of important commercial traits is the ability
to introduce DNA into the cells by transformation using methods that are
convenient and reliable (11). Essentially, transformation involves the modification of the genotype of a cell by introducing DNA from another source,
via the uptake of free DNA. Electroporation is a transformation technique
whereby host cells are concentrated and mixed with the DNA to be transformed. An electric voltage is passed through the host cells causing the
pores in the cell membrane to transiently open, and DNA can then be taken
up. The first *Lactobacillus* strain to be efficiently and reproducibly transformed was a strain of *L. casei*, by Chassy and Flickinger (12). Since then,
many other strains of lactobacilli have been successfully transformed (Table
6.1) with different electroporation protocols, which due to the heterogeneity of the group need to be optimized for each species and within the
species for each strain. Many of these procedures are based on weakening
the cell wall with either glycine, sucrose, or magnesium, or polyethylene
glycol (PEG) in the electroporation buffer (13). In general, electroporation
frequencies vary from 10^2 to 10^6 transformants/µg DNA. Recently, two
studies have examined the optimization of this for lactobacilli, and parameters that have been found to influence the efficiency of transformation
include strains used, vectors, and buffers (14, 15). Treatment of recipient
lactobacilli with lysozyme, glycine, or penicillin improved electrotransformation efficiencies up to 480-fold (14). A postelectroporation recovery time
of 2 to 3 h proved to be a critical step in achieving efficient and reproducible
electrotransformation of clinical lactobacilli with the plasmids pSA3 and

pNZ17. While pSA3 transformants also benefited from the use of subinhibitory concentrations of antibiotics in the selective plates, good transformation efficiencies with pNZ17 could be achieved when plated directly onto selective concentrations of chloramphenicol. Also, significant differences in electrotransformation efficiencies were reported between guinea pig vaginal *Lactobacillus* isolates (maximum of 4.8×10^4 transformants/µg pNZ17 DNA) and the human *L. casei* strain ATCC 393 (3.7×10^6 transformants/µg pNZ17 DNA).

Plasmids are autonomous, self-replicating, extrachromosomal pieces of circular DNA that can be used to introduce foreign DNA into bacterial cells. The plasmid vectors most widely used for lactobacilli are reported to be of three types (16). The first of these are plasmids based on rolling circle replication (RCR) replicons; these vectors have two origins of replication, one for *Esherichia coli* and Gram-positive bacteria, or alternatively these *Lactobacillus* vectors will have a second replication origin for Gram-negative bacteria. Many of these vectors are those initially constructed for *L. lactis* such as the many derivatives of pWV01 (17), e.g., pGK12, which can successfully replicate in lactococci, *E. coli*, *L. Bacillus*, *subtilis*, and a number of lactobacilli and carries erythromycin and chloramphenicol resistance markers. A second class of vectors are those of the pSA3 type, which have two origins, one of which is functional in *E. coli* and a second in Gram-positive bacteria. The third vector type is those which encode selectable markers and replication origins from Gram-negative bacteria on native *Lactobacillus* plasmids, which have *Lactobacillus* RCR replicons and encode erythromycin-chloramphenicol resistance, lacZ or xyl catabolism marker genes, and which can replicate not only in lactobacilli but also in other Gram-positive bacteria, *E. coli* and *Bacillus*. An important consideration when developing cloning and expression systems for lactobacilli is that promoter activity levels can be strain dependent (18), and that replication efficiencies and plasmid copy numbers may differ. In addition, *Lactobacillus* strains show varied preferences for codon usage (19), which could influence the efficiency of translation of a given protein.

Genetic transformation of *Bifidobacterium* strains poses more of a challenge because little is known about the genetics of this genus. A limited number of systems for the efficient and reproducible genetic transformation of bifidobacteria and for broad host transformation into 10 different species have been described in Table 6.1. The system described by Argnani and coworkers (20) relies on preincubation of the bacteria at low temperatures in the electroporation buffer and on the use of plasmid vectors with a replicon from *Actinomycetaceae*. Interestingly, AT-rich plasmids from *Lactococcus* and *Lactobacillus* do not replicate in *Bifidobacterium animalis* and presumably in other bifidobacteria (22). The development of food-grade cloning and expression vectors for the *Bifidobacterium* genus may be improved by the use of *E. coli-Bifidobacteria* shuttle vectors constructed from three fully sequenced bifidobacteria plasmids, pMB1, pCIBb1, and pKJ50 (23). In contrast to lactococci and lactobacilli, genetic research with bifidobacteria has been very limited, with only a few genes sequenced, e.g., *lacZ*, *xfp*, β-*gal*, β-d-glucosidase, bile salt hydrolase (24–29), and a few plasmids, e.g., pMB1 and pKJ50 having been isolated from the genus (30, 31).

6.2.2   Food-Grade Markers

For selection purposes and to be firmly maintained in the host, transformation procedures require selection markers. In order to be food-grade, transferable antibiotic markers may not be used. LAB, which have a long history of safe use, are a typical source of these food-grade markers. Food-grade markers have been grouped into three classes (10) that include the markers based on sugar utilization, auxotrophic markers, and markers that confer resistance or immunity. Food-grade markers that have been successfully used for lactobacilli include the genes for xylose fermentation (32) and lactose fermentation (33, 34), inulin fermentation (35), nisin resistance (36), and lacticin F immunity (37). In the latter case, the genes that confer immunity to lacticin F *(lacF)* do so specifically in lactobacilli, and therefore this system has certain limitations for use as a food-grade marker. Subsequently, the genes involved in immunity to lacticin 3147 *(ltnI)* from the lactococcal plasmid pMRC01 have been identified and heterologously overexpressed when cloned behind a nisin-inducible promoter (38). If adapted for use in strains of *Lactobacillus* and bifidobacteria, this broad host-range system may confer immunity to lacticin 3147, which could then be used to select transformants in the absence of antibiotic selection.

Inducible promoter systems adapted for use in lactobacilli are quite limited in number. The most efficient systems to date include *xylR* promoter from *L. pentosus*, α-amylase from *L. amylovorus*, *p*-coumarate decarboxylase promoter from *L. plantarum*, the powerful nisin-controlled expression (NICE) system in *L. plantarum* (39) and more recently phage φFSW in *L. casei* S1 (40). In this system, the thermoinducible promoter-repressor cassette from *L. casei* phage φFSW was adjoined to promoterless *gus*A reporter gene in *L. casei* and successfully directed *gus*A transcription there.

Interesting, systems recently developed for lactococci that allow food-grade cloning, and secondly food-grade engineering of native plasmids, could prove to be useful resources in the development of vectors suitable for use in probiotic bacteria. The first system designed by Sorensen and coworkers (41) employs an amber suppressor, *supD*, as selectable marker on a food-grade cloning vector pFG200 that consists entirely of *Lactococcus* DNA, with the exception of a small polylinker region. *supD* encodes an altered tRNA^{ser} and suppresses only amber codons, which are stop-codons in approximately 10% of *Lactococcus* genes. Selection for the vector is based on the suppression of pyrimidine auxotrophs generated by introduction of an amber codon in the chromosomal *pyrF* gene. If this efficient food-grade cloning system was applied in probiotic strains, it would allow food-grade genetic modification of these bacteria and overexpression of a host of relevant genes. A novel system, recently developed by Cotter and coworkers (42), that has been originally developed for *Lactococcus*, combines RepA⁺ temperature-sensitive helper plasmid and RepA⁻ cloning vector to create a system that allows functional analysis and modification of native lactococcal plasmids. If applied to novel plasmids of probiotic strains, this system would allow the removal or replacement of undesirable plasmid encoded genes, permit functional analysis of genes, and open the door to creation of a wealth of native probiotic food-grade vectors.

6.2.3 Manipulation of the Probiotic Genome through Homologous Recombination

The integration of foreign genes into the genome constitutes an interesting option for securely maintaining cloned genes without the need for selective markers. This approach has been used successfully in a number of cases including the integrative food-grade vector system successfully designed for *L. acidophilus*, containing β-galactosidase gene (β-*gal*) from *L. bulgaricus*, which was used to combine with the host chromosome, allowing selection and stable expression of β-*gal* (43). The transformant strain demonstrated β-galactosidase activity comparable with that of *Streptococcus thermophilus* and *L. bulgaricus*, strains which have been used for their high -galactosidase activity to alleviate symptoms of lactose intolerance. A second integration system for *L. casei* employs two different types of site-specific recombinases to achieve the stable integration of vector DNA into the host genome. This system was used successfully to make a stable food-grade *L. casei* strain completely immune to phage A2 infection during milk fermentation (44). An integrative expression vector developed for *L. casei* that has great potential for the food and health industries was developed by Gosalbes and coworkers (45). This system allowed the selection of stable mutants capable of expressing foreign genes under the tight control of the *lac* operon promoter. Russell and Klaenhammer (46) have also designed the first efficient and effective integration strategy for thermophilic probiotic lactobacilli such as *L. acidophilus* and *L. gasseri*. This system employs two pWV01-based plasmids, one is the integration plasmid, and the second is a helper plasmid with *RepA*. By a homologous recombination event, this system was successfully used to disrupt *lacL* in *L. acidophilus* and *gusA* in *L. gasseri*. It is expected that this system could be used in a number of Gram-positive bacteria because of the broad host range of pWV01-based plasmids. This system is not, however, suitable for single-plasmid integration experiments due to the moderate stability of the pWV01 replicon above 42°C. A system has recently been described by Neu and Henrich (47) in *L. gasseri* using the new delivery vehicle pTN1 whose replication is efficiently shut down at 42°C. A novel complementation marker that has recently been used in lactobacilli is the *alr* gene from *L. lactis* together with the nisin-regulatory genes in *L. plantarum* (48).

It has recently been reported that IS (insertion sequence) elements and *attP*-integrase systems are the most useful emerging technologies as integration systems for lactobacilli (45). The use of IS elements may, however, be limited due to their tendency for additional transposition events, in comparison with phage-based integration systems that are stable, occur as single copies, and are capable of integrating large fragments of DNA at specific, nonessential sites on the host chromosome.

An interesting approach toward manipulating the genetic material of bacteria is the use of classical chemical mutagenesis, which was employed by Ibrahim and O'Sullivan (49) in a study aimed at overproducing β-galactosidase from probiotic bacteria. Chemical mutagenesis can be considered food grade because it does not involve the introduction of heterologous DNA or the manipulation of existing DNA by recombinant means (49). Indeed β-gal production in this study was successfully increased in *L. delbrueckeii* and *B. longum* by 137% and 222%, respectively.

6.3   Genomic Dissection of Probiotics

The establishment of a large bank of microbial genome sequences will provide the inspiration to guide physiological studies, mutant selection, and the use of genetic and protein engineering for probiotic strain development. The accumulation of this sequence data is increasing at a rapid rate due to improvements in large-scale sequencing techniques. Indeed, entire bacterial genome sequences can now be determined in days (50). The genomic sequence of a number of lactobacilli have already been completed or are in the course of completion. These include strains isolated from the human gastrointestinal tract (GIT) such as *L. plantarum*, WCFS (see Reference 51), *L. johnsonii*, Nestle (see Reference 52), *L. acidophilus* (see http://www. calpoly.edu/~ rcano/ Eric_LAB7), *L. gasseri,* and *L. casei* (see http://www. jgi.doe.gov/). Species of the genus *Bifidobacterium* isolated from the human GIT that are currently undergoing sequencing include *B. breve* (see Reference 52) and two strains of *B. longum* (see Reference 52). For a comprehensive review on bifidobacteria and *Lactobacillus* sequencing projects, the reader is referred to Klaenhammer et al. (52) and Renault (53). The rapidly emerging genome sequence information is invaluable to the future course of genetic engineering, laying the foundations for acquiring and analyzing biological data by transcriptome, proteome, and metabolome analyses, as well as structural genomics, which together constitute the era of functional genomics. A possible concern is the discovery of unfavorable genes and metabolic capabilities that are present on mobile genetic elements, (i.e., antibiotic resistance genes, virulence genes), which raises the possibility that it should be a requirement that all microorganisms used in food and feeds be sequenced (54). A quick and simple screening technique designed for genetic testing for antimicrobial resistance genes

TABLE 6.2

Examples of Studies Showing the Genetic Engineering of Lactobacilli and Bifidobacteria

Organism	Genetic Manipulation	Main Objective	Ref.
L. plantarum	Express choA	Aid cholesterol breakdown in food	(59)
L. plantarum	Express model antigen M6-gp41E	Vaginal immunization for HIV	(119)
L. plantarum	Express cholera toxin B	Protection against cholera toxin B	(120)
L. casei	Express E7 of HPV 16	Vaginal immunization for papillomavirus	(121)
L. casei	Express hMBP	Prevention of infectious diseases or treatment of autoimmune disease	(67)
L. johnsonii	Ldh mutant	Produce a safer strain	(57)
L. plantarum	Express Frag C	Protection against tetanus toxin	(69)
L. plantarum	Express green fluorescent protein	Aid tracking in the GIT	(86)
B. adolescentis	Express endostatin	Inhibit tumor angiogenesis	(64)

has already begun on *Campylobacter* and *Staphylococcus,* as reported by Westin and coworkers (55).

6.4 Genetic Engineering for Design of Improved Probiotic Cultures

6.4.1 Accentuate Functionality

The improvement and/or modulation of health-promoting properties of probiotics through genetic manipulation (Table 6.2 for review) is still at an early stage of investigation. This is partly to do with the fact that molecular mechanisms for health promotion are not well understood. An exception to this is the ability of certain probiotic strains to alleviate the symptoms of lactose intolerance in individuals who cannot tolerate dietary lactose. This condition is caused by a deficiency of β-gal in the small intestine, an enzyme which hydrolyzes the main carbohydrate, lactose, in milk, to form glucose and galactose. The consumption of live bifidobacteria and lactobacilli improves lactose digestion and alleviates the symptoms of lactose intolerance (56). Chemical mutagenesis, which is considered food grade (49), has successfully been used to generate β-gal overproducing mutants of bifidobacteria and lactobacilli, thereby improving the potential of these probiotic cultures for treating lactose intolerance in humans.

One of the most well-known examples of a probiotic strain that has been genetically engineered for human consumption is the *L. johnsonii* La1 mutant (57). When cultured in milk, the native *L. johnsonii* La1 strain ferments lactose to L- and D-lactate, the latter being potentially harmful because it is linked with D-lactate acidosis and encephalopathy in patients with short bowel syndrome or intestinal failures (58). The genetically engineered *L. johnsonii* La1 mutant, however, has a single *in vitro* deletion in the *ldhD* gene, which effectively results in the production of L-lactate from pyruvate instead of the potentially harmful D-lactate. Another example of where the health-promoting properties of a probiotic strain have been enhanced is the expression of the heterologous gene product cholesterol oxidase (*choA*) from *Streptomyces*, in probiotic *L. plantarum* (59). Cholesterol oxidase is the first enzyme in the degradation of cholesterol to hydrogen peroxide and ketosteroids. This expansion of the metabolic activity of the probiotic strain potentially leads the way to the generation of probiotic bacteria capable of changing the composition of dairy foods, such as lowering their cholesterol content (60).

A major hurdle in gene therapy for cancer is the specific delivery of anticancer gene products to the solid tumor (60). Bifidobacteria have been found to germinate and proliferate in solid tumors (61), which has led to the construction of genetically engineered bifidobacteria useful as specific gene-delivery vectors for cancer gene therapy (62). In a study by Li and coworkers (62), *B. adolescentis* was used to transport endostatin gene to solid tumors in tumor-bearing mice. Endostatin has been shown to be an effective inhibitor of tumor angiogenesis and growth (63), and when delivered to the site of local tumors by the *B. adolescentis* vector, tumor growth was inhibited (64).

The need for improved methods for oral delivery of antigen at low cost together with concerns over the safety of attenuated live delivery systems have led to an interest in the use of LAB as vaccine-delivery vehicles. Lactobacilli of human origin can potentially colonize or at least survive transit to the gut mucosa, which makes them extremely attractive as vaccine vectors for the delivery of heterologous antigens (Figure 6.2) (64), and similarly streptoccocci are naturally found to colonize the oral cavity. The use of live LAB vectors such as *Lactobacillus* and *Streptococcus* to deliver antigens including fragment C of tetanus toxin model antigen M6-gp41E of HIV gp41 protein and E7 protein of HPV type 16, at the mucosal surface are just a few examples from the number of potential LAB that are currently being engineered for therapeutic purposes (53, 66). Interestingly, parenteral immunization with a recombinant *L. casei* strain that can secrete human myelin basic protein (hMBP) has been proposed for use in oral tolerance induction for patients with multiple sclerosis (67). In this study, a series of *L. casei* expression vectors were constructed to secrete hMBP or hMBP as a fusion protein with *E. coli* β-glucuronidase. The successful expression and secretion of these heterologous proteins by *L. casei* and their ability to induce peripheral T-cell tolerance makes them ideal candidates for the development of delivery vehicles for the prevention of infectious diseases or treatment of autoimmune diseases.

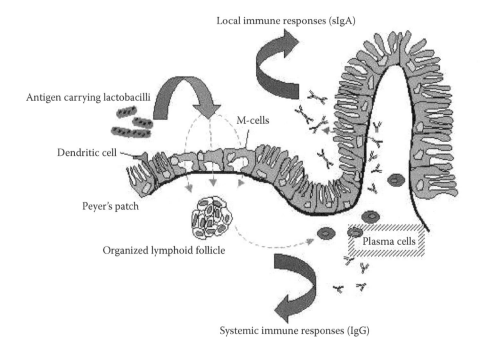

FIGURE 6.2
Lactobacilli as vaccine delivery vehicles. Lactobacilli are engineered to express an antigen and present it at the gastrointestinal tract where it is detected by cells of the mucosal immune system. This is done directly by dendritic cells or by M-cells of the Peyer's patches, which then transport the cells to antigen-presenting cells (APCs). APCs, depending on the signal, will activate CD4[+] T-cells, resulting in the stimulation of plasma cells to produce antigen-specific immunoglobulins. (Adapted from Reference 68.)

For the delivery of therapeutic molecules, it is essential that dose-response curves be established for the targeted biological effect (39). This may be particularly important for lactobacilli, because in comparison with other LAB vectors used as vaccine delivery vehicles, these may not elicit an immune response (68). However, in a recent comparative study by Grangette and coworkers (69), of the immunogenicity of recombinant *L. plantarum* and *L. lactis* producing equivalent amounts of the tetanus toxin fragment C (TTFC), the *Lactobacillus* strain used, i.e., *L. plantarum* NCIMB8826, was successful in eliciting the strongest ELISA TTFC-specific and protective humoral response. This would suggest that for future vaccine development, the ability of the bacterial vector to persist in the GIT will impact on the immunogenicity and the level of protection induced. An essential factor that may limit the use of lactobacilli in future vaccine vector development is the number of gaps in a comprehensive understanding of the molecular genetics underlying probiotic effects in this species (66). Thus, the potential for promoting health using well-characterized recombinant LAB such as lactococci holds much appeal (will be discussed later).

6.4.2 Improved Technological Suitability

Fermented dairy foods, including milk and yogurt, are among the most accepted food carriers for delivery of viable probiotic cultures to the human GIT. It has been recommended that foods containing such bacteria (i.e., probiotic functional foods) should contain at least 10^7 live microorganisms per gram or per milliliter (70) at the time of consumption, in order to benefit the consumer. Technological challenges associated with the introduction and maintenance of such high numbers of probiotic cultures in foods include the form of the probiotic inoculant, ability of the probiotic culture to retain viability in the environment of the food matrix, and maintenance of probiotic characteristics in the food product through to the time of consumption (71). Spray-dried and freeze-dried cultures are useful means of introducing probiotic culture into these food systems. However, the use of such approaches in preparing cultures may impair viability and probiotic functionality (71). Indeed, during the production of probiotic functional foods, substantial loss of culture viability may be experienced. Culture viability may be reduced in all steps from producer to site of action in the host mainly as a result of structural and physiological injury to the bacterial cell. Survival of probiotic bacteria through the contrasting and changing environments of food and GIT and the ability to react to minor fluctuations in either of these environments is controlled by the stress-responsive elements of the genome. Inherent resistance to stress obviously differs among strains and species of probiotic bacteria. For example, survival of *L. acidophilus* and *B. bifidum* varied in commercial yogurt preparations over a 5-week period, and the rate of decline due to acid stress was more rapid for bifidobacteria strains compared with lactobacilli during storage (72). Similarly, screening of a wide range of LAB species revealed that a larger number of bifdobacteria were inherently more sensitive to bile stress than lactobacilli (73). Indeed, lactobacilli also demonstrate strain-to-strain variation in resistance to stress (74). Two probiotic *Lactobacillus* strains, *L. paracasei* NFBC338 and *L. salivarius* UCC118, dried under identical conditions, demonstrated a 100-fold difference in survival during drying, with *L. paracasei* NFBC338 proving to be the more technologically robust strain (74). This study also revealed that *L. paracasei* maintained constant viability, whereas *L. salivarius* declined by approximately 1 log during 2 months of powder storage.

A well-characterized phenomenon among LAB is that sublethal stress can lead to an elevated state of resistance (75–80), which in the case of probiotic bacteria may be exploited to determine preconditioning treatments that can improve survival during food processing, storage, and GI transit (78). The use of genomics and proteomics has allowed the identification of the genes involved in *Lactobacillus* and bifidobacterial stress responses such as *F1F0-ATPase* (acid stress) (81), *groESL* and *dnaK* (heat stress) (80), or *Osp* (oxygen stress) (83). The regulatory mechanisms, however, involved in the adaptive response of LAB remain poorly understood (84). The mechanisms responsible for improved survival of heat- and salt-adapted probiotic *L. rhamnosus*

HN001 during storage in dried form were examined using two-dimensional gel electrophoresis, which revealed that the level of expression of classical heat shock proteins GroEL and DnaK were upregulated in the stressed cultures when compared with a nonstressed control (80). Interestingly, this study also revealed changes in the synthesis of some glycolytic enzymes, triose phosphate isomerase, glyceraldehyde-3-phosphate dehydrogenase, phosphoglycerate kinase, and enolase, raising the possibility that these enzymes are subject to the same regulatory mechanisms as those of the heat and salt stress responses in lactobacilli. This insight into the proteomics responsible for increased resistance to stress could in the future be applied to the generation of more technologically robust probiotic strains, by over-expressing the protective genes using food-grade molecular tools.

In cheddar cheese-making, the proteolytic activities of the various cheese flora determine the rate of ripening of the cheese and the flavor in the final product. In a study by Christensson et al, (85), attempts were made to overexpress a proteolytic enzyme PepO from the probiotic nonstarter strain *L. rhamnosus* HN001. Overexpression of the enzyme was very low from its own promoter, but with a more efficient system, the active PepO from *L. rhamnosus* HN001 could benefit cheese-making technology by controlling cheese bitterness.

As a means of tracking and monitoring the physiology of bacteria in different environments, including the GIT, studies have been successfully undertaken to fuse the green fluorescent protein (GFP) reporter to the *nis*A promoter, which is then controlled by extracellular nisin via signal transduction (86). When introduced into *L. plantarum*, this system allows the successful tracking of the strain both *in vitro* and *in vivo*. The food-grade generation of silent mutations in the pepX gene of *L. helveticus* CNRZ32 was proposed as an effective genetic labeling technique for detection of probiotic lactobacilli (87).

6.4.3 Modification of Nonprobiotic Strains toward Improving Health

To date, much of the research in engineering strains to have health benefits has been directed at the development of new vaccination techniques and at therapies to rectify enzyme malfunctions in the digestive tract using innocuous bacteria such as *L. lactis,* which does not colonize the digestive tract of man or animals (88, 89). The first report of an *L. lactis* strain being used as antigen carrier was by Iwaki and coworkers in BALB/c mice (90), when the recombinant strain expressed a *Streptococcus mutans* surface protein antigen and successfully stimulated antigen-specific IgA and IgG immune responses following three consecutive doses and one booster dose of 10^9 formalin-killed cells. Since then, recombinant *L. lactis* cells have been modeled to express a range of proteins, such as tetanus toxin fragment C, which is capable of protecting mice from lethal challenge with tetanus toxin (91). In more recent studies, *L. lactis* cells have also been modeled to express an *L7/L12* protein to protect against *Brucella abortis* infection (92) and urease subunit B antigen to protect against *Helicobacter pylori* infection (93).

The coexpression of stimulatory molecules as a means of further directing the immune response is an interesting development being made with live vaccine vectors. In particular, this approach has been successful in the treatment of irritable bowel disease (IBD) and Crohn's disease using IL-10-secreting *L. lactis* (94). Fourteen daily intragastric doses of *L. lactis* (LL-mIL10) secreting biologically active murine IL-10 (mIL-10) at 2×10^7 and control *L. lactis* were administered to mice in which chronic colitis had been induced using dextran sodium sulfate (DSS). This study showed that mice treated with *LL-mIL10* demonstrated close to 50% decrease in pathological symptoms, and 4 of 10 mice in the *LL-mIL10*-treated group had the histological characteristics of healthy mice. Adenomas in mice treated with DSS are common; however, in the mice treated with *LL-mIL10,* no adenomas were evident. Furthermore, mIL-10 was also being synthesized *in situ* by *LL-mIL-10* in the colon. Although the *LL-mIL10* strain was detected in other parts of the GIT, mIL-10 was not detectable there, which may be due to protein degradation in these areas (94). When compared with the performance of standard antiinflammatory methods, the *LL-mIL10* model worked equally well; however, a much lower dose of mIL10 from *LL-mIL10* was required to demonstrate the same effect. This is most likely due to the rapid delivery of mIL-10 at its therapeutic target by *LL-mIL10* cells producing the protein in the lumen where it can diffuse to the target site, or alternatively the lactococci may be taken up by M cells and secrete mIL-10 into the intestinal lymphoid tissue. This *LL-mIL10* model system where the therapeutic agent is synthesized by food-grade bacteria is a major breakthrough in the development of localized delivery vehicles and has been approved for use in human clinical trials.

An exciting new development in the use of LAB for vaccination has been that the repeat domain in a major muramidase in *L. lactis*, Acm-A, can bind to bacterial cells when added to these cells from the outside (95). Foreign antigens can thus be presented on the surface of *L. lactis* in order to elicit a mucosal immunogenic response by connecting this repeat domain to other proteins by genetic engineering. This system has been effectively used to express the MSA2 protein of *Plasmodium falciparum* by nasal and intragastric immunization (96). Different approaches such as engineering leaky lactococcal strains are currently being designed to increase the efficiency and safety of LAB vaccine vectors (97). However, to date, examples of detailed immunization studies are limited, with most being conducted in mice, and a paper has reported immunization in monkeys (98). The development of effective LAB vaccines for use in humans will depend on elucidating a full understanding of the mucosal immune system and a thorough assessment of the efficacy and safety of these live vectors (66). In this respect, improvement of local immunization, efficiency, analysis of *in vivo* antigen production, unraveling of the *Lactobacillus* colonization mechanisms, and the development of biologically contained strains are just some of the future areas of research that need to be addressed (97).

LAB can also be used to deliver enzymes of medical importance, such as lactase (99) and lipase (100). The potential use of LAB to supplement the human supply of lactase in the gut in cases of deficiency has been studied extensively (99), and, as mentioned earlier, has proven successful with the use of bifidobacteria and lactobacilli. An excellent example of where LAB can be used to deliver enzymes is the use of genetically modified lactococci as a vector to deliver lipase cloned from *Staphylococcus hyicus* to the GIT (100). This system aided lipid digestion in an animal model of pancreatic insufficiency and may be used in the treatment of this condition in humans (100). Research on the development of LAB as cell factories for production of biologically active compounds and nutrients such as vitamins and low-calorie sugars is an emerging area of importance in the development of foods or food components that have a claimed health-medical benefit (101).

In the future, a major factor in the development of bacteria that promote health in the host will undoubtedly be the improvement of genetic tools. In particular, the systems employed for the secretion and anchoring of proteins and peptides are often inefficient, resulting in inadequate levels of antigen being secreted and exposed on the vector surface. However, the design of new protein-targeting systems for LAB are improving. For example, a study by Dieye and coworkers (102) designed an expression export system that enabled the targeting of a reporter protein to specific locations in *L. lactis* by modifying the signals at the N- and C-termini of the reporter protein.

6.5 Functional Genomics and Proteomics

The era of functional genomics has revolutionized the rapid development in our understanding of the molecular and adaptive responses in bacterial cells. For probiotic bacteria, this critical research involves the correlation of genotypes and phenotypes that impact probiotic functionality. The availability of probiotic genome sequences will greatly accelerate the success of functional genomics technologies such as microarrays and proteomics. DNA microarrays operate on the convention of DNA hybridization and can be used to quantitate levels of gene expression as well as to detect gene mutations. For DNA microarrays, gene probes are spotted onto glass slides by a high precision robot, followed by hybridization to sample material containing either fluorescently labeled reverse transcribed mRNA (for gene expression analysis) or genomic DNA (for mutation analysis). The fluorescent signal is then read by a laser scanner, and the extracted data may then be analyzed using computer tools. Microarray analysis offers the possibility of assessing coordinated gene expression across the varying conditions to which probiotic bacteria are exposed, and arrays are also being developed to examine changes in the composition of the gut flora (103).

The proteome describes the entire protein complement of a cell, and the term refers to functional genomics at the protein level. There are two facets to proteomics research, the first, i.e., expression proteomics, is the study of global changes in protein expression; and the second, i.e., cell map proteomics, involves the study of protein-protein interactions. O'Farrell (104) and Klose (105) originally described a two-dimensional gel electrophoresis technique that could separate proteins on the basis of isoelectric point and molecular weight. With advances in matrix-assisted laser desorption and ionization time of flight (MALDI-TOF) mass spectroscopy, and the parallel growth in protein and expressed sequence tag (EST) databases for rapid identification of proteins (106), the combination of these two techniques is a powerful approach. To date, a few proteomic maps have been completed for LAB; however, much of the research is biased toward the study of specific functions or situations encountered during processing (107). Future proteomic research for LAB, and probiotic bacteria in particular, is certain to be applied to biomedical studies, i.e., strain identification, mechanisms of action, and drug resistance.

Bioinformatic systems to date have allowed for cluster analysis of transcriptome data (i.e., all RNAs synthesized in the cell) to provide cellular process information and also to assign function to unknown genes (50). The integrative E-Cell model building of "minimal" cells to predict their dynamics and allowing "*in silico*" experimenting can readily be used for modeling of LAB. This software package enables modeling of metabolic pathways and other higher-order cellular processes such as protein synthesis and signal transduction, and this system can continuously be updated as more genomic information becomes available. A second valuable program is the "Virtual cell" that can be used to simulate cellular processes. For this system, the information is provided as biochemical and electrophysiological data of the cell. This era of functional genomics, composed of microarrays, proteomics, and advances in bioinformatics, will set the course for genome mining and functional analysis of biological systems by readouts of tens of thousands of genes and their products simultaneously (50).

6.6 Safety Legislation and Risk Assessment

It is crucial that probiotic cultures are safe for human use; this is especially important in the case of immunocompromised hosts. The majority of microorganisms used as probiotics are LAB, which have GRAS (generally recognized as safe) status. A LABIP workshop organized to discuss the safety of LAB recognized that the major concern regarding their use of Lactic Acid Bacteria Industrial Platform as probiotics is their intrinsic resistance to antibiotics (108). This is particularly a problem when considering the use of *Enterococci,* which are inherently vancomycin resistant (108). Lactobacilli and bifidobacteria, on the other hand, generally do not carry resistances that are transmissible, and in many cases are sensitive to clinically used antibiotics (108). Many probiotic cultures have undergone safety testing. This is particularly true in the case of *Lactobacillus GG,* which has a considerable bank of data to demonstrate its safety (109). Genetically modified probiotic bacteria must undergo even more rigorous safety assesment prior to use for human consumption. Genetically modified organisms (GMOs) are currently defined as organisms that contain genetic material that has been altered in a manner that does not occur by natural mutation, mating, or recombination event (53). The introduction of GMOs to the food chain is regulated in the U.S. by the Food and Drug Administration (FDA) and in Europe by the Novel Foods Regulation (258/97/EC). The FDA defines a food additive as any substance "not generally recognized as safe" by qualified experts. Most LAB used in foods are perceived as having GRAS status, due to a long history of safe use, thus food producers do not need FDA approval prior to marketing of LAB-containing products. In the case of probiotics, the U.S. Dietary Supplement Health and Education Act (DSHEA) (1994) states that a new ingredient in dietary supplement does not need FDA approval prior to release on the market, but the FDA must be notified at least 75 days prior to the intended market date (110). In Europe, the Novel Foods Regulation defines a novel food or food ingredient as "food and food ingredients which have not hitherto been used for human consumption to a significant degree within the community," making this regulation applicable to newly discovered strains or GMOs. Where microorganisms are used as vehicles for production, both the safety of the microorganism and final product must be determined. Because many LAB contain plasmid or transposon-encoded antibiotic resistance genes, a critical factor that must be determined is the stability of the novel microorganism, in particular the ability of the microorganism to transfer genetic material to other microorganisms within the host (110). On a global basis, these regulations determine that for GMOs to reach the marketplace, their novel genetic combinations should be selected, tightly maintained, and expressed using food-grade systems that are safe, stable, and sustainable (111).

The purpose of risk assessment for the deliberate release of GMOs is to identify and evaluate the risks associated with their release. This is controlled in the U.S. by the Environmental Protection Agency (EPA) under the Toxic Substances Control Act Biotechnology Program (112). Other organizations involved in assessing the risk of releasing GMOs include the European Commission and BATS (the Biosafety Research and the Assessment of Biotechnology Impacts of the Swiss Priority Program Biotechnology), and EFB (the Working Party on Safety in Biotechnology of the European Federation Biotechnology) (112). In the EU, Directive 90/220/EEC is the principal legislation governing experimental and commercial release of GMOs (53); this was updated in October 2002. Self-cloning (10) is an exception to the EU Directive on the contained use of GMOs (CEC219, 1990). By adhering to the terms of these organizations, it is possible to achieve approval for the use of GMOs in the EU (100). In Europe, the time and cost associated with acquiring legal approval and a subsequent market for GMOs will possibly dictate that there will be a limited number of these products. It is more likely that major markets for these types of products will develop in the U.S. and the rest of the world (54). With the advent of better documentation of the probiotic effect and clearly described health benefits, it is possible that European consumers will accept GMO-probiotics.

6.7   Conclusion and Future Developments

Molecular biology has provided the means to manipulate probiotic bacteria for both novel and specialized tasks, either in food or in the intestinal tract of the host. Genetic engineering of probiotic strains may, in the future, be directed at meeting general consumer demands such as reduction of heart disease, cancer, osteoporosis, high blood pressure, stress, tooth decay, and obesity, and products that provide energy and enhance athletic performance (113). More specific requirements of future probiotic strains, such as production of riboflavin, folate, B-vitamins, and low-calorie sugars, will also need to be addressed. Food-grade bacteria are already being engineered to produce medicinal enzymes such as cholesterol oxidase, lipase, and β-galactosidase, and this may be expanded to bioconversion capabilities for improving the nutritional quality of foods. For example, engineered *Lactobacillus* strains may have the capacity of some bifidobacteria isolates (114) to convert linoleic acid to the beneficial fatty acid *cis*-9, *trans*-11 conjugated linoleic acid (CLA). Targets for microbiologists engineering probiotic bacteria in the future are likely to be centered around metabolic engineering, linking cultures with protective prebioitics, colonization, delivery of digestive enzymes, and vaccine developments (115). No doubt the delicate balance between advances in molecular biology and bioinformatics, safety risk and assessment, market demands and consumer opinion, and finally economics will determine the future success of genetically engineered probiotic bacteria.

Acknowledgments

This work was funded by the Irish Government under the National Development Plan 2000-2006, the European Research and Development Fund, and by EU Project QLK1-CT-2000-30042. C. Desmond is in receipt of a Teagasc Walsh Fellowship.

References

1. FAO/WHO. Evaluation of Health and Nutritional Properties of Powder Milk with Live Lactic Acid Bacteria. Proceedings of FAO/WHO Expert Consultation, Cordoba, Argentina, 2001.
2. AC Huggett, B Schilter. Research needs for establishing the safety of functional foods. *Nutr. Rev.*, 54: S143–S148, 1996.
3. C Stanton, C Desmond, G Fitzgerald, R Ross. Probiotics-health benefits — reality or myth. *Aust. J. Dairy Tech.*, 58: 107–113, 2003.
4. E Isolauri, M Kaila, H Mykkanen, WH Ling, S Salminen. Oral bacteriotherapy for viral gastroenteritis. *Dig. Dis. Sci.*, 39: 2595–2600, 1994.
5. M Kalliomaki, S Salminen, H Arvilommi, P Kero, P Koskinen, E Isolauri. Probiotics in primary prevention of atopic disease: a randomised placebo-controlled trial. *Lancet*, April: 1076–1079, 2001.
6. P Gionchetti, F Rizzello, U Helwig, A Venturi, KM Lammers, P Brigidi, B Vitali, G Poggioli, M Miglioli, M Campieri. Prophylaxis of pouchitis onset with pro-biotic therapy: A double-blind, placebo-controlled trial. *Gastroenterology*, 124: 1202–1209, 2003.
7. M Cross. Immunoregulation by probiotic lactobacilli: pro-Th1 signals and their relevance to human health. *Clin. Appl. Immunol.*, 3: 115–125, 2002.
8. WP Hammes, H Hertel. Research approaches for pre- and probiotics: challenges and outlook. *Food Res. Int.*, 35: 165–170, 2002.
9. A Bolotin, P Wincker, S Mauger, O Jaillon, K Malarme, J Weisenbach, SD Ehrlich, A Sorokin. The complete genome sequence of the lactic acid bacterium *Lactococcus lactis* ssp *lactis* IL1403. *Genome Res.*, 11: 731–753, 2001.
10. W de Vos. Safe and sustainable systems for food-grade fermentations by ge-netically modified lactic acid bacteria. *Int. Dairy J.*, 9: 3–10, 1999.
11. J Sambrook, D Russell. *Molecular Cloning. A Laboratory Manual*, 3rd ed, Cold Spring Harbor Laboratory Press, Cold Spring Harbor, New York, 2001.
12. BM Chassy, J Flickinger. Transformation of *Lactobacillus casei* by electroporation. *FEMS Microbiol. Lett.*, 44: 173–177, 1987.
13. TW Aukrust, MB Brurberg, IF Nes. Transformation of *Lactobacillus* by electropo-ration. *Methods Mol. Biol.*, 47: 201–208, 1995.
14. MQ Wei, CM Rush, JM Norman, LM Hafner, RJ Epping, P Timms. An improved method for the transformation of *Lactobacillus* strains using electroporation. *J. Microbiol. Methods*, 21: 97–109, 1995.
15. MT Aymerich, M Hugas, M Garriga, RF Vogel, J Monfort. Electroporation of meat lactobacilli. Effect of several parameters on their efficiency. *J. Appl. Bacte-riol.*, 75: 320–325, 1993.
16. MJ Kullen, TR Klaenhammer. Genetic modification of intestinal lactobacilli and bifidobacteria. *Curr. Issues Mol. Biol.*, 2: 41–50, 2000.
17. J Kok, MB Jos, M van der Vossen, G Venema. Construction of plasmid cloning vectors for lactic streptococci which also replicate in *Bacillus subtilis* and *Escher-ichia coli*. *Appl. Environ. Microbiol.*, 48: 726–731, 1984.

18. A McCracken, P Timms. Efficiency of transcription from promoter sequence variants in *Lactobacillus* is both strain and context dependent. *J. Bacteriol.*, 181: 6569–6572, 1999.

19. PH Pouwels, JA Leunissen. Divergence in codon usage of *Lactobacillus* species. Nucleic Acids Res., 22: 929–936, 1994.

20. A Argnani, RJ Leer, N van Luijk, PH Pouwels. A convenient and reproducible method to genetically transform bacteria of the genus *Bifidobacterium*. *Microbiology*, 142 (Pt 1): 109–114, 1996.

21. M Rossi, P Brigidi, D Matteuzzi. An efficient transformation system for *Bifidobacterium* spp. *Lett. Appl. Microbiol.*, 24: 33–36, 1997.

22. V Scardovi. Genus *Bifidobacteriumus*. In: PHA Sneath, NS Mair, ME Sharpe, JG Holt, Eds. *Bergey's Manual of Systematic Bacteriology, vol. 2.* Baltimore, MD: The Williams & Wilkins Co., 1986, 1418–1434.

23. MJ van der Werf, K Venema. Bifidobacteria: genetic modification and the study of their role in the colon. *J. Agric. Food Chem.*, 49: 378–383, 2001.

24. MN Hung, Z Xia, NT Hu, BH Lee. Molecular and biochemical analysis of two beta-galactosidases from *Bifidobacterium infantis* HL96. *Appl. Environ. Microbiol.*, 67: 4256–4263, 2002.

25. N Nunoura, O Kohji, T Keiko, T Hisanori, Y Toshihiro, I Masayuki, Y Hideaki, Y Kenji, K Hidehiko. Cloning and nucleotide sequence of the beta-D-glucosidase gene from *Bifidobacterium breve* clb and expression of beta-D-glucosidase activity in *Escherichia coli*. Biosci. Biotechnol. Biochem., 60: 2011–2018, 1996.

26. M Rossi, L Altomare, VRA Gonzalez, P Brigidi, D Matteuzzi. Nucleotide sequence, expression and transcriptional analysis of the *Bifidobacterium longum* MB 219 lacZ gene. Arch. Microbiol., 174: 74–80, 2000.

27. H Tanaka, H Hashiba, J Kok, I Mierau. Bile salt hydrolase of *Bifidobacterium longum*—biochemical and genetic characterization. *Appl. Environ. Microbiol.*, 66: 2502–2512, 2000.

28. L Meile, LM Rohr, TA Geissmann, M Herensperger, M Teuber. Characterization of the D-xylulose 5-phosphate/D-fructose 6-phosphate phosphoketolase gene (xfp) from *Bifidobacterium lactis*. *J. Bacteriol.*, 183: 2929–2936, 2001.

29. MI Trindade, VR Abratt, S Reid. Induction of sucrose utilization genes from *Bifidobacterium lactis* by sucrose and raffinose. *Appl. Environ. Microbiol.*, 69: 24–32, 2003.

30. MS Park, DW Shin, KH Lee, G Ji. Sequence analysis of plasmid pKJ50 from *Bifidobacterium longum*. *Microbiology*, 145 (Pt 3): 585–592, 1999.

31. M Rossi, P Brigidi, A Gonzalez Vara y Rodriguez, D Matteuzzi. Characterization of the plasmid pMB1 from *Bifidobacterium longum* and its use for shuttle vector construction. Res. *Microbiol.*, 147: 133–143, 1996.

32. M Posno, PT Heuvelmans, MJ van Giezen, BC Lokman, RJ Leer, P Pouwels. Complementation of the inability of *Lactobacillus* strains to utilize D-xylose with D-xylose catabolism-encoding genes of *Lactobacillus pentosus*. *Appl. Environ. Microbiol.*, 57: 2764–2766, 1991.

33. H Hashiba, R Takiguchi, K Jyoho, K Aoyama. Establishment of a host-vector system in *Lactobacillus helveticus* with beta-galactosidase activity as a selection marker. *Biosci. Biotechnol. Biochem.*, 56: 190–194, 1992.

34. TM Takala, PE Saris, S Tynkkynen. Food-grade host/vector expression system for *Lactobacillus casei* based on complementation of plasmid-associated phospho- beta-galactosidase gene lacG. *Appl. Microbiol. Biotechnol.*, 60: 564–570, 2003.

35. E Wanker, RJ Leer, PH Pouwels, H Schwab. Expression of *Bacillus subtilis* levanase gene in *Lactobacillus plantarum and Lactobacillus casei. Appl. Microbiol. Biotechnol.*, 43: 297–303, 1995.

36. TM Takala, P Saris. A food-grade cloning vector for lactic acid bacteria based on the nisin immunity gene nisI. *Appl. Microbiol. Biotechnol.*, 59: 467–471,

37. GE Allison, T Klaenhammer. Functional analysis of the gene encoding immunity to lactacin F, lafI, and its use as a *Lactobacillus*-specific, food-grade genetic marker. *Appl. Environ. Microbiol.*, 62: 4450–4460, 1996.

38. O McAuliffe, C Hill, RP Ross. Identification and overexpression of ltnI, a novel gene which confers immunity to the two-component lantibiotic lacticin 3147. *Microbiology*, 146 (Pt 1): 129–138, 2000.

39. S Pavan, P Hols, J Delcour, MC Geoffroy, C Grangette, M Kleerebezem, A Mercenier. Adaptation of the nisin-controlled expression system in *Lactobacillus plantarum*: a tool to study *in vivo* biological effects. *Appl. Environ. Microbiol.*, 66: 4427–4432, 2000.

40. B Binishofer, I Moll, B Henrich, U Blasi. Inducible promoter-repressor system from the *Lactobacillus casei* phage phiFSW. *Appl. Environ. Microbiol.*, 68: 4132–4135, 2002.

41. KI Sorensen, R Larsen, A Kibenich, MP Junge, E Johansen. A food-grade cloning system for industrial strains of *Lactococcus lactis. Appl. Environ. Microbiol.*, 66: 1253–1258, 2000.

42. PD Cotter, C Hill, R Ross. A food-grade approach for functional analysis and modification of native plasmids in *Lactococcus lactis. Appl. Environ. Microbiol.*, 69: 702–706, 2003.

43. MY Lin, S Harlander, D Savaiano. Construction of an integrative food-grade cloning vector for *Lactobacillus acidophilus. Appl. Microbiol. Biotechnol.*, 45: 484–489, 1996.

44. MC Martin, JC Alonso, JE Suarez, M Alvarez. Generation of food-grade recombinant lactic acid bacterium strains by site-specific recombination. *Appl. Environ. Microbiol.*, 66: 2599–2604, 2000.

45. MJ Gosalbes, CD Esteban, JL Galan, G Perez-Martinez. Integrative food-grade expression system based on the lactose regulon of *Lactobacillus* casei. *Appl. Environ. Microbiol.*, 66: 4822–4828, 2000.

46. WM Russell, T Klaenhammer. Generating site specific integrations in the genomes of thermophilic, probiotic lactobacilli via homologous recombination. Proceedings of the 101st General Meeting of the American Society for Microbiology, Orlando, FL, 2001.

47. T Neu, B Henrich. New thermosensitive delivery vector and its use to enable nisin-controlled gene expression in *Lactobacillus gasseri. Appl. Environ. Microbiol.*, 69: 1377–1382, 2003.

48. PA Bron, MG Benchimol, J Lambert, E Palumbo, M Deghorain, J Delcour, WM De Vos, M Kleerebezem, P Hols. Use of the alr gene as a food-grade selection marker in lactic acid bacteria. *Appl. Environ. Microbiol.*, 68: 5663–5670, 2002.

49. SA Ibrahim, D O'Sullivan. Use of chemical mutagenesis for the isolation of food grade beta-galactosidase overproducing mutants of bifidobacteria, lactobacilli and *Streptococcus thermophilus*. J. Dairy Sci., 83: 923–930, 2000.

50. OP Kuipers. Genomics for food biotechnology: prospects of the use of high-throughput technologies for the improvement of food microorganisms. *Curr. Opinion Biotechnol.*, 10: 511–516, 1999.

51. M Kleerebezem, J Boekhorst, R van Kranenburg, D Molenaar, OP Kuipers, R Leer, R Tarchini, SA Peters, HM Sandbrink, MW Fiers, W Stiekema, RM Lankhorst, PA Bron, SM Hoffer, MN Groot, R Kerkhoven, M de Vries, B Ursing, WM de Vos, RJ Siezen. Complete genome sequence of *Lactobacillus plantarum* WCFS1. *Proc. Natl. Acad. Sci. U. S. A.*, 100: 1990–1995, 2003.

52. T Klaenhammer, E Altermann, F Arigoni, A Bolotin, F Breidt, J Broadbent, R Cano, S Chaillou, J Deutscher, M Gasson, M van de Guchte, J Guzzo, A Hartke, T Hawkins, P Hols, R Hutkins, M Kleerebezem, J Kok, O Kuipers, M Lubbers, E Maguin, L McKay, D Mills, A Nauta, R Overbeek, H Pel, D Pridmore, M Saier, D van Sinderen, A Sorokin, J Steele, D O'Sullivan, W de Vos, B Weimer, M Zagorec, R Siezen. Discovering lactic acid bacteria by genomics. *Antonie Van Leeuwenhoek*, 82: 29–58, 2002.

53. P Renault. Genetically modified lactic acid bacteria: applications to food or health and risk assessment. *Biochimie*, 84: 1073–1087, 2002.

54. EB Hansen. Commercial bacterial starter cultures for fermented foods of the future. *Int. J. Food Microbiol.*, 78: 119–131, 2002.

55. L Westin, C Miller, D Vollmer, D Canter, R Radtkey, M Nerenberg, J O'Connell. Antimicrobial resistance and bacterial identification utilizing a microelectronic chip array. *J. Clin. Microbiol.*, 39: 1097–1104, 2001.

56. T Jiang, A Mustapha, D Savaiano. Improvement of lactose digestion in humans by ingestion of unfermented milk containing *Bifidobacterium longum*. *J. Dairy Sci.*, 79: 750–757, 1996.

57. B Mollet. Genetically improved starter strains: opportunities for the dairy industry. *Int. Dairy J.*, 9: 11–15, 1999.

58. G Bongaerts, J Tolboom, T Naber, J Bakkeren, R Severijnen, H Willems. D-lactic acidemia and aciduria in pediatric and adult patients with short bowel syndrome. *Clin. Chem.*, 41: 107–110, 1995.

59. P Kiatpapan, M Yamashita, N Kawaraichi, T Yasuda, Y Murooka. Heterologous expression of a gene encoding cholesterol oxidase in probiotic strains of *Lactobacillus plantarum and Propionibacterium freudenreichii* under the control of native promoters. *J. Biosci. Bioeng.*, 92: 459–465, 2001.

60. DI Pereria, GR Gibson. Effect of consumption of probiotics and predotics on seum lipid levels in humans. *Crit. Rev. Brochem. Mol. Biol.* 37:259–281, 2002.

61. K Yazawa, M Fujimori, J Amano, Y Kano, S Taniguchi. *Bifidobacterium longum* as a delivery system for cancer gene therapy: selective localization and growth in hypoxic tumors. *Cancer Gene Ther.*, 7: 269–274, 2000.

62. X Li, GF Fu, YR Fan, WH Liu, XJ Liu, JJ Wang, G Xu. Bifidobacterium adolescentis as a delivery system of endostatin for cancer gene therapy: selective inhibitor of angiogenesis and hypoxic tumor growth. *Cancer Gene Ther.*, 10: 105–111, 2003.

63. MS O'Reilly, T Boehm, Y Shing, N Fukai, G Vasios, WS Lane, E Flynn, JR Birkhead, BR Olsen, J Folkman. Endostatin: an endogenous inhibitor of angiogenesis and tumor growth. *Cell*, 88:277–285, 1997.

64. N Ibnar, Zerki, S Blum, EJ Schifrin, T von der Weid, Divergent patterns of colonization and immune response elected from two intestinal *Lactobacillus* strains that display similar preperties *in vitro*. *Infect. immun.* 71:428–436, 2003.

65. PH Pouwels, RJ Leer, M Shaw, MJ Heijne den Bak-Glashouwer, TD Tielen, E Smit, B Martinez, J Jore, PL Conway. Lactic acid bacteria as antigen delivery vehicles for oral immunization purposes. *Int. J. Food Microbiol.*, 41: 155–167, 1998.

66. G Perdigon, R Fuller, R Raya. Lactic acid bacteria and their effect on the immune system. *Curr. Issues Intest. Microbiol.*, 2: 27–42, 2001.
67. CB Maassen, JD Laman, MJ den Bak-Glashouwer, FJ Tielen, JC van Holten-Neelen, L Hoogteijling, C Antonissen, RJ Leer, PH Pouwels, WJ Boersma, DM Shaw. Instruments for oral disease-intervention strategies: recombinant *Lactobacillus casei* expressing tetanus toxin fragment C for vaccination or myelin proteins for oral tolerance induction in multiple sclerosis. *Vaccine*, 17: 2117–2128, 1999.
68. JF Seegers. Lactobacilli as live vaccine delivery vectors: progress and prospects. *Trends Biotechnol.*, 20: 508–515, 2002.
69. C Grangette, H Muller-Alouf, MGeoffroy, D Goudercourt, M Turneer, A Mercenier. Protection against tetanus toxin after intragastric administration of two recombinant lactic acid bacteria: impact of strain viability and *in vivo* persistence. Vaccine, 20:3304–3309, 2002.
70. N Ishibashi, S Shimamura. Bifidobacteria: research and development in Japan. Food Technol., 46: 126 – 135, 1993.
71. C Stanton, C Desmond, M Coakley, JK Collins, G Fitzgerald, RP Ross. Challenges facing development of probiotic-containing functional foods. In: Farnworth E., Ed., *Challenges Facing Development of Probiotic-Containing Functional Foods*, CRC Press, Boca Raton, Florida, 2003, Chap 11.
72. NP Shah, WEV Lankaputhra, ML Britz, W Kyle. Survival of *Lactobacillus acidophilus* and *Bifidobacterium bifidum* in commercial yoghurt during refrigerated storage. *Int. Dairy J.*, 5: 515–521, 1995.
73. G Kociubinski, P Perez, G De Antoni. Screening of bile resistance and bile precipitation in lactic acid bacteria and bifidobacteria. *J. Food Prot.*, 62: 905–912, 1999.
74. GE Gardiner, E O'Sullivan, J Kelly, MA Auty, GF Fitzgerald, JK Collins, RP Ross, C Stanton. Comparative survival rates of human-derived probiotic *Lactobacillus paracasei* and *L. salivarius* strains during heat treatment and spray drying. *Appl. Environ. Microbiol.*, 66: 2605–2612, 2000.
75. NP Shah. Probiotic bacteria: selective enumeration and survival in dairy foods. *J. Dairy Sci.*, 83: 894–907, 2000.
76. HK Park, JS So, T Heo. Acid adaptation promotes survival of Bifidobacterium breve against environmental stress. *Food Biotechnol.*, 4: 226–230, 1995.
77. G Schmidt, R Zink. Basic features of the stress response in three species of bifidobacteria: *B. longum, B. adolescentis,* and *B. breve. Int. J. Food Microbiol.*, 55: 41–45, 2000.
78. C Desmond, C Stanton, GF Gitzgerald, K Collins, RP Ross. Environmental adaptation of probiotic lactobacilli towards improved performance during spray drying. *Int. Dairy J.*, 11: 801 – 808, 2001.
79. WS Kim, L Perl, JH Park, JE Tandianus, N Dunn. Assessment of stress response of the probiotic *Lactobacillus acidophilus. Curr. Microbiol.*, 43: 346–350, 2001.
80. J Prasad, P McJarrow, P Gopal. Heat and osmotic stress responses of probiotic *Lactobacillus rhamnosus* HN001 (DR20) in relation to viability after drying. *Appl. Environ. Microbiol.*, 69: 917–925, 2003.
81. MJ Kullen, T Klaenhammer. Identification of the pH-inducible, proton-translocating F1F0-ATPase (atpBEFHAGDC) operon of *Lactobacillus acidophilus* by differential display: gene structure, cloning and characterization. *Mol. Microbiol.*, 33: 1152–1161, 1999.

82. M Ventura, R Zink, GF Fitzgerald, D von Sindern. Gene structure and transcriptional organization of the dnaK operon of *Bifidobacterium breve* UCC 2003 and application of the operon in bifidobacterial tracing. *Appl. Environ. Microbiol.* 71: 487–500, 2005.

83. JB Ahn, HJ Hwang, J Park. Physiological responses of oxygen tolerant anaerobic *Bifidobacterium longum* under oxygen. *J. Microbiol. Biotechnol.*, 11: 443–451, 2001.

84. M van de Guchte, P Serror, C Chervaux, T Smokvina, SD Ehrlich, E Maguin. Stress responses in lactic acid bacteria. *Antonie Van Leeuwenhoek*, 82: 187–216, 2002.

85. C Christensson, H Bratt, L LJ Collins, T Coolbear, R Holland, MW Lubbers, PW O'Toole, J Reid. Cloning and expression of an oligopeptidase, PepO, with novel specificity from *Lactobacillus rhamnosus* HN001 (DR20). *Appl. Environ. Microbiol.*, 68: 254–262, 2002.

86. EE Vaughan, B Mollet, WM deVos. Functionality of probiotics and intestinal lactobacilli: light in the intestinal tract tunnel. *Curr. Opin. Biotechnol.*, 10: 505–510, 1999.

87. E Malinen, LR Laitinen, A Palva. Genetic labelling of lactobacilli in a food grade manner for strain-specific detection of industrial starters and probiotic strains. *Food Microbiol.*, 18: 309–317, 2001.

88. M Gruzza, M Fons, MF Ouriet, Y Duval-Iflah, R Ducluzeau. Study of gene transfer *in vitro* and in the digestive tract of gnotobiotic mice from *Lactococcus lactis* strains to various strains belonging to human intestinal flora. *Microb. Releases*, 2: 183–189, 1994.

89. N Klijn, AH Weerkamp, WM de Vos. Genetic marking of *Lactococcus lactis* shows its survival in the human gastrointestinal tract. *Appl. Environ. Microbiol.*, 61: 2771–2774, 1995.

90. M Iwaki, N Okahashi, I Takahashi, T Kanamoto, Y Sugita-Konishi, K Aibara, T Koga. Oral immunization with recombinant *Streptococcus lactis* carrying the *Streptococcus mutans* surface protein antigen gene. *Infect. Immun.*, 58: 2929–2934, 1990.

91. JM Wells, PW Wilson, PM Norton, MJ Gasson, RWL Page. *Lactococcus lactis*: high-level expression of tetanus toxin fragment C and protection against lethal challenge. *Mol. Microbiol.*, 8: 1155–1162, 1993.

92. LA Ribeiro, V Azevedo, Y Le Loir, SC Oliveira, Y Dieye, JC Piard, A Gruss, P Langella. Production and targeting of the *Brucella abortus* antigen L7/L12 in *Lactococcus lactis*: a first step towards food-grade live vaccines against brucellosis. *Appl. Environ. Microbiol.*, 68: 910–1006, 2002.

93. MH Lee, Y Roussel, M Wilks, S Tabaqchali. Expression of *Helicobacter pylori* urease subunit B gene in *Lactococcus lactis* MG1363 and its use as a vaccine delivery system against *H. pylori* infection in mice. *Vaccine*, 19: 3927–3935, 2001.

94. L Steidler, W Hans, L Schotte, S Neirynck, F Obermeier, W Falk, W Fiers, E Remaut. Treatment of murine colitis by *Lactococcus lactis* secreting interleukin-10. *Science*, 289: 1352–1355, 2000.

95. WN Konings, J Kok, OP Kuipers, B Poolman. Lactic acid bacteria: the bugs of the new millennium. *Curr. Opin. Microbiol.*, 3: 276–282, 2000.

96. K Leenhouts, G Buist, J Kok. Anchoring of proteins to lactic acid bacteria. *Antonie Van Leeuwenhoek*, 76: 367–376, 1999.

97. A Mercenier, U Wiedermann, H Breiteneder. Edible genetically modified microorganisms and plants for improved health. *Curr. Opin. Biotechnol.*, 12: 510–515, 2001.

98. S Di Fabio, D Medaglini, CM Rush, F Corrias, GL Panzini, M Pace, P Verani, G Pozzi, F Titti. Vaginal immunization of Cynomolgus monkeys with *Streptococcus gordonii* expressing HIV-1 and HPV 16 antigens. *Vaccine*, 16: 485–492, 1998.

99. SE Gilliland, HS Kim. Effect of viable starter culture bacteria in yogurt on lactose utilization in humans. *J. Dairy Sci.*, 67: 1–6, 1984.

100. S Drouault, C Juste, P Marteau, P Renault, G Corthier. Oral treatment with *Lactococcus lactis* expressing *Staphylococcus hyicus* lipase enhances lipid digestion in pigs with induced pancreatic insufficiency. *Appl. Environ. Microbiol.*, 68: 3166–3168, 2002.

101. J Hugenholtz, W Sybesma, MN Groot, W Wisselink, V Ladero, K Burgess, D van Sinderen, JC Piard, G Eggink, EJ Smid, G Savoy, F Sesma, T Jansen, P Hols, M Kleerebezem. Metabolic engineering of lactic acid bacteria for the production of nutraceuticals. *Antonie Van Leeuwenhoek*, 82: 217–235, 2002.

102. Y Dieye, S Usai, F Clier, A Gruss, J-C Piard. Design of a protein-targeting system for lactic acid bacteria. *J. Bacteriol.*, 183: 4157–4166, 2001.

103. AL Hart, AJ Stagg, M Frame, H Graffner, H Glise, P Falk, MA Kann. The role of the gut flora in health and disease, and its modification as therapy. *Ailment Pharmacol. Ther.* 16: 1383–1393, 2000.

104. PH O'Farrell. High resolution two-dimensional electrophoresis of proteins. *J. Biol. Chem.*, 250: 4007–4021, 1975.

105. J Klose. Protein mapping by combined isoelectric focusing and electrophoresis of mouse tissues. A novel approach to testing for induced point mutations in mammals. *Humangenetik*, 26: 231–243, 1975.

106. WP Blackstock, M Weir. Proteomics: quantitative and physical mapping of cellular proteins. *Trends Biotechnol.*, 17: 121–127, 1999.

107. MC Champomier-Verges, E Maguin, MY Mistou, P Anglade, JF Chich. Lactic acid bacteria and proteomics: current knowledge and perspectives. *J. Chromatogr. B Analyt. Technol. Biomed. Life Sci.*, 771: 329–342, 2002.

108. S Salminen, A von Wright, L Morelli, P Marteau, D Brassart, WM de Vos, R Fonden, M Saxelin, K Collins, G Mogensen, SE Birkeland, T Mattila-Sandholm. Demonstration of safety of probiotics — a review. *Int. J. Food Microbiol.*, 44: 93–106, 1998.

109. M Saxelin. *Lactobacillus GG*-a human probiotic strain with thorough clinical documentation. *Food Rev. Int.*, 13: 293–313, 1997.

110. J Feord. Lactic acid bacteria in a changing legislative environment. *Antonie Van Leeuwenhoek*, 82: 353–360, 2002.

111. WM deVos. Safe and sustainable systems for food-grade fermentations by genetically modified lactic acid bacteria. *Int. Dairy J.*, 9: 3–10, 1999.

112. O Doblhoff-Dier, H Bachmayer, A Bennett, G Brunius, K Burki, M Cantley, C Stepankova, G Tzotzos, K Wagner, R Werner. Safe biotechnology 9: values in risk assessment for the environmental application of microorganisms. The Safety in Biotechnology Working Party of the European Federation of Biotechnology. *Trends Biotechnol.*, 17: 307–311, 1999.

113. M Hilliam. The market for functional foods. *Int. Dairy J.*, 6: 349–353, 1998.

114. M Coakley, RP Ross, M Nordgren, G Fitzgerald, R Devery, C Stanton. Conjugated linoleic acid biosynthesis by human-derived *Bifidobacterium* species. *J. Appl. Microbiol.*, 94: 138–145, 2003.

115. TR Klaenhammer, MJ Kullen. Selection and design of probiotics. *Int. J. Food Microbiol.*, 50: 45–57, 1999.

116. JB Luchansky, PM Muriana, TR Klaenhammer. Application of electroporation for transfer of plasmid DNA to *Lactobacillus, Lactococcus, Leuconostoc, Listeria, Pediococcus, Bacillus, Staphylococcus, Enterococcus and Propionibacterium. Mol. Microbiol.*, 2: 637–646, 1988

117. H Hashiba, R Takiguchi, S Ishii, K Aoyama. Transformation of *Lactobacillus helveticus* subsp. *jugurti* with plasmid pLHR by electroporation. *Agric. Biol. Chem.*, 54: 1537–1541, 1990.

118. P Serror, T Sasaki, SD Ehrlich, E Maguin. Electrotransformation of *Lactobacillus delbrueckii* (subsp.*) bulgaricus* and *L. delbrueckii (subsp.) lactis* with various plasmids. *Appl. Environ. Microbiol.*, 68: 46–52, 2002.

119. P Hols, P Slos, P Dutot, J Reymund, P Chabot, B Delplace, J Delcour, A Mercenier. Efficient secretion of the model antigen M6-gp41E in *Lactobacillus plantarum* NCIMB 8826. *Microbiology*, 143 (8): 2733–2741, 1997.

120. P Slos, P Dutot, J Reymund, P Kleinpeter, D Prozzi, MP Kieny, J Delcour, A Mercenier, P Hols. Production of cholera toxin B subunit in *Lactobacillus*. *FEMS Microbiol. Lett.*, 169: 29–36, 1998.

121. D Medaglini, MR Oggioni, G Pozzi. Vaginal immunization with recombinant Gram-positive bacteria. *Am. J. Reprod. Immunol.*, 39: 199–208, 1998.

7

Immunochemical Methods for Detection of Probiotics

Analía G. Abraham, María Serradell, Graciela L. Garrote,
Alberto C. Fossati, and Graciela L. De Antoni

CONTENTS

7.1 Antigenic Molecules and Structures Associated with Probiotic Characteristics

A probiotic is a microbial dietary adjuvant that beneficially affects the host physiology by modulating mucosal and systemic immunity, as well as improving nutritional and microbial balance in the intestinal tract (1).

Among bacterial probiotics for human gut, lactic acid bacteria (LAB) and bifidobacteria are widely used in food and pharmaceutical products. However, the evidence for their health benefits has not been rigorously demonstrated in all cases. For that reason, in 1996, a multicenter European research project began to study several commercial probiotics and demonstrated that some strains can positively affect human health in rigorously conducted human clinical studies (2). Common criteria used for isolating, characterizing, and selecting probiotic microorganisms include the following: strains of human origin, safe for oral administration (GRAS status), stable in the gut environment, and specific health-related effects (3).

About the role of probiotics in health and disease, there is a bibliography in which a compilation of commercial strains has been published (4). Probiotic action can be ascribed to several mechanisms such as competitive exclusion, production of antimicrobial substances, modulation of the immune response, alteration of intestinal bacterial metabolic activity, alteration of microecology in human intestine, and inhibition of bacterial translocation (4). Association of probiotic action with surface bacterial properties can be considered in almost all mechanisms mentioned. Some probiotic microorganisms have the ability to interact with different substrata like gut epithelial cells, intestinal mucus, pathogenic microorganisms, toxins, and mutagenic compounds. Through this interaction, probiotic strains exert their action of toxic and pathogenic inhibition. Interaction of probiotics and substratum has different degrees of specificity. In general, the putative probiotic characteristic is strain specific as revealed by *in vitro* experiments (5, 6).

To exert their probiotic effects, maintenance of LAB in the gastrointestinal tract is necessary to prevent their rapid removal by contraction of the gut. The ability of LAB to adhere to mucosal surfaces could confer a competitive advantage. Clearly, the ability to adhere to the intestinal mucus or epithelial cells is required for long-term persistence in the gut. Under these conditions probiotic bacteria can produce antagonist effects on pathogens by different mechanisms such as competition for nutrients, release of antimicrobial substances, and prevention of the subsequent attachment of pathogens by competitive exclusion. Different studies have suggested that adhesive probiotic bacteria can prevent the attachment of pathogens and stimulate their removal from the infected intestinal tract (7). Laboratory models using human intestinal cell lines such as Caco-2 and intestinal mucus have been developed to study the adhesion of probiotic bacteria and their competitive exclusion of pathogenic bacteria (8). Some adhesive strains of *Lactobacillus* showed a strong inhibition of the adhesion of food-borne pathogens by a combined effect of both bactericidal activity and competition for attachment sites (9). Results obtained with *Lactobacillus* and *Bifidobacterium* indicate that surface-mediated properties, such as aggregation, can have a role in adhesion and colonization (5, 10). Commercial probiotic strains such as *Lactobacillus* GG are able to slightly reduce S-fimbria-mediated adhesion of *Escherichia coli* to glycoproteins extracted from feces. In addition, adhesion of *Salmonella typhimurium* was significantly inhibited by probiotic *L. johnsonii* LJ1 and *L. casei* Shirota (11).

The association of probiotic bacteria with the intestinal mucosa is the first step in the development of the immune response. Studies conducted with fluorescent LAB and transmission electron microscopy in mouse revealed that some strains of lactobacilli can interact with Peyer's patches, M cells, and also epithelial cells of the small and large intestine. This interaction was shown to be host specific for the genus *Bifidobacterium* but not for *Lactobacillus* (12). These interactions are probably regulated by specific molecules on the bacterial surface.

Given the binding capacity of their surface envelopes, probiotic bacteria can be used as scavengers of mutagenic substance (13, 14), toxic compounds like bile acids (15, 16), and pathogenic microorganisms (by coaggregation) (17, 18). These abilities are desirable in probiotic strains that are unable to interact with the gut epithelium.

Frequently, adhesion is highly specific and depends on the presence of a certain receptor on the substratum and adhesins on the probiotic surface. Adhesins can be attached to the bacterial wall, to appendages such as fimbriae, and to structures like the S-layer. Molecules such as teichoic acids, lipoteichoic acids, proteins, glycoproteins, and polysaccharides participate in the adhesion phenomena. Among adhesins, lectins are oligomeric or multimeric proteins or glycoproteins with binding specificities for a particular carbohydrate structure (19).

Some probiotic strains produce antagonistic metabolites including organic acids, diacetyl, and a wide range of bacteriocins, some of which have activity against food pathogens such as *Listeria monocytogenes* and *Clostridium botulinum*. The bacteriocin nisin has been used as an effective biopreservative in some dairy products for decades, whereas a number of more recently discovered bacteriocins demonstrate increasing potential in food applications (20–22).

7.2 Application of Immunochemical Techniques to Potentially Detect Probiotic Bacteria

Because of their high specificity and sensitivity, immunological methods are useful tools for the identification and detection of microorganisms. There are a number of formats available today that employ antibodies to specifically detect certain food-borne pathogens. Detection systems include the enzyme-linked immunosorbent assay (ELISA), particle agglutination tests, precipitation tests (immunoimmobilization), direct immunostaining, antibody-coated paramagnetic beads, and antibody-based biosensors (23). ELISA is the most widespread and useful technique for the detection of food-borne pathogens because of its simplicity and ability to analyze large numbers of samples at one time.

Specifically, there are commercially available immunoassays that use either polyclonal antibodies or monoclonal antibodies to detect *Salmonella*, *Listeria*, *E. coli* O157:H7, *Staphylococcus aureus*, *Campylobacter* (23), enterotoxic *B. cereus*, and other organisms and their toxins (24). Immunoassays are also used in the food industry to detect spore-forming and nonspore-forming bacteria, and even to differentiate vegetative cells from spores (25). In addition, several species-specific and genus-specific ELISAs have been developed using monoclonal or polyclonal antibodies against exoantigens and heat-stable polysaccharides from a wide range of fungi, including *Aspergillus*, *Penicillium*, and *Fusarium* (26).

Application of immunoassays to detect LAB or their metabolites is still in an initial developmental stage. LAB enzymes as peptidases (27) or malolactic enzyme (28) have been detected by immunoblotting after sodium dodecyl sulfate–polyacrylamide gel electrophoresis (SDS-PAGE).

A summary of putative probiotic components detected by immunochemical methods are shown in Tables 7.1 and 7.2. Specific surface proteins (29–31) and teichoic acid (Antigen E) (32) are some of the surface molecules recognized by antibodies.

Antibodies produced against surface molecules of *Lactobacillus brevis* give us an example of the importance of antigen selection. Several antibodies have been obtained against *L. brevis* that allow the detection or identification of isolates at different levels (Table 7.1). An antibody against the S-layer allows the grouping of *L. brevis* strains according to the homology of their S-layer reactivity (33), while a second antibody against an antigen common to all *L. brevis* strains is used to identify the species (32). An antibody against specific antigens present in *L. brevis* that causes spoilage to beer is used to detect strains that may impair the quality of beer (30).

Selection of a method also determines the specificity of the assay. Antiserum against the S-layer of *L. kefir* recognize specifically the S-layer of the same species under a competitive ELISA. However, immunoblot or dot blot assays performed with the same antiserum show cross-reaction with the S-layer of *L. parakefir* (29).

The detection limit of an immunoassay depends on the format of the test. As was reported, the detection limit for pediocin A1 was found to be 2.5 $\mu g/ml$ by immunodotting, 1 $\mu g/ml$ in the NCI-ELISA, 0.025 $\mu g/ml$ in the CI-ELISA, and the sensitivity was even enhanced in the CD-ELISA (34).

An antiserum against the S-layer of *L. kefir* can detect 10^4 CFU/ml of different strains of *L. kefir* by CI-ELISA in kefir-fermented milk containing a

TABLE 7.1

Application of Immunoassay to Surface Protein Detection

Antigen	Antibody Obtained	Assay	Recognize	Application	Ref.
S-layer of *L. kefir*	Antiserum	Competitive ELISA	S-layer of *L. kefir*	Determination of *L. kefir* in kefir fermented milk	(29)
S-layer of *L. kefir*	Antiserum	Dot blot Western blot	S-layer of *L. kefir* and cross-reaction with S-layer of *L. parakefir*	Characterization of antiserum reactivity	(29)
Whole cell of *L. brevis* (S-layer deficient strain)	Antiserum adsorbed with cell wall of the same strain	Agglutination Precipitation	Antigen E in *L. brevis* strain and cross reactivity with lactobacilli (serological groups E and D)	Determination of serological group	(32)
Whole cell of *L. brevis* 578 with S-layer	Antiserum adsorbed with cell wall of *L. brevis* L47	Precipitation Agglutination	Surface molecules beneath S-layer	Determination of *L. brevis* beer spoilage strain	(30)
Whole cell of *L. brevis* strains with S-layer	Antiserum	SDS-PAGE and Western blot	S-layer of the homologous strain	*L. brevis* classification by immunological properties of S-layer	(33)
Whole cell of *L. kefir* or *L. kefiranofaciens*	Antiserum adsorbed with broken cells	Immunofluorescence	Structures present in whole cell of *L. kefir* or *L. kefiranofaciens*	Distribution of *L. kefir* and *L. kefiranofaciens* in kefir grains	(42)
High molecular mass cellular surface protein of *L. reuteri* from supernatant and bacterial surface	Antiserum adsorbed with nonadhering *L. reuteri*	ELISA Immunofluorescence	High-molecular-weight protein	Select clones that express the protein. Adhesion of protein to mucus	(31)
Whole cell of *L. casei* Shirota	Monoclonal antibodies	ELISA	*L. casei* strain Shirota	Identification of *L. casei* Shirota isolated from feces	(36)
Whole cell of *Veillonella, Enterococcus*	Monoclonal antibodies	Competitive ELISA	*Veillonella, Enterococcus*	Detection of poultry probiotic bacteria	(35)

TABLE 7.2

Application of Immunoassay to Bacteriocin Detection

Antigen	Antibody Obtained	Assay	Recognize	Application	Ref.
C-terminal fragment of pediocin PA1 and purified pediocin PA1	Antiserum	NCI ELISA CI ELISA CD ELISA	Pediocin PA1	Screening of Pediocin PA1 producing strain	(34)
Nisin	Antiserum	ELISA	Lantibiotics and nisin	Cross-reactivity with nisin	(37)
Nisin		ELISA	Nisin-like bacteriocins	Screening of *L. lactis* strains producing nisin-like bacteriocins	(38)
Nisin	Antiserum	Anti-nisin antibodies on magnetic beads	Nisin	Rapid purification of nisin	(41)
Nisin	Monoclonal antibodies	Immunoaffinity chromatography	Nisin	Purification of nisin	(40)
Nisin	Monoclonal antibodies	Flow ELISA	Nisin A	Quantification of nisin	(39)

Note: NCI ELISA: noncompetitive indirect ELISA. CI ELISA: competitive indirect ELISA. CD ELISA: competitive direct ELISA.

complex mixture of lactobacilli, lactococci, and yeasts at a concentration of 10^8 to 10^9 CFU/ml (29). The same detection limits were described by Durant et al. (35) in the detection of probiotic strains for chicken in fecal content. For lower bacterial concentration, an enrichment step is necessary; however, subsequent identification is quite rapid when compared with standard culture methods. Monoclonal antibodies against *L. casei* Shirota were used to determine the survival of this strain in the gastrointestinal tract of individuals after consumption of fermented milk containing this probiotic strain. The strain was isolated from feces using a selective method prior to identification using monoclonal antibodies (36).

Antibodies have also been used to study surface molecules on *L. reuterii* involved in adhesion to host epithelium and to analyze possible mechanisms of action (30). Polyclonal antibodies were used for the detection of pediocin PA-1 (pedA1). Antibodies were also used to determine structural similarities between different bacteriocins and nisin (37) and for the isolation of a bacteriocin-producer *Lactoccoccus* strain (38). Development of a flow method based on the affinity of antibodies allowed the on-line determination of nisin (39).

Antibodies against bacteriocins can be used for the purification of bacteriocin by immunoaffinity chromatography (40) or by one-step immunomagnetic separation (41).

7.3 Methodology for Immunochemical Detection of Probiotics

7.3.1 Preparation of Antigens

7.3.1.1. *Preparation of Bacterial Antigens*

Bacterial cells contain numerous components, which are potential antigens. Components from cell wall and membrane and cytoplasmic proteins can induce an immune response during a bacterial infection and, therefore, can be used as immunogens. Because of the dissimilarity of components of different bacterial species, few purification methods are widely applicable (43).

A first question to be aware of is that culture conditions can be critical for recovering antigens of interest. In some cases, liquid media are the best choice (i.e., recovery of extracellular soluble products like enzymes or bacteriocins), wheras solid media are more appropriate for other purposes.

In order to isolate different antigenic fractions of microorganisms, many different techniques can be used. The extracellular proteins can be isolated and concentrated from filtered culture supernatant by saline precipitation using a series of saturation steps with ammonium sulfate.

If the aim is to obtain intracellular antigens, different methods can be used to disrupt bacterial cells. These methods can be separated into relatively simple procedures, such as grinding frozen material in a mortar and pestle and methods that require specialized instruments (43). This latter category may include:

- Ultrasonic disruption: some bacterial species are more susceptible than others to ultrasonic disruption. Optimal time of exposure and frequency depend on the volume and density of bacterial culture, the bacterial species, and the physiologic state of the cells.

- Pneumatic press disruption: in this case, the bacterial cells in the form of a frozen paste are compressed and passed through a small orifice into a second chamber, where expansion and cell rupture occur.

- Mechanical disruption: a concentrated suspension of cooled bacteria is mixed with an equal volume of glass beads, and the mixture is shaken for 5 to 15 min. The small beads are separated by filtration and can be reused after washing with concentrated nitric acid and distilled water.

Additionally, physical properties of bacterial cells can be exploited in other methods such as heat disruption, freezing and thawing cycles, or osmotic lysis.

After disruption, the cell wall fraction has to be separated from intracellular components. In order to do this, two methods are usually used (43):

- Centrifugation: conditions normally depend on the size of the cells of the particular species. Unbroken cells are generally removed by centrifugation at 2500 to 3500 g for 10 to 15 min, whereas cell walls are removed at 8000 to 12,000 g for 20 to 30 min.

- Density gradient centrifugation: this method usually employs a sucrose gradient.

7.3.1.2   Purification of Bacterial Proteins

Purified antigens are frequently used for developing serological methods for the diagnosis of infectious diseases or to quantify and identify microorganisms. Purification of antigens involves diverse techniques, including chromatography and preparative electrophoresis.

Liquid chromatography can be classified according to stationary phase characteristics. Liquid-solid chromatography is the technique more frequently used for purification of molecules of interest in immunology. Several procedures based on different principles can be used, such as

- Gel filtration: separation is based on different size and shape of molecules. The stationary phase is composed of a matrix with pores of different and controlled sizes. Therefore, the smaller the protein, the greater the volume of buffer needed to elute it. This procedure is frequently employed to determine the molecular weight of proteins (43, 44).

- Ion exchange chromatography: in this case, proteins with different charges on their surface interact differentially with particular charged groups and are adsorbed to the stationary phase. Conditions of binding can be modified by changing the saline concentration or the pH of the buffer used as the eluting reagent. This procedure offers a variety of possibilities because of the diversity of matrices and charged groups that can be used.

- Hydrophobic interaction chromatography: proteins dissolved in the mobile phase establish hydrophobic interactions with the stationary phase, and the elution of proteins is achieved by reducing the saline concentration of the eluting buffer. Matrices used consist of nonpolar ligands bound to hydrophilic gels. Hydrophobicity of ligands, saline concentration of buffer, and temperature affect the binding (44).

- Affinity chromatography: this method is based on the interaction between antigenic proteins and specific antibodies. In order to purify a particular antigen, specific antibodies (polyclonal or monoclonal) are covalently bound to beads, and after incubation of beads with antigen solution, unbound antigens are removed by washing. Bound antigens are ultimately eluted using diverse eluting reagents (see Section 7.3.2.2) (45, 46).

Preparative electrophoresis is another useful procedure (44). If samples are run under denaturing conditions, they can be separated according to their molecular weight. That is the case of preparative SDS-PAGE, an adaptation of the corresponding analytical procedure. Samples are prepared in a buffer solution containing SDS (usually 2% w/v), and consequently they become negatively charged. The pore size of the gel depends on the percentage of acrylamide and *bis*-acrylamide, and it determines the size range of proteins that can be separated. Components are automatically collected in separated vials.

7.3.2 Production of Specific Antibodies

7.3.2.1 *Production of Polyclonal Antibodies (Antisera)*

Antibodies generated in a natural immune response or after immunization in a laboratory are a heterogeneous mixture of antibodies of different specificities and affinities. These antibodies constitute the antiserum. Usually, the animals selected for the production of antisera are rabbits or mice.

TABLE 7.3

Routes and Doses Most Frequently Used in Laboratory Immunization of Mice or Rabbits Experiments

Form of Antigen	Examples	Mice		Rabbit		Adjuvant
		Route	Dose	Route	Dose	
Soluble proteins	Enzymes, carrier proteins conjugated with peptides, immune complexes	ip, sc	5–50 μg	sc, im, id, iv	50–1000 μg	Yes
Particulate proteins	Bacterial cells Structural proteins	ip, sc	5–50 μg	sc, im, id	50–1000 μg	Yes
Insoluble proteins	Proteins from inclusion bodies, bound to beads	ip, sc	5–50 μg	sc, im, id	50–1000 μg	Yes
Carbohydrates	Polysaccharides, glycoproteins	ip, sc	10–50 μg	sc, im, id, iv	50–1000 μg	Yes

Note: ip: intraperitoneal; sc: subcutaneous; im: intramuscular; id: intradermal; iv: intravenous.

Because of the dynamics of the immune system, composition of the antiserum is continuously changing along the immunization, making each antiserum irreproducible (47). In most cases, more than one dose of antigen is needed to generate a strong immune response. Generally, three to five immunizations 3 to 4 weeks apart are carried out, and it is usually necessary to inject the antigen mixed with an adjuvant. An adjuvant is any substance that enhances the immunogenicity of components mixed with it. There are many different adjuvants (45), but the most employed in regular protocols consist of an oil-in-water emulsion known as incomplete Freund's adjuvant (IFA). Complete Freund's adjuvant (CFA) is the emulsion added with muramyldipeptide (component of *Mycobacterium* spp.). Usually, CFA is used for the first challenge, while IFA is employed in the following immunizations, in both cases mixed in a 1:1 ratio with the antigen solution.

There are many options to immunize the animals, depending basically on the antigen and the animal. The route, dose, and form in which the antigen is administered can profoundly affect the type of response produced. Different routes can be used to inject the antigen: subcutaneous, intradermal, intramuscular, intraperitoneal, or intravenous injection. In certain situations, an oral or intranasal administration can be used. Table 7.3 shows the routes and doses most frequently used in a laboratory immunization of mice or rabbits using different types of bacterial antigens (45).

7.3.2.2   Preparation and Production of Monoclonal Antibodies

Monoclonal antibodies have many other advantages over polyclonal antibodies in addition to specificity. They also provide reproducibility and unlimited supply.

The clone selection theory, presented by Burnet in 1959 (48), proposed that each B lymphocyte produces only a specific antibody in response to stimulation. This idea inspired Georges Köhler and César Milstein to devise a technique for producing a homogeneous population of antibodies of known antigen specificity. They fused spleen cells from an immunized mouse to cells of a murine myeloma and obtained hybrid cells that proliferated indefinitely and secreted antibody specific for the antigen employed to immunize the donor (49). The spleen cells provide the ability to make antibodies, whereas the myeloma cells provide the ability to proliferate continuously in culture.

Immunization options were shown in the previous section, so we will not discuss that issue here again. There are many myeloma cell lines that can be used to fuse to spleen cells. They can proliferate indefinitely *in vitro* and usually do not produce antibodies (i.e., NSO, SP2/0*, FO**). They are cultured in a carbonate-buffered medium (i.e., RPMI 1640, pH = 7.4) containing 10 to 20% of bovine calf serum, glutamine, pyruvate, and antibiotics (streptomycin and penicillin) (RPMI-BCS). Usually, splenocytes and myeloma cells are used in a ratio ranging from 1:1 to 10:1. In our experience, ratios of 2:1

or 3:1 provide good results. Polyethyleneglycol (PEG) at 50% is generally used to induce the fusion, and it must be added slowly with shaking. Because of its toxicity, PEG must be removed by washing with increasing volumes of a saline or buffer solution after 2 min of contact. The whole procedure is performed at 37°C. An example of a fusion protocol is shown below.

1. Mix spleen cells with myeloma cells (approx. 1×10^8 cells) in a ratio of 1:1 to 5:1.
2. Centrifuge the mixture at $300 \times g$ for 5 min and discard the supernatant.
3. Place the pellet on a bath at 37°C, and add 1 ml of PEG 50% slowly during 1 min with constant shaking.
4. Shake for 1 min.
5. Add 2 ml of buffer solution during 2 min, with shaking.
6. Add 5 ml of buffer solution during 2 to 3 min, with shaking.
7. Add 10 ml of buffer solution and complete volume to 50 ml with the same buffer.
8. Centrifuge at $300 \times g$ for 5 min, and discard the supernatant.
9. Add 50 ml of HAT medium, and distribute in 5 to 10 culture microplates (100 µl per well).

After the fusion, cells are grown in a selective medium in which only the hybridomas are able to survive. Because myeloma cells are defective in the enzyme hypoxanthine guanine phosphoribosyltransferase (HPGRT), the selective medium contains aminopterin (to inhibit the *de novo* synthesis of purines and thymidine), and hypoxanthine and thymidine (HAT medium). HAT medium can be prepared by adding HAT mixture to RPMI-BCS medium.

After 10 to 15 days of fusion, cells are screened for the production of specific antibodies. The screening has to be performed by a rapid method, capable to detect as low as nanograms per milliliter or micrograms per milliliter of antibody and to allow the selection of cells producing antibodies with the desired characteristics. Indirect ELISA is the procedure frequently employed (see Section 7.3.4.2.), although other techniques can be used.

Once specific hybridomas have been selected, they are cloned by limit dilution or in semisolid agar, and clones are frozen until they are used. Selected monoclonal antibodies can be produced *in vivo* by producing ascitic fluid in mice (generally 2 to 20 mg/ml of antibody can be recovered), or *in vitro*, by growing hybridomas in bioreactors (allows the production of gram-to-kilogram quantities of antibody) (44). Antibodies can be purified by chromatographic procedures from the supernatant of hybridoma cultures or from ascitic fluid. One of the available techniques is anion-exchange chromatography, in which the stationary phase is usually diethylaminoethyl–cellulose (DEAE-cellulose). Antibody solutions and eluting buffer are adjusted to pH

8.0 for purification of IgG subtypes (45). Another option is affinity chromatography, which can be based on antigen–antibody interactions or on the capacity of certain bacterial proteins (proteins A and G) to bind to the Fc region of immunoglobulins. In the first case, antigen is usually covalently bound to inert beads, which are then mixed with the antibody solution. After the specific binding of antibodies to the antigen, the unbound substances in the sample are removed by washing. The beads are then treated with an elution buffer, and the purified antibody is released. There are several elution reagents, such as ionic detergents, dissociating agents, chaotropic agents, organic solvents, high salt buffers, or high or low pH buffers. Notwithstanding, the systems using proteins A or G are frequently the most recommended techniques. Each of these proteins binds to a different spectrum of antibodies. In general, protein A is recommended for mouse monoclonal antibodies from the IgG2a, IgG2b, and IgG3 subclasses, whereas protein G is used for purifying IgG1. Usually, a low pH solution (pH = 3.0) is used to elute the bound antibody (45).

A simple scheme of the whole experiment is shown in Figure 7.1.

7.3.3 Labeling of Antibodies

7.3.3.1 Labeling Antibodies with Fluorochromes

Fluorochromes are among the most useful labels that can be attached to antibodies. Two fluorochromes are the most commonly used: fluorescein and rhodamine. There are problems associated with the use of fluorochrome, such as over- or underlabeling. Therefore, for each antibody there is an optimal ratio of fluorochromes to antibody. For labeling antibodies with the isothiocyanate derivatives of these fluorochromes, the antibody solution should not contain additional molecules with free amino groups (45).

1. Prepare an antibody solution of at least 2 mg/ml, and adjust it to pH 9.0 (i.e., in sodium borate or carbonate buffer).
2. Dissolve the fluorescein isothiocyanate or tetramethylrhodamine isothiocyanate in anhydrous DMSO at 1 mg/ml.
3. Add 50 μl of fluorochrome solution per milliliter of antibody solution very slowly. Protein solution has to be continuously mixed.
4. Leave the reaction in the dark for 1 h or overnight at 4°C.
5. Add NH_4Cl to 50 mM and incubate for 2 h at room temperature.
6. Separate the unbound dye from the conjugate by gel filtration.
7. The ratio of fluorochrome to protein can be estimated by measuring the absorbance at 495 nm and 280 nm for fluorescein, or 575 nm and 280 nm for rhodamine. The ratio should be between 0.3 and 1.0 for fluorescein, or between 0.3 and 0.7 for rhodamine.

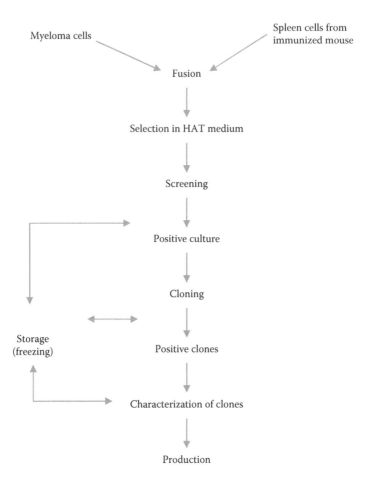

FIGURE 7.1
Schematic chart flow of monoclonal antibodies production.

7.3.3.2 *Labeling Antibodies with Biotin*

Antibodies can be easily labeled with biotin. The biotin groups can act as binding groups for avidin or strepavidin, which are commercially available, coupled to enzymes, fluorochromes, or iodine. The affinity of biotin for avidin or streptavidin is exceptionally high (Ka 10^{15} liter mol^{-1}); thus, the interaction is essentially irreversible and can be used as an amplification system to reveal antigen-antibody reactions (45). The technique for the covalent coupling of biotin to antibodies is very simple and normally does not have any adverse effect on the antibody.

1. Adjust the antibody solution (1 to 3 mg/ml) to pH 8.8 in 0.1 *M* sodium borate buffer.

2. Prepare a solution of *N*-hydroxysuccinimide biotin at 10 mg/ml in anhydrous DMSO.

3. Add the biotin ester to the antibody at a ratio of 25 to 100 µg of ester to 1 mg of antibody.

4. Mix and incubate at room temperature for 4 h.

5. Incubate with ammonium chloride solution (20 ml of 1 M per 250 mg of ester) for 10 min at room temperature.

6. Dialyze against the desired buffer, and purify by gel filtration.

7.3.3.3   Labeling Antibodies with Enzymes

The most employed enzymes in immunochemical methods are horseradish root peroxidase and alkaline phosphatase. There are two general methods for coupling these enzymes to antibodies. The first is a chemical method involving covalent binding between both proteins. In this case, the pH of the buffer used to carry out the conjugation is crucial to enhance the enzyme–antibody interaction. The most frequently used coupling agents are glutaraldehyde or sodium periodate. The latter is commonly used for conjugation of peroxidase to IgG immunoglobulins and allows the binding of up to 90% of the enzyme. The final step of these procedures involves the purification of conjugated antibody by ion exchange, followed by affinity chromatography (using concanavalin A), and previous precipitation with 35% ammonium sulfate. A protocol for this chemical conjugation is shown below.

1. Dissolve 2.5 mg of peroxidase in 0.25 ml of 100 mM NaHCO$_3$.

2. Add 0.25 ml of 12 mM NaIO$_4$ and incubate for 2 h at room temperature in the dark.

3. Dialyze the antibody solution against 100 mM carbonate buffer pH = 9.

4. Mix the antibody solution with peroxidase and incubate for 3 h at room temperature in the dark.

5. Add 1/20 in volume of NaBH$_4$ solution (5 mg/ml in NaOH 0.1 mM) and incubate for 30 min at 4°C.

6. Add 1 volume of NaBH4 solution to 10 volumes of the mixture, and leave at 4°C for 1 h.

7. Precipitate with 50% saturated ammonium sulfate for 1 h, centrifuge at 6000 × g for 15 min, and discard the supernatant.

8. Wash with 50% saturated ammonium sulfate.

9. Dissolve the pellet in a 0.1 M phosphate buffer pH = 7.2, and dialyze against the same solution.

10. Pass the sample through a column of Concanavalin A Sepharose 4B™ (conjugated antibody will bind to the column). Elute with α-methyl mannose (10 to 100 mM in 0.1 M phosphate, 0.1 M NaCl, pH = 7.2).

The other coupling method is based on the interaction between the enzyme and an antienzyme antibody. Enzyme–antienzyme complexes are formed to reveal immunoenzymatic reactions. This procedure was optimized to detect immunoglobulins from different species with peroxidase (PAP method) or alkaline phosphatase (APAAP method).

7.3.4 Methods for Detection

Numerous immunochemical techniques can be appropriate to detect specific antibodies or to quantify a specific component in a sample. In this section, we will discuss only a few different options that can be particularly helpful in bacteriology.

7.3.4.1 Indirect Immunofluorescence

This method is particularly useful to study the interaction between micro-organisms and specific cells and for the detection of microorganisms in tissues. Although the sensitivity of the technique is high, which is advantageous, its specificity can be low. Consequently, an experienced operator is crucial to avoid misreading a result. Indirect antigen detection using unlabeled primary antibodies and fluorochrome-coupled secondary antibodies involves a simple series of incubations. Both primary and fluorochrome-conjugated antibodies should be diluted in a protein-containing solution (i.e., phosphate buffer with 3% bovine serum albumin [BSA]). Dilutions of antibodies vary according to the characteristics of the sample (45).

1. Place coverslips, slides, or plates containing the sample on a flat surface.
2. Add the primary antibody, and incubate for at least 30 min at room temperature.
3. Wash three times with phosphate buffered saline pH 7.4 (PBS).
4. Apply the fluorochrome-coupled secondary antibody, and incubate for at least 20 min at room temperature (incubation times greater than 1 h have to be avoided).
5. Wash three times with PBS. Mount using buffered glycerol.
6. Observe under fluorescence microscope.

7.3.4.2 Noncompetitive Indirect ELISA

This is a two-step method. It consists of the immobilization of the antigen (i.e., proteins, whole bacterial cells, etc.) on a solid phase followed by a reaction with specific antibodies and finally an incubation with an enzyme-conjugated antibody (51). Two enzymes are mainly used: horseradish root peroxidase (HRP) or alkaline phosphatase. They catalyze the oxidation of a

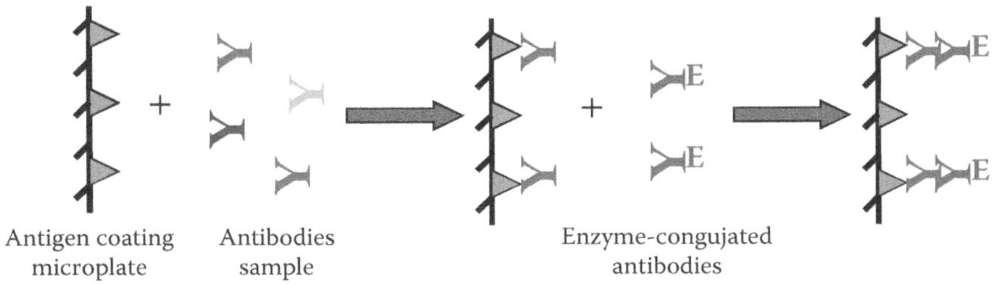

FIGURE 7.2
Noncompetitive indirect ELISA.

substrate, which produces a colored soluble product. Color intensity is read in a spectrophotometer.

This method can be used to identify microorganisms, metabolic products, specific antigens, etc. in different samples. A simple representation of the assay is shown in Figure 7.2.

To improve the binding of antigens to the microplate, antigen solutions can be adjusted to pH 9.0 (generally with sodium carbonate buffer 0.1 *M*). Usually, concentrations ranging from 0.5 to 100 µg/ml of protein are used to coat the plate. To block the free sites on the plate (those not occupied by the antigen), incubation with a solution of unrelated proteins (1% BSA, gelatin, and nonfat milk at 1 to 5%) is performed. Generally, samples where specific antibodies are to be detected are rabbit or mice antisera, human sera, hybridoma culture supernatant or ascites fluid. Thus, reaction parameters such as dilution and time and temperature of incubation depend on the characteristics of a particular sample. Usually these samples are diluted in blocking solution. A washing step with PBS containing 0.05% Tween 2 is needed after each incubation.

1. Apply 100 µl of the antigen solution to each well, and incubate at 37°C for 1 or 2 h or overnight at 4°C.
2. Wash three times with washing solution.
3. Add 200 µl of blocking solution to each well, and incubate for 1 or 2 h at 37°C or overnight at 4°C.
4. Wash three times with washing solution.
5. Apply 100 µl of antibody sample to each well and incubate under the chosen conditions.
6. Wash three times with washing solution.
7. Add 100 µl of the HRP-conjugated anti-immunoglobulin antibody to each well, and incubate for 1 h at 37°C.
8. Wash three times with washing solution.

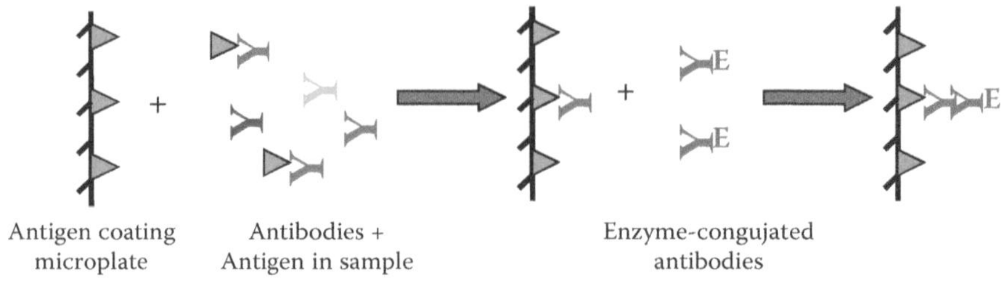

FIGURE 7.3
Competitive sequential ELISA.

9. Add 100 µl of the substrate solution (2 mg/ml *ortho*-phenylenediamine and 1 µl/ml H_2O_2 30%, in 0.1 *M* citric acid, 0.1 *M* $Na_2HPO_4.2H_2O$, pH = 5). Leave for 10 to 15 min in the dark.

10. Read absorbance in a spectrophotometer at 490 nm.

7.3.4.3   *Competitive Sequential ELISA*

This method is mainly useful to quantify a specific component in a sample (i.e., bacterial cell proteins, etc.) and detection limit is as low as nanograms per milliliter of protein.

This technique is based on the competition for the antibody binding between a soluble antigen present in the sample and antigens bound to a microplate (50). The scheme shown in Figure 7.3 depicts one of the many different forms to perform this type of assay.

In this procedure, a curve performed with standard antigen solutions of different concentration is used as a reference. Diluting, washing, time, and temperature conditions are similar to those of the indirect ELISA method.

1. Apply 100 µl of the antigen solution to each well, and incubate at 37°C for 1 or 2 h or overnight at 4°C.

2. Wash three times with washing solution.

3. Add 200 µl of blocking solution to each well, and incubate for 1 or 2 h at 37°C or overnight at 4°C, and then wash three times with washing solution.

4. In parallel, mix a volume of antigen standard solution (at different concentrations) or the unknown samples with the same volume of a chosen dilution of the antibody solution in separated vials (preincubated mixture) and incubate under the chosen conditions.

5. Apply 100 µl of preincubated mixture to each well, and incubate under the chosen conditions.

6. Wash three times with washing solution.

7. Add 100 µl of the HRP-conjugated anti-immunoglobulin antibody to each well, and incubate for 1 h at 37°C.

8. Wash three times with washing solution.
9. Add 100 μl substrate solution (2 mg/ml *ortho*-phenylenediamine and 1 μl/ml H_2O_2 30%, in 0.1 M citric acid, 0.1 M $Na_2HPO_4.2H_2O$, pH = 5). Leave for 10 to 15 min in the dark.
10. Read absorbance in a spectrophotometer at 490 nm.

7.3.4.4   Capture ELISA

This is another available procedure for the quantification of particular substances or cells in a sample (Figure 7.4). In this type of assay, the capture antibody is directly coupled to the polystyrene microplate (52). The "sandwich ELISA" outlined as follows is one of many different ways in which this test can be performed.

1. Apply 100 μl of the coating antibody solution to each well, and incubate at 37°C for 1 or 2 h or overnight at 4°C.
2. Wash three times with washing solution.
3. Add 200 μl of blocking solution to each well, and incubate for 1 or 2 h at 37°C or overnight at 4°C, and then wash three times with washing solution.
4. Apply 100 μl of antigen solution (standard or sample) in suitable concentrations and incubate for 1 or 2 h at 37°C or overnight at 4°C.
5. Wash three times with washing solution.
6. Add 100 μl of the second antibody solution, and incubate at chosen conditions.
7. Wash three times with washing solution.
8. Add 100 μl of the HRP-conjugated anti-immunoglobulin antibody to each well, and incubate for 1 h at 37°C.
9. Wash three times with washing solution.
10. Add 100 μl of the substrate solution (2 mg/ml *ortho*-phenylenediamine and 1 μl/ml H_2O_2 30%, in 0.1 M citric acid, 0.1 M $Na_2HPO_4.2H_2O$, pH = 5). Leave for 10 to 15 min in the dark.
11. Read absorbance in a spectrophotometer at 490 nm.

7.3.4.5   Immunoblotting or Western Blot

The Western blot technique combines the high resolution of the electrophoretic methods with the sensitivity and specificity of immunochemical assays. It consists of the transfer of the components previously separated by electrophoresis from the gel to a membrane (usually nitrocellulose or modified nylon sheets) and the detection of the antigenic fractions using specific antibodies (45). This procedure allows determining the specificity of particular antibodies to detect a desired component in a sample.

Electrophoresis can be performed in agarose or polyacrylamide gels, and the SDS-PAGE is the most commonly used procedure. In the latter case, the transfer can be performed employing electric current (taking advantage of the negative charge of the molecules in the gel) or by diffusion. The time required for transfer depends on the acrylamide concentration in the gel and the molecular size of the proteins. Once transfer is done, blocking of free sites is performed, and membranes are sliced for the following analysis. Incubations with primary and enzyme-conjugated antibodies should be carried out in blocking solution. The assay can be revealed using chromogenic or chemiluminescent substrates; the latter option is generally more sensitive (0.1 to 0.5 ng) than the first (1 to 10 ng). The chromogenic substrates have to produce water-insoluble colored products, which precipitate on the membrane.

1. Prepare the antigen sample, and run an SDS-PAGE.
2. Soak the membrane in the transfer buffer (25 mM Tris- (hydroxymethyl) aminomethane, 192 mM glycine, 20% methanol, pH = 8.3). Fill the tank with the same buffer.
3. Place three or four sheets of filter paper covering both sides of the electrotransfer cassette.
4. Place the gel and the nitrocellulose membrane on the electrotransfer cassette (gel on the cathode side and the membrane on the anode side). Run for 45 to 60 min at 300 mA or at 35 to 40 V for 1 to 3 h.
5. Stain the membrane with a 3% Ponceau S solution for 5 min; remove the dye with distilled water (optional).
6. Slice the membrane, and incubate with blocking solution for 2 h at 37°C or overnight at 4°C.
7. Wash three times with washing solution.
8. Apply antibody samples, and incubate at chosen conditions.
9. Wash three times with washing solution.
10. Add the HRP-conjugated anti-immunoglobulin antibody, and incubate for 1 h at 37°C.
11. Wash three times with washing solution.
12. Add an adequate volume of the substrate solution (15 ml of Tris Buffered Saline [TBS], 9 mg of 4-chloronaphthol, 3 ml methanol, 12 μl of H_2O_2, 30%).

7.3.4.6   Dot Blot

In this assay, antigenic samples are directly applied onto a nitrocellulose sheet. The rest of the procedure is analogous to that of the previously described method for immunoblotting. It has the advantage of being a rapid and very simple method for detection of specific antigen–antibody reactions.

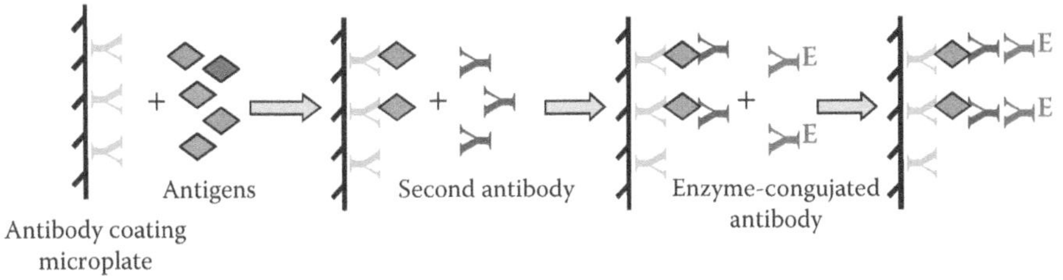

Antigens

Antibody coating
microplate

Second antibody

Enzyme-congujated
antibody

FIGURE 7.4
Capture ELISA.

7.3.4.7 Flow Cytometry

This technique allows individual cells to be identified by their surface antigens and to be counted. Individual cells within a mixed cell population are first tagged by treatment with specific monoclonal antibodies to cell-surface proteins labeled with fluorescent dyes. The labeled cells are forced through a nozzle, creating a fine stream of liquid containing cells individually separated at constant intervals. As each cell passes in front of a laser beam, it scatters the laser light, and dye molecules bound to the cell are excited and emit fluorescence (47). Therefore, this is a very important procedure to study the expression of different antigenic components on the bacterial surface.

7.4 Conclusions

Both adhesins and bacteriocins are molecules involved in probiotic action and are useful as antigens for specific antibody production. Regarding the use of immunochemical methods for probiotic detection, it is important to point out that probiotic action is strain specific. Therefore, once the probiotic strain has been isolated and characterized, it is possible to determine a specific characteristic only present in that strain and absent in other strains of the same species.

For the screening of probiotic strains among a large collection of new isolates, it might be possible to search for the presence of specific molecules involved in probiotic action by the use of immunochemical methods, for example, by exploring the presence of bacteriocins or molecules responsible for mucus adhesion or coaggregation.

Specific antibodies will be used in the future to identify and classify probiotics isolates (ELISA, dot blot, and agglutination), to detect and localize probiotic strains in complex communities like the intestine (immunoblotting and immunofluorescence), to quantify probiotic strains in food products and feces (competitive ELISA and colony blot), to define the mechanism of probiotic action (interaction with epithelium or mucus or other bacteria), to detect extracellular probiotic metabolites (bacteriocins and adhesions), and to study adhesion mechanisms and the role of determined epitopes in surface properties of probiotic microorganisms.

Development of immunochemical methods for probiotics could be applied to the rational selection and design of novel probiotics as well as to the microbiological control of commercial probiotic products and the detection of probiotics in the gastrointestinal tract. Although an important collection of probiotic strains used in food and pharmaceutical products is available, the development of immunochemical methods could facilitate the search for other strains with superior characteristics. Such methods could be used for the detection and study of new strains present in complex microbiological communities, like those present in kefir granules and in the gut of man and animals. The use of immunochemical methods for pursuing these goals requires, for example, knowing the chemical composition of the epitope of an adhesion having a determined probiotic action, so that a specific antibody can be obtained. It must be kept in mind that the probiotic character is strain dependent; as reflected in the bibliography, only certain strains within the *Lactobacillus* and *Bifidobacterium* genera have probiotic characteristics.

Acknowledgment

To Diana Velasco for bibliographic technical assistance.

References

1. AS Naidu, WR Bidlack, and RA Clemens. Probiotic spectra of lactic acid bacteria (LAB). *Crit. Rev. Food Sci. Nutr.*, 38 (1): 13–126, 1999.
2. T Mattila-Sandholm, S Blum, JK Collins, R Crittenden, W de Vos, C Dunne, R Fonden, G Grenov, E Isolauri, B Kiely, P Marteau, L Morelli, A Ouwehand, R Reniero, M Saarela, S Salminen, M Saxelin, E Schiffrin, F Shanahan, E Vaughan, and A von Wright. Probiotics: towards demonstrating efficacy. *Trends Food Sci. Technol.*, 10: 393–399, 1999.
3. JHJ Huis In't Veld and C Shortt. Selection criteria for probiotic microorganisms. *R. Soc. Med. Int. Congr. Symp. Ser.*, 219: 27–36, 1996.
4. YK Lee, K Nomoto, S Salminen, and SL Gorbach. *Handbook of Probiotics*. Wiley-Interscience, New York, 1999.
5. PF Pérez, J Minnaard, EA Disalvo, and GL De Antoni. Surface properties of Bifidobacterial strains of human origin. *Appl. Environ. Microbiol.*, 64: 21–26, 1998.
6. A Gomez Zavaglia, G Kociubinski, PF Pérez, and GL De Antoni. Isolation and characterization of *Bifidobacterium* strains for probiotic formulation. *J. Food Prot.*, 61: 865–873, 1998.

7. I Benno and T Mitsuoka. Impact of *Bifidobacterium longum* on human fecal microflora. *Microbiol. Immun.*, 36: 683–694, 1992.

8. YK Lee, CY Lim, WL Teng, AC Ouwehand, EM Tuomola, and S Salminen. Quantitative approach in the study of adhesión of lactic acid bacteria to intestinal cells and their competition with enterobacteria. *Appl. Environ. Microbiol.*, 66: 3692–369, 2000.

9. K Todoriki, T Mukai, S Sato, and T Toba. Inhibition of adhesion of food-borne pathogens to Caco-2 cells by *Lactobacillus* strains. *J. Appl. Microbiol.*, 91: 154–159, 2001.

10. C Cesena, L Morelli, M Alander, T Siljander, E Tuomola, S Salminen, T Mattila-Sandholm, T Vilpponen-Salmela, and A von Wright. *Lactobacillus crispatus* and its nonaggregating mutant in human colonization trials. *J. Dairy Sci.*, 84: 1001–1010, 2001.

11. EM Tuomola, AC Ouwehand, and SJ Salminen. The effect of probiotic bacteria on the adhesion of pathogens to human intestinal mucus. *FEMS Inmunol. Med. Microbiol.*, 26: 137–142, 1999.

12. G Perdigon, M Medina, E Vintini, and JC Valdez. Intestinal pathway of internalisation of lactic acid bacteria and gut mucosal immunostimulation. *Int. J. Immunopathol. Pharmacol.*, 13: 141–150, 2000.

13. O Sreekumar and A Hosono. The heterocyclic amine binding receptors of *Lactobacillus gasseri* cells. *Mutat. Res.*, 421: 65–72, 1998.

14. PR Lo, RC Yu, CC Chou, and HY Tsai. Antimutagenic activity of several probiotic bifidobacteria against benzo(a)pyrene. *J. Biosci. Bioeng.*, 94: 148–153, 2002.

15. A Gomez Zavaglia, G Kociubinski, PF Pérez, EA Disalvo, and GL De Antoni. Effect of bile on lipid composition and the surface properties of bifidobacteria. *J. Appl. Microbiol.*, 93: 794–799, 2002.

16. G Kociubinski, A Gómez Zavaglia, PF Pérez, EA Disalvo, and GL De Antoni. Effect of bile components on the surface properties of bifidobacteria. *J. Dairy Res.*, 69: 293–302, 2002.

17. AH Richard, P Gilbert, NJ High, PE Kolenbrander, and PS Handley. Bacterial coaggregation: an integral process in the development of multi-species biofilms. *Trends Microbiol.*, 11: 94–100, 2003.

18. M Rinkinen, K Jalava, E Westermarck, S Salminen, and AC Ouwehand. Interaction between probiotic lactic acid bacteria and canine enteric pathogens: a risk factor for intestinal *Enterococcus faecium* colonization? *Vet. Microbiol.*, 92: 111–119, 2003.

19. Ofek I and RJ Doyle. *Bacterial Adhesion to Cells and Tissues.* Chapman & Hall, New York, 1994.

20. J Cleveland, TJ Montville, IF Nes, and ML Chikindas. Bacteriocins: safe, natural antimicrobials for food preservation. *Int. J. Food Microbiol.*, 71(1): 1–20, 2001.

21. RP Ross, S Morgan, and C Hill. Preservation and fermentation: past, present and future. *Int. J. Food Microbiol.*, 79: 3–16, 2002.

22. JM Rodriguez, MI Martinez, N Horn, and HM Dodd. Heterologous production of bacteriocins by lactic acid bacteria. *Int. J. Food Microbiol.*, 80(2): 101–16, 2003.

23. BJ Robison. Immunodiagnostics in the detection of foodborne pathogens. In *Food Microbiological Analysis*. New Technologies. IFT Basic Symposium series. M Tortorelo and S Gendel, Editors. Marcel Dekker Inc. NY. 1997.

24. T Hawryluk and I Hirshfield. A superantigen bioassay to detect staphylococcal enterotoxin A. *J. Food Prot.*, 65 (7): 1183–1187, 2002.

25. N Charni, C Perissol, J Le Petit, and N Rugani. Production and characterisation of monoclonal antibodies against vegetative cells of *Bacillus cereus*. *Appl. Environ. Microbiol.*, 66: 2278–2281, 2000.

26. S Li, RR Marquardt, and D Abramson. Immunochemical detection of molds: a review. *J. Food Prot.*, 63: 281–291, 2000.

27. H Laan, RE Haverkort, L De Leij, and WN Konings. Detection and localization of peptidases in *Lactococcus lactis* with monoclonal antibodies. *J. Dairy Res.*, May 63(2): 245–256, 1996.

28. L Labarre, JF Cavin, C Divies, and J Guzzo. Using specific polyclonal antibodies to study the malolactic enzyme from *Leuconostoc oenos* and other lactic acid bacteria. *Lett. Appl. Microbiol.*, 26(4): 293–296, 1998.

29. L Garrote, M Serradell, A Abraham, MC Añón, G De Antoni, and A Fossati. Detection of heterofermentative lactobacillus in kefir fermented milk. Submitted to *Food and Agricultural Immunology*, 2004.

30. T Yasui and K Yoda. Purification and partial characterisation of an antigen specific to *Lactobacillus brevis* strains with beer spoilage activity. *FEMS Microbiol. Lett.*, 151: 167–176, 1997.

31. Roos S and H Jonsson. A high molecular mass cell surface protein from *Lactobacillus reuteri* 1063 adheres to mucus components. *Microbiology*, 148: 433–442, 2002.

32. T Yasui and K Yoda. The group E antigen is masked by the paracrystalline surface layer in *Lactobacillus brevis*. *J. Ferment. Bioeng.*, 1: 35–40, 1997.

33. T Yasui, K Yoda, and T Kamiya. Analysis of S-layer proteins of *Lactobacillus brevis*. *FEMS Microbiol. Lett.*, 133: 181–186, 1995.

34. M I Martinez, AM Suarez, C Herranz, P Casaus, LM Cintas, JM Rodriguez, and PE Hernandez. Generation of polyclonal antibodies of predetermined specificity against pediocin PA-1. *Appl Environ. Microbiol.*, 64(11): 4536–4545, 1998.

35. JA Durant, CR Young, DJ Nisbet, LH Stanker, and SC Ricke. Detection and quantification of poultry probiotic bacteria in mixed culture using monoclonal antibodies in an enzyme linked immunosorbent assay. *Int. J. Food Microbiol.*, 38: 181–189, 1997.

36. N Yuki, K Watanabe, A Mike, Y Tagami, R Tanaka, M Ohwaki, and M Morotomi. Survival of a probiotic, *Lactobacillus casei* strain Shirota, in the gastrointestinal tract: selective isolation from faeces and identification using monoclonal antibodies. *Int. J. Food Microbiol.*, 48(1): 51–57, 1999.

37. MB Falahee and MR Adam. Cross-reactivity of bacteriocins from lactic acid bacteria and lantibiotics in a nisin bioassay and ELISA. *Lett. Appl. Microbiol.*, 15(5): 214, 1992.

38. CM Franz, M Du Toit, A von Holy, U Schillinger, and WH Holzapfel. Production of nisin-like bacteriocins by *Lactococcus lactis* strains isolated from vegetables. *J. Basic Microbiol.*, 37(3): 187–196, 1997.

39. R Nandakumar, MP Nandakumar, and B Mattiasson. Quantification of nisin in a flow-injection immunoassay. *Biosensors Bioelectronics*, 15: 241–247, 2000.

40. AM Suarez, JM Azcona, JM Rodríguez, B Sanz, and P Hernández. One step purification of nisin A by immunoaffinity chromatography. *Appl. Environ. Microbiol.*, 63: 4990–4992, 1997.

41. G Prioult, C Turcotte, L Labarre, C Lacroix, and I Fliss. Rapid purification of Nisin Z using specific monoclonal antibody-coated magnetic beads. *Int. Dairy J.*, 10: 627–633, 2000.

42. K Arihara, T Toba, and S Adachi. Immunofluorescence microscopic studies on distribution of *Lactobacillus kefiranofaciens* and *Lb kefir* in kefir grains. *Int. J. Food Microbiol.*, 11: 127–134, 1990.
43. LA Herzenberg, DM Weir, *Handbook of Experimental Immunology*. Blackwell Science Publications, Oxford, UK, 1979.
44. A Johnstone and R Thorpe. *Immunochemistry in Practice*. Blackwell Science, Cambridge, MA, 1997.
45. E Harlow and D Lane. *Antibodies: A Laboratory Manual*. Cold Spring Harbor Laboratory Press, Cold Spring Harbor, New York, 1988.
46. FA Golbaum, CP Rubbi, and CA Fossati. Removal of LPS from a *Brucella* cytoplasmic fraction by affinity chromatography with an anti-LPS monoclonal antibody as immunosorbant. *J. Med. Microbiol.*, 40: 174–178, 1994.
47. CA Janeway, P Travers, M Walport, and M Shlomchik. *Immunobiology: The Immune System in Health And Disease*. Garland Publishing, New York, 2001.
48. FM Burnet. *The Clonal Selection Theory of Acquired Immunity*. London, Cambridge, University Press, 1959.
49. G Kohler and C Milstein. Continuous cultures of fused cells secreting antibody of predefined specifiity. *Nature*, 7, 256 (5517): 495–497, 1975.
50. FG Chirdo, MC Añón, and CA Fossati. Development of high-sensitive enzyme immunoassay for gliadin quantification using the streptavidin-biotin amplification system. *Food Agric. Immunol.*, 10: 143–155, 1998.
51. PC Baldi, MM Wanke, ME Loza, and CA Fossati. *Brucella abortus* cytoplasmic proteins used as antigens in an ELISA potentially useful for the diagnosis of canine brucellosis. *Veterinary Microbiol.*, 41: 127–134, 1994.
52. FA Golbaum, CA Velicovsky, PC Baldi, S Mörtl, A Bacher, and CA Fossati. The 18-kDa cytoplasmic protein of *Brucella* species — an antigen useful for diagnosis — is a lumazine synthase. *J. Med. Microbiol.*, 48: 833–839, 1999.

8

Molecular-Based Methods Directed toward Probiotics

Marlene E. Janes and Nigel Cook

CONTENTS

8.1 Introduction

The word probiotic is derived from the Greek meaning "for life". Probiotics are mainly consumed by eating foods or supplements that contain a mono- or mixed culture of live bacteria. The microorganisms exist in a viable state in foods like dairy products or as lyophilized forms in supplements. The beneficial effects attributed to the consumption of probiotic bacteria include improving intestinal tract health, enhancing the immune system, synthesizing and enhancing the bioavailability of nutrients, reducing symptoms of lactose intolerance, decreasing the prevalence of allergy in susceptible individuals, and reducing the risk of chronic disorders like ulcerative colitis and colorectal cancers (1–9).

The probiotic bacteria that can be attributed to improving health are mainly from two genera: *Lactobacillus* (e.g., *Lactobacillus acidophilus*, *L. casei*, *L. bulgaricus*, *L. plantarum*, *L. salivarius*, *L. rhamnosus*, and *L. reuteri*) and *Bifidobac-*

terium (e.g., *Bifidobacterium bifidum, B. longum,* and *B. infantis*) (10). Several molecular methods have been used for the identification of *Lactobacillus* and *Bifidobacterium* species. These methods are important for efficient detection and discrimination of strains in products as well as in feeding trials involving human subjects — so as to better understand the effects probiotics may have on the improvement of health (5, 11–14).

Part of the improvement in the health of individuals who consume probiotic bacteria can be attributed to the production of organic compounds such as lactic and acetic acids in the human intestinal tract or food products. These organic compounds can inhibit the growth of infectious bacteria that cause different types of diarrhea by decreasing pH. Probiotics also promote good digestion and boost immune function (9, 15). A reduction in the gut pH by lactic acid probiotic bacteria may improve intestinal motility and relieve constipation. In this context, the consumption of *Bifidobacterium* and *Lactococcus* strains by elderly patients with constipation improved intestinal motility and bowel behavior (15–17).

Probiotics can also produce bacteriocins that have antimicrobial activity against several foodborne bacteria (18–20). Bacteriocin-producing lactic acid bacteria are naturally found on the surfaces of raw produce, in milk, and in fermented food products (18–20). Generally, bacteriocins have antimicrobial activities against Gram-positive bacteria, but when they are combined with agents that alter the bacterial cell wall, they have been shown to reduce colony counts of both *Salmonella* species and *Escherichia coli* in broth systems (21, 22).

Probiotic bacteria can improve the quantity, availability, and digestibility of some dietary nutrients due to enzymatic reactions (7, 9). Fermentation of food products with some of the lactic acid probiotic bacteria increases bioavailability of protein and fat, B vitamins, especially folic acid and biotin, and the production of free amino acids (7, 9). Because the B vitamins are biocatalysts that increase digestive enzymes, probiotic bacteria can improve the overall digestive process and help to alleviate symptoms of intestinal malabsorbtion (7).

One of the most common health benefits of probiotic bacteria is their ability to alleviate symptoms of lactose intolerance, which affects around half of the world's population, by the production of lactase during the fermentation of food products or in the intestinal tract or stomach of humans who consume the probiotic cultures (9).

Molecular methods have been used to detect and monitor in time the lactic acid bacterial population of ripening cheese products (23, 24). These studies indicated that the predominant bacterial strains in the cheese products were *Lactobacillus* species and *Streptococcus thermophilus* (23, 24). Commercial yogurts analyzed by polymerase chain reaction-denaturing gradient gel electrophoresis (PCR-DGGE) contained the typical species associated with yogurt production, *Streptococcus thermophilus* and *Lactobacillus delbrueckii* subsp. *bulgaricus,* along with *Bifidobacterium lactis* (25).

In order to decrease the incidence of gastric or intestinal illness, humans should consume probiotic bacteria at a level of 10^7 to 10^{11} per day (2, 4, 5). A clinical trial with 60 patients who had inflammatory bowel diseases who consumed a daily 400-ml drink containing 5×10^7 CFU/ml of *Lactobacillus plantarum* showed a reduction in pain and flatulence after 4 weeks of intake (26). PCR analysis, using species-specific primers, of fecal samples from 10 patients affected by inflammatory bowel diseases who consumed a commercial probiotic supplement containing 3×10^{11} CFU/g of viable lyophilized bacteria for 2 months showed a similar bacterial colonization pattern as in healthy subjects (13, 14).

Several studies using molecular techniques have evaluated the retention of bacterial strains in feces of individuals consuming commercially available probiotic supplements that contain the following lyophilized bacterial strains: *S. thermophilus, B. longum, B. infantis, B. breve, L. acidophilus, L. casei, L. delbrueckii* subsp. *bulgaricus, L. plantarum* (13, 14, 26, 27). The results of these studies indicated that persistence in the human gut is dependent upon the specific strain of probiotic and the microflora of the individuals being evaluated (13, 14).

With the development of cultural-independent molecular techniques, a more detailed analysis of the intestinal tract microflora can be evaluated. In addition, molecular techniques will allow a more comprehensive understanding of the interaction of probiotic bacteria with the other microflora in the intestinal tract. There is an extensive library of DNA primers available for the detection of probiotic bacteria that are genus or species specific and even strain specific (28–34). These primers were developed from the most variable region of the bacterial DNA sequence, the region that codes for the 16S or 23S ribosomal RNA (rRNA).

8.2 Molecular-Based Identification Methods

A variety of DNA-based methods can be employed to identify probiotic microorganisms. They are based mainly on restriction enzyme analysis or PCR, sometimes in combination.

8.2.1 Isolation of DNA

The first step in the molecular identification of probiotic bacteria is the isolation of the DNA from pure cultures, fecal material, or food products. Klijn et al. (33) developed a method to isolate small amounts of DNA from a single colony. The authors found that this method of isolation produced reliable results that could be used to characterize isolates from starter cultures and environmental samples (Protocol 8.1).

PROTOCOL 8.1

Isolation of DNA from a Single Colony[a]

Cell Lysis:

1. Suspend one colony into 50 µl of 10 mM Tris HCl buffer (TE buffer) (pH 8.0) containing 400-µl of lysozyme.
2. Incubate at 37°C for 40 min.
3. Add 50 µl of 10% SDS (sodium dodecyl sulfate) and 250 µl of the TE buffer.
4. Incubate for 30 min at 50°C, and then centrifuge at 12,000 g for 3 min at 4°C.
5. Pipette supernatant into a fresh 1.5-ml tube.

DNA Precipitation:

1. Add 60 µl of 3 M sodium acetate and 1 ml of 96% ethanol to the supernatant, and place on ice for 20 min.
2. Centrifuge at 12,000 g for 20 min at 4°C.
3. Dissolve the DNA pellet in 50 µl of the TE buffer.

---

[a] Protocol from Klijn et al. (33).

A more simple method used in our laboratory for bacterial cell lysis is to suspend a single colony into 400-µl dH$_2$O in a microcentrifuge tube and then place the tube into a 95°C water bath for 5 min. After heating, the tubes are immediately placed into ice that causes the bacterial cells to lyse. This method can be used for the detection and identification of pure bacterial cultures with the use of PCR primers.

A more detailed procedure has to be followed for the isolation of DNA from mixed microbial communities such as those found in fecal material and food products (Protocol 8.2). Mainly this is due to the enzymes found in these mixed environments that may interfere with the reagents used during PCR such as the Taq polymerase. Recently, a rapid method for detection and enumeration of *Streptococcus thermophilus* and *Bifidobacterium* strains from human fecal material was developed that does not require time-consuming DNA isolation and purification (Protocol 8.3). This method uses centrifugation and a washing procedure to separate bacteria from the enzyme contaminants of fecal samples.

8.2.2 Primers for Identification of Probiotic Bacteria

Primers can be used to study microbial habitats and ecology, or as rapid methods for the detection or identification of a particular microorganism. A variety of DNA primers have been sequenced for most probiotic bacteria (Tables 8.1 and 8.2). These primers were sequenced from the 16S and 23S ribosomal DNA regions of the gene that specifically codes for the V1 and V2 variable regions. The primers developed from the V1 and V2 regions of

---

PROTOCOL 8.2

Isolation of DNA from Fecal and Food Samples Using Mini Bead Beater

---

Cell Lysis:

1. Add 50 to 100 mg (wet weight) of feces or food products to a 2-ml screw-cap tube (keep tubes on ice).
2. Add 500 µg of lysis buffer (200 mM NaCl, 200 mM TrisBase, 2 mM sodium citrate, 10 mM CaCl$_2$, 50 mM EDTA), 20 µl of poly A (10 mg/ml), 20 µl of 10% pyrophosphate, and 30 µl of lysozyme (100 mg/ml).
3. Incubate for 40 min at 37°C.
4. Add 10 µl of 20% SDS and 500 µl of phenol:CHCl3:IAA (24:24:1) and 0.3 g acid-washed zirconium beads. Mix on low setting for 2 min in the Mini Bead Beater.
5. Spin at 12,000 g for 3 min at 4°C. Pipette supernatant into fresh 1.5-ml tube.

DNA Precipitation:

1. Extract samples with one volume of phenol:CHCl3:IAA (24:24:1). Mix by inversion and centrifuge at 12,000 g for 2 min at 25°C. Remove aqueous upper layer and put in new tube.
2. Precipitate nucleic acids by adding an equal volume of isopropanol and 0.1 volume 3 M NaAcetate, pH 5.2. Put on ice for 20 min, next spin at 12,000 g for 20 min at 4°C.
3. Remove the supernatant, and rinse pellet with 500 µl of cold 70% ethanol. Invert tube and allow to air dry for 30 min. Resuspend pellets in 20 µl of TE buffer.

---

[a] Protocol from Reference 68.

PROTOCOL 8.3

Isolation of DNA from Fecal Samples by Centrifugation[a]

---

1. Add 1 g of feces to 9 ml of sterile 0.05 M phosphate-buffered saline (PBS) (pH 7.4), and mix by vortexing for 5 to 10 min.
2. Centrifuge the sample at 2450 × g for 1 min, and collect the supernatant collected; repeat three times.
3. Collect the bacterial pellet by centrifuging the supernatant at 8000 × g for 3 min.
4. Wash the bacterial pellet four times with 2.5 ml of PBS, and then wash with 1 ml of sterile water.
5. Resuspend the bacterial pellet with 0.3 ml of Triton X-100.
6. Heat the sample at 100°C for 5 min, and immediately cool in ice; repeat five times.

---

[a] Protocol from Reference 14.

TABLE 8.1

Primers Used for the Detection and Identification of Probiotic Lactobacilli Species

Species	Primer Sequences	PCR Fragment Length (bp)	PCR Annealing Temp (°C)	MgCl$_2$	Ref.
L. delbrueckii	ACGGATGGATGGAGAGCAG GCAAGTTTGTTCTTTCGAACTC	200–300	72	3.0	64
L. acidophilus	AGCTGAACCAACAGATTCAC ACTACCAGGGTATCTAATCC	200–300	62	1.5	35
L. acidophilus	TCTAAGGAAGCGAAGGAT CTCTTCTCGGTCGCTCTA	200–300	72	3.0	64
L. rhamnosus	CTATTTAGTAATCACAGAAAA TAACAGCAGTCTCCAAATGG	595	72	3.0	65
L. rhamnosus	CAGACTGAAAGTCTGACGG GCGATGGCGAATTTCTATTATT	200–300	58	2.0	35
L. casei	CAGACTGAAAGTCTGACGG GCGATGCGAATTTCTTTTTC	290	55	2.0	35
L. plantarum	GCCGCCTAAGGTGGGACAGAT TTACCTAACGGTAAATGCGA	200–300	55	2.0	35
L. fermentum	GCCGCCTAAGGTGGGACAGAT CTGATCGTAGATCAGTCAAG	200–300	55	3.0	35
L. sharpeae	GATAATCATGTAAGAAACCGC ATATTGTTGGTCGCGATTCG	200–300	58	1.5	35
L. curvatus	GCTGGATCACCTCCTTTC TTGGTACTATTTAATTCTTAG	250	72	1.0	24
L. pentosus	GCTGGATCACCTCCTTTC GTATTCAACTTATTAGAACG	250	72	1.0	24

TABLE 8.2

Primers Used for the Detection and Identification of Probiotic Bifidobacterial Species

Species	Primer Sequences	PCR Fragment Length (bp)	PCR Annealing Temp (°C)	Ref.
B. adolescentis	CTCCAGTTGGATGCATGTC CGAAGGCTTGCTCCCAGT	279	72	66
B. bifidum	CCACATGATCGGCATGTGATTG CCGAAGGCTTGCTCCCAA	278	72	66
B. breve	CCGGATGCTCCATCACAC ACAAAGTGCCTTGCTCCCT	340	72	66
B. infantis	CCATCTCTGGGATCGTCGG TATCGGGGAGCAAGCGTGA	565	72	55
B. longum	GTTCCCGACGGTCGTAGAG GTGAGTTCCCGGCATAATCC	153	72	67
B. lactis	GTGGAGACACGGTTTCCC CACACCACACAATCCAATAC	680	95	36

the 16S and 23S rDNA are specific enough that bacteria can be identified at the genus, species, and even subspecies level. The use of these primers will give us a better understanding of the dosage of probiotic bacteria needed for persistence within the intestine, if particular probiotic bacteria can colonize the intestine and how long it takes for the probiotic bacteria to be eluted from the gut. Several biotechnology companies produce the primers and can be located on the Internet.

8.2.3   PCR-Based Identification Methods

PCR is a rapid method for the amplification of DNA fragments from a known sequence using two primers to generate the PCR fragment (see Tables 8.1 and 8.2). The main instrument needed to perform PCR is a thermocycler that is available through most biotechnology companies. The basic procedure for preparing the samples for generating the PCR fragments is explained in the protocols throughout this article. The thermocycler is programmed according to the manufacturer's instructions. The basic cyclic parameters for amplification of *Lactobacillus* species rDNA by PCR are as follows: initial denaturation at 94°C for 2 to 4 min; 30 to 35 cycles of 94°C for 30 sec (denaturation), 56 to 61°C for 30 to 60 sec (amplification), and 68° for 1 min (annealing); final extension 68°C for 7 min (27, 31, 35). The cyclic parameters for *Bifidobacterium* species are as follows: initial denaturation at 95°C for 5 min; 40 cycles of 95°C for 1 min (denaturation), 53°C for 30 sec (amplification), and 72°C for 5 min (annealing); final extension 68°C for 7 min (Protocol 4). When low numbers of the indicator bacteria are expected, the number of PCR cycles should be increased. One study using a pure culture of *B. lactis* indicated that after 25 cycles of PCR, 100 bacteria/ml could be detected, and by increasing to 50 cycles, 1 to 10 bacteria/ml were detected (36).

Sometimes it can be advantageous to identify simultaneously several different species of bacteria within one reaction tube, and this can be done by Multiplex PCR. During Multiplex PCR, several primer sets are added to a single reaction tube along with mixed DNA templates. This method has several advantages; mainly it can reduce the cost and time associated with identifying single bacterial species. Henegariu et al. (37) developed a step-by-step protocol for conducting Multiplex PCR that also included all possible pitfalls of the procedure.

Theoretically, PCR should, by judicious choice of the amplified sequence, be able to identify a target simultaneously with its detection, i.e., the presence of a PCR signal should be completely diagnostic for the desired target. However, it is not always easy to find sequences that are suitably variant among species or strains such that they can be used to mediate discrimination between them in a single PCR. Amplification of random or repetitive sequences can be used as approaches to identify closely related strains, and have been reported as useful in this regard for probiotic microorganisms.

Random amplification of polymorphic DNA (RAPD) is a PCR-based method in which a pattern of amplicons is produced through the simulta-

---

PROTOCOL 8.4

Identification of Probiotic Bacteria Using Polymerase Chain Reaction (PCR)[9]

1. To a 0.5- or 0.2-ml microcentrifuge tube (depending on the thermal cycler) on ice add:

2. 26.5 μl of water (double-distilled)

3. 5 μl of 10 *Taq* buffer (0.5 M KCl, 0.1 M Tris – HCl pH 9, 1% Triton X – 1.5 mM $MgCl_2$[a])

4. 5 μl of mixed dNTPs, each at 200 μM (dCTP + dGTP + dATP + dTTP)

5. 6 μl of forward primer (50 μg/ml)

6. 6 μl of reverse primer (50 μg/ml)

7. 1 μl of substrate DNA (1 ng/μl)

8. 0.5 μl of Taq (or similar) DNA polymerase (5 units/μl)

9. Mix, then place in the thermal cycle, and initiate PCR. See references in Tables 8.1 and 8.2 for the PCR conditions for the specific primers.

10. Remove 5 μl, and run on a gel to verify that the specific DNA was present.

---

[a] The concentration of $MgCl_2$ varies for the primer sets used by the lactobacilli species (see Table 8.1).

---

[a] Protocol from Reference 69.

---

neous amplification of many chromosomal sequences mediated by annealing of short oligonucleotide primers. The primers are composed of random nucleotide sequences and are not designed to match any specific sequence in the target organism's DNA, but by using a low annealing temperature in the reaction will allow them to bind and mediate amplification. If cycling conditions are maintained, then the pattern produced by RAPD should be reproducible for an individual species or strain. In actuality, the method may be affected by variability in laboratory conditions or equipment. Furthermore, it can require some effort to find suitable primers to produce specific RAPD patterns for particular strains of interest (38). Nonetheless, RAPD has been shown to be capable of differentiating between *L. acidophilus* group strains (39) and has been useful for monitoring introduced and indigenous lactobacilli in the intestinal tract (40). RAPD analysis of yeast isolates from feta cheese provided reliable identification at species level and good discrimination at strain level (12).

Amplification of repetitive chromosomal sequences can produce distinctive patterns between closely related species or strains. Ventura and Zink (41) used enterobacterial repetitive intergenic consensus PCR to explore phylogenetic relationships between *B. animalis* and *B. lactis* isolates; their findings led them to support redefining the classification of *B. lactis* as a subspecies of *B. animalis*.

PCR and restriction enzyme analysis can be used in combination to facilitate identification of microorganisms, and this approach has been used for the study of probiotic organisms. Digestion of amplified 16S ribosomal gene sequences can produce group- or species-distinctive patterns; it has been employed to identify *Lactobacillus* spp. (42). A variation of this approach, which compared the size of fragments produced after restriction of labeled 16S amplicons from fecal samples, has been used to monitor bacterial populations in the digestive tract after ingestion of *L. acidophilus* (43). This technique, terminal restriction fragment (TRF) pattern analysis, can allow several species or strains to be visualized simultaneously and can be employed directly on fecal or environmental samples, without the necessity of prior culturing (44).

Other culture-independent techniques, which can mediate the identification of individual bacterial species or strains, include DGGE (45). DGGE analysis is based on the differences in electrophoretic mobility of partially denatured amplicons in polyacrylamide gels containing a linear gradient of chemical denaturants (urea and formamide) or temperature. Molecules with different sequences may have different melting behaviors and therefore migrate differently; this property can be used to separate otherwise indistinguishable amplicons from different species or strains. DGGE has been used to detect *Lactobacillus* and other species in the human gastrointestinal tract (46), to monitor *Bifidobacterium* populations (47), and to study the development of the *Lactobacillus*-like community (28) in the gastrointestinal tract of human volunteers; and to verify the bacterial species present in commercial probiotic products (24, 25, 48).

Many of the studies mentioned above have found that the identity of probiotic strains used in commercial products does not always conform to that which is designated in the product. Klein et al. (49) strongly recommended that the taxonomy of probiotic microorganisms should be clarified, and considered that molecular techniques should be used in combination with culture-based techniques to achieve this. They affirmed that knowledge of taxonomy would provide the basis for quality assurance schemes, as required by national legislation and consumer interest. Harmonized approaches are required to elucidate fully the taxonomy of probiotic organisms and produce validated methods for determination of strain identity. The acquisition of full taxonomic knowledge and production of widely accepted identification methods should ensure that probiotics are safely and beneficially used.

8.2.4   Restriction Enzyme Analysis-Based Identification Methods

Restriction enzymes cut, or digest, DNA at specific sequences. The resulting fragments are separated by size using electrophoresis through agarose gels, producing a pattern of bands after staining with a fluorescent dye. The banding pattern can be representative of a particular species or strain. The banding patterns may be quite difficult to interpret without sophisticated computer-aided analysis (50), and care must be taken in the choice of enzymes used. Identification methods based on restriction enzyme analysis, which have been used to identify probiotic microorganisms, include ribotyping and pulsed field gel electrophoresis.

Ribotyping involves typing species and strains through characteristic variations in their ribosomal (r)RNA genes. Chromosomal DNA is digested by restriction enzymes, and after electrophoresis the DNA is transferred to a membrane by Southern blotting. The membrane-bound DNA is hybridized to probes, which are specific for 16S and 23S rRNA genes. The probes are labeled to allow visualization of the pattern produced, which is used to determine the identity of the original bacterium. In general, ribotyping produces easily interpretable fingerprint patterns and can be particularly useful in species recognition (50). Ribotyping has been used to confirm the identity of probiotic *Lactobacillus* strains inoculated into foods (51) and to identify *Lactobacillus* isolates from human colonic biopsy samples (52).

When electrophoresis of digested DNA is performed with periodic changes in the orientation of the electric field, it can facilitate the separation of high-molecular-weight fragments. This is the basis of pulsed field gel electrophoresis, or PFGE. Commonly, rare-cutting restriction enzymes are used; these act at uncommon DNA sequences and thus produce large fragments that can be several hundred kilobases in length. The technique has high discriminatory power. Roussel et al. (53) used it to differentiate closely related strains of *L. acidophilus*, and Ferrero et al. (54) used it to characterize *L. casei* strains. Roy et al. (55) differentiated commercially available bifidobacterial strains by PFGE, and it has also been used to determine the identity of *Lactobacillus* and *Bifidobacterium* strains that had probiotic properties (56). PFGE was shown to be more efficient than ribotyping for differentiation of *Enterococcus faecalis* strains (57). Lund et al. (58) and Gelsomino et al. (59) used the technique to confirm the establishment in the intestine of enterococcal species following the daily consumption of dairy products containing them. RAPD has also been used to confirm the identity of *Pediococcus* spp. (60).

8.3 Future Directions

With the burgeoning developments in PCR technology and the increasing use of methods based on PCR for detection of microorganisms, the need for international standardization has become apparent (61, 62). With regard to probiotics, regulatory measures to ensure the identity of a product may be underpinned by standard detection and identification procedures. The provision of these standards will require harmonized approaches to development and evaluation, among those laboratories involved in method development.

However, PCR-based detection does not unequivocally indicate that the target was viable. An alternative molecular detection technique, nucleic acid sequence-based amplification (NASBA), may be able to provide rapid detection of viable microorganisms (63–67) and could be worth consideration for future probiotics testing.

8.4   Conclusion

Molecular techniques can be useful tools for evaluating the effectiveness of probiotic bacteria in the prevention and treatment of gastrointestinal disorders. Now that a large set of rDNA primers are available for the identification of probiotic bacteria, more extensive research needs to be conducted to determine the most effective probiotic bacteria for improving the health of humans. Mainly, more work needs to be done to determine if probiotic bacteria are capable of persisting within the human intestinal tract and what conditions will improve or enable them to colonize therein.

References

1. S Bengmark. Colonic food: pre- and probiotics. *Am. J. Gastroenterol.*, 95: S5–S7, 2000.
2. L Sherwood, MD Gorbach. Probiotics and gastrointestinal health. *Am. J. Gastroenterol.*, 95:S2–S4, 2000.
3. J Levy. The effects of antibiotic use on gastrointestinal function. *Am. J. Gastroenterol.*, 95: S9–S10, 2002.
4. M Pochapin. The effect of probiotics on *Clostridium difficile* diarrhea. *Am. J. Gastroenterol.*, 95: S11–S13, 2000.
5. J Saavedra. Probiotics and infectious diarrhea. *Am. J. Gastroenterol.*, 95: S16–S18, 2000.
6. M Schultz, RB Sartor. Probiotics and inflammatory bowel diseases. *Am. J. Gastroenterol.*, 95: S19–S21, 2000.
7. S Cunningham-Rundles, S Ahrne, S Bengmark, R Johann-Liang, F Marshall, L Metakis, C Califano, AM Dunn, C Grassey, G Hinds, J Cervia. Probiotics and immune response. *Am. J. Gastroenterol.*, 95: S22–S25, 2000.
8. Vanderhoof JA. Summary. *Am. J. Gastroenterol.*, 95: S27, 2000.
9. Kopp-Hoolihan L. Prophylactic and therapeutic uses of probiotics: a review. *J. Am. Diet. Assoc.*, 101: 229–241, 2001.
10. GW Tannock, A Tilsala-Timisjarvi, S. Rodtong, J NG, K Munro. Identification of *Lactobacillus* isolates from the gastrointestinal track, silage and yoghurt by 16S-23S gene intergenic spacer region sequence comparisons. *Appl. Environ. Microbiol.*, 65: 4264–4267, 1999.

11. GW Tannock. Molecular methods for exploring the intestinal ecosystem. *Br. J. Nutr.*, 87: 199–201, 2002.
12. E Psomas, C Andrghetto, E Litopoulou-Tzanetaki, A Lombardi, N Tzanetakis. Some probiotic properties of yeast isolates from infant faeces and Feta cheese. *Int. J. Food Microbiol.*, 69: 125–133, 2001.
13. P Brigidi, B Vitali, E Swennen, L Altomare, M Rossi, D Matteuzzi. Specific detection of *Bifidobacterium* strains in a pharmaceutical probiotic product in human feces by polymerase chain reaction. *Syst. Appl. Microbiol.*, 23: 391–399, 2000.
14. P Brigidi, E Swennen, B Vitali, M Rossi, D Matteuzzi. PCR detection of Bifidobacterium strains and *Streptococcus thermophilus* in fecus of human subjects after oral bacteriotherapy and yogurt consumption. *Int. J. Food Microbiol.*, 81: 203–209, 2003.
15. A Bennet, KG Eley. Intestinal pH and propulsion: an explanation of diarrheas in lactase deficiency and laxation by lactulose. *J. Pharm. Pharmacol.*, 28: 192–195, 1976.
16. M Seki, T Igarashi, Y Fukuda, S Simamura, T Kaswashima, K Ogasa. The effect of *Bifidobacterium* cultured milk on the "regularity" among an aged group. *Nutr. Foodstuff*, 31: 379–387, 1978.
17. L Motta, G Blancato, G Scornavacca, M De Luca, E Vasquez, MR Gismondo, A lo Bue, G Chisari. Study on the activity of a therapeutic bacterial combination in intestinal motility disorders in the aged. *Clin. Ter.*, 138: 27–35, 1991.
18. ME Janes, R Nannapaneni, MG Johnson. Identification and characterization of two bacteriocin producing bacteria isolated from garlic and ginger root. *J. Food Prot.*, 62: 899–904, 1999.
19. Z Yildirim, MG Johnson. Characterization and antimicrobial spectrum of bifidocin B, a bacteriocin produced by *Bifidobacterium bifidum* NCFB. *J. Food Prot.*, 61: 47–51, 1998.
20. Z Yildirim, MG Johnson. Detection and characterization of a bacteriocin produced by *Lactococcus lactis* subsp. *cremoris* isolated from radish. *Lett. Appl. Microbiol.*, 26: 297–304, 1998.
21. CK Cutter, GR Siragusa. Treatments with nisin and chelators to reduce *Salmonella* and *Escherichia coli* on beef. *J. Food Prot.*, 57: 1028–1030, 1995.
22. KA Stevens, NA Klapes, BW Sheldon, TR Klaenhammer. Nisin treatment for inactivation of *Salmonella* species and other Gram-negative bacteria. *Appl. Environ. Microbiol.*, 57: 3613–3615, 1991.
23. L Mannu, R Giovanni, R Comunia, MC Fozzi, MF Scintu. A preliminary study of lactic acid bacteria in whey starter culture and industrial Pecorino Sardo Ewes' milk cheese: PCR-identification and evolution during ripening. *Int. Dairy J.*, 12: 17–26, 2002.
24. F Berthier, E Beuvier, A Dasen, R Grappin. Origin and diversity of mesophilic lactobacilli in Comte cheese, as revealed by PCR with repetitive and species-specific primers. *Int. Dairy J.*, 11: 293–305, 2001.
25. S Fasoli, M Marzotto, L Rizzotti, F Rossi, F Dellaglio, S Torriani. Bacterial composition of commercial probiotic products as evaluated by PCR-DGGE analysis. *Int. J. Food Microbiol.*, 82: 59–70, 2003.
26. S Nobaek, ML Johansson, G Molin, S Ahrn's, B Jeppsson. Alteration of intestinal microflora is associated with reduction in abdominal bloating and pain in patients with irritable bowel syndrome. *Am. J. Gastroenterol.*, 95: 1231–1238, 2000.

27. RM Satokari, EE Vaughan, ADL Akkermans, M Saarela, WM DE Vos. Polymerase chain reaction and denaturing gradient gel electrophoresis monitoring of fecal *Bifidobacterium* populations in a prebiotic and probiotic feeding trial. *Syst. Appl. Microbiol.*, 24: 227–231, 2001.

28. HGHJ Heilig, EG Zoetendal, EE Vaughan, P Marteau, ADL Akkermanns, WM de Vos. Molecular diversity of *Lactobacillus* spp. and other lactic acid bacteria in the human intestine as determined by specific amplification of 16S ribosomal DNA. *Appl. Environ. Microbiol.*, 68: 114–123, 2002.

29. J Sui, S Leighton, F Busta, L Brady. 16S ribosomal DNA analysis of the fecal lactobacilli composition of human subjects consuming a probiotic strain *Lactobacillus acidophilus* NCFM. *J. Appl. Microbiol.*, 93: 907–912, 2002.

30. M Drake, CL Small, KD Spence, BG Swanson. Rapid detection and identification of *Lactobacillus* spp. in dairy products by using the polymerase chain reaction. *J. Food Prot.*, 59: 1031–1036, 1996.

31. MI Castellanos, A Chauvet, A Deschamps, C Barreau. PCR methods for identification and specific detection of probiotic lactic acid bacteria. *Curr. Microbiol.*, 33: 100–103, 1996.

32. MJ Kullenm, RB Sanozky-Dawes, DC Crowell, TR Klaenhammer. Use of the DNA sequence of variable regions of the 16S rRNA gene for rapid and accurate identification of bacteria in the *Lactobacillus acidophilus* complex. *J. Appl. Microbiol.*, 89: 511–516, 2000.

33. N Klijn, AH Weerkamp, WM de Vos. Identification of mesophilic lactic acid bacteria by using polymerase chain reaction-amplified variable regions of 16S rRNA and specific DNA probes. *Appl. Environ. Microbiol.*, 57: 3390–3393,1991.

34. S Torriani, G Zapparoli, F Dellagho. Use of PCR-based methods for rapid differentiation of *Lactobacillus delbrueckii* subsp. *bulgaricus* and *L. delbrueckuii* subsp. *lactis*. *Appl. Environ. Microbiol.*, 65: 4351–4356, 1999.

35. J Walter, GW Tannock, A Tilsala-Timisjarvi, S Rodtong, DM Loach, K Munro, T Alatossava. Detection and identification of gastrointestinal *Lactobacillus* species by using denaturing gradient gel electrophoresis and species-specific PCR primers. *Appl. Environ. Microbiol.*, 66: 297–303, 2000.

36. M Ventura, R Reniero, F Zink. Specific identification and targeted characterization of *Bifidobacterium lactis* from different environmental isolates by a combined multiplex-PCR approach. *Appl. Environ. Microbiol.*, 67: 2760–2765, 2001.

37. O Henegariu, NA Heerema, SR Dlouhy, GH Vance, PH Vogt. Multiplex PCR: Critical parameters and step-by-step protocol. *BioTechniques*, 23: 504–511, 1997.

38. K Brandt, A Tilsala-Timisjarvi, T Alatossava. Phage-related DNA polymorphism in dairy and probiotic *Lactobacillus*. *Micron*, 32: 59–65, 2001.

39. EM Du Pleiss, LMT Dicks. Evaluation of random amplified polymorphic DNA (RAPD)-PCR as a method to differentiate *Lactobacillus acidophilus*, *Lactobacillus crispatus*, *Lactobacillus amylovorus*, *Lactobacillus gallinarum*, *Lactobacillus gasseri*, and *Lactobacillus johnsonii*. *Curr. Microbiol.*, 31: 114–118, 1995.

40. GE Gardiner, C Heinemann, ML Baroja, AW Bruce, D Beuerman, J Madrenas, G Reid. Oral administration of the probiotic combination *Lactobacillus rhamonosus* GR-1 and *L. fermentum* for human intestinal applications. *Int. Dairy J.*, 12: 191–196, 2002.

41. M Ventura, R Zink. Rapid identification, differentiation, and proposed new taxonomic classification of *Bifidobacterium lactis*. *Appl. Environ. Microbiol.*, 68: 6429–6434, 2002.

42. D Roy, S Sirois, D Vincent. Molecular discrimination of Lactobacilli used as starter and probiotic cultures by amplified ribosomal DNA restriction analysis. *Curr. Microbiol.*, 42: 282–289, 2001.
43. M Ferrero, C Cesena, L Morelli, G Scolari, M Vescovo. Molecular characterization of *Lactobacillus casei* strains. *FEMS Microbiol. Lett.*, 140: 215–219, 1996.
44. BG Clement, LE Kehl, KL DeBord, CL Kitts. Terminal restriction fragment patterns (TRFPs), a rapid, PCR-based method for the comparison of complex bacterial communities. *J. Microbiol. Methods*, 31: 135–142, 1998.
45. G Muyzer. DGGE/TGGE a method for identifying genes from natural ecosystems. *Curr. Opin. Microbiol.*, 2: 317–322, 1999.
46. J Walter, C Hertel, GW Tannock, CM Lis, K Munro, WP Hammes. Detection of *Lactobacillus, Pediococcus, Leuconostoc,* and *Weissella* species in human feces by using group-specific PCR primers and denaturing gradient gel electrophoresis. *Appl. Environ. Microbiol.*, 67: 2578–2585, 2001.
47. RM Satokari, EE Vaughan, ADL Akkermans, M Saarela, WM De Vos. Bifidobacterial diversity in human feces detected by genus specific PCR and denaturing gradient gel electrophoresis. *Appl. Environ. Microbiol.*, 67: 504–513, 2001.
48. R Temmerman, I Scheirlink, G Huys, J Swings. Culture-independent analysis of probiotic products by denaturing gel electrophoresis. *Appl. Environ. Microbiol.*, 69: 220–226, 2003.
49. G Klein, A Pack, C Bonaparte, G Reuter. Taxonomy and physiology of probiotic lactic acid bacteria. *Int. J. Food Microbiol.*, 41: 103–125, 1998.
50. WP Charteris, PM Kelly, L Morelli, DK Collins. Selective detection, enumeration and identification of potentially probiotic *Lactobacillus* and *Bifidobacterium* species in mixed bacterial populations. *Int. J. Food Microbiol.*, 35: 1–27, 1997.
51. S Erkkilä, ML Suihko, S Eerola, E Petäjä, T Mattila-Sandholm. Dry sausage fermented by *Lactobaccillus rhamnosus* strains. *Int. J. Food Microbiol.*, 64: 205–210, 2001.
52. P Kontula, ML Suihko, T Suortti, M Tenkanen, T Mattila-Sandholm, A von Wright. The isolation of lactic acid bacteria from human colonic biopsies after enrichment on lactose derivatives and rye arabinoxylo-oligosaccharides. *Food Microbiol.*, 17: 13–22, 2000.
53. Y Roussel, C Colmin, JM Simonet, B Decaris. Strain characterization, genome size and plasmid content in the *Lactobacillus acidophilus* group (Hansen and Mocquot). *J. Appl. Bacteriol.*, 74: 549–556, 1993.
54. M Ferrero, C Cesena, L Morelli, G Scolari, M Vescovo. Molecular characterization of *Lactobacillus casei* strains. *FEMS Microbiol. Lett.*, 140: 215–219, 1996.
55. D Roy, P Ward, G Champagne. Differentiation of bifidobacteria by use of pulsed-field gel electrophoresis and polymerase chain reaction. *Int. J. Food Microbiol.*, 29: 11–29, 1996.
56. J Prasad, H Gill, J Smart, PK Gopal. Selection and characterisation of *Lactobacillus* and *Bifidobacterium* strains for use as probiotics. *Dairy J.*, 8: 993–1002, 1998.
57. ME Gordillo, KV Singh, BE Murray. Comparison of ribotyping and pulsed-field gel electrophoresis for subspecies differentiation of strains of *Enterococcus faecalis*. *J. Clin. Microbiol.*, 31: 1570–1574, 1993.
58. B Lund, I Adamsson, C Edlund. Gastrointestinal transit survival of an *Enterococcus faecium* probiotic administered with or without vancomycin. *Int. J. Food Microbiol.*, 77: 109–115, 2002.

59. R Gelsomino, M Vancanneyt, TM Cogan, J Swings. Effect of raw-milk cheese consumption on the enterococcal flora of human feces. *Appl. Environ. Microbiol.*, 69: 312–319, 2003.
60. PJ Simpson, C Stanton, GF Fitzgerald, RP Ross. Genomic diversity within the genus *Pediococcus* as revealed by randomly amplified polymorphic DNA, PCR and pulsed-field gel electrophoresis. *Appl. Environ. Microbiol.*, 68: 765–771, 2002.
61. J Hoorfar, N Cook. Critical aspects of standardization of PCR. In: K Sachse, J Frey, Eds., *Methods in Molecular Biology: PCR Detection of Microbial Pathogens.* Humana Press, Totowa, NJ, 2002, pp. 51–64.
62. B Malorny, PT Tassios, P Rådström, N Cook, M Wagner, J Hoorfar. Standardization of diagnostic PCR for the detection of foodborne pathogens. *Int. J. Food Microbiol.*, 83: 39–48, 2003.
63. N Cook. The use of NASBA for the detection of microbial pathogens in food and environmental samples. *J. Microbiol. Methods*, 53: 165–174, 2003.
64. A Tilsal-Timisjarvi, T Alatossava. Development of oligonucleotide primers from the 16S-23S rRNA intergenic sequences for identifying different dairy and probiotic lactic acid bacteria by PCR. *Int. J. Food Microbiol.*, 35: 49–56, 1997.
65. A Tilsala-Timisjarvi, T Alatossava. Strain-specific identification of probiotic *Lactobacillus rhamnosus* with randomly amplified polymorphic DNA-derived PCR primers. *Appl. Environ. Microbiol.*, 64: 4816–4819, 1998.
66. T Matsuki, K Watanabe, R Tanaka, H Oyaizu. Rapid identification of human intestinal bifidobacteria by 16S rRNA-targeted species- and group-specific primers. *FEMS Microbiol. Lett.*, 167: 113–121, 1998.
67. RF Wang, WW Cao, CE Cerniglia. PCR detection and quantitation of predominant anaerobic bacteria in human and animal fecal samples. *Appl. Environ. Microbiol.*, 62: 1242–1247, 1996.
68. RG Kok, A Dewaal, F Schut, GW Wellins, G Wenk, KJ Hellingwerf. Specific detection and analysis of probiotic *Bifidobacterium* species in infant feces. *Appl. Environ. Microbiol.*, 62: 3668–3672, 1996.
69. JR Johnston. Molecular *Genetics of Yeast: A Practical a Practical Approach.* IRL Press of Oxford University Press, Oxford, UK., 1994.

9

Application of Repetitive Element
Sequence-Based (rep-) PCR and Denaturing
Gradient Gel Electrophoresis for the
Identification of Lactic Acid Bacteria in
Probiotic Products

Robin Temmerman, Liesbeth Masco, Geert Huys, and Jean Swings

CONTENTS

9.1 Introduction

Although the first recordings on the potential health-promoting effects of lactic acid bacteria (LAB) in fermented milk products date from the beginning of the 20th century, it is only since the 1990s that LAB-containing products have been linked to the health and physical condition status of the modern consumer. The strong expansion of the human probiotic market has also led to a diversification of probiotic species (1). Although the genera *Lactobacillus* and *Bifidobacterium* are among the most commonly used probiotics, strains belonging to other Gram-positive genera such as *Enterococcus, Lactococcus, Pediococcus,* and *Bacillus* have also appeared on the market as well as yeasts such as *Saccharomyces boulardii.*

Next to their functional and technological properties, the safety aspect of probiotics is of major importance both to the producer and the consumer. Because the current state of evidence suggests that probiotic effects are mostly strain specific (2), correct identification of probiotics is crucial to link a strain to a specific health effect as well as to enable surveillance and epidemiological studies (3). Over the past decade, there were several major attempts by the scientific community to clarify the classification and identification of probiotic microorganisms (4). Furthermore, it is also assumed that probiotic effects are dose dependent, although there is currently no consensus on the quantitative estimation of a daily probiotic dose (5, 6). A reliable quantification of probiotic bacteria is regarded of equal importance to their identification and should therefore be performed on a frequent basis using widely acknowledged techniques. In a number of independent studies, it has been demonstrated that the identity and the number of recovered species do not always correspond to the label information of the probiotic product itself (4, 7–9).

The need for legislation in which guidelines are defined for quality control is urgent and of immediate relevance. The Thematic Priority "Food Quality and Safety" of the European Sixth Framework Program, clearly underlines the need for this type of research through the availability of a budget of $900 million (10). It is generally agreed that quality control guidelines should be based on the use of standardized and harmonized methods to enable comparisons among products on an international scale.

Most research studies report on the identification of isolates obtained from probiotic products, rather than the identification of pure cultures or starter organisms obtained from the manufacturer itself (4, 7–9). Hence, there is a strong need for efficient and validated techniques in order to perform such product analyses. Although the majority of these techniques still rely on culture-dependent approaches, the use of culture-independent methods for taxonomic characterization is expected to increase in the near future. This chapter will focus mainly on identification strategies for LAB included in probiotic products. As a culture-dependent approach, the use of selective isolation and cultivation media coupled with genotypic fingerprinting and identification using repetitive element sequence-based (rep-) polymerase chain reaction (PCR) will be thoroughly discussed. The denaturing gradient gel electrophoresis (DGGE) method represents one of the most promising methodologies among the currently available culture-independent techniques.

9.2 Culture-Dependent Analysis

Despite the availability of molecular techniques, enumeration and detection of probiotic bacterial species is still widely performed using culture media. These media need to support growth of all bacteria that may occur in probiotic products. However, the specific nutritional demands of certain probiotics may sometimes hamper the development of a cultivation medium with the desired selectivity and electivity. Because at present no fully optimized culture-independent alternatives exist for the quantitative microbial analysis of probiotic products, it is very important that each step of the protocol be performed as standardized as possible in order to reduce the biases possibly encountered when performing culture-dependent analysis (11).

The following section discusses all points involved in the culture-dependent microbial analysis of probiotic products, from various sample treatments to methodologies for identification of the obtained isolates with special focus on rep-PCR.

9.2.1 Sample Treatment

Depending on the type of product sample (e.g., dairy product or freeze-dried food supplement), different approaches may be required for enumeration and identification of probiotic bacteria. The labels of most probiotic products usually mention the estimated amount of bacteria present in the product on the expiration date, which is also the recommended date for microbial analysis. In comparative studies, it is especially relevant to perform analyses of products at the same date in relation to the time of production. For dairy products, the shelf life is usually limited to 1 month, whereas that of freeze-dried products may range from 1 to 5 years. In the latter case, it is obvious that quantification at the expiration date may be impractical to perform.

9.2.1.1   Dairy Products

- Prior to the actual enumeration, it is important to verify if the products were kept refrigerated from the point of purchase or shipment up to the start of the actual analysis. In case that analysis should be postponed until the expiry date, products need to be stored at a maximum of 6°C.

- Although the risk of air contamination is considered to be minimal, it is recommended that all subsequent steps be carried out in a laminar flow cabinet.

- Using a sterile pipette, 1 ml of product is sampled from the middle of the product recipient. The intake of air is avoided because certain probiotic bifidobacteria are highly sensitive even to the slightest trace amount of oxygen due to their strictly anaerobic metabolism. For viscous dairy products, air-free sampling can be achieved by slightly moving the pipette around in the matrix while taking up the sample.

- One milliliter of product is transferred to a tube containing 9 ml of sterile physiological peptone solution (PPS) (NaCl: 8.5 g/l; neutralized bacteriological peptone: 1 g/l), after which the solution is homogenized through gentle shaking. The suspension is set to rest for approximately 30 min, and for recovery of bifidobacteria, reducing agent (e.g., 0.5 g/l of cysteine-HCl) needs to be added to the PPS.

- A 10-fold dilution series is prepared in PPS, each time using a new pipette in order to prepare the next dilution. Dilutions should be plated within 30 min after preparation.

9.2.1.2   Freeze-Dried Products

- In contrast to dairy products, most freeze-dried products (e.g., powders, capsules, and tablets) can be stored at room temperature. Because of the much longer shelf life (mostly 1 to 3 years) of freeze-dried products, analysis at the expiry date is often difficult to achieve in practice.

- Powder: In case the freeze-dried powder is contained in a box or sachet, approximately 100 mg of powder is weighed under sterile conditions and dissolved in 9 ml of PPS.

- Capsules: The average net weight of one capsule is between 50 and 150 mg. In order to avoid the weighing of exactly 100 mg of powder from a capsule, it is recommended to empty a complete capsule in 9 ml of PPS. If the exact content (in milligrams) of such a capsule is known, it is possible to calculate the equivalent to 100 mg. However, if no information about the weight is available, the weight of the capsule should be determined before and after emptying.

- Tablets: Hard tablets should be crushed to powder using a sterile mortar, after which the powder is dissolved in 9 ml of PPS. The equivalent to 100 mg can be calculated when the weight of one tablet is known.
- The PPS solution containing the freeze-dried product is homogenized through shaking and set to rest for 1 hour, in order to dissolve small remaining lumps. Again, PPS containing cysteine-HCl has to be used for the recovery of bifidobacteria.
- After a second homogenization through gentle shaking, a 10-fold dilution is prepared as outlined for dairy products.

9.2.2   Plating

Although most LAB display good growth on deMan, Rogosa, and Sharpe (MRS) (12) medium, enumeration of specific bacterial groups such as enterococci, streptococci, and bifidobacteria requires the use of specially designed selective culture media. On the other hand, the selectivity of certain culture media may lead to an underestimation of the actual number of targeted organisms present in the original sample. In this regard, reliable enumeration of bifidobacteria remains difficult to achieve because of their (strict) anaerobic character and high nutritional demands (11). Therefore, the most optimal approach for culture-dependent quantitative analysis of probiotic products includes the use of a range of culture media with overlapping selective or elective properties. Next to the choice of these culture media, the plating method should also be carefully chosen. Table 9.1 presents the most commonly used culture media with their corresponding incubation parameters.

9.2.2.1   Culture Media

- deMan, Rogosa, and Sharpe (MRS) (12): MRS agar and broth were designed to allow growth of LAB including the genera *Lactobacillus, Streptococcus, Pediococcus,* and *Leuconostoc*. Depending on culture

TABLE 9.1

Overview of Culture Media and Incubation Parameters Used for Enumeration and Isolation of Probiotic LAB

Bacterial Group	Medium	Incubation Parameters
Lactobacillus, Pediococcus, Streptococcus, Lactococcus, Leuconostoc	MRS	24–48 h, 37°C, MA
Streptococcus	M17	48 h, 37°C, MA
Enterococcus	KAAAB	24 h, 37°C, AE
Bifidobacterium	Acidified MRS, MRS + TTC, BSM	72–96 h, 37°C, AN

Note: MA = microaerophilic, AE = aerobic, AN = anaerobic. For more information, see 9.2.2.

conditions, enterococci and bifidobacteria can also be isolated on this medium. Selectivity can be enhanced by addition of specific agents such as cycloheximide (antibiotic inhibiting growth of yeasts) or sorbic acid (inhibiting growth of yeasts by pH reduction of the medium from 6.2 to 5.7) (13).

- M17: A well-buffered medium for the growth of the nutritionally fastidious lactic streptococci (14). Shankar and Davies (15) recommended M17 agar for the enumeration of *Streptococcus thermophilus* from yogurt. Meanwhile, the medium has also demonstrated its use for analysis of other probiotic products.

- Bifidobacteria: The majority of the media that were developed for the enumeration of bifidobacteria are in fact either too selective or inhibit insufficiently the growth of nonbifidobacteria. Because bifidobacteria have variable physiological requirements for growth, none of the currently described media are suitable for the recovery of all species. As far as *Bifidobacterium*-containing probiotic dairy products are concerned, an overview of the currently available media and methods is presented by Roy (11). For routine enumeration of bifidobacteria from pure cultures, commercially available non-selective media such as reinforced clostridial agar (RCA) and MRS supplemented with cysteine-HCl are recommended. These media provide an excellent recovery of bifidobacteria and are time and cost effective. Furthermore, the addition of certain short-chain fatty acids (e.g. propionic acid) or bifidogenic factors (e.g., lactulose) to these non-selective media might improve their elective properties. For selective enumeration of bifidobacteria in dairy products, existing medium bases are often supplemented with elective or selective agents, e.g., Columbia agar base media supplemented with lithium chloride and sodium propionate plus raffinose; Columbia agar with propionic acid and dicloxacillin or MRS medium supplemented with neomycin, paromycin, nalidixic acid, and lithium. In a recent study by Leuschner and coworkers (17), spread plating on four different agars including MRS, acidified MRS, MRS with triphenyl tetrazolium chloride (TTC) and a selective bifidobacteria medium (BSM) was evaluated for the selective enumeration of probiotic bifidobacteria used as feed additives. For routine analysis of products containing only bifidobacteria, the use of MRS agar supplemented with cysteine-HCL is recommended. In presence of either enterococci or pediococci and lactobacilli, acidified MRS and MRS supplemented with TTC is recommended, respectively. In cases where bifidobacteria are less predominant in the presence of other microorganisms, the use of BSM for selective enumeration was reported to be the most optimal choice (17).

- KAAAB: The kanamycin aesculin azide agar base medium was originally designed by Mossel and coworkers (16) for the selective isolation of enterococci from food samples. Round, white or gray colonies about 2 mm in diameter, surrounded by black zones of at least 1 cm in diameter is the typical morphology of enterococci on KAAAB.

9.2.2.2   Different Plating Techniques

Especially in the case of bifidobacteria, it is important to find a suitable combination of culture medium and plating technique. In order to obtain statistically significant quantitative data, all dilutions are usually plated in triplicate.

- With the spread plate method, 50 µl of each decimal dilution is transferred and spread on the appropriate medium using a sterile Drigalsky spatula. More advanced (automated) systems such as spiral plating can be used to perform a more accurate enumeration.
- With the pour plate method, 1 ml of decimal dilution is transferred to an empty petri dish after which the appropriate medium is poured into the plate and gently mixed to obtain a homogeneous distribution. Make sure the medium has cooled down to an acceptable temperature of about 50°C before pouring.

9.2.3   Incubation

The incubation parameters are crucial in any isolation procedure because they can largely influence the growth capacity of the organisms. Besides the length of incubation, the atmosphere and temperature should be well chosen. Table 9.1 lists the incubation parameters for a number of culture media discussed earlier.

Although most LAB are aerotolerant, incubation under microaerophilic conditions (e.g., 5% O_2, 3.5% CO_2, 7.5% H_2, 84% N_2) greatly enhances their growth after 24 to 48 h. The incubation temperature is usually set at 37°C, although growth of thermophilic species may require incubation at higher temperatures, such as 42°C for *Lactobacillus acidophilus*.

Regardless of the medium used, bifidobacteria always require anaerobic incubation. Because of their complex physiological demands, these organisms are relatively slow growers, and for this reason incubation at 37°C lasts for 72 to 96 h. In order to enhance the degree of anaerobicity in the (vented) petri dishes, a medium containing cysteine-HCl (0.1% w/v) is preferred.

9.2.4   Enumeration and Isolation

9.2.4.1   Enumeration

- In order to obtain statistically significant counts, dilutions are usually plated in triplicate, from which an average value is calculated.

- Plates containing between 30 and 300 colonies are considered for counting. Based on the initial volume of 50 µl that has been plated and the corresponding dilution factor, the number of colony forming units (CFU) per gram or milliliter of product can be calculated.
- Dairy products: The initial solution prepared (1 ml of product in 9 ml of PPS) is considered as the first decimal dilution.
- Freeze-dried products: In case 100 mg of product has been diluted in 9 ml of PPS, this solution is considered as the second decimal dilution because CFU values of freeze-dried products are expressed per gram.

9.2.4.2   Isolation and Strain Storage

At this point of the product analysis, only quantitative data are obtained. In order to reveal the taxonomic composition of the product, identification of a batch of representative isolates has to be performed. Because of practical limitations, it is often not possible or simply not necessary to identify all colonies obtained on the culture medium. Therefore, a statistically significant balance has to be found between the amount of isolates selected for subsequent identification and the workload involved.

- Select those petri dishes that contain a representative number (approximately 100) of well-separated colonies.
- Depending on the selectivity of the medium used, up to 10 colonies can be picked using a sterile loop. The selection of these colonies is mainly based on their morphology, which in case of small colony sizes is determined using a microscope.
- Except for bifidobacteria, all presumptive LAB colonies can be purified on MRS agar using the following incubation parameters: 37°C, 48 h, aerobic (or microaerophilic). Bifidobacterial colonies need to be subcultured on media sustaining their growth, such as RCA or MRS supplemented with cysteine-HCl, and incubated anaerobically for 72 h at 37°C.
- If required, several repetitive inoculations have to be performed in order to obtain a pure culture.
- Isolates can be preserved for long-time storage at –80°C, either in medium supplemented with glycerol, or by using a bead storage system (e.g., MicroBank, Pro-Lab Diagnostics®).

9.2.5   Identification–Repetitive Element Sequence-Based (rep-) PCR

Especially in the field of food production and gastrointestinal microbiology, there is a huge interest in the identification and characterization of LAB, mainly because of (i) the association of these organisms with health-promoting properties; (ii) their application in numerous food products as probiotics; and (iii) the requirements of legislative and industrial bodies, as well as the consumer, with respect to safety, labeling, and strain integrity (18). Next to the improvement of the technological and functional aspects of probiotics, safety and quality control are equally important and should ideally be performed on a frequent basis (19). At present, a broad range of techniques is available for the identification of LAB with varying degrees of discriminatory power, reproducibility, and workload (20). At present, the most commonly used techniques for LAB identification include 16S rDNA sequencing and the phenotypic API50CHL system (Biomérieux), mainly because of the availability of (on-line sequence) databases and the good exchangeability of generated data. However, both methods have insufficient taxonomic resolution to discriminate among certain groups of LAB species.

A technique combining high taxonomic resolution (subspecies level) with acceptable reproducibility and low workload, also allowing dereplication of the set of isolates, is rep-PCR. This genotypic technique is based on the use of outwardly facing oligonucleotide PCR primers complementary to interspersed repetitive sequences, which enable the amplification of differently sized DNA fragments lying between these elements. Examples of evolutionarily conserved repetitive sequences are BOX, ERIC, REP, and (GTG)$_5$ (21). rep-PCR fingerprinting is considered to be a valuable tool for classifying and typing a wide range of bacterial genera (21, 22). Within the LAB, this method has been evaluated for identification of *Lactobacillus* (23) and *Bifidobacterium* species (24).

9.2.5.1 Bacterial Strains and Cultivation

Similar to the use of other DNA fingerprinting methods, a taxonomical framework of reference strains has to be constructed in order to be able to perform an identification of unknown isolates with rep-PCR. Using a representative set of well-characterized type and reference strains obtained from international bacteria collections, such as the BCCM™/LMG Bacteria Collection (Ghent University, Belgium) or DSMZ (Braunschweig, Germany), a reliable identification database can be generated.

Presumptive LAB strains should be grown overnight at 37°C on MRSA, whereas bifidobacteria are usually grown overnight at 37°C under anaerobic conditions (80% N_2, 10% H_2, 10% CO_2) on Modified Columbia Agar comprising 23 g of special peptone, 1 g of soluble starch, 5 g of NaCl, 0.3 g of cysteine-HCl-H_2O, 5 g of glucose, and 15 g of agar dissolved in 1 l of distilled water.

9.2.5.2 Total DNA Extraction

Extraction of total bacterial DNA is based on the method of Pitcher and coworkers (25) with slight modifications regarding the concentration of lysozyme and an additional step involving RNase at the end of the procedure.

- From a pure overnight-grown culture, half a loop of cells is harvested and washed in 500 µl TE-buffer (1 mM EDTA pH 8.0; 10 mM Tris-HCl pH 8.0) after which the cell suspension is centrifuged for 2 min at 13,000 rpm.

- Following the removal of the supernatant, the resulting pellet is frozen at –20°C for at least 1 h to facilitate the rupture of the Gram-positive cell wall.

- The thawed pellet is resuspended in 150 µl of lysozyme-solution (5 mg of lysozyme in 150 µl of TE buffer), followed by an incubation step at 37°C during 40 min.

- Subsequently, 500 µl of GES reagents (600 g/l guanidium-thiocyanate; 200 ml/l of 0.5M EDTA [pH 8] and 10 g/l of sarkosyl, sodium lauryl sarcosinate) is added, and the solution is gently mixed until a transparent solution is obtained, which is allowed to stand on ice for 10 min.

- To enhance the precipitation of proteins, 250 µl of ammonium-acetate (7.5 M) is added and carefully mixed with the lysed cell suspension. Keep on ice for 10 min.

- To separate the precipitated proteins from the nucleic acids, 500 µl of chloroform/isoamyl alcohol (24:1) is added. The mixture is thoroughly shaken until a homogeneous one-phase solution is obtained, after which it is centrifuged for 20 min at 13,000 rpm. The isoamyl alcohol serves as an antifoaming agent.

- Following the centrifugation step, a three-phase solution is usually obtained: the upper aqueous phase (containing nucleic acids), the interphase (containing a layer of proteins and cell debris), and the lower phase (containing the chloroform fraction). Approximately 700 µl of the upper phase is transferred to a new Eppendorf tube containing 0.54 volumes (approximately 378 µl) ice-cold isopropanol. By means of careful mixing, the nucleic acids are precipitated as a white fluffy cloud.

- The nucleic acids are centrifuged for 1 min at 13,000 rpm, after which the supernatant is gently removed and the remaining pellet is washed two times (with 1 min centrifugation and supernatant removal in between) with 150 µl 70% ethanol to remove traces of isopropanol.

- The obtained semitransparent pellet is dried under vacuum and dissolved in 100 µl of TE buffer.

- The resulting DNA pellet is dissolved in 200 µl TE-buffer and kept overnight at 4°C.

- The RNase step can be performed once most of the pellet has solubilized. Remaining RNA is then digested by adding 2 µl of an RNase solution (10 mg of RNase dissolved in 1 ml of milli-Q water) followed by an incubation step of 90 min at 37°C.

- Quality of the DNA extracts is verified by spectrophotometric measurements at 260, 280, and 234 nm. The DNA is then diluted to a working concentration of 50 ng/µl. The integrity of the DNA is checked by electrophoresis on a 1% agarose gel in 1 × TAE (0.04 M Tris-acetate, 0.001 M EDTA) buffer and visualization under ultraviolet light after staining with ethidium bromide.

9.2.5.3 rep-PCR Fingerprinting

Several repetitive sequence-based oligonucleotide primers can be used for rep-PCR (22). Depending on the group of organisms and the desired taxonomic resolution, it may be necessary to evaluate several of these primers. A list of primers frequently used is presented in Table 9.2. For all rep-PCR assays, a universal PCR reaction mix can be used in which the primer is interchangeable. Annealing temperatures for the different primers are mentioned in Table 9.2.

- Each **25 µl PCR reaction** contains in milli-Q water:
  - 5 µl 5 × Gitschier buffer [83 mM $(NH_4)_2SO_4$, 335 mM Tris-HCl (pH 8.8), 33.5 mM $MgCl_2$, 32.5 µM EDTA (pH 8.8), and 150 mM β-mercaptoethanol]
  - 160 µg/ml BSA
  - 10% DMSO
  - 1.25 mM of each of 4 dNTPs (dATP, dCTP, dGTP, and dTTP)
  - 0.3 µg/µl oligonucleotide primer
  - 2 units of DNA polymerase
  - 50 ng of template DNA

- PCR amplification is performed in a DNA thermal cycler (e.g., Perkin Elmer 9600) through the following program:
  - Initial denaturation: 95°C for 7 min

TABLE 9.2

Sequence and Recommended Annealing Temperature of rep-PCR Primers

Designation	Sequence	Annealing Temperature (°C)
ERIC1R	5'-ATGTAAGCTCCTGGGGATTCAC-3'	52
ERIC2	5'-AAGTAAGTGACTGGGGTGAGCG-3'	52
BOXA1R	5'-CTACGGCAAGGCGACGCTGACG-3'	52
GTG$_5$	5'-GTGGTGGTGGTGGTG-3'	40
REP1R	5'-IIIICGICGICATCIGGC-3'	40
REP2I	5'-ICGICTTATCIGGCCTAC-3'	40

Note: I = inosine.

Source: Based on Versalovic et al., 1994.

- 30 cycles of denaturation (94°C, 1 min), 1 min of annealing at a temperature depending on the primer (see Table 9.2), and extension (65°C, 8 min)
- Final extension step (65°C, 16 min)

- The amplified fragments are fractionated on a 1.5% agarose gel during 16 h at a constant voltage of 55 V in 1 × TAE at 4°C
- The rep-PCR genomic fragments are visualized after staining with ethidium bromide under ultraviolet light, followed by digital capturing of the image using a CCD camera.
- Although simple visual comparison of obtained fingerprints with those of reference and type strains may be sufficient to determine the presence of a certain species, it is recommended to further process and compare the rep-PCR fingerprints with the user-generated identification database of reference strains using a software package such as BioNumerics V2.5 (Applied Maths, Kortrijk, Belgium).

An example of a classification based on a rep-PCR fingerprint of strains belonging to six bifidobacterial species used as probiotics is presented in Figure 9.1.

9.3 Culture-Independent Analysis

In the course of various independent studies, it has been demonstrated that the identity and number of microbial species recovered from probiotic products do not always correlate with the information stated on the product labels (4, 7, 8; see also *The safety of probiotics in foods in Europe and its legislation*; chapter 18). These studies mainly relied on the use of culture media to isolate the bacteria from the probiotic product, after which a selection of purified isolates is usually identified using one or more molecular techniques. However, because these cultivation-dependent approaches have proven limitations in terms of recovery rate and reproducibility, the set of recovered isolates may not always truly reflect the actual microbial composition of the product (11, 26, 27). Moreover, although certain products actually claim to contain only dead bacteria, other techniques are required that are able to detect both viable and nonviable bacteria.

Culture-independent methods such as DGGE represent a valuable alternative to detect and identify bacteria in probiotic products in a fast and reliable manner. Essentially, the complete DGGE methodology consists of three steps: (i) extraction of total bacterial DNA from the probiotic product, (ii) PCR amplification of specific regions of the 16S rDNA gene, and (iii)

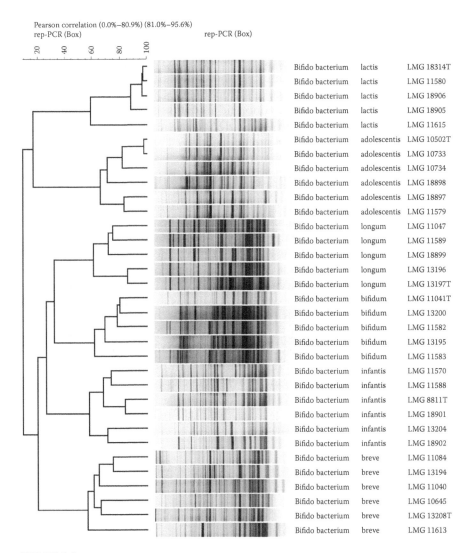

Pearson correlation (0.0%–80.9%) (81.0%–95.6%)
rep-PCR (Box)   rep-PCR (Box)

Bifido bacterium	lactis	LMG 18314T
Bifido bacterium	lactis	LMG 11580
Bifido bacterium	lactis	LMG 18906
Bifido bacterium	lactis	LMG 18905
Bifido bacterium	lactis	LMG 11615
Bifido bacterium	adolescentis	LMG 10502T
Bifido bacterium	adolescentis	LMG 10733
Bifido bacterium	adolescentis	LMG 10734
Bifido bacterium	adolescentis	LMG 18898
Bifido bacterium	adolescentis	LMG 18897
Bifido bacterium	adolescentis	LMG 11579
Bifido bacterium	longum	LMG 11047
Bifido bacterium	longum	LMG 11589
Bifido bacterium	longum	LMG 18899
Bifido bacterium	longum	LMG 13196
Bifido bacterium	longum	LMG 13197T
Bifido bacterium	bifidum	LMG 11041T
Bifido bacterium	bifidum	LMG 13200
Bifido bacterium	bifidum	LMG 11582
Bifido bacterium	bifidum	LMG 13195
Bifido bacterium	bifidum	LMG 11583
Bifido bacterium	infantis	LMG 11570
Bifido bacterium	infantis	LMG 11588
Bifido bacterium	infantis	LMG 8811T
Bifido bacterium	infantis	LMG 18901
Bifido bacterium	infantis	LMG 13204
Bifido bacterium	infantis	LMG 18902
Bifido bacterium	breve	LMG 11084
Bifido bacterium	breve	LMG 13194
Bifido bacterium	breve	LMG 11040
Bifido bacterium	breve	LMG 10645
Bifido bacterium	breve	LMG 13208T
Bifido bacterium	breve	LMG 11613

FIGURE 9.1
Dendrogram showing clustering analysis of digitized rep-PCR fingerprints generated using the BOX primer for type and reference strains of probiotic *Bifidobacterium* species. LMG: strain accession number for the BCCM™/LMG culture collection. Note: *B. lactis* changed to *B. animalis* subsp. *lactis* (Masco et al., 2004) (28).

electrophoresis of the resulting amplicons on a DGGE gel. At present, DGGE analysis is one of the most suitable and widely used methods to study complex bacterial communities in various environments (29). Moreover, the DGGE-based approach presented in the next section can also be used as a culture-independent identification method, provided that a predetermined database is available. In less than 30 h, a given probiotic product can be analyzed to verify the species composition stated on its label (9).

9.3.1 Principle

DGGE allows the sequence-dependent separation of a mixture of PCR-amplified DNA fragments, all identical in size, on an acrylamide gel containing a well-defined gradient of denaturing components (Figure 9.2). In conventional agarose or acrylamide gel electrophoresis, DNA fragments are separated by size, with the electrophoretic mobility of each fragment being disproportionate to its size. In DGGE, DNA fragments of the same size are separated by their denaturing profile, i.e., how the double-stranded DNA (dsDNA) becomes (partially) single stranded (ssDNA) when it is subjected to an increasingly denaturing environment. This physical denaturation of the dsDNA fragment proceeds gradually, during which discrete portions of the fragment will denature through so-called melting domains. As a result of this conformational change, the mobility of the DNA fragment through the acrylamide gel is gradually reduced until it comes to a complete halt. The position in a gel where the dsDNA fragment melts and becomes ssDNA is dependent on the nucleotide sequence and percentage G+C content of the fragment. Differences in DNA sequences will yield different origins of melting domains and will result in different positions in the gel where the DNA fragment is halted. For each type of application, the optimal denaturing gradient has to be prepared by means of mixing the desired volumes of a 100% and 0% denaturing acrylamide solution (100% denaturant is 40% formamide and 7 M urea), resulting for instance in the commonly used 35 to 70% denaturing gel (30). Originally being developed for mutation analysis,

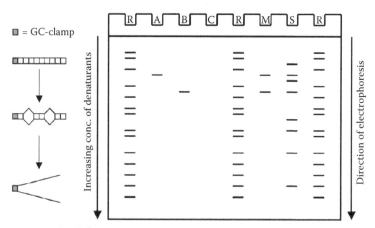

R = Reference pattern, A = Organism 1, B = Organism 2, C = Organism 3,
M = Mix of organisms 1, 2 and 3, S = unknown sample

FIGURE 9.2
Principle of DGGE. PCR amplicons of equal length are electrophoretically separated in a sequence-dependent manner. The increasing gradient of denaturing components along the gel confers the double-stranded amplicons into single-stranded DNA through melting domains. A GC-clamp attached to the 5′ end of one of both PCR primers prevents the amplicons from completely denaturing.

DGGE has been shown to detect differences in the denaturing behavior of small DNA fragments (200 to 700 bp) that differ by as little as 1 bp.

9.3.2 Method Description

9.3.2.1 DNA Extraction

Extraction of total bacterial DNA from probiotic products is based on the method described by Pitcher and coworkers (25) with slight modifications depending on the type of starting material.

- Dairy products: 1 ml of product is centrifuged for 10 min at 13,000 rpm followed by removal of the supernatant and resuspension of the pellet in 1 ml of Tris EDTA (TE) buffer. In case large fruit particles are present in the product, 50 ml of the product is centrifuged for 2 min at 1500 rpm, after which 1 ml of the top liquid is taken and centrifuged for another 10 min at 13,000 rpm. After removal of the supernatant, the remaining pellet is dissolved in 1 ml of TE buffer.

- Capsules and powders: The content of one capsule, usually approximately 100 mg, or 100 mg of powder is dissolved in 10 ml of sterile PPS and softly shaken until a homogeneous suspension is obtained. One milliliter of this suspension is transferred to an Eppendorf tube, which is then centrifuged for 10 min at 13,000 rpm. The supernatant obtained is removed, and the remaining pellet is suspended in 1 ml of TE buffer.

- Tablets: One tablet is crushed in a sterile mortar, and the powder obtained is dissolved in 10 ml of PPS. After homogenization, 1 ml of this suspension is centrifuged for 10 min at 13,000 rpm. The remaining pellet is finally dissolved in 1 ml of TE buffer.

- The subsequent steps in the DNA extraction procedure are identical to those described in the rep-PCR protocol (see Section 9.2.5.2).

9.3.2.2 PCR

A number of primer pairs that have been frequently used for DGGE analysis are listed in Table 9.3. When designing new DGGE primers, it should be kept in mind that the length of the amplicons obtained should not exceed 500 bp, which is the upper resolution limit of the technique. Furthermore, one or both primers should contain a GC-clamp (Table 9.3), which prevents the amplicons from becoming completely single stranded and running off the gel. An initial analysis of a probiotic sample is usually performed using universal primers amplifying the V3 or V6–V8 region of the 16S rDNA. More specific primers can be used to detect certain bacterial groups of interest, such as *Lactobacillus* (31) and *Bifidobacterium* (32). In order to further enhance

TABLE 9.3

Selection of Frequently Used 16S rDNA PCR Primers for DGGE Analysis of LAB Populations

Primer	Target	Position	Sequence	Ref.
F357-GC	Bacterial 16S rDNA V3 region	(341-357)	5'-GC-clamp-GCCTACGGGAGGCAGCAG-3'	28
518-R	Bacterial 16S rDNA V3 region	(518-534)	5'-ATTACCGCGGCTGCTGG-3'	28
U968F-GC	Bacterial 16S rDNA V6-V8 region	(968-985)	5'-GC-clamp-AACGCGAAGAACCTTAC-3'	32
L1401R	Bacterial 16S rDNA V6-V8 region	(1401-1418)	5'-GCGTGTGTACAAGACCC-3'	32
Bif164-GC-f	Bifidobacterial 16S rDNA	(146-164)	5'-GC-clamp-GGGTGGTAATGCCGGATG-3'	30
Bif662-r	Bifidobacterial 16S rDNA	(662-680)	5'-CCACCGTTACACCGGGAA-3'	30
Lac1-f	*Lb., Pd., Lc., Ws.* 16S rDNA	(333-352)	5'-AGCAGTAGGGAATCTTCCA-3'	29
Lac2GC-r	*Lb., Pd., Lc., Ws.* 16S rDNA	(661-679)	5'-GC-clamp-ATTICACCGCTACACATG-3'	29

Note: GC-clamp = 5'-CGCCCGCCGCGCGCGGCGGGCGGGGCGGGGGCACGGGGG-3'; *Lb., Lactobacillus; Pd., Pediococcus; Lc., Leuconostoc, Ws., Weisella;* I = inosine.

the taxonomic resolution of DGGE analysis, a combination of multiple primers can be applied (33).

- For the universal primers V3 and V6–V8, PCR reaction volumes of 50 μl consist of 6 μl of 10 × PCR buffer containing 15 m*M* MgCl₂, 2.5 μl of BSA, 2.5 μl of dNTPs (2 m*M* each), 2 μl of each primer (5 μ*M*), 1.25 units of *Taq* polymerase, 33.75 μl of sterile milli-Q, and 1 μl of 10-fold diluted DNA solution.

- The following PCR program is used: initial denaturation at 94°C for 5 min; 30 cycles of denaturation at 94°C for 20 sec, annealing at 55°C for 45 sec, and extension at 72°C for 1 min; final extension at 72°C for 7 min followed by cooling to 4°C.

- The PCR product is checked by loading a mix of 8 μl of PCR product and 2 μl of loading dye on a 2% (w/v) agarose gel, which is electrophoresed for 30 min at 100 V.

9.3.2.3 DGGE Analysis

The following protocol for DGGE analysis is based on Muyzer and coworkers (30), with modifications according to Temmerman and coworkers (9). A

TABLE 9.4

Volumes of the 0 and 100% Denaturing Acrylamide
Solution to Be Mixed in order to Obtain the Desired
Denaturing Solutions for Preparation of a 160 x 160 x 1
mm DGGE Gel

Desired Gradient	0% Solution (ml)	100% Solution (ml)
0	12.0	0.0
10	10.8	1.2
20	9.6	2.4
30	8.4	3.6
35	7.8	4.2
40	7.2	4.8
45	6.6	5.4
50	6.0	6.0
55	5.4	6.6
60	4.8	7.2
65	4.2	7.8
70	3.6	8.4
80	2.4	9.6
90	1.2	10.8
100	0.0	12.0

Note: The most commonly used gradient for bacterial detection
is 35–70%.

DGGE gel consists of an 8% acrylamide gel containing a gradient of denaturing components (ureum and formamide). By diluting a 100% denaturing
polyacrylamide solution (containing 7 M urea and 40% formamide) with a
0% denaturing polyacrylamide solution (containing no denaturing components), the polyacrylamide solutions with the desired denaturing percentage
can be obtained (Table 9.4). For analysis of probiotic bacteria, two types of
denaturing gradients are recommended: (i) a 35 to 70% gradient for the
detection of all probiotic LAB and (ii) a 55 to 75% gradient for the detection
of bifidobacteria. Furthermore, a list of all solutions needed and instructions
for their preparation are presented in Table 9.5.

- The 24-ml gradient gels (160 × 160 × 1 mm) are cast using a gradient
  former and a pump set at a constant speed of 5 ml/min.
- Denaturing gels are allowed to polymerize for 3 h at 20°C.
- A 5-ml nondenaturing stacking gel (5 ml of the 0% denaturing acrylamide solution) is poured on top of the separation gel.
- After insertion of the comb, the gel is allowed to polymerize for 1 h.
- Subsequently, approximately 20 µl of PCR sample is loaded into the
  wells, and electrophoresis is performed for 16 h at 70 V in a 1× TAE
  buffer at a constant temperature of 60°C.

TABLE 9.5

Solutions Needed for DGGE Analysis

- 10% (w/v) APS solution in milliQ: store at -20°C in aliquots of 1 ml.
- TAE (50×): 242 g tris base, 57 ml of acetic acid, and 100 ml of 0.5*M* EDTA (pH 8): add milliQ up to 1000 ml.
- 0.5 *M* EDTA (pH 8): 186.1 g EDTA and 20 g NaOH in 800 ml milliQ; set the pH at 8.0 by adding NaOH grains.
- Running Buffer (= Tank Buffer) = 1× TAE: 140 ml 50× TAE in 6860 ml milliQ
- Water-saturated isobutanol: 9.5 ml butanol in 100 ml milliQ
- Staining solution: 500 ml 1x TAE + 50 μl EtBr.
- 100% Denaturing acrylamide solution (200 ml): 54 ml acrylamide solution (30% w/v acrylamide, 0.8% w/v *bis*-acrylamide [37,5:1]) + 84 g ureum + 80 ml 100% formamide + 4 ml 50× TAE (becomes 1× TAE) + milliQ up to 200 ml
- 0% acrylamide solution (200 ml): 54 ml 30% acrylamide-solution + 4 ml 50× TAE + milliQ up to 200 ml

---

- Gels are stained with ethidium bromide or SYBR Green for 30 min, followed by visualization of DGGE band profiles under ultraviolet light.
- Digital capturing is performed using a CCD camera.

9.3.2.4   *Processing of DGGE Gels*

In order for DGGE to be used as a direct identification method, a reference pattern has to be designed consisting of different type strain amplicons. By running this reference pattern every six lanes on each DGGE gel, it becomes possible to digitally normalize the gel patterns by comparison with a standard pattern using the BioNumerics (BN) software package version 3.50 (AppliedMaths, St.-Martens-Latem, Belgium, http://www.applied-maths.com). This normalization enables comparison of different DGGE gels provided that they consist of the same denaturing gradient and that they were run under standardized conditions. Before performing DGGE analysis of unknown probiotic samples, band positions of the type and other reference strains of the probiotic species of interest have to be determined and stored in a BN identification database. Subsequently, the amplicons obtained from probiotic products are run on a DGGE gel, and after normalization based on the standard reference pattern of the BN database, individual bands in the product band pattern can be identified by comparison with this database. Amplicons of previously identified isolates or reference strains, corresponding to the species claimed on the product label, can be run next to the amplicon of the probiotic product itself as an additional confirmation for identification (Figure 9.3).

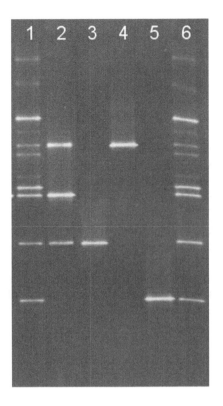

FIGURE 9.3

Example of a probiotic product analysis using V3-DGGE. Lanes 1 and 6: reference pattern for normalization; Lane 2: probiotic product amplicons; Lane 3: *Lactobacillus rhamnosus*; Lane 4: *Lactobacillus acidophilus*; and Lane 5: *Bifidobacterium lactis*. It can be seen that the product contains a species (*Streptococcus thermophilus* after comparison with the database) that is not claimed on the label (middle band in Lane 2). Furthermore, the claimed species *B. lactis* is not detected in the product (the band of Lane 5 should also be present in Lane 2).

9.4 Conclusions and Future Perspectives

It is beyond any doubt that an accurate (molecular) identification is highly important in the development of probiotic foods in terms of product safety and integrity, and of functional monitoring upon product consumption. Although the number of papers dealing with microbial analyses of probiotic products is rather limited, it is a general observation of all these studies that a significant percentage of the products contain an insufficient number of living organisms, and/or suffer from mislabeling regarding their species composition (4, 7, 8). Because the majority of these studies are performed using culture-dependent approaches, their results may not truly reflect the actual composition of the product due to biases introduced by the use of culture media and unreliable identification of the recovered isolates (11, 27).

The development and optimization of culture-independent methods such as DGGE has significantly increased the speed and sensitivity of microbial quality control of probiotic products (9), although a reliable quantification and the distinction between dead or live bacteria remains difficult. The development of real-time PCR is a major step forward in the direction of a culture-independent quantification, although this methodology still needs to be optimized for the analysis of probiotic bacteria (34). Besides the identity and number of probiotic bacteria, metabolic activity is an important factor in relation to the functionality of these organisms. The combination of reverse transcriptase PCR with real-time technology (35) or fluorescent *in situ* hybridization (FISH) (36) is expected to simultaneously reveal data on the number, the activity, and the identity of bacteria in a product. Finally, technologies such as microarrays (37) or mass spectrometry (38) will contribute to the complete microbial analysis of various bacterial communities, although these approaches suffer from a significant cost. In the long run, it is expected that this technological optimization will also find applications in clinical and functional trials of new probiotics.

References

1. C Stanton, G Gardiner, H Meehan, K Collins, G Fitzgerald, PB Lynch, RP Ross. Market potential for probiotics. *Am. J. Clin. Nutr.*, 73:476S–483S, 2001.
2. S Gilliland. Technological and commercial applications of lactic acid bacteria; health and nutritional benefits in dairy products. http://www.fao.org/es/ESN/food/food andfood_probio_en.stm 2002.
3. G Reid, M Araya, L Morelli, ME Sanders, C Stanton, M Pineiro, P Ben Embarek. Joint FAO/WHO Working Group Report on Drafting Guidelines for the Evaluation of Probiotics in Food. London, Ontario, Canada, April 30 and May 1, 2002.
4. JMT Hamilton-Miller, S Shah, JT Winkler. Public health issues arising from microbiological and labelling quality of foods and supplements containing probiotic microorganisms. *Public Health Nutr.*, 2:223–229, 1999.
5. R Foschino, I Cafaro, G Ottogalli. Study on viability of probiotic bacteria in retail samples of fermented milks. *Ann. Microbiol. Enzimol.*, 47:151–164, 1997.
6. ME Sanders, J Huis in't Veld. Bringing a probiotic-containing functional food to the market: microbiological, product, regulatory and labeling issues. *Antonie Van Leeuwenhoek Int. J. Gen. Mol. Microbiol.*, 76:293–315, 1999.
7. WH Holzapfel, P Haberer, J Snel, U Schillinger, JHJ Huis in't Veld. Overview of gut flora and probiotics. *Int. J. Food Microbiol.*, 41:85–101, 1998.
8. R Temmerman, B Pot, G Huys, J Swings. Identification and antibiotic susceptibility of bacterial isolates from probiotic products. *Int. J. Food Microbiol.*, 81(1):1–10, 2003a.

9. R Temmerman, I Scheirlinck, G Huys, J Swings. Culture-independent analysis of probiotic products using denaturing gradient gel electrophoresis (DGGE). *Appl. Environ. Microbiol.*, 69(1):220–226, 2003b.

10. J Lucas. EU-funded research on functional foods. *Br. J. Nutr.*, 88(S):S131–S132, 2002.

11. D Roy. Media for the isolation and enumeration of bifidobacteria in dairy products. *Int. J. Food Microbiol.*, 69:167–182, 2001.

12. JC DeMan, M Rogosa, ME Sharpe. A medium for the cultivation of lactobacilli. *Appl. Bacteriol.*, 23:130–135, 1960.

13. G Reuter. Elective and selective media for lactic acid bacteria. *Int. J. Food Microbiol.*, 2:55–68, 1985.

14. AW Anderson, PR Elliker. The nutritional requirements of lactic streptococci isolated from starter cultures. Growth in a synthetic medium. *J. Dairy Sci.*, 36:161–167, 1953.

15. PA Shankar, FL Davies. Associative bacterial-growth in yoghurt starters — Initial observations on stimulatory factors. *J. Soc. Dairy Technol.*, 30:28–30, 1977.

16. RGK Leuschner, J Bew, P Simpson, PR Ross, C Stanton. A collaborative study of a method for the enumeration of probiotic bifidobacteria in animal feed. *Int. J. Food Microbiol.*, 83(2):161–170, 2003.

17. DAA Mossel, PGH Bijker, I Eelderink. Streptococci of Lancefields Group-D in foods — Their significance, enumeration and control. *Arch. Lebensmittelhyg.*, 29:121–127, 1978.

18. AL McCartney. Application of molecular biological methods for studying probiotics and the gut flora. *Br. J. Nutr.*, 88:S29–S37, 2002.

19. M Saarela, G Mogensen, R Fondén, J Mättö, T Matilla-Sandholm. Probiotic bacteria: safety, functional and technological properties. *J. Biotech.*, 84:197–215, 2000.

20. J Rafter. Lactic acid bacteria and cancer: mechanistic perspective. *Br. J. Nutr.*, 88(S1):S89–S94, 2002.

21. J Rademaker, B Hoste, FJ Louws, K Kersters, J Swings, L Vauterin, P Vauterin, FJ de Bruijn. Comparison of AFLP and rep-PCR genomic fingerprinting with DNA-DNA homology studies: *Xanthomonas* as a model system. *Int. J. Syst. Evol. Microbiol.*, 50: 665–677, 2000.

22. J Versalovic, M Schneider, F de Bruijn, JR Lupski. Genomic fingerprinting of bacteria using repetitive sequence-based polymerase chain reaction. *Methods Mol. Cell Biol.*, 5:25–40, 1994.

23. D Gevers, G Huys, J Swings. Applicability of rep-PCR fingerprinting for identification of *Lactobacillus* species. *FEMS Microbiol. Lett.*, 205:31–36, 2001.

24. L Masco, G Huys, D Gevers, L Verbrugghen, J Swings. Identification of *Bifidobacterium* species using rep-PCR fingerprinting. *Syst. Appl. Microbiol.*, 26(4):557–563, 2003.

25. DG Pitcher, NA Saunders, RJ Owen. Rapid extraction of bacterial genomic DNA with guanidium thiocyanate. *Lett. Appl. Microbiol.*, 8:151–156, 1989.

26. F Ampe, N Omar, C Moizan, C Wacher, JP Guyot. Polyphasic study of the spatial distribution of microorganisms in Mexican pozol, a fermented maize dough, demonstrates the need for cultivation-independent methods to investigate traditional fermentations. *Appl. Environ. Microbiol.*, 65:5464–5473, 1999.

27. D Ercolini, G Moschetti, G Blaiotta, S Coppola. The potential of a polyphasic PCR-DGGE approach in evaluating microbial diversity of natural whey cultures for water-buffalo mozzarella cheese production: Bias of culture-dependent and culture-independent analyses. *Syst. Appl. Microbiol.*, 24(4):610–617, 2001.

28. L Masco, M Ventura, R Zink, G Huys, J Swings. Polyphasic taxonomic analysis of *Bifidobacterium animalis* and *Bifidobacterium lactis* reveals relationships at the subspecies level: reclassification of *Bifidobacterium animalis* as *Bifidobacterium animalis* subsp. animalis subsp. nov. and *Bifidobacterium lactis* as *Bifidobacterium animalis* subsp. *lactis* subsp. nov. Int. J. Syst. Evol. *Microbiol.* 54: 1137–1143, Part 4, 2004.

29. G Muyzer. DGGE/TGGE a method for identifying genes from natural ecosystems. *Curr. Opin. Microbiol.*, 2:317–322, 1999.

30. G Muyzer, EC de Waal, AG Uitterlinden. Profiling of complex microbial populations by denaturing gradient gel electrophoresis analysis of polymerase chain reaction-amplified genes coding for 16S rRNA. *Appl. Environ. Microbiol.*, 59:695–700, 1993.

31. J Walter, C Hertel, GW Tannock, CM Lis, K Munro, WP Hammes. Detection of *Lactobacillus, Pediococcus, Leuconostoc,* and *Weissella* species in human feces by using group-specific pcr primers and denaturing gradient gel electrophoresis. *Appl. Environ. Microbiol.*, 67(6):2578–2585, 2001.

32. RM Satokari, EE Vaughan, ADL Akkermans, M Saarela, WM De Vos. Bifidobacterial diversity in human feces detected by genus-specific PCR and denaturing gradient gel electrophoresis. *Appl. Environ. Microbiol.*, 67:504–513, 2001.

33. R Temmerman, L Masco, T Vanhoutte, G Huys, J Swings. Development and validation of a nested PCR-denaturing gradient gel electrophoresis method for taxonomic characterization of bifidobacterial communities. *Appl. Environ. Microbiol.*, 69: 6380–6385, 2003.

34. B Vitali, M Candela, D Matteuzzi, P Brigidi. Quantitative detection of probiotic *Bifidobacterium* strains in bacterial mixtures by using real-time PCR. *Syst. Appl. Microbiol.*, 26(2):269–276, 2003.

35. G Bleve, L Rizzotti, F Dellaglio, S Torriani. Development of reverse transcription (RT)-PCR and real-time RT-PCR assays for rapid detection and quantification of viable yeasts and molds contaminating yogurts and pasteurised food products. *Appl. Environ. Microbiol.*, 69(7):4116–4122, 2003.

36. D Sohier, A Lonvaud-Funel. Rapid and sensitive *in situ* hybridization method for detecting and identifying lactic acid bacteria in wine. *Food Microbiol.*, 15:391–397, 1998.

37. RF Wang, ML Beggs, LH Robertson, CE Cerniglia. Design and evaluation of oligonucleotide-microarray method for the detection of human intestinal bacteria in fecal samples. *FEMS Microbiol. Lett.*, 213 (2):175–182, 2002.

38. G Reid, BS Gan, YM She, S Weinberger, JC Howard. Rapid identification of probiotic *Lactobacillus* biosurfactant proteins by ProteinChip tandem mass spectrometry tryptic peptide sequencing. *Appl. Environ. Microbiol.*, 68(2):977–980, 2002.

10

Genetically Modified Probiotics

Farid E. Ahmed

CONTENTS

10.1 Introduction

Biological agents (biotherapeutic agents, or probiotics, derived from the Greek "for life") have been used to treat a variety of infections, especially those of the mucosal surfaces such as the gut and vagina. After the discovery and development of antibiotics, the value of these traditional treatments diminished. We are now being forced to look at alternatives to antibiotics because of the increasing number of resistant strains (1). The term probiotic was first used in 1965 to describe a live microbial supplement that beneficially affects the host by improving its microbial balance (2). Since then, research has looked at possible clinical uses for these agents, and in 1995 — when a greater understanding of their properties had been developed — the term biotherapeutic agents was proposed to describe microorganisms with specific therapeutic properties that also inhibit the growth of pathogenic bacteria. The bacteria that are found in probiotic products such as yogurt, kefir, and fermented vegetables are not normally found in the human intestine, as the gastrointestinal (GI) tract is often a hostile one for them. Thus, bacteria eaten in probiotic products do not normally colonize the intestine, but are flushed and eliminated from the body (3).

Prebiotics are foods or nutrients that are used by specific bacteria that can be added to the diet to increase the chances that these bacteria grow and thrive in the intestine (e.g., fructooligosaccharides); they — as well as members of the genera *Lactococcus* and *Lactobacillus* — are classified by the U.S. Food and Drug Administration (FDA) as GRAS (generally recognized as safe) for human use (4). The most frequently used probiotics are lactobacilli and bifidobacteria, owing to their recognition as members of the indigenous microflora of humans, their history of safe use, and the general body of evidence that supports their positive roles. The potential mechanisms of their action include competitive bacterial interactions, production of antimicrobial metabolites and nutraceuticals, immunomodulation, mucosal conditioning, constipation relief and preventing antibiotic-associated diarrhea, mercury detoxification, reduction in cardiovascular disease risk factors, prevention and alleviation of allergic diseases, management of atopic dermatitis, and reduction in carcinogens and mutagens (5–12). The emerging use of probiotics in several GI disorders (e.g., inflammatory bowel disease, or irritable bowel syndrome [IBS]) has led to increased interest in their use in patients with IBS (13). Universal acceptance of probiotics is, however, not without critics or risk, with over 11 case reports of *Salmonella boulardii* septicemia. It has been suggested that probiotic use may distract from proven interventions such as modification of antibiotic use (14).

This chapter will concentrate on Gram-positive, nonspore-forming, economically important lactic acid bacteria (LAB) that exist naturally in foods and the intestine since they have a long history of use for food production and preservation (15). LAB are a genetically diverse group of bacteria with G+C contents varying from 34 to 53%, which encompass rod-shaped bacteria such as lactobacilli, and cocci such as streptococci, lactococci, entococci, pedicocci, and leuconostoc (16). Although many representatives of LAB are safe and used for generation of food, some species such as streptococci are pathogens (17). LAB are widely used in starter cultures in fermented food processing, and a number of applications have been proposed to enhance their technological properties, increase the reliability of food-making processes, and improve product safety and quality. LAB can be engineered to function as cell factories to massively produce metabolites of interest such as food additives and aroma compounds. They have also been used to produce proteins with applications to health or the development of new vaccines. Knowledge of the interaction between certain LAB and the human host will allow exploiting their expected natural potential to improve health (18).

10.2 Cloning of LAB

A variety of genetic systems were developed to analyze and modify LAB, especially *L. lactis* and others of industrial interest. These systems may be classified as cloning, chromosome modification and expression systems. This review will concentrate on those introduced into food, referred to as "food grade" because of their safety.

10.2.1 Cloning Systems

The following plasmids using selective markers to allow selection and maintenance in the host have been developed: (a) the origin of replication of natural plasmids combined with food-grade selection markers such as the wide host-range pWV01 or pVS40 plasmids (19,20), the narrow host-range pCI305 (21) in *Lactococci*, or pFR18 in *Leuconostoc mesenteroides* (22); (b) interactive plasmid strategies such as *L. casei*, mv4 of *L. delbrueckii* and TP901-1 in *L. lactis* (23); and (c) systems based on homologous recombination by single crossover of nonreplicative plasmids as integrative vector by removing part of their replication function (24). Two kinds of markers have been utilized: (a) selectable that confers new phenotypes (e.g., sugar such as sucrose and xylose genes; bacteriocin resistance such as those conferring insensitivity to nisin or lactinin; or metal such as cadmium), and (b) those that restore an impaired function necessary for cell viability and that can be conditionally inactivated to produce auxotrophic mutations. Disadvantages of these markers are (a) the copy number of plasmids may vary, (b) plasmids may be lost in the absence of marker selection, (c) plasmids may be structurally unstable, and (d) vectors require the introduction of DNA in addition to that of the desired gene (18).

10.2.2 Chromosomal Modification Systems

Allelic replacement in the chromosome has several advantages over replicative or single crossover integration vectors as it allows stable DNA insertion or genetic modifications without leaving any foreign DNA other than that desired. Allelic replacement occurs by double crossover between two regions of homology flanking the modification and the corresponding regions on the chromosome. This procedure may occur spontaneously upon transformation using the natural competence machinery, such as those described in *Streptococcus pneumoniae* (25) and *Bacillus subtilis* (26). Alternatively, a thermosensitive plasmid-based system has been developed allowing gene replacement by a two-step procedure (Figure 10.1). A mutation in the gene encoding plasmid replication protein of the plasmid pWV01 has been selected that allows maintenance of the plasmid at 30°C,

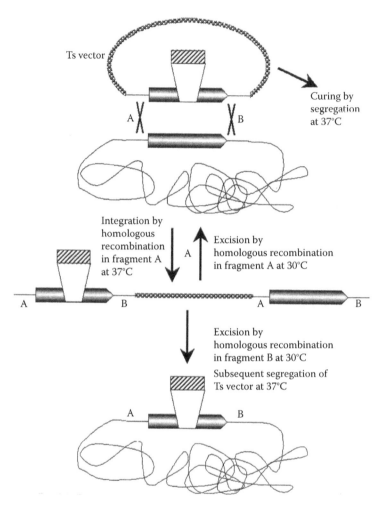

FIGURE 10.1
Food-grade allelic exchange and gene integration in the chromosome by a two-step procedure that employs a thermosensitive plasmid. From Reference 18; with permission.

but not at 37°C (27). This plasmid can direct homologous integration in the *L. lactis* chromosome when it carries a chromosomal DNA fragment of sufficient length (500 bp) (28), allowing sequential recombination in the two fragments flanking the central modification (nucleotide change, DNA deletion or insertion). The thermosensitive plasmid vector with selective markers can then be cured by growing the modified strain at 37°C for several generations to allow plasmid segregation once replication is blocked. This plasmid may also be used to select food-grade mutants containing a single insertion (IS) element as a new DNA fragment in the genome (29).

10.2.3 Expression Systems

In addition to cloning systems, gene expression systems have been developed in *Lactococcus lactis* that allow the controlled expression of homologous or heterologous genes such as those based on promoters controlled by sugar (lactose operon promoter) (30), by salt (*gadC* promoter) (31), by temperature upshift (*tec* phage promoter) (32), pH decrease (P170) (33), and phage infection (phi31-promoter) (34). A dose-dependent system of induction was developed using sublethal concentrations of nisin in *L. lactis* (35), and this system was extended to other LAB species (36). Sugar-dependent expression systems have also been developed in other LAB such as lactobacilli (37). Inducible systems are not always easy to manage under industrial conditions, especially if a constant level of production is required for metabolic control. In such a case, a well-defined constitutive promoter allowing the constitutive and defined level of expression of downstream genes has been created that could be applied to any bacterial species (38).

10.3 Construction of Starter LAB Strains

10.3.1 Production of Genetic Variants

Mutagenesis strategies using compounds such as ethyl methyl sulfonate (EMS), or *N*-methyl-*N'*-nitro-*N*-nitrosoguanidine have been employed (35). Additional mutations may, however, occur, necessitating careful testing for other important traits. To circumvent these problems, genetic engineering utilizing recombinant DNA technology was employed. For example, gene technology allowing the production of similar *ald*B mutants by direct allelic replacement using an appropriate thermosensitive vector has been carried out (36). This approach was also adopted to obtain food-grade mutants of *L. lactis* resistant to phages by the inactivation of the phage infection protein (*pip*) involved in phage adsorption and DNA injection (37). Similar approaches could also be used in other LAB species such as *Streptococcus thermophilus* where the inactivation of the phosphoglucomutase gene enhances polysaccharide production (38), and that of the urease gene reduces delay in the acidification in milks containing high amounts of urea (18). Specific mutants may also be isolated using an IS element (29) in which the final strains only contain ISS1 originating from *L. lactis* at a well-defined place in the chromosome.

10.3.2 Engineering Strains with Genes from Other LAB or Other Bacteria

Strains with increased proteolytic properties were constructed by transfer of genes coding PepN, PepC, PepX, and PepI peptidases of a highly proteolytic

L. helveticus strain (39), or PepI, PepL, PepW, and PepG from *L. delbrueckii* (40) into *L. lactis* using a food-grade cloning system. Another example of a gene transfer between LAB is provided by the construction of bacteriophage-resistant strains. However, phage resistance systems may interfere with phage adsorption, phage DNA injection, phage replication, transcription, RNA translation, protein assembly, and phage packaging. These mechanisms are often carried out by mobile elements such as plasmids and transposons, suggesting that lateral transfer of these genes occurs under pressure of phage infection. To improve phage resistance, mechanisms can be combined as a function of their target in phage development and of the phages present in the factories (41). In addition, targeting specific steps in phage development may introduce a further phage replication origin that competes with that of the phage (42). Another strategy is to induce the expression of a lethal gene upon phage infection (43), or to massively produce antisense mRNA against essential phage genes (44). A drawback of these systems is their narrow range of action. DNA shuffling, exploiting the properties of type I restriction enzymes, was also found to generate new restriction-modification mechanisms (45). New functions can also be provided by inter-LAB cloning. For example, an amylolytic *L. plantarum* silage strain with high starch-degrading ability was developed by expressing *L. amylovorus* amylase gene. This recombinant strain may have potential as a silage inoculant for crops such as alfalfa in which water-soluble carbohydrate levels are low, but which contain starch as an alternative carbohydrate source (46).

Genes can also be introduced from distant bacteria to produce genetically modified organisms (GMOs). For example, a heterologous catabolic glutamate dehydrogenase (GDH) gene from *Peptostreptococcus asaccharolyticus* was introduced into *L. lactis* to allow this organism to produce alpha-ketoglutarate from glutamate, an amino acid present at high levels in cheese (47). Another example is the metabolism engineering of LAB to produce high amounts of L-alanine not contaminated by the D-stereoisomer in *L. lactis* from pyruvate in a single step by alanine dehydrogenase (Figure 10.2). To redirect the carbon flux from pyruvate (which usually leads to lactate) to alanine, the *Bacillus sphaericus* alanine dehydrogenase gene (*ala*DH) was expressed in an L-LDH-deficient lactococcal strain. The constructed strain produced alanine as the sole end product. Finally, stereospecific production (>99%) of L-alanine could be achieved by inactivating the host-gene encoding alanine racemase (48). The use of LAB modified with an exogenous gene is currently limited in the European Union (EU) by a moratorium. Genetically modified LAB *S. thermophilus* have been used as improved biosensors for the detection of antibiotic residues in milk by introducing luciferase genes into this strain and optimizing their expression. This has been the only approved LAB under EU directive 90/220/EEC since December 1997. The bacterium, however, will not be found in food products as it is destroyed after the test has been attempted on a small sample (49).

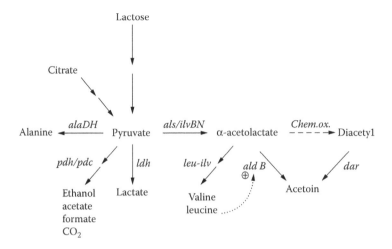

FIGURE 10.2
Pyruvate metabolism pathways leading to various amino acid aromatic compounds. From Reference 18; with permission.

10.4 Molecular Phenotyping of LAB

Two recent developments in molecular biology have advanced a number of fronts that have been historically problematic for probiotics. The first is the availability of molecular tools to properly recognize probiotic species and individual strains that have eliminated major confusion over strain identity and ancestry. Classical microbiological approaches (e.g., colony morphology, fermentation patterns, stereotyping, or a combination of them [50–52]) cannot classify a culture taxonomically, and phenotype characterization cannot distinguish closely related species such as those in the *Lactobacillus acidophilus* family occupying similar biological niches and are likely to play similar functions (Figure 10.3). On the other hand, phylogenetic analysis is a powerful taxonomic classification tool for bacterial cultures to accumulate data on ribosomal RNA sequences, and it has provided a means for comparative identification of probiotic cultures. Phylogenetic analysis can be conducted in varying degrees and combined with other characteristics (e.g., phenotypes) as needed to make definitive taxonomic classifications. Phylogenetics has now recognized 54 species of lactobacilli, 18 of which may be considered as probiotics; and 31 species of *Bifidobacterium*, 11 of which have been detected in human feces (53).

A number of sequence-based typing systems have been used to analyze conserved regions of the rRNA operon, or other conserved genes in probiotic cultures (54): (a) PCR amplification and sequencing of ~1500 bp of the 16S rRNA gene; (b) PCR amplification and sequencing of ~450 bp of the internal

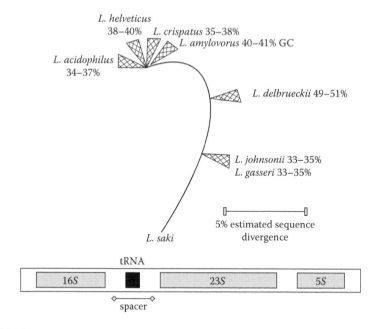

FIGURE 10.3
Phylogenetic relationships among members of the *Lactobacillus acidophilus* complex. From Reference 50; with permission.

transcribed spacer region; (c) PCR amplification and sequencing of alternative genes that are universally present and highly conserved (e.g., *recA* gene of bifidobacteria [55]); and (d) PCR amplification and sequencing of ~ 50 bp variable region of 16S rRNA to identify members of *L. acidophilus* complex (53). Although there are a number of alternative classification methods such as hybridization with species-specific probes (56), generation of profile PCR amplicons by species-specific primers (57), and PCR-ELISA (58), the outcomes are often variable and results are less definitive than direct sequencing of rRNA.

The second key developmental step in probiotics has been the availability of molecular fingerprinting methods (59) for identification and tracking of individual strains. These methods include ribotyping, amplified fragment length polymorphism (AFLP), pulsed-field gel electrophoresis (PFGE), and random amplified polymorphic DNA (RAPD) including multiplex PCR, arbitrary primed PCR, and triplicate arbitrary primed PCR (TAP-PCR) (53). Ribotyping (60) and PFGE (61) allow an overall clear identification potential at the strain level but are not suitable for routine use as they are tedious, slow, and repeatable. Alternative typing methods are those that employ PCR including RAPD analysis (62), by a TAP-PCR, or AFLP (63). TAP-PCR uses a specific primer targeting a highly conserved sequence within the 16S rRNA gene, and the resulting PCR amplicons can be used for molecular fingerprinting of a wide range of LAB (64). In the AFLP method, total genomic DNA is restriction digested followed by ligation of the synthetically generated adapters with DNA sequence corresponding to the sticky end of restriction products, then carrying out preselective PCR amplification using two primers complementary to the adapter-ligated ends with one preselected base at the 3′ end, and finally selective amplification generating the final product — by a second series of labeled primers that generally contain three tandemly oriented preselected bases at the 3′ end—is carried out. These amplified products are then separated by polyacrylamide gel electrophoresis (PAGE) on a sequencing gel (65).

The use of PCR primers complementary to the adapter and to the restriction site sequence yields strain-specific amplification patterns. AFLP and PFGE are reliable methods but are time-consuming and not suited for routine use. Techniques such as RAPD and TP-PCR show little reproducibility and require a complicated setup of experimental controls. Recently, enterobacterial repetitive intragenic consensus PCR (ERIC-PCR) and repetitive extragenic palindromic PCR (REP-PCR) methods that employed species-specific primers and raid PCR methods have been applied to molecular phenotyping of lactobacilli (59). REP elements and ERIC sequences are dispersed throughout bacterial genomes, and inter-REP and inter-ERIC distances or profiles are typical for given bacterial species and strains within a given species (66). Figure 10.4a shows an ERIC-PCR analysis on 16 *L. johnsonii* strains (Table 10.1) using primers targeting the 16S rRNA genes isolated from different environments. Multiple DNA fragments ranging from 4.072 kb to <0.298 kb were generated. A common intensive band of ~ 260 bp was found in 15 of the 16 strains, and only a weak signal was detected for *L. gasseri* (Lane 12 in Figure 10.4a). The 16 isolates were grouped into six different groups with distinguished ERIC-PCR patterns (E1 to E6) as seen in Table 10.1. A computer program employed for DNA pattern analysis produced dendograms that represented genetic relationships among the various strains (59). In REP-PCR genotyping, REP primers used for detection of *L. johnsonii* were made to contain the nucleotide isomer at an ambiguous position in the PCR consensus squence. These primers generated 2 to 11 fragments ranging in size from ~ 300 to 5000 bp (Figure 10.4b). A dominant band of ~ 1200 bp was found in all strains, except for *L. johnsonii* ATCC 11506 (Lane 18). REP amplification patterns were less complicated than ERIC patterns. Computer analysis of REP profiles generated only four groups (R1 to R4 in Table 10.1) (59). Results of other fingerprinting techniques on *L. johnsonii* are presented in Table 10.1. Not all typing methods resulted in the same genotype grouping because they look at different genotypical traits. Thus, REP- and ERIC-PCR that employ the complete bacterial genome in various microecological or GI lesions offer an alternate opportunity to characterize *Lactobacillus* strains because the interpretation power of rRNA-based data (i.e., the use of only one single gene or operon) in molecular typing has been questioned (59).

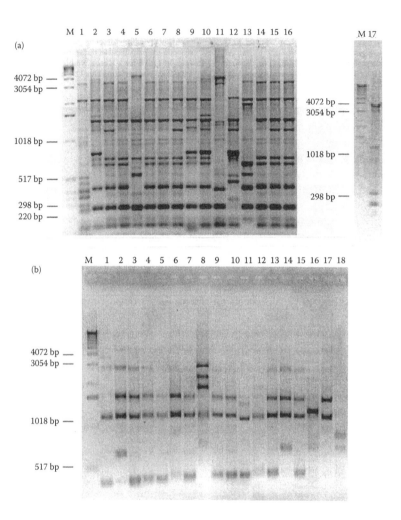

FIGURE 10.4
ERIC- and REP-PCR Fingerprints. (a) ERIC-PCR fingerprints of *Lactobacillus* strains: Lane 1, *L. johnsonii* ATCC 33200; Lane 2, *L. johnsonii* ATCC 332, Lane 3, *L. johnsonii* DSM 20553; Lane 4 *L. johnsonii* NCC 1669; Lane 5, *L. johnsonii* ATCC 11506; Lane 6, *L. johnsonii* NCC 1657; Lane 7, *L. johnsonii* NCC 1646, Lane 8, *L. johnsonii* NCC 1717; Lane 9, *L. johnsonii* JCM 8793; Lane 10, *L. johnsonii* NCC 1741; Lane 11; *L. johnsonii* JCM 8791, Lane 12, *L. gasseri* NCC 30604; Lane 13, *L. johnsonii* NCC 1703, Lane 14, *L. johnsonii* NCC 533; Lane 15, *L. johnsonii* NCC 1b; Lane 16, *L. johnsonii* NCC 1c; Lane 17, *L. johnsonii* NCC 1627; Lane M, 1-kb molecular size DNA ladder marker. From Reference 59; with permission. (b). REP-PCR profiles of *Lactobacillus* strains: Lane 1, *L. gasseri* NCC 30604; Lane 2, *L. johnsonii* NCC 1741, Lane 3, *L. johnsonii* NCC 1b; Lane 4 *L. johnsonii* NCC 533; Lane 5, *L. johnsonii* NCC 1c; Lane 6, *L. johnsonii* JCM 8793; Lane 7, *L. johnsonii* NCC 1669, Lane 8, *L. gasseri* DSM 20243; Lane 9, *L. johnsonii* NCCM 1646; Lane 10, *L. johnsonii* NCC 1717; Lane 11, *L. johnsonii* NCC 1703; Lane 12, *L. gasseri* JCM 8791; Lane 13, *L. johnsonii* DSM 20553, Lane 14, *L. johnsonii* ATCC 332; Lane 15, *L. johnsonii* NCC 1657; Lane 16, *L. johnsonii* ATCC 33200; Lane 17, *L. johnsonii* NCC 1627; Lane 18, *L johnsonii* ATCC 11506; Lane M, 1-kb molecular size DNA ladder marker. From Reference 59; with permission.

TABLE 10.1

Bacterial Identification by PCR Method and Grouping of *L. johnsonii* Strains According to the Overall Combination of All Achievable Typing Results[a]

Group and Strain	Species-specific Primer Identification	Genotype by:					Origin
		ERIC-PCR Type	REP-PCR Type	AFLP Type	TAP-PCR Type	PFGE Type	
Group 1[b]							
L. johnsonii NCC 1669	*L. johnsonii*	E1	R1	A1	C1	P1	Human Feces
L. johnsonii NCC 1657	*L. johnsonii*	E1	R1	A1	C1	P1	Human Feces
L. johnsonii NCC 533	*L. johnsonii*	E1	R1	A1	C1	P1	Human Feces
L. johnsonii NCC 1b	*L. johnsonii*	E1	R1	A1	C1	P1	Human Feces
L. johnsonii NCC 1c	*L. johnsonii*	E1	R1	A1	C1	P1	Human Feces
L. johnsonii NCC 1717	*L. johnsonii*	E1	R1	A1	C1	P1	Human Feces
L. johnsonii NCC 1646	*L. johnsonii*	E1	R1	A1	C1	P1	Human Feces
L. johnsonii NCC 20553	*L. johnsonii*	E1	R1	A1	C1	P1	Sour Milk
L. johnsonii NCC 332	*L. johnsonii*	E1	R1	A1	C2	P2	Unknown
L. johnsonii NCC 1741	*L. johnsonii*	E1	R1	A1	C2	P3	Human Feces
Group II							
L. johnsonii ATCC 11506	*L. johnsonii*	E2	R2	A2	C2	P2	Unknown
Group III							
L. johnsonii NCC 11506	*L. johnsonii*	E3	R3	A3	C1	P4	Cheese
Group IV							
L. johnsonii JCM 8791	*L. johnsonii*	E4	R1	A4	C2	P5	Mouse Feces
Ungroup Strains							
L. johnsonii ATCC 33200T	*L. johnsonii*	E6	R4	A5	C1	P6	Human Blood
L. johnsonii JCM 8793	*L. johnsonii*	E1	R1	A6	C1	P7	Pig Feces
L. johnsonii NCC 1627	*L. johnsonii*	E5	R1	A7	C1	P8	Unknown

[a] Types were determined by ERIC-PCR, REP-PCR, AFLP, PFDE, and TAP-PCR.

[b] Groups are based on linkage in the same cluster or are designed to the same cluster by at least three genotypical methods.

Source: Reference 59; with permission.

10.5 Gene Expression and Its Measurement

Genome sequencing and analysis of probiotic cultures identifies two major categories of genes: (a) those that will be required for survival and activity in grossly contrasting and changing environments (e.g., food vs. the GI tract), and (b) responsive gene systems that react to the varied stimuli encountered within the food-carrier or GI tract (Figure 10.5). Induced expression of critical genes (e.g., adherence, bacteriocin production, acid tolerance) is believed to be essential among many key functions that determine the survival, activity, and colonization potential of probiotic cultures. The roles of probiotics in these contrasting environments are distinct. Initially, it is survival with minimal energy through the food carrier that is apparent. Eventually through the GI tract, passage and survival among varying stresses (e.g., acid, bile) and functional activity (competitiveness and performance) at the targeted *in vivo* locations become the predominant stimuli (55).

Analysis of gene expression is important because changes in the physiology of an organism or a cell are accompanied by changes in patterns of gene expression. Techniques of gene expression (65) can be used to detect novel genes introduced into probiotics. New techniques for analyses of gene expression have been developed, which include (a) comprehensive open systems such as serial analysis of gene expression (SAGE), differential display (DD) analysis, RNA arbitrarily primed (RAP)-PCR, restriction endonucleolytic analysis of differentially expressed sequences (READS), amplified restriction fragment-length polymorphism (AFLP), total gene expression analysis (TOGA), and use of internal standard competitive template primers

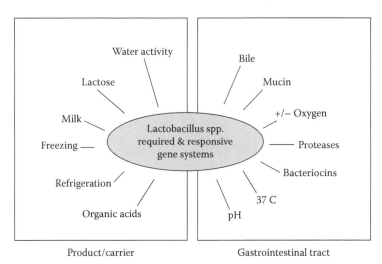

FIGURE 10.5
Stimuli encountered by probiotics in two key environments. From Reference 53; with permission.

(CTs) in a quantitative multiples RT-PCR method [StaRT-(PCR)], and (b) focused closed systems such as: high density cDNA filter hybridization (HDFCA) analysis, suppression subtractive hybridization (SHH), differential screening (DS), several forms of high-density cDNA arrays or oligonucleotide chips, and tissue microarrays. Sometimes, a combination of these systems is used to enhance the sensitivity and specificity of the assays. Although closed systems are excellent for the initial screening of large number of sequences, the value of the information generated is generally limited to an often arbitrarily chosen known sequence. On the other hand, only the open system platform has the potential to evaluate the expression patterns of tens of thousands of genes that have not yet been cloned or partially sequenced in a quantitative manner (65).

The study of gene expression by microarray technology is based on immobilizing cDNA or oligonucleotides on solid support. The introduction of differently labeled fluorescent probes for control and test samples to hybridize with DNA on arrays made miniaturization of arrays possible (67). Microanalysis offers the capability to examine differential gene expression across the entire genome and has been used for the identification of organisms in environmental, clinical, or food samples based on the presence of unique sequences, e.g., 16S rRNA, 23S rRNA, and key functional genes for the transgenic organisms (68) (Figure 10.6). The hindrance to utilization of microarrays on a wide scale in the biosafety arenas is due to the perceived initial large expense and the lack of standardization of the technology (65).

Employing functional genomic methods in probiotic screening will enhance existing traits by genetic modification of cultures. Targets for genetic modification and improvement include (a) immunomodulation and oral vaccine development, (b) antimicrobials and bacteriocins, (c) vitamin synthesis and production, (d) adhesion and colonization determinants, (e) production and delivery of digestive enzymes, and (f) metabolic engineering to alter products (e.g., polysaccharides, organic acids) or link cultures with specialty prebiotics designed to enhance the performance of probiotics *in vivo* (53).

10.6 Risk Assessment and Regulatory Controversies

Fermentation-based bioprocessors rely extensively on strain improvement for commercialization, and until recently strain (improvement) was based mainly on the selection of new natural strains. However, numerous improvements can be achieved by employing techniques of recombinant DNA, although no genetically modified (GM) LAB have been developed except for a strain utilized as a biosensor for detecting biocides in food (49).

GMOs are defined as organisms in which the genetic material has been altered in a way that does not occur naturally by mutation, mating, or natural recombination. GM foods have not gained worldwide acceptance because

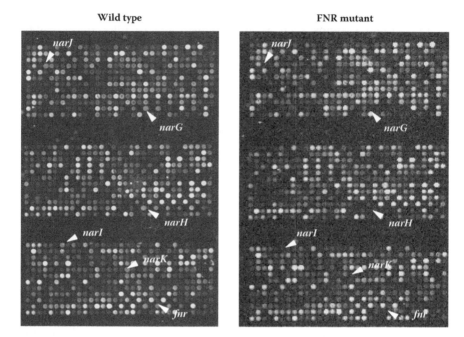

FIGURE 10.6

Illustration of a DNA microarray experiment in the bacterium *Bacillus subtilis*. This false-color image shows the induction of *nar* genes involved in nitrate reduction and their regulation by FNR protein under anaerobic conditions. Genes induced under anaerobic conditions are shown in red. From Reference 67; with permission.

of unmollified consumer suspicion resulting from earlier food and environmental concerns, transparent regulatory oversight, and mistrust in government bureaucracies, all factors which fueled debate about the environmental and public health safety issues of introduced genes (69); for example, potential gene flow to other organisms, the destruction of biological diversity, allergenicity, antibiotic resistance, and GI problems. Other economical and ethical issues pertaining to intellectual property rights added to these concerns. These uncertainties caused countries, such as those of the European Union (EU), to either restrict or require mandatory labeling the importing of bioengineered foods (70). Opposition to GMOs reached a peak during the mid-1990s in response to public pressure. In 1998, the EU introduced a *de facto* moratorium on the importation and production of GM foods, and in March 2003, the European Commission upheld the moratorium (71).

In the U.S., less stringent legislation did not stipulate mandatory labeling of GM foods, but has instead recommended a voluntary labeling of bioengineered foods and requested that companies notify the FDA of their intent to market GM foods at least 120 d before launch (70). The regulatory climate in the U.S. has been much less restrictive due to a different perception of the risk, and more of consumers' trust in regulatory authorities than in the EU. Social acceptance of GM foods, however, is not uniform in developed countries. Consumer concerns are based on ethical considerations (scientists playing God) or safety worries (more testing needs to be done) (69).

Risk assessments look specifically at how the GM product was developed and examine the risks associated with the gene products in the food (e.g., toxic or allergic responses). The determination of the overall risk of GMOs' strain improvement requires the identification of any characteristics of the GMOs which might lead to adverse effects, the evaluation of their potential consequences, and their possible occurrence. From these data, the risk can be estimated, and the application of strategies to manage these risks may be developed (18). The concept of "substantial equivalence" has been used for checking the safety of GM foods for human consumption. If the chemical composition of a modified food is found to be equivalent to its natural antecedent, then it is considered safe. However, some scientists in Europe argue that tests used to define this equivalence do not include all biological, toxicological, or immunological aspects of GM foods (69).

The definition of the GMO varies among EU countries. For example, in the diacetyl-overproducing strains, the same mutation inactivating the *addB* gene could be obtained in different ways (i.e., spontaneous mutation, induced mutagenesis, or genetic engineering). In Denmark, a spontaneous mutant isolated after a transient step in which the strain contained foreign DNA has been considered as a GMO, although this strain should be similar to that obtained by chemical mutagenesis (72). Also a mutant obtained by allelic replacement of *aldB* by a modified copy of the gene, even if it is a deletion similar to a natural event, would be considered as a GMO. This is because the EU regulations are considered to be relevant the way the modification was carried out, in contrast to the U.S. regulations that only take into account the final product. However, in terms of safety evaluation, the means by which a mutant has been constructed is irrelevant unless the technique employed introduced side effects. Very little is known about the latter, except that induced random mutagenesis often produces secondary mutations due to the lack of targeting. On the other hand, theoretically speaking, genetically derived mutants allow targeted modifications, and thus should avoid further mutations. The formal demonstration of the directness of the latter technique has not been proven, because until recently, side effects could not easily be determined due to lack of appropriate technology (18).

An EU-founded program "Express-Fingerprints" (QLK3-2001-01373) is currently testing potential side effects of the technology in *L. lactis* looking at *aldB* and *pip* mutants obtained by chemical mutagenesis and measuring levels of gene expression by two analytical methods: two-dimensional gel electrophoresis and DNA microarrays, which will allow the determination of relevant changes in the level of expression in at least 450 proteins and the transcription of over 2100 genes, larger than 89 bp. The Express-Fingerprints program should also give information about deregulation due to the desired change itself (18).

In addition to variants or mutants of technological relevance for improvement of product quality, strains expressing genes derived from closely or distantly related species could be developed. Here, three issues should be considered: (a) safety of the new gene product, (b) whether or not the new gene induces undesirable functions in the new host, and (c) whether or not any danger exists in case of transfer of the new gene. The first and third issues should be examined on a case-by-case basis, while the second one could be assessed by the same strategy employed by Express-Fingerprints. Various strains of *L. lactis* expressing heterologous proteins will be tested in order to evaluate the risk linked to the second issue. However, the main issue for GMOs, especially those of human origin, is the evaluation of human health risks of uncontrolled product expression following transfer of the transgene into commensal bacteria (18).

An additional issue that should be taken into consideration if the modification technology does not seem to influence safety is what percentages of gene could be inserted without having to label the product as containing a GMO? Thus, if it is assumed that many mutants (e.g., punctual, insertion of IS, deletions) will be substantially equivalent to naturally occurring strains if a specific mutation occurs at a very low rate (i.e., 10^{-9} per generation), more than 100 such cells will be present in a yogurt containing 10^9 viable cells per gram. The current labeling requirement in the EU specifies that the product must be labeled if it contains >0.9% GMOs. Since gene shuffling occurs on the chromosome (45), plasmids (73), transposons (74), or on phage-related elements (75), establishing the limit of the natural gene pool might then not be easy.

Transfers, including undesirable genes, occur naturally between bacteria from different species (76), and sequence data show that strains within the same species differ significantly in their genome (77). Gene transfer has been shown between pathogenic and commensal bacteria (78), and vehicles for these transfers include phages common between pathogens and food bacteria (79). Moreover, phages are known to be vehicles for pathogenicity islands in the former bacteria, all factors alluding to the potential of natural transfer of genes between pathogenic and natural isolates of food species. Considering that food bacteria are often derivatives of environmental and commensal bacteria, including probiotic strains isolated from intestinal flora (80), one has to wonder if a function isolated from a strain belonging to a bacterial species that has a long history of safe use will also be safe in another environment (18).

Assessment by global analyses such as that carried out in the Express-Fingerprints program on a set of natural strains underlies the variation in expression of genes and functions already characterized during the annotation of the genome and used to design the DNA microarray. A new DNA fragment and its expression product might not be detected, although it may encode new traits. GM microorganisms seem better characterized than new strains of food species. The new high throughput genomic technologies such as microarrays and proteomics, after perfection, will allow rapid characterization of inserted new elements (or their expression) in new strains, which could eventually be used to identify newly introduced GMO strains (18).

Technologies such as genome shuffling that use the natural cellular machinery, such as competence, are emerging (81). Strains produced by this technology may not be considered as a GMO, although the derived bacterium is less characterized than a GMO. Genome shuffling was applied to improve the acid tolerance of a poorly characterized industrial strain of *Lactobacillus* (82). New shuffled lactobacilli that grew at lower pH and produced more lactic acid than does the wild-type strain were identified, suggesting that genome shuffling is broadly useful for the evolution of tolerance and other complex phenotypes in industrial microorganisms. Genome analysis on food bacteria shows that the full set of genes necessary for this purpose is present in the genomes of *L. lactis* and *Streptococcus thermophilus*, a food bacteria closely related to the human commensal *S. salivarius* (83). Thus, the study of genome evolution promises to provide us with organisms where the genetic material has been naturally modified in many ways.

References

1. A D'Souza, C Rajkumar, J Cooke, C Bulpitt. Probiotic in prevention of antibiotic associated diarrhoea: meta analysis. *Br. Med. J.*, 324: 1361–1364, 2002.
2. RB Fuller. Probiotics in human medicine. *Gut*, 32: 439–442, 1991.
3. LV McFarland, GW Elmer. Biotherapeutic agents: past, present and future. *Microecol. Ther.*, 23: 46–73, 1995.
4. JE Teitelbaum, WA Walker. Nutritional impact of pre- and probiotics as protective gastrointestinal organisms. *Annu. Rev. Nutr.*, 22: 107–138, 2002.
5. S Houard, M Heinderyckx, A Bollen. Engineering of non-conventional yeasts for efficient synthesis of macromolecules: the methylotrophic genera. *Biochimie*, 84: 1089–1093, 2002.
6. C Krüger, Y Hu, Q Pan, H Marcotte, A Hultberg, D Delwar, PJ van Dalen, PH Pouwels, RJ Leer, CG Kelly, C van Dollenweerd, JK Ma, L Hammarström. *In situ* delivery of passive immunity by lactobacilli producing single-chain antibodies. *Nature Biotechnol.*, 20: 702–706, 2002.
7. C Ouwehand, H Lagstrom, T Suomalainen, S Salminen. Effect of probiotics on constipation, fecal azoreductase activity and fecal mucin content in the elderly. *Ann. Nutr. Metab.*, 46: 159–162, 2002.
8. MA Brudnak. Probiotics as an adjuvant to detoxfication protocols. *Med. Hypoth.*, 58: 382–385, 2002.
9. J Van Loo, N Jonkers. Evaluation in human volunteers of the potential anticarcinogenic activities of novel nutritional concepts: prebiotics, probiotics and synbiotics (the SYNCAN project QLK1-1999-00346). *Nutr. Metab. Cardiovasc. Dis.*, 11(4 Suppl): 87–93, 2001.
10. M Naruszewicz, ML Johansson, D Zapolska-Downar, H Bukowska. Effect of *Lactobacillus plantarum* 299v on cardiovascular disease risk factors in smokers. *Am. J. Clin. Nutr.*, 76: 1249–1255, 2002.
11. V Rosenfeldt, E Benfeldt, SD Nielsen, KF Michaelsen, DL Jeppesen, NH Valerius, A Paerregaard. Effect of probiotic *Lactobacillus* strains in children with atopic dermatitis. *J. Allerg. Clin. Immunol.*, 111: 389–395, 2003.

12. E Isolauri, H d C Ribiero, G Gibson, J Saavedra, S Saliminen, J Vanderheed, W Varavithya. Functional foods and probiotics: working group report of the First World Congress of Pediatric Gastroenterology and Nutrition. *J. Pediatr. Gastroenterol. Nutr.*, 35: S106–S109, 2002.

13. JA Madden, JO Hunter. A review of the role of the gut microflora in irritable bowel syndrome and the effects of probiotics. *Br. J. Nutr.*, 88 (Suppl 1): S67–S72, 2002.

14. J Beckly, S Lewis. Probiotics and antibiotic associated diarrhoea. The case for probiotics remains unproved. *Br. Med. J.*, 325: 902, 2002.

15. LL Mckay, KA Baldwin. Applications for biotechnology—present and future improvements in lactic acid bacteria. *FEMS Microbiol. Rev.*, 87: 3–14, 1990.

16. ME Stiles, WH Holzapfel. Lactic acid bacteria of foods and their current taxonomy. *Int. J. Food Microbiol.*, 36: 1–29, 1997.

17. MR Adams. Safety of industrial lactic acid bacteria. *J. Biotechnol.*, 68: 171–178, 1999.

18. P Renault. Generally modified lactic acid bacteria: applications to food or health risk assessment. *Biochimie*, 84: 1073–1087, 2002.

19. KJ Leenhouts, B Tolner, S Bron, J Kok, G Venema, JFML Seegers. Nucleotide sequence and characterization of the broad-host-range lactococcal plasmid pWVO1. *Plasmid*, 26: 55–66, 1991.

20. A von Wright, K Raty. The nucleotide sequence for the replication region of pVS40, a lactococcal food grade cloning vector. *Lett. Appl. Microbiol.*, 17: 25–28, 1993.

21. F Hayes, P Vos, GF Fitzgerald, WM de Vos, C Daly. Molecular organization of the minimal replicon of novel, narrow-host range, lactococcal plasmid pC1305. *Plasmid*, 25: 16–26, 1991.

22. F Biet, Y Cenatiempo, C Fremaux. Characterization of pFR18, a small cryptic plasmid from *Leuconostoc mesenteroides* ssp. Mesenteroides FR52, and its use as a food grade vector. *FEMS Microbiol. Lett.*, 179: 375–383, 1999.

23. L Brondsted, K Hammer. Use of the integration elements encoded by the temperate lactococcal bacteriophage TP901-1 to obtain chromosomal single-copy transcriptional fusion in *Lactococcus lactis*. *Appl. Environ. Microbiol.*, 65: 752–758, 1999.

24. E Emond, R Lavallee, G Drolet, S Moineau, G LaPointe. Molecular characterization of a theta replication plasmid and its use for development of a two-component food-grade cloning system for *Lactococcus lactis*. *Appl. Environ. Microbiol.*, 67: 1700–1709, 2001.

25. EA Campbell, SY Choi, HR Masure. A competence regulon in *Streptococcus pneumoniae* revealed by genomic analysis. *Mol. Microbiol.*, 27: 929–939, 1998.

26. D Dubnau, CM Lovett. Transformation and recombination. In: AL Sonenehein, JA Hoch, R Losicks, Eds., Bacillus subtilis *and Its Closest Relatives: From Gene to Cells*. American Society for Microbiology Press, Washington, DC, 2002, pp. 453–471.

27. E Maguin, P Duwat, T Hege, SD Ehrlich, A Gruss. New thermosensitive plasmid for gram-positive bacteria. *J. Bacteriol.*, 17: 5633–5638, 1992.

28. I Biswas, A Gruss, SD Ehrlich, E Maguin. High efficiency gene inactivation and replacement system for gram-positive bacteria. *J. Bacteriol.*, 175: 3628–3635, 1993.

29. E Maguin, H Prevost, SD Ehrlich, A Gruss. Efficient insertional mutagenesis in lactococci and other gram-positive bacteria. *J. Bacteriol.*, 178: 931–935, 1996.

30. J Payne, CA MacCormick, HG Griffin, MJ Gasson. Exploitation of a chromosomally integrated lactose operon for controlled gene expression in *Lactococcus lactis. FEMS Microbiol. Lett.,* 136: 19–24, 1996.

31. JW Sanders, G Venema, J Kok. A chloride-inducible gene expression casette and its use in induced lysis of *Lactococcus lactis. Appl. Environ. Microbiol.,* 63: 4977–4882, 1997.

32. A Nauta, D Vansinseren, H Karsens, E Smit, G Venema, J Kok. Inducible gene expression mediated by a repressor-operator system isolated from *Lactococcus lactis* bacteriophage rlt. *Mol. Microbiol.,* 19: 1331–1341, 1996.

33. SM Madsen, J Arnau, A Vrang, M Givskov, H Israelsen. Molecular characterization of the pH-inducible and growth phase-dependent promoter P170 of *Lactococcus lactis. Mol. Microbiol.,* 32: 75–87, 1999.

34. DJ O'Sullivan, SA Walker, SG West, TR Klaenhammer. Development of an expression strategy using a lytic phage to trigger explosive plasmid amplification and gene expression. *N.Y. Biotechnol.,* 14: 82–87, 1996.

35. H Boumerdassi, C Monnet, M Desmazeaud, G Corrieu. Isolation and properties of *Lactococcus lactis* subsp. *lactis* biovar diacetylactis CNRZ 483 mutants producing diacetyl and acetoin from glucose. *Appl. Environ. Microbiol.,* 63: 2293–2299, 1997.

36. SR Swindell, KH Benson, HG Griffin, P Renault, SD Ehrlich, MJ Gasson. Genetic manipulation of the pathway for diacetyl metabolism in *Lactococcus lactis. Appl. Environ. Microbiol.,* 62: 2641–2643, 1996.

37. MR Monteville, B Ardestani, BL Geller. Lactococcal bacteriophages require a host cell wall carbohydrate and a plasma membrane protein for adsorption and ejection of DNA. *Appl. Environ. Microbiol.,* 60: 3204–3211, 1994.

38. F Levander, M Svensson, P Radstrom. Enhanced exopolysaccharide production by metabolic engineering of *Streptococcus thermophilus. Appl. Environ. Microbiol.,* 68: 784–790, 2002.

39. V Joutsjoki, S Luoma, M Tamminen, M Kilpi, E Johansen, A Palva. Recombinant *Lactococcus* starters as a potential source of additional peptidolytic activity in cheese ripening. *J. Appl. Microbiol.,* 92: 1159–1166, 2002.

40. U Wegmann, JR Klein, I Drumm, OP Kuipers, B Henrich. Introduction of peptidase genes from *Lactococcus delbrueckii* subsp. lactis into *Lactococcus lactis* and controlled expression. *Appl. Environ. Microbiol.,* 65: 4729–4733, 1999.

41. S Moineau. Applications of phage resistance in lactic acid bacteria. *Antonie Van Leeuwenhoek,* 76: 377–382, 1999.

42. DJ O'sullivan, C Hill, TR Klaenhammer. Effect of increasing the copy number of bacteriophage origins of replication, in trans, on incoming-phage proliferation. *Appl. Environ. Microbiol.,* 59: 2449–2456, 1993.

43. GM Djordjevic, DJ O'Sullivan, SA Walker, MA Conkling, TR Klaenhammer. A triggered- suicide system designed as a defense against bacteriophages. *J. Bacteriol.,* 179: 6741–6748, 1997.

44. SG Kim, CA Batt. Antisense messenger RNA-mediated bacteriophage resistance in *Lactococcus lactis* subsp. *lactis. Appl. Environ. Microbiol.,* 57: 1109–1113, 1991.

45. C Schouler, M Gautier, SD Ehrlich, MC Chopin. Combinational variation of restriction modification specificities in *Lactococcus lactis. Mol. Microbiol.,* 28: 169–178, 1998.

46. A Fitzsimons, P Hols, J Jore, RJ Leer, M O'Connell, J Delcour. Development of an Amylotic *Lactobacillus plantarum* silage strain expressing the *Lactobacillus amylovorus* alpha-amylase gene. *Appl. Environ. Microbiol.*, 60: 3529–3535, 1994.

47. L Rijnen, S Bonneau, M Yvon. Genetic characterization of the major lactococcal aromatic aminotransferase and its involvement in conversion of amino acids to aroma compounds. *Appl. Environ. Microbiol.*, 65: 4873–4880, 1999.

48. P Hols, M Kleerebezem, AN Schanck, T Ferain, J Hugenholtz, J Delcour, WM de Vos. Conversion of *Lactococcus lactis* from homolactic to homoalanine fermentation through metabolic engineering. *Nat. Biotechnol.*, 17: 588–592, 1999.

49. MF Jacobs, S Tynkkynen, M Sibakov. Highly bioluminescent *Streptococcus thermophilus* strain for the detection of dairy-relevant antibiotics in milk. *Appl. Microbiol. Biotechnol.*, 44: 405–412, 1995.

50. K-H Schleifer, M Ehrmann, E Brockmann, W Ludwig, R Amann. Application of molecular methods for the classification and identification of lactic acid bacteria. *Int. Dairy J.*, 5: 1081–1094, 1995.

51. M Alander, R Korpela, M Saxelin, T Vilpponen-Salmela, T Mattila-Sandholm, A von Wright. Recovery of *Lactobacillus rhamnosus* GG from human colonic biopsies. *Lett. Appl. Microbiol.*, 24: 361–364, 1997.

52. S Ahrne, S Nobaek, B Jeppsson, I Adierberth, AE Wold, G Molin. The normal *Lactobacillus* flora of healthy human rectal and oral mucosa. *J. Appl. Microbiol.*, 85: 88–94, 1998.

53. TR Klaenhammer, MJ Kullen. Selection and design of probiotics. *Int. J. Food Microbiol.*, 50: 45–57, 1999.

54. DJ O'Sullivan. Methods for the analysis of the intestinal microflora. In: GW Tannock, Ed., *Probiotics: A Critical Review*. Horizon Scientific Press, Norfolk, UK, 1999, pp. 23–44.

55. MJ Kullen, MJ Brady, DJ O'Sullivan. Evaluation of using a short region of the *recA* gene for rapid and sensitive speciation of dominant bifidobacteria in the human large intestine. *FEMS Microbiol. Lett.*, 154: 377–383, 1997.

56. B Pot, C Hertel, W Ludwig, P Descheemaeker, K Kersters, KH Schleifer. Identification and classification of *Lactobacillus acidophilus*, *L. gasseri*, and *L. johnsonii* strains by SDS-PAGE and rRNA-targeted oligonucleotide probe hybridization. *J. Gen. Microbiol.*, 139: 513–517, 1993.

57. A Tilsala-Timisjarvi, T Alatossava. Development of oligonucleotide primers from the 16S-23S rRNA intergenic sequences for identifying different dairy and probiotic lactic acid bacteria by PCR. *Int. J. Food Microbiol.*, 35: 49–56, 1997.

58. R Laitinen, E Malinen, A Palva. I: Application to simultaneous analysis of mixed bacterial samples composed of intestinal species. *Syst. Appl. Microbiol.*, 25: 241–248, 2002.

59. M Ventura, R Zink. Specific identification and molecular typing of *Lactobacillus johnsonii* by using PCR-based methods and pulsed-field gel electrophoresis. *FEMS Microbiol. Lett.*, 217: 141–154, 2002.

60. S Rodtong, GW Tannock. Differentiation of *Lactobacillus* strains by ribotyping. *Appl. Environ. Microbiol.*, 59: 3480–3484, 1993.

61. HJ Busse, EBM Denner, W Lubitz. Classification and identification of bacteria — current approaches to an old problem — overview of methods used in bacterial systematics. *J. Biotechnol.*, 47: 3–38, 1996.

62. JD William, AR Kubelik, K Livak, JA Rafalski, SV Tingey. DNA polymorphism amplified by arbitrary primers are useful as genetic markers. *Nucleic Acids Res.*, 18: 6531–6535, 1990.

63. A Gancheva, B Pot, K Vanhonacker, B Hoste, K Kersters. A polyphasic approach towards the identification of strains belonging to *Lactobacillus acidophilus* and related species. *Syst. Appl. Microbiol.*, 22: 573–585, 1999.

64. SM Cusick, DJ O'Sullivan. Use of a single, triplicate arbitrarily primed-PCR procedure for molecular fingerprinting of lactic acid bacteria. *Appl. Environ. Microbiol.*, 66: 2227–2231, 2000.

65. FE Ahmed. Molecular techniques for studying gene expression in carcinogenesis. *J. Environ. Sci. Health*, C20: 77–116, 2002.

66. J Versalovic, T Koeuth, JR Lupski. Distribution of repetitive DNA sequences in eubacteria and application to fingerprinting of bacterial genomes. *Nucleic Acids Res.*, 19: 6823–6831, 1991.

67. RW Ye, T Wang, L Bedzyk, KM Croker. Application of DNA microassays in microbial systems. *J. Microbiol.*, 47: 257–272, 2001.

68. RJ Lipshutz, SP Fodor, TR Gingeras, DJ Lockhart. High density synthetic oligonucleotide arrays. *Nature Genet.*, 21: 30–24, 1999.

69. BEB Mosley. The safety and social acceptance of novel foods. *Int. J. Food Microbiol.*, 50: 25–31, 1999.

70. FE Ahmed. Detection of genetically modified organisms in foods. *Trends Biotechnol.*, 20: 215–223, 2002

71. A Hellemans. Consumer fear cancels European GM research. *Scientist*, 17: 52–54, 2003.

72. CM Henriksen, D Nilsson, S Hansen, E Johansen. Application of genetically modified microorganisms: gene technology at Chr Hnasen A/S. *Int. Dairy J.*, 9: 17–24, 1999.

73. M Nardi, P Renault, V Monnet. Duplication of the *pepF* gene and shuffling of DNA fragments on the lactose plasmid of *Lactococcus lactis*. *J. Bacteriol.*, 179: 4164–4171, 1997.

74. T Immonen, G Wahlsrom, T Takala, PE Saris. Evidence for a mosaic structure of the Tn5481 in *Lactococcus lactis* N8. *DNA Seq.*, 9: 245–261, 1998.

75. E Durmaz, TR Klaenhammer. Genetic analysis of chromosomal regions of *Lactococcus lactis* acquired by recombinant lytic phages. *Appl. Environ. Microbiol.*, 66: 895–903, 2000.

76. M Teuber, L Meile, F Schwarz. Acquired antibiotic resistance in lactic acid bacteria from food. *Antonie Van Leeuwenhoek*, 76: 115–137, 1999.

77. JC Smoot, KD Barbian, J Van Gospel, LM Smoot, MS Chausseu, GL Sylva, DE Sturdevant, SM Ricklefs, SF Porcella, LD Parkins, B Beres, DS Campbell, TD Smith, Q Zhang, V Kapur, JA Daly, LG Veasy, JM Musser. Genome sequence and comparative microarray analysis of serotype M18 group A *Streptococcus* strains associated with acute rheumatic fever outbreaks. *Proc. Natl. Acad. Sci. U.S.A.*, 99: 4668–4673, 2002.

78. A Kalia, MC Enright, BG Spratt, DE Bessen. Directional gene movement from human-pathogenic to commensal-like streptococci. *Infect. Immun.*, 69: 4858–4869, 2001.

79. F Desiere, WM McShan, D van Sinderen, JJ Ferreti, H Brussow. Comparative genomics reveals close genetic relationships between phages from dairy bacteria and pathogenic streptococci: evolutionary implications for prophage-host interactions. *Virology*, 288: 325–341, 2001.

80. S Salminen, A von Wright, L Morelli, P Marteau, D Brassart, WM de Vos, R Fonden, M Saxelin, K Collins, G Mogensen, SE Birkeland, T Mattila-Sandholm. Demonstration of safety of probiotics — a review. *Int. J. Food Microbiol.*, 44: 93–106, 1998.

81. YX Zhang, K Perry, VA Vinci, K Powell, WP Stemmer, SB del Cardayre. Genome shuffling leads to rapid phenotypic improvement in bacteria. *Nature*, 415: 644–646, 2002.

82. R Patnaik, S Louie, V Gavrilovic, K Perry, WP Stemmer, CM Ryan, S Del Cardayre. Genome shuffling of *Lactobacillus* for improved acid tolerance. *Nat. Biotechnol.*, 20: 707–712, 2002.

83. A Bolotin, P Wincker, S Mauger, O Jaillon, K Malarme, J Weissenbach, SD Ehrlich, A Sorokin. The complete genome sequence of the lactic acid bacterium *Lactococcus lactis* ssp. *lactis* IL403. *Genome Res.*, 11: 731–753, 2001.

11

Use of Probiotics in Preharvest Food Safety Applications

Francisco Diez-Gonzalez and Gerry P. Schamberger

CONTENTS

11.1 Introduction

The concept of probiotics has evolved since Lilly and Stillwell (1) used it in 1965 to refer to growth-promoting substances for microorganisms. There are many definitions for the term probiotic, but one of the most widely accepted definitions was coined by Fuller (2) in 1989, as "a live microbial supplement that beneficially affects the host by improving its intestinal microbial balance." This definition, however, restricts the benefits of probiotics to the organism that ingests them. A third definition of probiotics that might be more flexible is the one coined by Shaafsma (3) in 1996, who stated that "oral probiotics are living organisms which upon ingestion in certain numbers, exert health effects beyond inherent basic nutrition." In this definition, it is not clear who ingests the probiotics and who gets the health effects.

In 1998, Zhao et al. (4) used the term "probiotic bacteria" to refer to *Escherichia coli* strains that were fed to cattle with the purpose of reducing the fecal carriage of serotype O157:H7. While the authors were seeking an ultimate food safety benefit by feeding these strains, the well-being of the direct host was not improved, as *E. coli* O157:H7 is not normally a cattle pathogen. Researchers in the human probiotics field might disagree with the use of the term probiotic as a preharvest food safety strategy, but for the purpose of this chapter we suggest that the meaning of the term probiotic could be an expansion of Fuller's, as follows: "live microorganisms orally ingested that beneficially affect the host and/or the consumer of the host by improving the host's intestinal microbial balance."

There are two other terms applied to the use of microbial cultures in livestock with the purpose of reducing food-borne pathogens or improving their health: competitive exclusion and direct-fed microbial products. Competitive exclusion (CE) in its broader biological meaning is the "principle that if two species try to occupy the same ecological niche, a superior species will eventually emerge to replace the inferior one" (5). For preharvest food safety, CE is defined as the use of microbial cultures to prevent colonization of or to displace populations of pathogenic bacteria in the gastrointestinal (GI) tract of animals. Typical CE cultures are undefined cultures of microorganisms that are obtained from healthy adult animals and are used to inoculate young animals to prevent pathogen colonization (6). The reader is referred to Chapter 12 in this book that covers in more detail the use of CE cultures.

According to the Center of Veterinary Medicine (CVM) of the U.S. Food and Drug Administration, CE products cannot be considered "generally recognized as safe" (GRAS) substances, and they should be regulated as animal drugs. The concept of direct-fed microbial products (DFMPs) is a term coined by the CVM with the purpose of classifying and regulating probiotic bacteria for use in livestock (7). DFMPs are regulated in the U.S. as food according to the Compliance Policy Guide 689.100, as long as they are well-defined microorganisms belonging to one of the genera approved by the Association of American Feed Control Officials (AAFCO) (e.g., *Saccharomyces, Bacillus, Lactobacillus, Lactococcus, Bifidobacterium, Bacteroides, Pediococcus*) containing known viable numbers. DFMPs can be marketed as such, as long as no health or structure or function claim is made about their use.

This chapter will be focused on probiotic preparations consisting of defined cultures including DFMP and non-DFMP microorganisms. It will include sections on the use of probiotics for nonfood safety purposes in livestock, and their most recent applications to reduce the prevalence of food-borne pathogens.

11.2   History of Probiotic Use in Animals

The benefits of consuming live probiotic microorganisms by humans was recognized as early as 1907 by Elie Metchnikoff, the Russian father of immunology (8), but it was not until 1924 that the first experiment with livestock was reported. Eckles et al. (9) investigated the utilization of baker's yeast as a feed supplement of vitamin B. In that pioneering report, it was concluded that yeast-fed calves did not grow faster than control groups and that their health status was not improved. The first report that describes the use of colicinogenic *Escherichia coli* as a probiotic for pigs was published in 1961 (10). In this study Tadd and Hurst could not eliminate hemolytic *E. coli* strains from the GI tract and did not reduce the incidence of disease in weaned pigs. However, they were able to show that the colicinogenic *E. coli* could get established in the large intestine.

Despite the fact that lactic acid bacteria (LAB) were recognized as a human probiotic early in the 1900s, the documented use of *Lactobacillus* strains for animals was not attempted until 1925, when *L. acidophilus* was fed to chickens to decrease cecal pH as a potential coccidiosis preventive measure (11). In 1965, a milk culture of *L. acidophilus* was given to piglets to treat enteritis in a swineherd (12). After a 2-month feeding of the probiotic culture, the incidence of diarrhea was markedly reduced, and the strain was recovered more than a year later. In one of the first reports that studied the effect of probiotics on animal productivity, an *L. acidophilus* preparation was given to pigs with the purpose of determining its influence on average daily gain (ADG) and feed efficiency (12). As a result of giving the *L. acidophilus* preparation for 5 d at the beginning of the feeding period, the live-weight gain per week increased an average of 7%, and the amount of feed per increase in live weight was reduced approximately 4%. These pioneering studies clearly indicated that probiotics could have an extraordinary potential for utilization in livestock.

In 1973, Tortuero (13) reported that *L. acidophilus* given to chickens via drinking water failed to increase the ADG or feed efficiency. After this study, a significant number of researchers investigated the effects of probiotics on poultry productivity (14), and in 1977 Fuller laid out some of the principles of the potential application of defined probiotic cultures to reduce potentially pathogenic bacteria (15). He reported that lactobacilli were capable of displacing *E. coli* and streptococci from chick crops and suggested that these microorganisms could also prevent the colonization of *Salmonella*. Ellinger et al. (16) provided some of the earliest evidence that indicated that probiotics could also be used in ruminants, as they observed that feeding *L. acidophilus* could reduce the levels of intestinal coliforms in calves. Since those early studies, the preferred probiotic has been lactobacilli, but a few reports such as the one written by Linton et al. (17) also explored the possibility of using *E. coli* to displace antibiotic-resistant bacteria.

11.3   Use of Probiotics for Animal Production

Probably the most important event that fostered the interest in probiotics for livestock production was Swann's report on the use of antibiotics in animal husbandry and veterinary medicine in 1969 (18). This study recommended the segregation of antibiotics used for therapeutic purposes from those used as feed supplements. Swann's report fueled intensive research on the use of probiotics as an alternative to antibiotics. Probiotics have been extensively tested and fed to cattle, swine, and poultry for the purposes of promoting growth, increasing feed efficiency, enhancing milk and egg production, improving meat and milk quality, and preventing infectious diseases (18). For all of these objectives, the ultimate goal is to increase animal productivity.

The mechanisms that result in the above-referenced benefits have not been fully elucidated, but it has been suggested that probiotics prevent or reduce the colonization of microorganisms that might be detrimental for animal growth, compete for nutrients, affect the animal immune system, and damage animal cells and tissues. In this section, a summary of the most relevant nonfood safety probiotic applications in livestock is described. For more detailed discussions on these probiotic applications the reader is encouraged to consult the references by Huber (19), Fuller (18), and Nousiainen and Setälä (20).

The diversity of probiotic microorganisms for livestock is broader than those intended for human use. A variety of *Lactobacillus* and *Bifidobacteria* species similar to those used as human probiotics have been tested and are currently marketed, but other bacterial species belonging to the genera *Streptococcus*, *Enterococcus*, *Bacillus*, *Clostridium*, and *Pediococcus* have also been fed to a range of farm animals (18). Not only bacterial probiotics are fed to livestock, but also yeast and fungi. Among these, *Saccharomyces cerevisiae* and *Aspergillus oryzae* probably account for the largest volume of probiotics currently marketed for cattle (19). Amaferm® (Biozyme, Inc., St. Joseph, MO) is an extract of an *A. oryzae* culture that is widely sold as a cattle supplement but is arguably a probiotic as it does not contain live organisms.

In calves, a number of reports have shown that different *Lactobacillus*, *Enterococcus,* and yeast increased feed intake and weight gain, as well as decreased diarrhea (19). In adult cattle, fungal cultures have been observed to increase feed efficiency and ADG. In lactating cows, slight stimulation of milk production has been reported by a number of researchers. The utilization of probiotic cultures in piglets has targeted the reduction of pathogenic *E. coli* to prevent infection and to increase animal performance (19, 20). The genera of bacteria most frequently used in swine are *Lactobacillus, Enterococcus,* and *Bacillus*. In poultry, defined cultures of *Lactobacillus* strains have been used to reduce and prevent colonization by *Salmonella* (14). A number of studies have shown little or no effect of feeding probiotics in farm animals, and these contradictory results appear to be due to a variety of factors such as age of animals, diet, culture viability, dose frequency, degree of stress, and colonization (18).

11.4   Infectious Disease Prevention

The utilization of probiotic cultures for preharvest food safety is relatively recent and remains a largely undeveloped field. However, some of the underlying mechanisms of inhibition of food-borne pathogens could be based on the same principles of displacement of animal pathogens. A number of authors have summarized the mechanisms of inhibition by probiotic bacteria as the following: nutrient competition, production of inhibitory compounds (e.g., organic acids, bacteriocins), immunostimulation, and competition for binding sites (21). The demonstration of the specific mechanisms of any given probiotic strain is still quite difficult, and sometimes *in vitro* inhibition does not always translates to *in vivo* effects (10).

The inhibition of potentially pathogenic bacteria in poultry has been actively studied since the 1970s, targeting largely *E. coli* and *Salmonella* (14). In 1982, feeding *L. acidophilus* was found to reduce the mortality of gnotobiotic chicks 20-fold after inoculation with pathogenic *E. coli* O2:K1 (22). Similarly the *Salmonella*-caused mortality of newly hatched chicks was also markedly prevented by feeding mixtures of *L. acidophilus* strains prophylactically (14). A number of reports have indicated that feeding lactobacilli to chickens can reduce the colonization by *Salmonella* strains (14, 23), but some researchers have frequently reported a complete lack of effect of these types of probiotics (24, 25). Recently, lactobacilli have also been investigated for prevention of coccidiosis, and Dalloul et al. (26) reported that this effect was mediated by stimulation of the immune system. Because *Salmonella* is both a human and poultry pathogen, the advances in bird infection prevention can be directly applicable for food safety.

Probiotic microorganisms have also been utilized in other bird species, but published reports are relatively rare. In turkey poultries, feeding of viable *Enterococcus faecium* appeared to stimulate the population of LAB, but no attempts were made to relate probiotic feeding to the bird's health status (27). Ehrmann et al. (28) developed the first combination of *Lactobacillus* strains isolated from ducks and reported the colonization of inoculated fowl. In another study, a proprietary probiotic culture was fed to ducklings in combination with antibodies against *Salmonella enteritidis*, and the number of infected birds was significantly lower than controls in at least half of the observations (29). In a unique application, Khajarern et al. (30) recently reported the use of a probiotic-fermented proprietary product to protect ducks from the toxicity of mycotoxins.

Postweaning diarrhea is an important cause of mortality in swine, and most of the research on the use of probiotics has been directed to control its etiological agent, enterotoxigenic *E. coli* serotype K88 (25). In 1993, Mulder et al. (25) used piglets that had been inoculated with *E. coli* K88 to investigate the effect of a mixture of lactobacilli, and they reported that the severity of the symptoms and the number of deaths in the control group were markedly greater than in those animals that were given the probiotic cultures. Treatments with *Enterococcus faecium* cultures have also shown to significantly reduce the morbidity and mortality of K88-inoculated gnotobiotic piglets (31), and this effect appears to be mediated by the ability of this lactic acid bacterium to inhibit the adhesion of serotype K88 to intestinal lining cells (32). In another study, combinations of *E. faecium* and *Bacillus toyoi*, however, had little effect to prevent diarrhea, despite the observation that *E. faecium* was associated with the intestinal mucosa (25). Despite the variability on the effects reported in the literature, there are several preparations that are currently marketed to reduce diarrhea in piglets.

In cattle, the prevention of infections by using probiotic microorganisms has been directed to treat scouring in newborn calves as an alternative to antibiotic administration. Feeding *Lactobacillus* and *Saccharomyces cerevisiae* cultures individually or in combination into milk-replacer formula has been reported to be effective in reducing the number of diarrheic calves and scouring symptoms (33–35). A mixture of LAB, however, was not effective in protecting calves against *Cryptosporidium parvum* infection (36). *Enterococcus faecium* cultures have also been related to scouring control in at least five separate studies (19), and there are at least four commercial brands of calf milk replacers supplemented with this bacterium (e.g., Calf Advantage® Support Formula, Newmilk, Inc., Stockton, CA; Micromanager® Microbials, PharmTech, Inc., Des Moines, IA.; Provita Protect®, Provita Eurotech, Ltd., Tyrone, Northern Ireland). In addition to treatment of scours, the use of *E. faecium* and *Propionibacterium* has been investigated to reduce rumen acidosis in adult cattle (37).

Infectious diseases caused by *Vibrio* and *Aeromonas* species are a major concern in aquaculture, and the ability of probiotic bacteria to control these pathogens in fish and shrimp has been extensively studied (38). *Bacillus* strains have been used in shrimp with relative success to reduce mortality caused by pathogenic bacteria (38), but a recent report indicated that this genus can also have deleterious effects on the quality of shrimp (39). *Pseudomonas fluorescens* strains have been given to rainbow trout and other fish varieties, and this treatment has resulted in as much as 46% reduction in mortality rates of trout inoculated with *Vibrio angillarum* (38, 40). *Aeromonas salmonicida* is one of the most deleterious bacterial pathogens for Atlantic salmon, but the number of deaths was significantly decreased when salmonids were fed a probiotic *Carnobacterium* K strain (41). Typical human probiotics such as *Lactobacillus rhamnosus*, *L. plantarum*, and *L. helveticus* have also been capable of reducing mortalities caused by *Aeromonas* and *Vibrio* pathogens (38). Ecuadorian shrimp farms have been able to control larval diseases by providing a *Vibrio alginolyticus* probiotic (42). Some of these infectious disease-control technologies could be very promising to control human pathogens transmitted via shellfish.

11.5 Food Safety Applications

11.5.1 Cattle

The treatment of cattle with probiotics to reduce the presence of *E. coli* O157:H7 is considered one of the most promising preharvest food safety interventions that can contribute to the eradication of this food-borne pathogen. Serotype O157:H7 is a highly virulent food-borne pathogen, and many studies have revealed that a large number of cattle asymptomatically harbor and shed this pathogen in their feces (43–45). Diminishing the ability of *E. coli* O157:H7 to colonize cattle would reduce outbreaks of enterohemorrhagic colitis from meats, produce, and water contaminated with cattle manure. There have been a few reports that have investigated the effectiveness of probiotic bacteria to reduce O157:H7 carriage in cattle (Table 11.1). These reports have focused on the utilization of LAB and other *E. coli* strains.

Zhao et al. (4) identified several *E. coli* strains with the ability to inhibit *E. coli* O157:H7 and conducted the first study using probiotic bacteria to inhibit O157:H7 in cattle. The authors screened 1200 bacterial isolates from cattle for *in vitro* inhibitory activity against serotype O157:H7. Eighteen isolates, 17 *E. coli* and 1 *Proteus mirabilis*, were found to inhibit five strains of *E. coli* O157:H7. Two calves were then orally inoculated with this set of anti-O157:H7 isolates, and four of these *E. coli* strains were selected for further utilization on the basis of their recovery from calves 27 d after inoculation.

These dominant anti-O157:H7 *E. coli* were then tested in a calf-feeding study. Six calves were inoculated with cultures of the four probiotic *E. coli* strains, and after 2 d, the probiotic-fed calves and nine control calves were administered a mixture of *E. coli* O157:H7 strains. The results showed that five of the six treated calves stopped shedding serotype O157:H7 within 18 d. Thirty days after the start of this experiment, *E. coli* O157:H7 was only detected in one of the six probiotic-treated calves, but this pathogen was still present in the feces of all control animals. Three *E. coli* strains (strains 271, 786, 797) were included in the first anti-O157:H7 probiotic patent, and they were selected for commercial application (46).

In a study yet to be published, Doyle (47) presented the results of another cattle experiment that used adult animals and the patented strains (Figure 11.1). In this trial, 20 steers were orally inoculated with 10^{10} cells of five O157:H7 strains, and at 48 and 72 h, 10 of them were given oral doses of 10^{10} cells of a mixture of *E. coli* strains 271, 786, and 797. Fecal shedding of O157:H7 was determined during 30 d. In probiotic-treated cattle, most of the animals had undetectable levels of O157:H7 after 12 d, but 9 of 10 untreated steers had counts greater than 2 logs of O157:H7 by day 30. This probiotic mixture is now under commercialization trials by Alpharma, Inc.

TABLE 11.1

Summary of Animal Trials Focused to Test the Effect Of Probiotic Bacteria against Enterohemorrhagic *E. coli* (EHEC)

Animal	Target Serotype(s)	Probiotic Bacteria	Dose	Major Findings	Ref.
Beef cattle	O157:H7	*Lactobacillus acidophilus*	Daily 10^9 CFU/ steer	Reduced natural prevalence 50%	(54)
Calves	O157:H7	*L. crispatus* *L. gallinarum* *Streptococcus bovis*	Daily for 2 weeks, 20 g dried strains/calf	Fecal shedding (FS) reduction of inoculated calves	(52)
Calves	O157:H7, O26, O111	*E. coli* (3 strains)	Single dose 10^{10} CFU/ calf	Reduction of FS of O157:H7 and O111 after 8 d	(49)
Calves	O157:H7	*E. coli* (4 strains)	Single dose 10^{10} CFU/ calf	Reduction of FS and colonization of O157:H7	(4)
Neonatal calves	O157:H7, O26, O111	*E. coli* (3 strains)	Single dose 10^{10} CFU/ calf	Reduction of FS of O26 and O111	(48)
Lambs	O157:H7	*L. acidophilus* *Ent. faecium* *L. casei* *L. fermentum* *L. plantarum*	Daily 6×10^6 CFU/ steer	Mix of all strains reduced 4 log CFU/g feces after 7 weeks	(70)

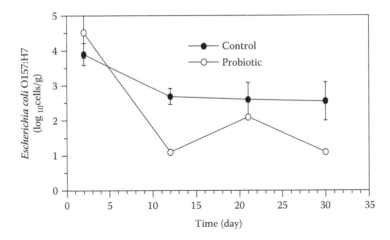

FIGURE 11.1

Effect of administration of probiotic *Escherichia coli* to adult steers on the fecal shedding of *E. coli* O157:H7 strains. A combination of three probiotics was orally given on days 2 and 3. Figure was drawn based on data from Reference 47.

Zhao et al. (48) have reported additional feeding studies assessing the ability of the same probiotic *E. coli* strains to prevent colonization and infection caused by *E. coli* O157:H7 and other enterohemorrhagic *E. coli* (EHEC) strains O26:H11 and O111:NM in neonatal calves. In that report, calves less than a week old were first inoculated with mixtures of the probiotic *E. coli*, and then 48 h later were administered a 5-strain mix of either O26, O111, or O157. Within 7 d of administering the EHEC strains, probiotic-treated calves were found to shed significantly less O26:H11 and O111:NM than the control animals. However, there was no significant reduction of O157:H7 between the control and probiotic-administered animals as the level of serotype O157:H7 sharply decreased in both groups. The authors hypothesized that the GI tract of milk-fed calves does not favor *E. coli* O157:H7 carriage, as compared to weaned calves.

In the most recent article using the three probiotic *E. coli* strains, a similar experiment was conducted with weaned calves inoculated with three types of EHEC (49). In that report, the authors confirmed the ability of the probiotic *E. coli* to reduce fecal shedding of serotype O157:H7 as there were statistical differences with the control group in half the measurements. The numbers of serotype O111:NM were also significantly reduced, but in only 25% of the sampling days. The fecal shedding of *E. coli* O26:H11 was, however, completely unaffected by the probiotic treatment, and this result may indicate that the probiotic treatment might be only effective in neonatal calves against this specific serotype.

Our laboratory has identified a collection of 24 colicinogenic *E. coli* that were selected on the basis of individual inhibition of over 20 strains of *E. coli* O157:H7 (50). This set of *E. coli* was identified from a collection of 540 recently isolated *E. coli* strains from humans and nine different animal species (cats, cattle, chickens, deer, dogs, ducks, horses, pigs, sheep). These 24 strains were selected for their ability to inhibit *E. coli* O157:H7 in solid and liquid culture due to the production of colicins, which are antimicrobial proteins produced by *E. coli* strains against other *E. coli* or enterobacteria. These strains were then cultivated in anaerobic chemostat experiments and challenged with four strains of O157:H7. Because the colicinogenic *E. coli* mixture was able to reduce O157:H7 five times faster than the calculated washout rate, we concluded that the O157:H7 strains were being killed in the culture vessel. Rep-PCR was used to identify that the colicinogenic strain F16, a cat isolate, was the predominant strain in these competition studies.

This collection of strains has been further characterized to assess their potential for probiotic applications (51). A total of 23 selected strains were first screened for the presence of known virulence factors, and six strains were eliminated from additional tests because of the presence of genes encoding for shiga toxins (*stx*), intimin (*eae*), or heat-stable toxin (STa). In addition, three strains were ruled out on the basis of multiple antibiotic resistance or single resistance to ampicillin. Colicins B, E1, E2/E7, E7, Ia/Ib, K, and M were detected in the remaining 14 strains, and four of those strains can produce three or four different colicins. A cattle trial is currently underway to test the effect of feeding a subgroup of colicin-producing strains on the fecal shedding of *E. coli* O157:H7 in experimentally inoculated calves.

LAB have also been investigated as potential probiotics to reduce serotype O157:H7 in cattle. Ohya et al. (52) evaluated the effect of feeding *Streptococcus bovis* and *Lactobacillus gallinarum* to calves on the fecal shedding of O157:H7. The authors first identified these two bacteria from the feces of adult cattle for their ability to produce lactic acid. The challenge study consisted of four calves being orally inoculated with a single strain of O157:H7. The probiotic was administered to the animals daily in two 1-week periods, days 7 to 13 and days 17 to 23 following the O157:H7 inoculation. *E. coli* O157 was not recovered in 3 of the calves by day 14. The fourth calf continued to shed O157 until day 28 when the pathogen level became undetectable. The results of this study, however, should be cautiously interpreted, as the investigators failed to include a control group.

Brashears et al. (53) have reported on the selection of lactobacilli for inhibition against *E. coli* O157:H7 as well as characterization of strains for desired properties. The authors first isolated over 600 LAB from cattle and screened the isolates for inhibition against 4 strains of O157:H7. Significant inhibitory activity against O157 was found in 52% of the strains, and 75 isolates with the best activity against serotype O157 were evaluated for their ability to tolerate acid and bile to predict if the strains would survive the harsh conditions of the GI tract. Nineteen isolates were then chosen for antibiotic susceptibility testing for six antibiotics. Only four isolates were susceptible to all of the antibiotics, whereas 13 were multiple antibiotic resistant. An isolate designated as strain M35 was selected as the best strain for use as a probiotic in cattle because it possessed the ability to significantly reduce O157:H7 in manure and rumen fluid. Using 16S rRNA sequence analysis, the authors identified M35 as being most closely related to *Lactobacillus crispatus*.

In another report, Brashears et al. (54) conducted a feeding study to investigate the effect of feeding LAB on the natural prevalence of *E. coli* O157:H7 in cattle. A total of 180 beef steers were used in this study, and groups of 60 animals were fed daily with either *Lactobacillus acidophilus* strain NPC747 or *Lactobacillus crispatus* strain NPC750. The researchers monitored the prevalence of natural O157:H7-positive animals throughout the course of the study as well as detecting O157:H7-positive hides at slaughter. The cattle supplemented with strain NPC747 were found to have as much as 50% less prevalence of fecal shedding of serotype O157:H7. When the animals were slaughtered, the prevalence of hides from lactobacilli-treated animals was at least 83% smaller than the prevalence from the control group. Moxley et al. (55) confirmed that this strain was capable of reducing the *E. coli* O157:H7 prevalence from 21 to 13%, but this difference was not statistically significant. This DFMP is currently being marketed as a beef performance additive and is being fed to approximately 1 million cattle in the U.S. (G. Ternus, personal communication, 2003). These results support the idea that probiotic bacteria have great potential as preharvest strategies against enterohemorrhagic *E. coli* in cattle populations.

11.5.2   Poultry

The control of pathogenic *Salmonella* in chickens by using probiotic cultures has been investigated extensively, and the early research was concentrated on preventing animal disease and reducing mortality. When poultry meat and eggs were recognized as vehicles of human salmonellosis, the application of probiotics as a tool for preventing food-borne disease was actively explored. A wide variety of microorganisms have been used as chicken probiotics, but apart from CE cultures, LAB bacteria have been the most frequently utilized group of probiotics. However, the effectiveness of LAB to reduce intestinal colonization of *Salmonella* reported in the literature varies significantly. This variability appears to be due to a variety of factors such as origin of strains, strain characteristics, experimental design, age of animals, diet, and stress, among others.

Initial work by Stavric et al. (24) and Hinton and Mead (56) indicated that feeding *Lactobacillus*, *Enterococcus*, and *Bifidobacterium* alone or in combination could not reduce the colonization of *Salmonella* in chicks. In another report, Stavric et al. (57) treated newly hatched chicks with different mixtures of as many as 50 different strains that belonged to the genera *Lactobacillus*, *Bacteroides*, *Escherichia*, *Strepococcus*, and six other genera, and they observed an increased inhibition of *Salmonella* colonization as the number of strains increased. Based on these results, Stavric et al. (24) argued that probiotics based on a few LAB strains would not likely be successful against *Salmonella*. However, Pascual et al. (23) have recently reported that a single strain of

Lactobacillus salivarius was capable of eliminating *Salmonella enteriditis* from the GI tract of 1-day-old chickens. The use of LAB in combination with *Salmonella*-specific avian antibodies has also been reported to significantly reduce *Salmonella typhimurium* counts, but this treatment was not capable of complete removal of the pathogen.

Escherichia coli strains have also been used as probiotics in chickens. Wooley and coworkers (58) developed a colicinogenic transformant capable of inhibiting *Salmonella* and *E. coli* O157:H7 *in vitro*. This probiotic strain (AvGOB18) was obtained by transferring the plasmid encoding for microcin 24 to a wild-type poultry *E. coli* isolate. Microcin 24 was originally identified in a uro-pathogenic *E. coli* and cloned into a pBR322 plasmid vector (59). When broiler chicks were given individual doses of strain AvGOB18, the level of artificially administered *Salmonella* was no different from that of controls, but if water sources contained approximately 10^6 cells/ml of probiotic strain, the count of *Salmonella typhimurium* reached almost undetectable levels after 3 weeks (Figure 11.2) (58). Strain AvGOB18 has been patented as a method to control food-borne pathogens in livestock (60).

Probiotics have also been applied as a preharvest food safety intervention for the inhibition of *Campylobacter jejuni* in chickens. *C. jejuni* is considered to be the major cause of bacterial food-borne illness in the U.S. (61). Studies have found that a large percentage of poultry is asymptomatically infected with *C. jejuni*, and this fact combined with inevitable carcass contamination during processing results in poultry products cited as a major source of infections (62). Because poultry is one of the most important natural reservoirs of *Campylobacter*, preharvest is a logical critical control point at which to focus eradication efforts. Researchers have explored the ability of LAB as

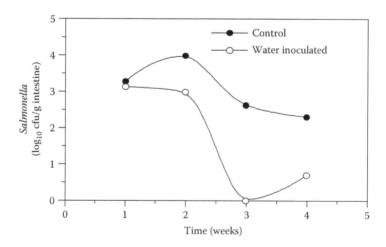

FIGURE 11.2

Effect of providing probiotic *Escherichia coli* strain AvGOB18 (10^6 cells/ml) in the drinking water to 1-day-old broiler chicks on the intestinal count of *Salmonella*. Figure was drawn with data taken from Reference 57.

well as bacteria selected for producing anti-*Campylobacter jejuni* compounds to reduce this food-borne pathogen in chickens.

Schoeni and Doyle (63) first reported on the selection of bacteria from chickens with the ability to produce anti-*C. jejuni* metabolites. Initially the authors identified only 8 White Leghorn hens among a total of 2320 birds to be negative for the presence of *C. jejuni*. Several bacteria were isolated from the mucus layer of the hens' cecums and screened for anti-*C. jejuni* activity. Nine isolates were capable of inhibiting *C. jejuni*, and they included *Pasteurella multocida*, *Micrococcus roseus*, *Citrobacter diversus*, *Proteus mirabilis*, *E. coli* O13:H-, two strains of *Klebsiella pneumoniae*, and two strains of CDC Group IIK. In five trials, 1-day-old chicks were administered the nine anti-*C. jejuni* isolates, and then 7 d later were inoculated with *C. jejuni*. The treated chicks had an average protection rate of 64% to colonization by *C. jejuni* when compared to the control groups.

In the same study (63), the researchers evaluated the nine probiotic strains for their capability of only using mucin to grow. The basis of this approach was that *C. jejuni* tends to be located in mucus-filled crypts in the cecum and mucin, a main component of mucus. After performing the mucin growth test on the nine isolates, three strains were selected (*K. pneumoniae*, *C. diversus*, and *E. coli*) for further feeding studies. These three strains provided an average colonization protection rate of 78% against *C. jejuni* in four feeding trials. The dominant isolate identified in these trials was the *E. coli* strain. However, when this *E. coli* strain was used as the sole probiotic in three feeding trials, it only provided a protection rate of 59%.

In an additional report, Schoeni and Wong (64) used the same three anti-*C. jejuni* isolates to investigate additional factors relevant to probiotic administration. The first factor was to compare the effectiveness of the strains when they were grown under aerobic or anaerobic conditions prior to animal inoculation. A previous article had reported on the need to grow CE bacteria under anaerobic conditions to maximize their effectiveness (65). Although chicks treated with anaerobically grown bacteria had lower colonization rates than chicks administered aerobically grown cells, no statistical difference was found between the groups.

The same report also investigated the dose frequency needed for the probiotic bacteria to prevent colonization of *C. jejuni*. This point is important to determine if a single dose of a probiotic is adequate to prevent pathogen colonization, or if an additional dose or doses are needed to establish the probiotic in the host. The authors inoculated chicks on day 1 and on day 4, but the additional dose did not appear to help reduce *C. jejuni* colonization. In the same study (64), the effect of supplementing probiotics with lactose, mannose, and fructooligosaccharides on the extent of inhibition of *C. jejuni* was determined. These carbohydrates alone were found to reduce *C. jejuni* colonization, as well as enhance the effectiveness of the probiotics.

Another research group evaluated the effectiveness of using LAB to reduce *C. jejuni* colonization in poultry. Morishita et al. (66) administered *Lactobacillus acidophilus* and *Streptococcus faecium* via water to chicks for 3 d. The chicks were also administered *C. jejuni* on day 1 after being given the probiotic. The authors monitored fecal shedding of the chicks until 40 d of age. The probiotic treatment in this study reduced the shedding of *C. jejuni* by 70% when compared to the control group.

11.5.3   Sheep

Escherichia coli O157:H7 is typically associated with cattle populations, but outbreaks of this pathogen have also been related to sheep (67). Mutton and lamb food products are not a frequent vehicle of contamination, but the fecal prevalence of this bacterium in sheep populations can be as high as 30% (68). A number of investigators have used lambs as a ruminant model to study preharvest interventions (69, 70). Lema et al. (71) investigated the use of probiotics to reduce fecal shedding of serotype O157:H7 in lambs. In this study, several LAB were fed to animals inoculated with *E. coli* O157:H7, and a decrease of more than 4 log CFU/g of feces was achieved after 7 weeks of feeding a mixture of *Lactobacillus acidophilus, Enterococcus faecium, L. casei, L. fermentum,* and *L. plantarum.* Feeding *E. faecium* alone reduced the level of EHEC by approximately 2 log CFU/g of feces, but treatment with *L. acidophilus* alone or mixed with *E. faecium* had almost no difference to control animals.

11.6   Selection Criteria for Preharvest Food Safety Probiotics

The selection of the specific probiotic bacteria intended to inhibit food-borne pathogens in livestock is probably the single most important step in the development of probiotics (4, 50, 52, 54, 63, 72). The primary criteria in the selection of probiotic strains are those directly affecting the potential reduction of the targeted pathogen, but other criteria must also be considered if the probiotic is to be successfully adopted in practice. Some of the primary selection factors relate directly to the ability of the probiotic to survive and get established in the animal's GI tract, such as acid tolerance or siderophore production. Other considerations include antibiotic resistance and possession of virulence factors, which are based largely on public health concerns. There might be some disagreement on what selection criteria are more important, but it is critical that a reasonable number should be considered to help predict the success and practicality of using the selected isolate(s).

The first step in developing a probiotic is the selection of potential strains. Most of the preharvest food safety published reports have isolated bacteria from the same type of animal for which the probiotic is intended (4, 52, 54, 72). The chosen isolates would theoretically be better adapted to colonize that particular GI tract environment. An arbitrary selection criterion that is set by the researchers is whether more than one specific group of bacteria would be considered (e.g., LAB, *E. coli*). A third criterion that has been reported by virtually all preharvest food safety reports, and perhaps one of the most important steps, is to determine that the isolated strains have the ability to inhibit the target pathogen *in vitro*. This antagonistic activity is typically tested against several strains of the pathogen, to detect any resistance variability among pathogen strains. The steps following the identification of inhibitory activity vary significantly among the different preharvest food safety reports.

The ability of bacteria to survive and proliferate in the GI tract is difficult to predict without an actual *in vivo* evaluation, but several assays can be used to assess the potential colonization traits of probiotic isolates. Bile and acid tolerance assays can be used to determine if strains can withstand harsh conditions of the GI tract. Brashears et al. (53) screened 75 probiotic LAB for acid and bile tolerance and found that the strains had a wide range of acid tolerance, whereas only one strain was sensitive to bile salts. The production of siderophores is another characteristic that can be tested to assess colonization potential. Siderophores are proteins synthesized by bacteria to sequester iron, which is a scarce nutrient in the GI tract (73). *E. coli* possessing siderophores have been found to have an increased advantage for intestinal colonization (74, 75).

The ability to adhere to intestinal epithelial cells can also be used to predict if the strain would establish in the GI tract (15). Garriga et al. (72) used high-adherence efficiency to chick epithelial cells as a criterion for selecting promising probiotic LAB for use in chicks. The most promising probiotic *E. coli* strains tested by Zhao et al. (4) had been originally isolated from intestinal and colonic tissues of cattle. Beneficial bacteria may also be chosen on their ability to exploit an ecological niche that allows them to proliferate. For example, Schoeni and Doyle (63) identified anti-*Campylobacter jejuni* isolates that could utilize mucin, a major component of the ecological niche that *C. jejuni* might occupy in chickens.

Virulence factors are undesirable attributes that potential probiotics should be screened for before their final selection. The ability of an intended probiotic to express a pathogen gene(s) could be detrimental for the intended animal. Even if the pathogenic organism does not affect the livestock receiving the probiotic, the virulence property may proliferate in the environment and also pose food- and water-borne risks to human populations. DebRoy and Maddox (76) recommended as many as 16 different virulence factors that included heat-labile toxin (LT), heat-stable toxins (STa, STb), shiga toxins (Stx1, Stx2), cytotoxic necrotizing factors (CNF1, CNF2), intimin (*eae*), and adhesins (K88, K99, 987P, F17, F18, F41, F42, F165) to be determined during the evaluation of *E. coli* isolated from various animal hosts. In our laboratory, we screened a collection of anti-O157 *E. coli* for nine different pathogen genes (STa, STb, LT, Stx1, Stx2, *eae*, enterohemolysin, K88, K99) with PCR and found that 6 of 23 possessed at least one virulence gene.

A major public health issue has been the concern over the emergence of antibiotic resistant-bacteria (77). In agriculture, the widespread use of antibiotics may lead to the selection of resistant strains that could proliferate and pass resistance factors to human pathogens (78). Consequently, the U.S. Food and Drug Administration (FDA) has proposed a criterion outline for assessing products designed for use in livestock (79). Antimicrobial agents are placed into one of three categories based on the product's significance to human health care. The utilization of a probiotic with antibiotic resistance in food-producing animals would be undesirable because the resistance genes could be transferred to other microbes in the GI tract and the environment (77, 80). When Brashears et al. (53) assessed their probiotic LAB for resistance, they found that 68% of their strains were multiple antibiotic resistant, and the final selected probiotic strain was sensitive to all antibiotics tested. Conversely, Garriga et al. (72) selected their probiotic strains for resistance to chloramphenicol, streptomycin, nalidixic acid, and erythromycin. The rationale for this selection scheme was that chicken feed regularly contains antibiotics that could inhibit supplemented beneficial bacteria that would be included in the diet.

As a last screening step, before the potentially probiotic isolates are given to livestock, a number of reports have introduced a variety of assays. Brashears et al. (53) assessed the inhibitory activity of LAB in manure and rumen fluid. Zhao et al. (4) initially fed their probiotic *E. coli* to only a couple of calves to identify which strains would predominate. Competition studies in continuous cultivation have been utilized in the development of defined mixed cultures (81). By using anaerobic continuous-culture chemostats, we were able to show that colicinogenic *E. coli* was capable of killing serotype O157:H7 and to identify which beneficial *E. coli* strain became predominant in the cultures (50).

An important aspect of using probiotic microorganisms is the mechanism of action that the probiotic utilizes to inhibit or exclude the pathogen. Of the reported selection and characterizations for food safety probiotics, few studies have elucidated the actual mode of action. Garriga et al. (72) found that the inhibitory activity of their LAB was not affected when treated with catalase; however, the inhibition was lost when the pH was adjusted to 6.5. The authors concluded that the source of the activity was not hydrogen peroxide; however, an acid may have caused or contributed to the inhibition. Brashears et al. (53) did not observe a pH reduction in the rumen and manure models that they used when assessing the ability of their LAB to inhibit *E. coli* O157:H7. The authors hypothesized that organic acids were not the cause of inhibition, but that the activity may have been the result of products such as hydrogen peroxide, bacteriocins, CO_2, or diacetyl. In our laboratory, the supernatant activity of our anti-O157:H7 *E. coli* was lost when treated with Proteinase K, indicating the inhibition was caused by a proteinaceous substance (50). The strains were then assessed with a bacteriophage assay, but the isolates were all negative. Further research using PCR confirmed that our probiotic *E. coli* possessed colicins.

Elucidation of the mode of action for probiotics may be useful to minimize the possible development of pathogen resistance as well as enhance pathogen reduction. It has long been recognized that antibiotic resistance can arise after exposure to antibiotics, and bacteria can also become resistant to the antimicrobial mechanisms of probiotic organisms (77). Food-borne pathogens such as *Listeria monocytogenes* and *Staphylococcus aureus* have been reported to spontaneously mutate with the ability to resist nisin (82). However, using a combination of different probiotic organisms with diverse mechanisms of action may reduce the risk of selecting resistant pathogens. For example, Hanlin et al. (83) found that the bacteriocin combination of nisin and pediocin AcH was more effective at reducing pathogen levels than when the bacteriocins were used singly. To reduce the possibility of pathogen resistance and improve pathogen reduction, it is advisable that a probiotic possesses multiple mechanisms of activity.

11.7 Conclusions

The utilization of probiotic microorganisms to prevent food-borne infections is one of the most promising preharvest intervention strategies. Several reports in the literature have provided conclusive evidence that many bacterial genera could be used to significantly reduce the colonization of livestock by food-borne pathogens. The fecal shedding of *Escherichia coli* O157:H7 in cattle could be reduced by feeding *Escherichia coli* and *Lactobacillus* strains. The reduction of *Salmonella* carriage in poultry could be prevented using lactic acid bacteria. *Lactobacillus, Klebsiella, Citrobacter*, and *E. coli* have also been effective against *Campylobacter jejunii* in poultry. It should be noted, however, that the long-term success of a probiotic culture is highly dependent on an initial thorough and systematic selection process to address environmental concerns and the potential for resistance development.

References

1. DM Lilly, RH Stillwell. Probiotics: growth promoting factors produced by microorganisms. *Science*, 147:747–748, 1965.
2. R Fuller. Probiotics in man and animals. *J. Appl. Bacteriol.*, 66:365–378, 1989.
3. G Schaafsma. State of the art concerning probiotic strains in milk products. *IDF Nutr. News Lett.*, 5:23–24, 1996.
4. T Zhao, MP Doyle, BG Harmon, CA Brown, POE Mueller, AH Parks. Reduction of carriage of enterohemorrhagic *Escherichia coli* O157:H7 in cattle by inoculation with probiotic bacteria. *J. Clin. Microbiol.*, 36:641–647, 1998.
5. C Morris. *Academic Press Dictionary of Science and Technology*. Academic Press, New York, 1992.

6. DJ Nisbet, DE Corrier, SC Ricke, ME Hume, JA Byrd II, JR Deloach. Maintenance of the biological efficacy in chicks of a cecal competititve-exclusion culture against *Salmonella* by continuous-flow fermentation. *J. Food Prot.*, 59:1279–1283, 1996.

7. CVM. CVM Policy on Competitive Exclusion. FDA Center for Veterinary Medicine, Rockville, MD, 1997.

8. J Schrezenmeir, M de Vrese. Probiotics, prebiotics, and synbiotics —approaching a definition. *Am. J. Clin. Nutr.*, 73:361S–364S, 2001.

9. CH Eckles, WM Williams, JW Wilbur, LS Palmer, M Harshaw. Yeast as a supplementary feed for calves. *J. Dairy Sci.*, 7:421–439, 1924.

10. AD Tadd, A Hurst. The effect of feeding colicinogenic *Escherichia coli* on the intestinal *E. coli* of early weaned pigs. *J. Appl. Bacteriol.*, 24:222–228, 1961.

11. JR Beach. The effect of feeding cultures of *Bacillus acidophilus*, lactose, dry skim milk, or whole milk on the hydrogen ion concentration of the contents of the ceca of chickens. *Hilgardia*, 1:145–166, 1925.

12. JOL King. *Lactobacillus acidophilus* as a growth stimulant for pigs. *The Vet.*, 5:273–280, 1968.

13. F Tortuero. Influence of implantation of *Lactobacillus acidophilus* in chicks on the growth, feed conversion, malabsorption of fats syndrome and intestinal flora. *Poultry Sci.*, 52:197–203, 1973.

14. PA Barrow. Probiotics for chickens. In: R Fuller, Ed., *Probiotics. The Scientific Basis.* Chapman & Hall, London, 1992.

15. R Fuller. The importance of lactobacilli in maintaining normal microbial balance in the crop. *Poultry Sci.*, 18:85–94, 1977.

16. DK Ellinger, LD Muller, PJ Glantz. Influence of feeding fermented colostrum and *Lactobacillus acidophilus* on fecal flora and selected blood parameters of young dairy calves. *J. Dairy Sci.*, 70(Suppl. 1):119, 1978.

17. AH Linton, K Howe, MH Richmond, HM Clemens, AD Osborne, B Handley. Attempts to displace the indigenous antibiotic resistant gut flora of chicken by feeding sensitive strains of *Escherichia coli* prior to slaughter. *J. Appl. Bacteriol.*, 45:239–247, 1978.

18. R Fuller. Probiotics for farm animals. In: GW Tannock, Ed., *Probiotics: A Critical Review,* Horizon Scientific Press, Whymodham, U.K., 1999.

19. JT Huber: Probiotics in cattle. In: R Fuller, Ed., *Probiotics 2: Applications and Practical Aspects.* Chapman & Hall, London, 1997.

20. J Nousiainen, J Setälä: Lactic acid bacteria as animal probiotics. In: S Salaminen, A Von Wright, Eds., *Lactic Acid Bacteria: Microbiology and Functional Aspects,* Marcel Dekker, New York, 1998.

21. JA Patterson, KM Burkholder. Application of prebiotics and probiotics in poultry production. *Poult. Sci.*, 82:627–631, 2003.

22. BA Watkins, BF Miller, DH Neil. *In vivo* effects of *Lactobacillus acidophilus* against pathogenic *Escherichia coli* in gnotobiotic chicks. *Poult. Sci.*, 61:1298–1308, 1982.

23. M Pascual, M Hugas, JI Badiola, JM Monfort, M Garriga. *Lactobacillus salivarius* CTC2197 prevents *Salmonella enteritidis* colonization in chickens. *Appl. Environ. Microbiol.*, 65:4981–4986, 1999.

24. S Stavric, TM Gleeson, B Buchanan, B Blanchfield. Experience on the use of probiotics for the control of *Salmonella* in poultry. *Lett. Appl. Microbiol.*, 14:69–71, 1992.

25. RWAW Mulder, R Havenaar, JHJ Huis in't Veld. Intervention strategies: the use of probiotics and competitive exclusion microfloras against contamination with pathogens in pigs and poultry. In: R Fuller, Ed., *Probiotics 2: Applications and Practical Aspects*. Chapman & Hall, London, 1997.
26. RA Dalloul, HS Lillehoj, TA Shellem, JA Doerr. Enhanced mucosal immunity against *Eimeria acervulina* in broilers fed a *Lactobacillus*-based probiotic. *Poult. Sci.*, 82:62–66, 2003.
27. W Vahjen, A Jadamus, O Simon. Influence of a probiotic *Enterococcus faecium* strain on selected bacterial groups in the small intestine of growing turkey poults. *Arch. Tierernaehr.*, 56:419–429, 2002.
28. MA Ehrmann, P Kurzak, J Bauer, RF Vogel. Characterization of lactobacilli towards their use as probiotic adjuncts in poultry. *J. Appl. Microbiol.*, 92:966–975, 2002.
29. RM Fulton, BN Nersessian, WM Reed. Prevention of *Salmonella enteritidis* infection in commercial ducklings by oral chicken egg-derived antibody alone or in combination with probiotics. *Poult. Sci.*, 81:34–40, 2002
30. JM Khajarern, S Khajarern, TH Moon, JH Lee. Effects of dietary supplementation of fermented chitin-chitosan (FERMKIT) on toxicity of mycotoxin in ducks. *Asian Aust. J. Anim. Sci.*, 16:706–713, 2003.
31. NR Underdahl. The effect of feeding *Streptococcus faecium* upon *Escherichia coli* induced diarrhea in gnotobiotic pigs. *Prog. Food Nutr. Sci.*, 7:5–12, 1983.
32. LZ Jin, RR Marquardt, X Zhao. A strain of *Enterococcus faecium* (18C23) inhibits adhesion of enterotoxigenic *Escherichia coli* K88 to porcine small intestine mucus. *Appl. Environ. Microbiol.*, 66:4200–4204, 2000.
33. TJ Bechman, JV Chambers, MD Cunningham. Influence of *Lactobacillus acidophilus* on performance of young dairy calves. *J. Dairy Sci.*, 60 (Suppl. 1):74, 1977.
34. G Bonaldi, L Buratto, G Darsie. Use of *Saccharomyces cerevisiae* and *Lactobacillus acidophilus* in veal calves. *Obiet Doc. Vet.*, 7:49–53, 1986.
35. N Agarwal, DN Kamra, LC Chaudhary, IAS Agarwal, NN Pathak. Microbial status and rumen enzyme profile of crossbred calves fed on different microbial feed additives. *Lett. Appl. Microbiol.*, 34:329–336, 2002.
36. JA Harp, P Jardon, ER Atwill, M Zylstra, S Checel, JP Goff, C De-Simone. Field testing of prophylactic measures against *Cryptosporidium parvum* infection in calves in a California dairy herd. *Am. J. Vet. Res.*, 57:1586–1588, 1996.
37. GR Ghorbani, DP Morgavi, KA Beauchemin, JA Leedle. Effects of bacterial direct-fed microbials on ruminal fermentation, blood variables, and the microbial populations of feedlot cattle. *J. Anim. Sci.*, 80:1977–1985, 2002.
38. A Irianto, B Austin. Probiotics in aquaculture. *J. Fish Dis.*, 25:633–642, 2002.
39. G Dalmin, K Kathiresan, A Purushothaman. Effect of probiotics on bacterial population and health status of shrimp in culture pond ecosystem. *Indian J. Exp. Biol.*, 39:939–942, 2001.
40. L Gram, J Melchiorsen, B Spanggaard, I Huber, TF Nielsen. Inhibition of *Vibrio anguillarum* by *Pseudomonas fluorescens* AH2, a possible probiotic treatment of fish. *Appl. Environ. Microbiol.*, 65:969–973, 1999.
41. PAW Robertson, C O'Dowd, C Burrells, P Williams, B Austin. Use of *Carnobacterium* sp. as a probiotic for Atlantic salmon (*Salmo salar* L.) and rainbow trout (*Oncorhynchus mykiss* Walbaum). *Aquaculture*, 185:235–243, 2000.

42. J Vandenberghe, L Verdonock, R Robles-Arozarena, G Rivera, A Bolland, M Balladares, B Gomez-Gil, J Calderon, P Sorgeloos, J Swings. Vibrios associated with *Litopenaeus vannamei* larvae, postlarvae, broodstock, and hatchery probionts. *Appl. Environ. Microbiol.*, 65:2592–2597, 1999.

43. AA Borczyk, MA Karmali, H Lior, LMC Duncan. Bovine reservoir for verotoxin-producing *Escherichia coli* O157:H7. *Lancet*, 1:98, 1987.

44. RO Elder, JE Keen, GR Siragusa, GA Barkocy-Gallagher, M Koohmaraie, WW Laegreid. Correlation of enterohermorrhagic *Escherichia coli* O157 prevalence in feces, hides, and carcasses of beef cattle during processing. *Proc. Natl. Acad. Sci. of U.S.A.*, 97:2999–3003, 2000.

45. D Smith, M Blackford, S Younts, R Moxley, J Gray, L Hungerford, T Milton, T Klopfenstein. Ecological relationships between the prevalence of cattle shedding *Escherichia coli* O157:H7 and characteristics of the cattle or conditions of the feedlot pen. *J. Food Prot.*, 64:1899–1903, 2001.

46. MP Doyle, T Zhao, BG Harmon, CA Brown. Control of enterohemorrhagic *E. coli* O157:H7 in cattle by probiotic bacteria and specific strains of *E. coli*. In Official Gazette, U.S. Patent Trademark Office, Patent U. S., 1999.

47. MP Doyle. Keeping Foodborne Pathogens Down on the Farm. U.S. Dept. Agric. Food Safety Inspection Service, 2001.

48. T Zhao, S Tkalcic, MP Doyle, BG Harmon, CA Brown, P Zhao. Pathogenicity of enterohemorrhagic *Escherichia coli* in neonatal calves and evaluation of fecal shedding by treatment with probiotic *Escherichia coli*. *J. Food Prot.*, 66:924–930, 2003.

49. S Tkalcic, T Zhao, BG Harmon, MP Doyle, CA Brown, P Zhao. Fecal shedding of enterohemorrhagic *Escherichia coli* in weaned calves following treatment with probiotic *Escherichia coli*. *J. Food Prot.*, 66:1184–1189, 2003.

50. GP Schamberger, F Diez-Gonzalez. Selection of recently isolated colicinogenic *Escherichia coli* strains inhibitory against *E. coli* O157:H7. *J. Food Prot.*, 65:1381–1387, 2002.

51. GP Schamberger, F Diez-Gonzalez. Characterization of colicinogenic *Escherichia coli* strains inhibitory to enterohemorrhagic *E. coli*. *J. Food Prot.*, (submitted), 67: 486–492, 2004.

52. T Ohya, T Marubashi, H Ito. Significance of fecal volatile fatty acids in shedding of *Escherichia coli* O157 from calves: experimental infection and preliminary use of a probiotic product. *J. Vet. Med. Sci.*, 62:1151–1155, 2000.

53. MM Brashears, D Jaroni, J Trimble. Isolation, selection, and characterization of lactic acid bacteria for a competitive exclusion product to reduce shedding of *Escherichia coli* O157:H7 in cattle. *J. Food Prot.*, 66:355–363, 2003.

54. MM Brashears, ML Galyean, GH Loneragan, JE Mann, K Killinger-Mann. Prevalence of *Escherichia coli* O157:H7 and performance by beef feedlot cattle given *Lactobacillus* direct-fed microbials. *J. Food Prot.*, 66:748–754, 2003.

55. RA Moxley, D Smith, TJ Klopfenstein, G Erickson, J Folmer, C Macken, A Hinkley, A Potter, B Finlay. Vaccination and feeding a competitive exclusion product as intervention strategies to reduce the prevalence of *Escherichia coli* O157:H7 in feedlot cattle. Proceedings of the 5th International Symposium on Shiga Toxin (Verocytotoxin)-producing *Escherichia coli* Infections, University of Aberdeen, Edinburgh, Scotland, 2003, p. 23.

56. M Hinton, GC Mead. Salmonella control in poultry: the need for the satisfactory evaluation of probiotics for this purpose. *Lett. Appl. Microbiol.*, 13:49–50, 1991.

57. S Stavric, TM Gleeson, B Blanchfield, H Pivnick. Competitive exclusion of *Salmonella* from newly hatched chicks by mixtures of pure bacterial cultures isolated from fecal and cecal contents of adult birds. *J. Food Prot.*, 48:778–782, 1985.

58. RE Wooley, PS Gibbs, EB Shotts. Inhibition of *Salmonella typhimurium* in the chicken intestinal tract by a transformed avirulent avian *Escherichia coli*. *Avian Dis.*, 43:245–250, 1999.

59. GJ O'Brien, HK Mahanty. Colicin 24, a new plasmid-borne colicin from a uropathogenic strain of *Escherichia coli*. *Plasmid*, 31:288–296, 1994.

60. RE Wooley, EB Shotts Jr. Biological control of food pathogens in livestock. In Official Gazette U.S. Patent Trademark Office Patent U.S.., 2000.

61. PS Mead, L Slutsker, V Dietz, LF McCaig, JS Bresee, C Shapiro, PM Griffin, RV Tauxe. Food-related illness and death in the United States. *Emerg. Infect. Dis.*, 5:607–625, 1999.

62. MJ Blaser. *Campylobacter jejuni* and food. *Food Technol.*, 36:89–92, 1982.

63. JL Schoeni, MP Doyle. Reduction of *Campylobacter jejuni* colonization of chicks by cecum-colonizing bacteria producing anti-*C. jejuni* metabolites. *Appl. Environ. Microbiol.*, 58:664–670, 1992.

64. JL Schoeni, ACL Wong. Inhibition of *Campylobacter jejuni* colonization in chicks by defined competitive exclusion bacteria. *Appl. Environ. Microbiol.*, 60:1191–1197, 1994.

65. JS Bailey. Factors affecting microbial competitive exclusion in poultry. *Food Technol.*, 41:88–92, 1987.

66. TY Morishita, PP Aye, BS Harr, CW Cobb, JR Clifford. Evaluation of an avian-specific probiotic to reduce the colonization and shedding of *Campylobacter jejuni* in broilers. *Avian Dis.*, 41:850–855, 1997.

67. NJ Strachan, DR Fenlon, ID Ogden. Modeling the vector pathway and infection of humans in an environmental outbreak of *Escherichia coli* O157. *FEMS Microbiol. Lett.*, 203:69–73, 2001.

68. IT Kudva, PG Hatfield, CJ Hovde: *Escherichia coli* O157:H7 in microbial flora of sheep. *J. Clin. Microbiol.*, 34:431–433, 1996.

69. M Lema, L Williams, L Walker, DR Rao. Effect of dietary fiber on *E. coli* O157:H7 shedding in lambs. *Small Rumin. Res.*, 43:249–255, 2002.

70. IT Kudva, CW Hunt, CJ Williams, UM Nance, CJ Hovde. Evaluation of dietary influences on *Escherichia coli* O157:H7 shedding by sheep. *Appl. Environ. Microbiol.*, 63:3878–3886, 1997.

71. M Lema, L Williams, DR Rao. Reduction of fecal shedding of enterohemorrhagic *Escherichia coli* O157:H7 in lambs by feeding microbial feed supplement. *Small Rumin. Res.*, 39:31–39, 2001.

72. M Garriga, M Pascual, JM Monfort, M Hugas. Selection of lactobacilli for chicken probiotic adjuncts. *J. Appl. Microbiol.*, 84:125–132, 1998.

73. K Hantke, V Braun. The art of keeping low and high iron concentrations in balance. In: G Storz, R Hengge-Aronis, Eds., *Bacterial Stress Responses*. American Society of Microbiology Press, Washington D.C., 2000, pp. 275–288.

74. M Der Vartanian, B Jaffeux, M Contrepois, M Chavarot, JP Girardeau, Y Bertin, C Martin. Role of aerobactin in systemic spread of an opportunistic strain of *Escherichia coli* from the intestinal tract of gnotobiotic lambs. *Infect. Immun.*, 60:2800–2807, 1992.

75. I Stojiljkovic, M Cobeljic, Z Trgovcevic, E Salaj-Smic. The ability of rifampin-resistant *Escherichia coli* to colonize the mouse intestine is enhanced by the presence of a plasmid-encoded aerobactin-iron(III) uptake system. *FEMS Microbiol. Lett.*, 90:89–94, 1991.

76. C DebRoy, CW Maddox. Identification of virulence attributes of gastrointestinal *Escherichia coli* isolates of veterinary significance. *Anim. Health Res. Rev.*, 1:129–140, 2001.

77. IM Gould. A review of the role of antibiotic policies in the control of antibiotic resistance. *J. Antimicrob. Chemother.*, 43:459–465, 1999.

78. L Tollefson, SF Altekruse, ME Potter. Therapeutic antibiotics in animal feeds and antibiotic resistance. *Rev. Sci. Tech. Int. Off. Epiz.*, 2:709–715, 1997.

79. FDA. A proposed framework for evaluating and assuring the human safety of the microbial effects of antimicrobial new animal drugs intended for use in food-producing animals. U.S. Food and Drug Administration, Rockville, MD, 2001.

80. AA Salyers, NB Shoemaker. Broad host range gene transfer: plasmids and conjugative transposons. *FEMS Microbiol. Ecol.*, 15:15–22, 1994.

81. D Nisbet. Defined competitive exclusion cultures in the prevention of enteropathogen colonisation in poultry and swine. *Antonie van Leeuwenhook*, 81:481–486, 2002.

82. X Ming, MA Daeschel. Nisin resistance of foodborne bacteria and the specific resistance responses of *Listeria monocytogenes* Scott A. *J. Food Prot.*, 56:944–948, 1993.

83. MB Hanlin, N Kalchayanand, P Ray, B Ray. Bacteriocins of lactic acid bacteria in combination have greater antibacterial activity. *J. Food Prot.*, 56:252–255, 1993.

12

Preharvest Food Safety Applications of Competitive Exclusion Cultures and Probiotics

Robin C. Anderson, Kenneth J. Genovese, Roger B. Harvey,
Todd R. Callaway, and David J. Nisbet

CONTENTS

12.1 Introduction

Beneficial microorganisms have long been used in postharvest food safety applications. For instance, the culture of cheeses, sauerkrauts, and other fermented foods has been used to preserve these foods in the days before refrigeration. Moreover, the administration of beneficial microorganisms as probiotics or functional foods to humans or animals to enhance health, gut function, or a variety of other purported medical purposes is not new and such use has been reviewed extensively by others (1–4). Traditionally, the term "probiotic" (meaning "for life") began being used to describe microbial supplements for animal feeds, and the following definition was subsequently formalized: "live microbial feed supplement which beneficially affects the host animal by improving intestinal microbial balance" (5). Since then, the definition has come to include application of such supplements to humans as well. Schrezenmeir and De Vrese (6) have proposed an expanded definition, "preparation of or a product containing viable, defined microorganisms in sufficient numbers, which alter the micro-flora (by implantation or colonization) in a compartment of the host and by that exert beneficial health effects in this host," so as to include the usage of products containing microbes (i.e., fermented dairy products, etc.) and not necessarily restrict the effect to colonization per se. In the present chapter, we review the more recent application of such microbial preparations, as well as preparations of mutualistic or commensal bacteria, to live animals for the specific purpose of enhancing food safety by reducing carriage and (or) shedding of zoonotic pathogens or as alternatives to growth-promoting antibiotics.

12.2   Rationale for Preharvest Intervention

Food-borne pathogens such as *Salmonella*, enterohemorrhagic *Escherichia coli*, and *Campylobacter* reside in the gastrointestinal tracts of food animals, often colonizing the animals at a very young age before full acquisition of a mature gut flora. Shedding of these pathogens in the feces during on-farm production contaminates the environment, which thus serves as a source for continuous reinoculation. Furthermore, shedding of these pathogens during transport, lairage, and at slaughter is considered a critical risk for contamination of meat and meat products. For instance, as many as 3.9 million human non-O157 shiga toxin-producing *E. coli*, *Salmonella*, and *Campylobacter* infections are estimated to occur in the U.S. each year (7), at a cost exceeding $4.5 billion (8). Consequently, strategies are sought to reduce the carriage of these pathogens on the farm and especially before entry into the abattoir. Many, if not most, of these human infections may be attributed to consumption of meat and poultry products contaminated during slaughter and processing (7, 9). More recently, concern over the emergence of resistant bacterial populations has put pressure on today's livestock producers to find alternatives to subtherapeutic administration of antibiotics. Thus, alternative technologies such as probiotics and competitive exclusion are sought to alleviate the need for antibiotics, particularly in the very young animal (10).

In healthy animals, the establishment of a mature gut flora consisting of numerous diverse genera of strict anaerobes may take up to 2 weeks or longer to occur, and not until then do populations of *Enterobacteriaceae*, including *Salmonella* and *E. coli*, which can be acquired as early as 8 h of age, cease being predominant members of the gut flora (11–14). Consequently, it is thought that interventions designed to expedite a mature colonization of the gastrointestinal tract with beneficial microorganisms would help preclude the establishment of certain zoonotic pathogens. The need for such interventions becomes even more apparent considering that consequences caused by modern agricultural production practices often delay or perturb the natural establishment of a healthy gut flora. For instance, washing and hatching of eggs in clean incubators essentially removes the hen's fecal material from the egg, thus preventing the chick from contacting its maternal source of inoculum. Likewise, hygienic sanitation procedures intended to wash away pathogenic bacteria from farrowing crates, nurseries, or elsewhere concurrently washes away inocula containing commensal and mutualist microbes as well. Young swine also experience a major dietary change upon weaning that can also perturb the maturing gut ecosystem, as evidenced by changes in volatile fatty acid accumulation, and this may also predispose the pig to certain enteric infections (15). The goal of intentionally administering beneficial microorganisms to food animals is to harness the protective effect of a healthy gut microflora.

12.3 Competitive Exclusion Cultures and Probiotics

Proof that the intentional administration of beneficial microbes to food animals could enhance food safety was elegantly presented by Nurmi and Rantala (16). They reported a dramatic increase in resistance to *Salmonella* colonization in chickens that had very early in life been administered a preparation of gut bacteria originating from healthy adult chickens. This concept has subsequently been referred to as "competitive exclusion" (4, 17, 18). Competitive exclusion cultures in the most technical sense are restricted in composition to microorganisms originating from an animal of the intended host species and are expected to colonize when administered to the neonate. Therefore, at least in theory, competitive exclusion cultures attempt to conserve and take advantage of synergies acquired during coevolution of host and microorganism. In contrast, probiotic preparations generally consist of individual species or mixtures of lactic acid bacteria or yeasts that are not necessarily of animal origin. The distinction between the two terminologies becomes confused, however, because the term "competitive exclusion" is also used to define the act for which the technology has been named. For instance, as the name implies, competitive exclusion cultures are thought to enhance the host organism's ability to resist colonization by pathogenic bacteria by facilitating the establishment of a mature, healthy population of gut bacteria capable of outcompeting the pathogens for limited nutrients, attachment sites, or niches within the ecosystem (2, 4, 17, 18). Whereas these modes of action have also been proposed for probiotics, long-term colonization of the target animal by the probiotic bacteria is often not claimed. Traditionally, competitive exclusion cultures had been developed to control *Salmonella* colonization in poultry via application to very young animals possessing undeveloped, immature gut environments. Interest now exists, however, in expanding the use of competitive exclusion technology to other animal species for the preharvest control of a variety of pathogens and as an environmentally compatible alternative to antibiotic use in agriculture.

12.3.1 Competitive Exclusion Cultures

Since the report by Nurmi and Rantala (16) on the protective effect of inoculating young chicks with gut bacteria obtained from healthy *Salmonella*-free adult chickens, numerous other studies have shown that similar treatments with defined or undefined mixtures of healthy gut bacteria provide a measure of protection to newly hatched avian species (19–24). Moreover, their extensive use outside the U.S. to enhance colonization resistance of avian species to *Salmonella* has been well documented (2, 4, 17, 18). A beneficial effect of competitive exclusion treatment against the colonization of swine by *Salmonella* has also been reported (25–30). Concerns by regulatory agencies regarding the potential of undefined cultures to harbor unknown or undetected pathogens or undesired genetic elements such as transferable antibiotic-resistance genes casts doubts as to the safety of undefined cultures (31). Subsequently, competitive exclusion cultures are classified as drugs by the Center for Veterinary Medicine (32). Only recently has an avian-derived competitive exclusion product of known microbial composition, trade named PREEMPT™, been approved by the U.S. Food and Drug Administration (33).

Whereas competitive exclusion cultures have traditionally been developed as interventions for very young animals with naive gut environments, considerable interest exists in developing similar technologies to control foodborne pathogens in more mature animals. For instance, although weaning of suckling pigs has been proposed as a critical control point for *Salmonella* control (34), evidence suggests that other critical control points may also exist during growing and finishing phases of production (35). A limitation of the traditional technology, however, is that it is less effective when applied to mature animals with preexisting *Salmonella* infections (36). Consequently, it is thought that the technology needs to be strengthened in order to be effective when administered to mature animals. One potential method may be to link the technology to specific substrates that are selectively or preferentially utilized by the competitive exclusion bacteria. For instance, lactose-adapted competitive exclusion cultures have been developed previously for poultry, and these have shown great promise in controlling *Salmonella* infections (37–39). In those studies, dietary lactose enhanced colonization resistance of molting hens and market age broilers to *Salmonella enteritidis* by stimulating growth of beneficial anaerobes at the expense of the nonlactose-fermenting *Salmonella*. *Salmonella* infections in U.S. poultry operations are generally asymptomatic, and thus combining the complementary use of lactose-adapted competitive exclusion cultures with effective quantities of lactose was not considered cost effective due to lack of productivity gains. *Salmonella* infections in swine often result in increased morbidity and mortality, particularly infections caused by the host-adapted serotype Choleraesuis (40), which means that lactose-adapted competitive exclusion cultures administered with effective quantities of bypass-lactose are likely be cost effective in controlling *Salmonella* infections of swine. Numerous other substrates, most notably fructooligosaccharides, are being investigated for their potential to synergistically enhance the biological activity of competitive exclusion and probiotic preparations and as such are referred to as synbiotics (5, 41–43).

The efficacy of competitive exclusion could be further enhanced by enriching the activity of constituent bacteriocin-producing microbes contained in the cultures. The concept here is that stronger competitive exclusion cultures containing only select bacterial species capable of exerting specific control against a particular pathogen or group of pathogens could ultimately be developed. Numerous strains of gut lactobacilli as well as other gut anaerobes are known to produce bacteriocins (44–51), and Duncan et al. (52) recently reported bacteriocin production by ruminal *Pseudomonas aeruginosa* strains isolated from sheep. Bacteria belonging to several of these genera are contained within the defined avian competitive exclusion culture PRE-EMPT™ (53) and are thus potential producers of bioactive compounds *in vivo*. Recently, a competitive exclusion culture utilizing several nonpathogenic, colicin-producing *E. coli* strains to specifically exclude *E. coli* O157:H7 from cattle has been developed (54, 55).

Because of the ability of competitive exclusion cultures to preclude colonization by enteropathogens, particularly in young animals, they are considered an attractive alternative to the subtherapeutic administration of antibiotics to food animals (10). A porcine-derived competitive exclusion culture is currently being tested under field conditions (56–58).

12.3.2 Probiotics

Whereas probiotics have enjoyed considerable use as feed supplements to enhance animal health and productivity (1, 2), their application as preharvest food safety interventions per se has been researched less so. Keen and Elder (59) reported little if any decrease in fecal shedding of *E. coli* O157:H7 in cattle when animals were fed either of two unnamed, yet commercially available, probiotic preparations for 2 d. However, their posttreatment measurements were taken 6 d after administration of the probiotic, and that short time frame may have limited the authors' ability to observe potential differences. In contrast, reports exist of reduced shedding of *E. coli* O157:H7 in lambs fed preparations of probiotic bacteria for 7 weeks (60) or calves fed probiotic bacteria at intervals over a 2-week period (61). More recently, Brashears et al. (62, 63) reported a 50% decrease in incidence of *E. coli* O157:H7 in cattle fed *Lactobacillus acidophilus* during a 60-day feedlot trial.

Like competitive exclusion cultures, the use of probiotics in lieu of subtherapeutic levels of antibiotics would seem to be a natural and attractive extension of their purported performance and animal health-enhancing activity (51); however, their use in this specific application has yet to be thoroughly researched.

12.4 Practical Issues in Culture Development and Application

Development of competitive exclusion cultures typically involves obtaining cecal or fecal material from healthy, pathogen-free animals (4, 17, 18). These contents are then subjected to various batch or continuous-flow culture protocols and screened to ensure the absence of pathogens. In the case of batch culture methodology, steps are often incorporated to dilute to extinction any potential pathogens that may be present. A disadvantage of numerous repeated transfers can result in the cultures losing efficacy in their ability to enhance colonization resistance *in vivo* (17, 18). This obviously limits the commercial lifespan of the respective cultures such that new cultures must frequently be redeveloped from freshly collected cecal or fecal contents. In contrast, the adaptation of continuous-flow culture methodology for development and maintenance of competitive exclusion cultures was a major improvement as these cultures retain their efficacy for extended periods of time (64).

Probiotic bacteria may or may not have been isolated from an animal and are often screened for preferred characteristics such as tolerance to acid and (or) bile, ability to adhere to gut epithelial cells, bacteriocin production, and for the ability to compete against other intestinal bacteria (65–68). Tests for biological activity of competitive exclusion cultures or probiotics against specific pathogens *in vitro* can be accomplished by appropriate selection methods (69), cell culture challenges (70), or by use of mixed gut populations grown in continuous-flow or sequencing fed-batch culture (71, 72). Establishment of competitive exclusion culture bacteria within the animal gut can be assessed experimentally by measuring increased colonization of the gut by total culturable anaerobes (73) that, as seen by electron microscopy, readily colonize the gut epithelium (74). With respect to the defined competitive exclusion culture PREEMPT™, culture establishment was also found to be highly correlated with an increased accumulation of propionic acid in cecal contents of treated birds, and this was further correlated to culture efficacy (75). However, determination of culture efficacy must ultimately rely on tests of their ability to enhance colonization resistance of target animals against challenge by relevant pathogens. With competitive exclusion cultures, *in vivo* challenge models are often performed with young chicks or pigs that had been either treated or not on their first day of life with the respective competitive exclusion cultures (17, 25–30, 56, 57, 76). Early treatment of very young animals with the competitive exclusion cultures is important as treatments applied to older weaned animals provided little, if any, protection (36, 77). On the other hand, probiotics are typically administered over a more extended period of time (51, 59–63, 78, 79). Cultures or probiotics are considered efficacious if colonization and (or) shedding of a test pathogen in treated animals is significantly less, whether by a qualitative (i.e., incidence) and (or) quantitative (i.e., concentration) measurement, in untreated controls (18, 80).

Probiotic preparations are often administered to food animals in their feed or water. Competitive exclusion cultures have typically been administered to chicks by spraying or in drinking water, whereas administration to piglets is done via oral gavage. Whereas probiotic preparations are usually, but not always, relatively aerotolerant, competitive exclusion cultures often contain a number of strictly anaerobic bacteria that are sensitive to exposure to oxygen. Thus, it is important to determine experimentally the efficacious dose regardless of administration procedure.

12.5 Future Outlook for Competitive Exclusion and Probiotics in Preharvest Food Safety

Development of competitive exclusion cultures and probiotics has yielded technologies capable of reducing carriage and shedding of food-borne pathogens in food-producing animals. With respect to *Salmonella* control, there is at present little economic incentive for certain food animal industries to accept and implement these technologies. For instance, with *Salmonella* levels maintained below allowable levels, the poultry industry's demand for the only FDA approved competitive exclusion culture, PREEMPT™, was not enough to sustain production (81). On the other hand, the classification of *E. coli* O157:H7 as an adulterant by the U.S. Food Safety Inspection Service continues to drive the beef industry's preharvest research efforts to control or eliminate this pathogen from slaughter animals. Moreover, economic and political pressures to reduce antibiotic usage in animal agriculture (82) may be the most powerful motivator to continue research and development of competitive exclusion and probiotic technologies for all industries involved in food animal production.

References

1. R Fuller. Probiotics for farm animals. In: GW Tannock, Ed., *Probiotics: A Critical Review*. Horizon Scientific Press, Wymondham, UK, 1999, pp. 15–22.

2. S Stavric, ET Kornegay. Microbial probiotics for pigs and poultry. In: RJ Wallace, A Chesson, Eds., *Biotechnology in Animal Feeds and Animal Feeding*. VCH, New York, 1995, pp. 205–230.

3. GW Tannock. Probiotics: time for a dose of realism. *Curr. Issues Intest. Microbiol.,* 4:33–42, 2003.

4. GC Mead. Prospects for "competitive exclusion" treatment to control salmonellas and other foodborne pathogens. *Vet. J.,* 159:111–123, 2000.

5. R Fuller. Probiotics in man and animals. *J. Appl. Bacteriol.,* 66:365–378, 1989.

6. J Schrezenmeir, M De Vrese. Probiotics, prebiotics, and synbiotics-approaching a definition. *Am. J. Clin. Nutr.,* 73(Suppl): 361–364, 2001.

7. PS Mead, L Slutsker, V Dietz, LF McCaig, JS Bresee, C Shaprio, PM Griffin, RV Tauxe. Food-related illness and death in the United States. *Emerg. Infect. Dis.,* 5:607–625, 1999.

8. ERS/USDA. In 2000, ERS estimated foodborne disease costs at $6.9 billion per year. http://www.ers.usda.gov/briefing/FoodborneDisease/features.htm (accessed May 26, 2003), 2001.

9. AC Baird-Parker. Foodborne illness. *Lancet,* 336:1231–1235, 1990

10. C Wray, RH Davies. Competitive exclusion — an alternative to antibiotics. *Vet. J.,* 159:107–108, 2000.

11. E Barnes. The avian intestinal flora with particular reference to the possible ecological significance of the cecal anaerobic bacteria. *Am. J. Clin. Nutr.,* 25:1475–1479, 1972.

12. KS Muralidhara, GG Sheggeby, PR Elliker, DC England, WE Sandine. Effect of feeding lactobacilli on the coliform and *Lactobacillus* flora of intestinal tissue and feces from piglets. *J. Food Prot.,* 40:288–295, 1977.

13. RW Schaedler. The relationship between the host and its intestinal microflora. *Proc. Nutr. Soc.,* 32:41–47, 1973

14. HW Smith, WE Crabb. The faecal bacterial flora of animals and man: its development in the young. *J. Pathol. Bacteriol.,* 82:53–66, 1961.

15. AG Mathew, AL Sutton, AB Scheidt, JA Patterson, DT Kelly, KA Meyerholtz. Effect of galactan on selected microbial populations and pH and volatile fatty acids in the ileum of the weanling pig. *J. Anim. Sci.,* 71:1503–1509, 1993.

16. E Nurmi, M Rantala. New aspects of *Salmonella* infection in broiler production. *Nature,* 241:210–211, 1973.

17. S Stavric. Defined cultures and prospects. *Int. J. Food Microbiol.,* 55:245–263, 1992.

18. S Stavric, J-Y D'Aoust. Undefined and defined bacterial preparations for competitive exclusion of *Salmonella* in poultry. *J. Food Prot.,* 56:173–180, 1993.

19. EM Barnes, CS Impey, BJH Stevens. Factors affecting the incidence and anti-*Salmonella* activity of the anaerobic cecal flora of the young chick. *J. Hyg.,* 82:263–283, 1979.

20. DE Corrier, DJ Nisbet, CM Scanlan, AG Hollister, DJ Caldwell, LA Thomas, BM Hargis, T Tomkins, JR DeLoach. Treatment of commercial broiler chickens with a characterized culture of cecal bacteria to reduce salmonellae colonization. *Poultry Sci.,* 74:1093–1101, 1995.

21. DJ Nisbet, DE Corrier, JR DeLoach. Effect of mixed cecal microflora maintained in continuous culture and of dietary lactose on *Salmonella typhimurium* colonization in broiler chicks. *Avian Dis.,* 37:528–535, 1994.

22. AB Lloyd, RB Cumming, RD Kent. Prevention of *Salmonella typhimurium* infection in poultry by pretreatment of chickens and poults with intestinal extracts. *Aust. Vet. J.*, 53:82–87, 1977.
23. CR Reid, DA Barnum. The effects of treatments of cecal contents on the protective properties against *Salmonella* in poults. *Avian Dis.*, 29:1–11, 1984.
24. OM Weinack, GH Snoeyenbos, CF Smyser, AS Soerjadi. Reciprocal competitive exclusion of *Salmonella* and *Escherichia coli* by native intestinal microflora of the chicken and turkey. *Avian Dis.*, 26:585–595, 1982.
25. RC Anderson, LH Stanker, CR Young, SA Buckley, KJ Genovese, RB Harvey, JR DeLoach, NK Keith, DJ Nisbet. Effect of competitive exclusion treatment on colonization of early weaned pigs by *Salmonella* serovar Choleraesuis. *Swine Health Prod.*, 7:155–160, 1999.
26. PJ Fedorka-Cray, JS Bailey, NJ Stern, NA Cox, SR Ladely, M Musgrove. Mucosal competitive exclusion to reduce *Salmonella* in swine. *J. Food Prot.*, 62:1376–1380, 1999.
27. KJ Genovese, RC Anderson, RB Harvey, TR Callaway, TL Poole, TS Edrington, PJ Fedorka-Cray, DJ Nisbet. Competitive exclusion of *Salmonella* from the gut of pigs. *J. Food Prot.*, 66:1353–1359, 2003.
28. DJ Nisbet, RC Anderson, SA Buckley, PJ Fedorka-Cray, LH Stanker. Effect of competitive exclusion on *Salmonella* shedding in swine. Proceedings of the 2nd International Symposium on Epidemiology and Control of *Salmonella* in Pork. Copenhagen, 1997, pp. 176–178.
29. DJ Nisbet, RC Anderson, RB Harvey, KJ Genovese, JR DeLoach, LH Stanker. Competitive exclusion of *Salmonella* serovar Typhimurium from the gut of early weaned pigs. Proceedings of the 3rd International Symposium on the Epidemiology and Control of *Salmonella* in Pork, Washington DC, 1999, pp. 80–82.
30. W Pullen, PJ Fedorka-Cray, D Reeves, JS Bailey, NJ Stern, NA Cox, SR Ladely. Mucosal competitive exclusion to reduce *Salmonella* in swine. Proceedings of the 30th Annual Meeting of the American Swine Practioners Association. St. Louis, 1999, pp. 17–18.
31. MS Nawaz, BD Erickson, AA Khan, SA Khan, JV Pothuluri, F Rafii, JB Sutherland, RD Wagner, CE Cerniglia. Human health impact and regulatory issues involving antimicrobial resistance in the food animal production environment. *Regulatory Res. Perspect.*, 1:1–10, 2001.
32. CVM Update. CVM Policy on Competitive Exclusion Products. Issued by: FDA, Center for Veterinary Medicine, Office of Management and Communications, Rockville, MD, February 21, 1997.
33. Federal Register. Vol 63, No 88:25163, 1998.
34. PJ Fedorka-Cray, DL Harris. Elimination of *Salmonella* species in swine by isolated weaning — the first critical control point. ISU Swine Res. Rep. 1994, Iowa State University, Ames, Iowa, pp. 146–149, 1995.
35. PR Davies, FGEM Bovee, JA Funk, WEM Morrow, FT Jones, J. Deen. Isolation of *Salmonella* serotypes from feces of pigs raised in a multiple-site production system. *J. Am. Vet. Med. Assoc.*, 212:1925–1929, 1998.
36. DE Corrier, JA Bryd II, ME Hume, DJ Nisbet, LH Stanker. Effect of simultaneous or delayed competitive exclusion treatment on the spread of *Salmonella* in chicks. *J. Appl. Poultry Res.*, 7:132–137, 1998.
37. DE Corrier, BM Hargis, A Hinton Jr, JR DeLoach. Protective effect of used poultry litter and lactose in the feed ration on *Salmonella enteritidis* colonization of leghorn chicks and hens. *Avian Dis.*, 37:47–52, 1993.

38. DE Corrier, DJ Nisbet, BM Hargis, PS Holt, JR DeLoach. Provision of lactose to molting hens enhances resistance to *Salmonella enteriditis* colonization. *J. Food Prot.*, 60:10–15, 1997.

39. DE Corrier, DJ Nisbet, CM Scanlan, G Tellez, BM Hargis, JR DeLoach. Inhibition of *Salmonella enteritidis* cecal and organ colonization in leghorn chicks by a defined culture of cecal bacteria and dietary lactose. *J. Food Prot.*, 56:377–381, 1994.

40. K Schwartz. Salmonellosis in swine. *Compend. Contin. Educ. Pract. Vet.*, 13:139–147, 1991.

41. A Bomba, R Nemcová, S Gancariková, R Hrich, R Kastel. Potentiation of the effectiveness of *Lactobacillus casei* in the prevention of *E. coli* induced diarrhea in conventional and gnotobiotic pigs. In: PS Paul, DH Francis, Eds., *Mechanisms in the Pathogenesis of Enteric Diseases 2*. Kluwer Academic/Plenum Publishers, New York, 1999, pp. 185–190.

42. A Bomba, R Nemcová, D Mudroová, P Guba. The possibilities of potentiating the efficacy of probiotics. *Trends Food Sci. Technol.*, 13:121–126, 2002.

43. JGM Houdijk, MW Bosch, MWA Verstegen, HJ Berenpas. Effects of dietary oligosaccharides on the growth and faecal characteristics of young growing pigs. *Anim. Feed Sci. Technol.*, 71:35–48, 1998.

44. WG Iverson, NF Mills. Bacteriocins of *Streptococcus bovis*. *Can. J. Microbiol.*, 22:1040–1047, 1976.

45. RW Jack, JR Tagg, B Ray. Bacteriocins of Gram-positive bacteria. *Microbiol. Rev.*, 59:171–200, 1995.

46. ML Kalmokoff, F Bartlett, RM Teather. Are ruminal bacteria armed with bacteriocins? *J. Dairy Sci.*, 79:2297–2306, 1996.

47. WJ Kim. Bacteriocins of lactic acid bacteria: their potentials as food biopreservative. *Food Rev. Int.*, 9:299–313, 1993.

48. TR Klaenhammer. Bacteriocins of lactic acid bacteria. *Biochimie*, 70:337–349, 1988.

49. A Laukova, M Marckova. Antimicrobial spectrum of bacteriocin-like substances produced by ruminal staphylococci. *Folia Microbiol.*, 38:74–76, 1993.

50. JA Southern, W Katz, DR Woods. Purification and properties of a cell-bound bacteriocin from a *Bacteroides fragillis* strain. *Antimicrob. Agents Chemother.*, 25:253–257, 1984.

51. M Wiemann. How do probiotic feed additives work? *Int. Poultry Prod.*, 11:7–9, 2003.

52. SH Duncan, CJ Doherty, JRW Govan, S Neogrady, P Galfi, CS Stewart. Characteristics of sheep-rumen isolates of *Pseudomonas aeruginosa* inhibitory to the growth of *Escherichia coli* O157:H7. *FEMS Microbiol. Lett.*, 180:305–310, 1999.

53. DE Corrier, DJ Nisbet, CM Scanlan, AG Hollister, JR DeLoach. Control of *Salmonella typhimurium* colonization in broiler chicks with a continuous-flow characterized mixed culture of cecal bacteria. *Poultry Sci.*, 74:916–924, 1995.

54. T Zhao, MP Doyle, BG Harmon, CA Brown, POE Mueller, AH Parks. Reduction of carriage of enterohemorrhagic *Escherchia coli* O157:H7 in cattle by inoculation with probiotic bacteria. *J. Clin. Microbiol.*, 36:641–647, 1998.

55. T Zhao, MP Doyle, JP Peters, KL Lechtenberg. Reduction of the carriage and fecal excretion of *E. coli* O157:H7 in beef cattle by competitive *E. coli*. 4th International Symposium and Workshop on Shiga Toxin (Verocytoxin)-producing *Escherichia coli* Infections. Kyoto, 2000, pp. 92.

56. KJ Genovese, RC Anderson, RB Harvey, DJ Nisbet. Competitive exclusion treatment reduces the mortality and fecal shedding associated with enterotoxigenic *Escherichia coli* infection in nursery-raised pigs. *Can. J. Vet. Res.*, 64:204–207, 2000.

57. KJ Genovese, RB Harvey, RC Anderson, DJ Nisbet. Protection of suckling neonatal pigs against an infection with an enterotoxigenic *Escherichia coli* expressing 987P fimbriae infection by the administration of a bacterial competitive exclusion culture. *Microb. Ecol. Health Dis.*, 13:223–228, 2001.

58. RB Harvey, KJ Genovese, RC Anderson, DJ Nisbet. Mixed culture of commensal bacteria reduces *E. coli* in nursery pigs. Proceedings of the 5th International Symposium on the Control and Epidemiology of Foodborne Pathogens in Pork. Hersonissos, 2003, pp. 151–153.

59. J Keen, R Elder. Commercial probiotics are not effective for short-term control of enterohemorrhagic *Escherichia coli* O157 infection in beef cattle. 4th International Symposium and Workshop on Shiga Toxin (Verocytoxin)-producing *Escherichia coli* Infections. Kyoto, 2000, pp. 92.

60. M Lema, L Williams, DR Rao. Reduction of fecal shedding of enterohemorrhagic *Escherichia coli* O157:H7 in lambs by feeding microbial feed supplement. *Small Ruminant Res.*, 39:31–39, 2001.

61. T Ohya, T Marubashi, H Ito. Significance of fecal volatile fatty acids in shedding of *Escherichia coli* O157 from calves: experimental infection and preliminary use of a probiotic product. *J. Vet. Med. Sci.*, 62:1151–1155, 2000.

62. MM Brashears, ML Galyean, GH Loneragan, JE Mann, K Killinger-Mann. Prevalence of *Escherichia coli* O157:H7 and performance by beef feedlot cattle given *Lactobacillus* direct-fed microbials. *J. Food Prot.*, 66:748–754, 2003.

63. Check-Off News. Checkoff-funded Research Shows Dramatic Results of Preharvest *E. coli* O157:H7 Interventions. http://www.beef.org/dsp/dsp_content.cfm?locationId=45&contentTypeId=2&contentId=2278. September 2003 (accessed October 27, 2003).

64. DJ Nisbet, DE Corrier, SC Ricke, ME Hume, JA Byrd II, JR DeLoach. Maintenance of the biological efficacy in chicks of a cecal competitive-exclusion culture against *Salmonella* by continuous flow fermentation. *J. Food Prot.*, 59:1279–1283, 1996.

65. MM Brashears, D Jaroni, J Trimble. Isolation, selection, and characterization of lactic acid bacteria for a competitive exclusion product to reduce shedding of *Escherichia coli* O157:H7 in cattle. *J. Food Prot.*, 66:355–363, 2003.

66. DJ O'Sullivan. Screening of intestinal microflora for effective probiotic bacteria. *J. Agric. Food Chem.*, 49:1751–1760, 2001.

67. AC Ouwehand, PV Kirjavainen, C Shortt, S Salminen. Probiotics: mechanisms and established effects. *Int. Dairy J.*, 9:43–52, 1999.

68. AMP Gomes, Malcata FX. *Bifidobacterium* spp. and *Lactobacillus acidophilus*: biological, biochemical, technological and therapeutical properties relevant for use as probiotics. *Trends Food Sci. Technol.*, 10:139–157, 1999.

69. LR Bielke, AL Elwood, DJ Donoghue, AM Donoghue, LA Newberry, NK Neighbor, BM Hargis. Approach for selection of individual enteric bacteria for competitive exclusion in turkey poults. *Poultry Sci.*, 82:1378–1382, 2003.

70. RD Wagner, M Holland, CE Cerniglia. An *in vitro* assay to evaluate competitive exclusion products for poultry. *J. Food Prot.*, 65:746–751, 2002.

71. DJ Nisbet, RC Anderson, DE Corrier, RB Harvey, LH Stanker. Modeling the survivability of *Salmonella typhimurium* in the chicken cecae using an anaerobic continuous-culture of chicken cecal bacteria. *Microb. Ecol. Health Dis.*, 12:42–47, 2000.

72. PWJJ van der Wielen, LJA Lipman, F van Knapen, S Biesterveld. Competitive exclusion of *Salmonella* serovar Enteriditis by *Lactobacillus crispatus* and *Clostridium lactatifermentans* in a sequencing fed-batch culture, *Appl. Environ. Microbiol.*, 68:555–559, 2002.

73. DJ Nisbet, SC Ricke, CM Scanlan, DE Corrier, AG Hollister, JR DeLoach. Inoculation of broiler chicks with a continuous–flow derived bacterial culture facilitates early cecal bacterial colonization and increases resistance to *Salmonella typhimurium. J. Food Prot.*, 57:12–15, 1994.

74. RE Droleskey, DE Corrier, DJ Nisbet, JR DeLoach. Colonization of mucosal epithelium in chicks treated with a continuous flow culture of 29 characterized bacteria: confirmation by scanning electron microscopy. *J. Food Prot.*, 58:837–842, 1995.

75. DJ Nisbet, DE Corrier, SC Ricke, ME Hume, JR DeLoach. Cecal propionic acid as an indicator of the early establishment of a microbial ecosystem inhibitory to *Salmonella* in chicks. *Anaerobe*, 2:345–350, 1996.

76. DJ Nisbet. Defined competitive exclusion cultures in the prevention of enteropathogen colonization in poultry and swine. *Antonie van Leeuwenhoek*, 81:481–486, 2002.

77. RC Anderson, DJ Nisbet, SA Buckley, KJ Genovese, RB Harvey, JR DeLoach, NK Keith, LH Stanker. Experimental and natural infection of early weaned pigs with *Salmonella choleraesuis. Res. Vet. Sci.*, 64:261–262, 1998.

78. J Tournut. Applications of probiotics to animal husbandry. *Rev. Sci. Tech. Off. Int. Epiz.*, 8:551–566, 1989.

79. SC Kyriakis, VK Tsiloyiannis, J Vlemmas, K Sarris, AC Tsinas, C Alexopoulos, L Jansegers. The effect of probiotic LSP 122 on the control of post-weaning diarrhea syndrome of piglets. *Res. Vet. Sci.*, 67:223–238.

80. GC Mead, PA Barrow, MH Hinton, F Humbert, CS Impey, C Lahellec, RWAW Mulder, S Stavric, NJ Stern. Recommended assay for treatment of chicks to prevent *Salmonella* colonization by competitive exclusion. *J. Food Prot.*, 52:500–502, 1989.

81. S Muirhead. MS Bioscience to halt production of salmonella product. *Feedstuffs*, 74:27, 2003.

82. McDonald's Global Policy on Antibiotic Use in Food Animals. http://www.mcdonalds.com/corporate/social/marketplace/antibiotics/global/media/antibiotics_policy.pdf, June 2003 (accessed October 27, 2003). This Web address is no longer valid. The information may be accessed at http://www.mcdonalds.com/news/2003/conpr06192003.html, which was last accessed by me on May 6, 2005.

13

Probiotics as Biopreservatives for Enhancing Food Safety

Ipek Goktepe

CONTENTS

13.1 Introduction

Probiotic bacteria are defined as "live microbial feed supplement" that beneficially affect the host by improving intestinal microbial balance (1). Most of the probiotic strains are natural inhabitants of the intestines of humans and animals (2). The potential health benefits of probiotics have produced considerable interest in the U.S. with respect to the use of probiotics in food, pharmaceutical, and feed product applications (3–8). Currently, most probiotic bacteria are cultivated from the genera *Lactobacillus* and *Bifidobacterium*

(9), but *Enterococcus* spp., *Bacillus* spp., and *Streptococcus* spp. also have great potential.

Probiotics have traditionally been consumed as constituents of fermented dairy products such as yogurt but have also been incorporated into drinks like yakult. Today there are more than 70 lactic acid bacteria-containing products worldwide, including sour cream, yogurt, powdered milk, buttermilk, and frozen deserts (10). Probiotics are increasingly being marketed as dietary supplements in tablet, capsule, and freeze-dried preparations.

Recent clinical trials have demonstrated the beneficial health effects of probiotic bacteria that have been incorporated in foods and animal feed. These effects include the prevention of diarrhea, balancing of intestinal microflora, stimulation of the immune system, modulation of systemic immune homeostasis, antitumor properties, and lactose intolerance (11–20). In addition to these beneficial health effects, researchers have demonstrated that probiotic bacteria produce bacteriocins and organic acids that act as antimicrobials to inhibit the growth of pathogens as well as improve functions of intestinal epithelial cells (21–29). The focus of this chapter will be on the antimicrobial use of selected probiotics in controlling and inhibiting food pathogens and their potential use to extend the shelf life of food products.

13.2   Probiotic Bacteria in Food Safety Applications

Today's consumers expect and demand a safer food supply than ever before. Yet, despite the improvements in food safety, the number of food-borne illnesses continues to rise. It is estimated that food-borne related illnesses cause 4.2 million cases of sickness and over 1200 deaths in the U.S. every year (30). Ensuring food safety is largely dependent on minimizing initial microbial contamination and preventing microbial growth during processing, distribution, and storage of foods. Although techniques such as refrigeration, pasteurization, canning, packaging, and the use of antimicrobial chemicals in foods are well-established methods for preserving food quality and safety, outbreaks of food-borne diseases continue to pose significant public health concerns. Recent trends in consumer behavior over the safety of foods and demand for "natural foods" have spurred the search for alternative antimicrobials, such as biopreservatives. A variety of biopreservatives have been investigated, including bacteria with antimicrobial activity, plant extracts, and bacteriocins produced by lactic acid bacteria. Lactic acid bacteria have been used to ferment or culture foods such as meat, fish, milk, cereals, tubers, and vegetables for at least 4000 years (10).

Lactic acid bacteria produce biopreservatives such as lactic acid, hydrogen peroxide, and bacteriocins that are used to retard both spoilage and the growth of pathogenic bacteria (4). Among these biopreservatives, bacteriocins are the most widely used in the food industry. One of the most studied bacteriocins is nisin, which is used as a food additive in more than 48 countries to preserve the quality of processed cheese, dairy products, and canned foods. Another form of food biopreservative that is gaining widespread use involves the incorporation of live cultures, such as bifidobacteria, lactobacillus, and enterobacter (probiotics) in foods. The remainder of this chapter will review the antimicrobial properties of major probiotic species and their potential use as food biopreservatives.

13.2.1 Bifidobacteria

Bacteria representing the genus *Bifidobacterium* were first discovered by Henry Tissier (1899, 1900) when studying the organism in the feces of infants. Tissier initially referred to these organisms as *Bacillus,* and it was not until 1924 that the genus *Bifidobacterium* was proposed to classify these unique organisms that exhibited a characteristic morphology and cell-wall constituents (31). The morphology of bifidobacteria is rods of various shapes that are often in a "V" shaped or "bifid" form. They are nonmotile, nonsporeforming, and are strictly anaerobic (32). Of the more than 500 bacteria that inhabit the human body, bifidobacteria are the most abundant microorganisms in the human gut. Ten different species of bifidobacteria have been isolated from the human body. Among these species, *B. longum, B. bifidum,* and *B. infantis* are the most commonly known.

13.2.1.1 *Antimicrobial Activity of Bifidobacteria against Pathogenic Microorganisms*

The ability of bifidobacteria to inhibit pathogenic bacteria has been attributed to the production of lactic and acetic acids and a variety of other antimicrobial compounds, such as bacteriocins. Bacteriocins are known to inhibit the growth of pathogenic bacteria, such as *Escherichia coli* and *Salmonella* (21). A number of the bifidobacterial strains have been used as antimicrobials (33–39) due to their ability to inhibit pathogenic organisms, both *in vivo* and *in vitro*. These include *Salmonella, Shigella, Clostridium, Bacillus cereus, Staphylococcus aureus, Candida albicans*, and *Campylobacter jejuni* (33, 40, 41). A *B. bifidum* culture has also been found to show inhibitory effect against *Shigella dysenteriae in vitro* (22).

The potential of bifidobacteria and their by-products to controll pathogens and spoilage microorganisms in food systems has been evaluated by several researchers. Gagnon et al. (42) tested the ability of five bifidobacterial strains to inhibit *E. coli* O157:H7 *in vitro*. Two of the isolates were found to be significantly effective in inhibiting *E. coli* O157:H7 on nalidix acid lithium chloride agar. O'Riordan and Fitzgerald (39) screened 22 *Bifidobacterium* strains for antimicrobial activity and found that 12 of the strains exhibited antagonistic activity against Gram-positive and Gram-negative bacteria, including *Pseuedomonas* spp., *Salmonella* spp, *E. coli, Lactobacillus* spp., *Strep-*

tococcus spp., and *Listeria innocua*. Lievin et al. (43) found human isolates of *Bifidobacterium* spp. CA1 and F9 to have a protective effect against *S. enterica* pathovar Typhimurium C5 infection. Similarly, *B. lactis* HN109 has been shown to provide antiinfectious activity against *E. coli* O157:H7 in a murine model (44). Gagnon et al. (42) reported that the isolates with the greatest inhibiting effect on *E. coli* O157:H7 also had significant potential for reducing their adhesion to Caco-2 cells. Competitive binding to intestinal cells has also been reported between *Bifidobacterium* sp. BL2928 and *E. coli* O157:H7 (45). In a study by Gibson and Wang (37), the antimicrobial activity of eight different strains of bifidobacteria was studied *in vitro* on a range of pathogenic bacteria, including *Listeria monocytogenes*, *Clostridium perfringens*, *Salmonella* spp., *Vibrio cholerae*, and *E. coli*. The most effective inhibition toward *E. coli* and *C. perfringens* occurred with *B. infantis* and *B. longum*. Further studies showed that six strains of bifidobacteria could variously excrete bacteriocins with significant antimicrobial activities against the pathogenic bacteria tested.

Studies involving infants fed with a formula including bifidobacteria have found a lower incidence of hospital-induced diarrhea and lower rates of rotavirus compared to infants fed with a standard formula (16). A similar study conducted by Rinne et al. (46) showed that infants fed with formula supplemented with a mixture of fructo- and galactooligosaccharides or breast-fed mothers who had been given probiotics had higher bifidobacteria microbiota composition than those of control groups, indicating the antagonistic activity of probiotics against competing bacteria, such as pathogenic *E. coli*.

Unlike *Lactobacillus* species, very limited information is available on the production of bacterocins by bifidobacteria. Bifidin, a bacteriocin produced by *B. bifidum* NCDC 1452, showed inhibitory activity against *E. coli*, *Bacillus cereus*, *Staphylococcus aureus*, *Micrococcus flavus*, and *Pseudomonas fluorescens* (33). *B. bifidum*'s bacteriocin activity was also demonstrated on *Streptococcus thermophilus*, *Lactobacillus acidophilus*, and *Pediococcus* spp. (23). In another study, the inhibitory activity of bifidocin B, a bacteriocin produced by *Bifidobacterium bifidum* NCFB 1454, was tested against Gram-positive and Gram-negative bacteria (21). Bifidocin B showed inhibitory activity against Gram-positive food pathogens; however, it did not show inhibitory activity against any of the Gram-positive or Gram-negative bacteria tested. Crude bifidocin B was highly active against food-borne pathogens and food spoilage bacteria including most *Listeria* species, *Enterococcus faecum*, *Enterococcus faecalis*, *Bacillus cereus*, and *Lactobacillus*, *Leuconstoc*, *Micrococcus*, and *Pediococcus* species (21). The adsorption of bifidocin B to the Gram-positive bacterial cell surface was attributed to the presence of nonspecific binding sites on these bacteria (21).

13.2.1.2 *Use of Bifidobacteria in Food Preservation*

Over the last decade, bifidobacteria have gained considerable attention in terms of their use as a biopreservative in the food industry. Demonstrated benefits of the inclusion of bifidobacteria into foods include increased shelf life, replacement of artificial preservatives, and enhanced food functionality. Dairy products containing bifidobacteria are especially gaining popularity with consumers due to their claimed health benefits. The use of bifidobacteria as antimicrobial surface treatments in foods was reported by several investigators (Table 13.1). Kim et al. (47) used a combination of *B. infantis* ATCC 15697, *B. longum* ATCC 15707, *B. adolescentis* 9H Martin, and sodium acetate to increase the shelf life of catfish fillets. The combination of *B. infantis* and 0.5% of sodium acetate extended the shelf life of catfish fillets by 3 d. Fish treated with *B. infantis* and 0.5% of sodium acetate were highly comparable to fresh fillets in terms of odor and appearance for up to 6 d. In another study, *B. infantis* NCFB 2255 and *B. breve* NCFB 2258 were incorporated into cottage cheese to inhibit spoilage microorganisms, e.g., *Pseudomonas* spp. (39). Both bifidobacterial strains exerted inhibitory activity against *Pseudomonas* and extended the shelf life of cottage cheese by 14 d, which is beyond the recommended shelf life. Microbiological and sensory qualities of shrimp treated with 1.5% potassium sorbate and *B. breve* NCFB 2258 were evaluated by Al-Dagal and Bazaraa (48). Surface treatment of potassium sorbate and *B. breve* NCFB 2258 reduced the growth of psychrotropic bacteria and extended the shelf life of shrimp by 3 d, compared to the control shrimps. The shelf life of fresh camel meat was also extended by 12 d after treatment with 10% sodium acetate and *B. breve* NCFB 2258 (49). The combination treatment suppressed the growth of spoilage bacteria, maintained good surface color, and prevented off-odor in fresh camel meat. The combined effect of 2 or 4% sodium acetate and *B. bifidum* ATCC 29521 on storage quality of raw chicken was studied by Goktepe et al. (50). The combination treatment extended the shelf life of refrigerated raw chicken by 3 d over the standard shelf life. Chickens treated with bifidobacteria in combination with sodium acetate had higher texture values (indicating firmness), lighter color (closest to fresh chicken color), and little or no off-odor for up to 9 d of storage. Such shelf-life extension was not observed in control (untreated) chicken samples,

TABLE 13.1

Bifidobacteria with Antimicrobial Activity

Species	Active Against	Examples of Food Application	Ref.
B. adolescentis	*L. innocua, Staphylococcus* spp., *C. tyrobutyricum*	Catfish	(39, 47)
B. bifidum	*E. coli, B. cereus, S. aureus, M. flavus, P. fluorescens*	Chicken	(33, 50)
B. breve	*L. innocua, Staphylococcus* spp., *C. tyrobutyricum*	Cottage cheese, shrimp, camel meat	(39,48–49)
B. infantis	*L. innocua, C. tyrobutyricum*	Catfish, cheddar-like cheese, cottage cheese	(39, 47)
B. longum	*L. innocua, Staphylococcus* spp., *C. tyrobutyricum*	Catfish	(47)

chicken treated with bifidobacteria alone and packaged in modified atmosphere, or those treated with sodium acetate alone at any of the concentrations used. These studies suggest that antimicrobial activity of bifidobacteria is enhanced when combined with an organic acid salt, such as sodium acetate. Such a combination treatment may have a potential to prevent food spoilage and enhance the quality and safety of treated food products.

13.2.2 *Lactobacillus*

The *Lactobacillus* genus contains approximately 60 species, including such organisms as *L. acidophilus*, *L. plantarum*, *L. casei*, and *L. rhamnosus*. Many *Lactobacillus* species used to be classified simply as *L. acidophilus*, resulting in the term "acidophilus" being used almost synonymously with probiotic organisms. Over the centuries people used *Lactobacillus* species to produce cultured foods with improved preservation properties and with characteristic flavors and textures different from the original food. These additional characteristic flavors and aromas are the result of metabolic products of lactobacilli. For example, acetaldehyde provides the characteristic aroma of yogurt, whereas diacetyl imparts a buttery taste to other fermented milks. The preservative action of *Lactobacillus* is explained by the presence of lactic acid, hydrogen peroxide, and bacteriocins. The use of *Lactobacillus* and/or lactobacilli-derived antimicrobials as biopreservatives in foods for the control of spoilage and pathogenic organisms has been extensively studied.

13.2.2.1 *Antimicrobial Properties of* Lactobacillus

Antimicrobial activities of *Lactobacillus* strains have been studied *in vitro* and *in vivo*. Such activity is attributed to the ability of *Lactobacillus* species to produce antimicrobial compounds and their capacity to interfere with the adhesion of pathogenic bacteria onto intestinal cells (Table 13.2). *In vitro*

TABLE 13.2

Lactobacillus Species with Antimicrobial Activity against Pathogens

Species	Active Against	Ref.
L. acidophilus	*E. coli* O157:H7, *S. enterica* serovar Typhimurium, *Y. pseudotuberculosis*, *L. monocytogenes*	(17, 25, 52)
L. casei subsp. rhamnosus	*E. coli* O157:H7, *K. pneumoniae*, *H. pylori*, *S. typhimurium*, *A. hydrophila*, *S. aureus*, *L. monocytogenes*	(25, 27, 51, 53–54)
L. fermentum	*S. typhimurium*, *S. pullorum*	(52)
L. delbrueckii subsp. bulgaricus	*Bacillus cereus*	(53)
Lactobacillus spp.	*C. jejuni/coli*	(26)
L. salivaris	*H. pylori*	(55)
L. johnsonii	*S. enterica* serovar Typhimurium,	(56)
L. rhamnosus	*S. enterica* serovar Typhimurium, *E. coli* O157:H7	(57–58)

studies using cultured human intestinal cells have demonstrated the adhesiveness and the bacterial interference effect of *Lactobacillus* species. Coconnier et al. (17) found that *L. acidophilus* LB inhibited adhesion of *E. coli* O157:H7, *Salmonella enterica* serovar Typhimurium, *Yersinia pseudotuberculosis*, and *Listeria monocytogenes* onto Caco-2 cells. Similarly, Forestier et al. (51) reported significant inhibitory activity of *L. casei* subsp. *rhamnosus* Lcr35 against *E. coli* O157:H7 and *Klebsiella pneumoniae* on Caco-2 cells. These investigators also found that *L. casei* subsp. *rhamnosus* Lcr35 inhibited colonization of the Caco-2 cells by *E. coli* O157:H7 and *Klebsiella pneumoniae*. Human *Lactobacillus* strains including *L. acidophilus* and *L. fermentum* have been shown to produce exclusion of and competition with pathogenic bacteria in cultured intestinal cells infected with *Salmonella pullorum* and *S. typhimurium* (52). Six of eight tested strains of *L. acidophilus* and one strain of *L. casei* tested *in vitro* on *Helicobacter pylori* NCTC 11637 inhibited the growth of *H. pylori* NCTC 11637 (25).

The inhibitory effect of *Lactobacillus* strains is believed to be due to lactic acid and bacteriocins production as discussed later. The inhibitory activity of *L. delbruecki* ssp. *bulgaricus* CFR 2028 was tested against a toxigenic strain of *Bacillus cereus* F 4810 (53) and found to have a pronounced antimicrobial activity against *B. cereus* at 37°C between late logarithmic and early stationary phases. The principle of the observed inhibitory activity was due to the production of antimicrobial compounds and pH-lowering effect of the *Lactobacillus* strain. In a study by Vescovo et al. (54), the antimicrobial effect of five strains of *L. casei* was investigated against *Aeromonas hydrophila*, *L. monocytogenes*, *S. typhimurium*, and *S. aureus in vitro*. This study showed that *L. casei* IMPCLC34 was the most effective strain in reducing the growth of *Aeromonas hydrophila*, *S. typhimurium*, and *S. aureus*, whereas no inhibitory effect was observed against *L. monocytogenes*. The antagonistic effect of *Lactobacillus* spp. (P93) isolated from chicken was investigated on 10 strains of *Campylobacter jejuni/coli* (26). *Lactobacillus* spp. (P93) showed bactericidal activity against all tested *Campylobacter* strains. The bactericidal effect was characterized by the production of organic acids in combination with the production of an anti-*Campylobacter* protein.

The antagonistic effects of *Lactobacillus* against other pathogens have also been demonstrated *in vivo* using murine models. In an *H. pylori*-infected gnotobiotic murine model, *L. salivaris* completely inhibited the growth of *H. pylori* and reduced the *H. pylori*-induced inflammatory response (55). In another study using conventional mice, oral treatment with acidophilus conferred protection against infection by *H. felis* and *H. pylori* (17). *L. johnsonii* La1 and GG strains developed antibacterial activity when gnotobiotic C3H/He/Oujco mice were orally infected with *S. enterica* serovar Typhimurium C5 (56). Similarly, *L. acidophilus* and *L. rhamnosus* HN001 have the ability to confer protection on BALB/c mice orally infected by *S. enterica* serovar Typhimurium (57, 58). Mice challenged with *E. coli* O157:H7 also exhibited lower morbidity and higher immune responses when fed with *L. rhamnosus* HN001 (58). Administration of *L. casei* to mice infected by *L. monocytogenes* showed suppression of *L. monocytogenes* colonization. Similarly, ingestion of viable *L. casei* significantly reduced the growth of *L. monocytogenes* in the tissues of orally infected Wistar rats. It was concluded that the enhanced resistance to *L. monocytogenes* infection induced by *L. casei* is probably due to increased cell-mediated immunity (27).

On the basis of evidence from the *in vitro* and *in vivo* studies discussed above, it is apparent that the antimicrobial activity of *Lactobacillus* is due to the pH-lowering effect caused by the production of organic acids and boosting of host immunity. Additionally, recent reports have revealed that antimicrobial effect of *Lactobacillus* strains might be the production of bacteriocins as defense and survival against invading microorganisms.

13.2.2.2 *Bacteriocins Produced by* Lactobacillus

Interest in the bacterocins produced by *Lactobacillus* species increased in recent years due to the potential application of *Lactobacillus* as starter bacteria for the fermentation of vegetables, meat, and fish products. Since the first *Lactobacillus* bacteriocin identification in 1975 (Lactocin 27, [59]), many other bacteriocins have been characterized and identified. Some of these are currently being used commercially to inhibit the growth of pathogenic bacteria in food products, such as fish, dairy, and meat products.

Up to now, a variety of bacteriocins from *Lactobacillus* species have been identified (Table 13.3). The most common strains known to produce bacteriocins are *L. sakei* and *L. curvatus*. Sakacin A, M, P, 674, K, and T, all produced by *L. sakei*, possess potent activity against *L. monocytogenes* (60–66). Factors influencing the antimicrobial activity of sakacin A and P in food systems were investigated by many authors, and the possibilities of direct application of these bacteriocins in food products were demonstrated (67). *Lactobacillus casei* CRL705 secretes a bacteriocin called lactocin 705 (68). Lactocin 705, a small hydrophobic and positively charged antimicrobial peptide that consists of 33 amino acids, possesses inhibitory effects against *L. monocytogenes*, *Streptococcus pyogenes*, and *S. aureus* (68, 69). Bacteriocins of *L. curvatus* are curvacin A, curvaticin 13, and curvaticin FS47 (62, 70). Curvacin A produced by *L. curvatus* LTH 1174 has an identical amino acid sequence as sakacin K from *L. sakei* CTC494 and sakacin A from *L. sakei* Lb 706 (66, 71).

Other bacteriocins obtained from *L. plantarum* strains include plantaricin BN, D, LP84, C19, C, and F and have shown an inhibitory activity toward some Gram-positive and Gram-negative bacteria (72–74). Zamfir et al. (75) identified a bacteriocin called acidophilin 801 produced by *L. acidophilus* IBB 801. Acidophilin 801 displayed a narrow inhibitory activity against Gram-positive and Gram-negative bacteria. *L. plantarum* ALC 01 was reported to secrete the bacteriocin pediocin AcH, also produced by *Pediococcus acidilactici*, with strong antilisteral activity (76).

Recently, several new bacteriocins produced by *Lactobacillus* were identified and characterized. Buchnericin LB produced by *L. buchneri* inhibited the growth of many Gram-positive bacteria including *Listeria*, *Bacillus*, *Staphylo-*

TABLE 13.3

Bacteriocins Produced by *Lactobacillus* Species and Their Activity against Pathogens

Bacteriocin(s)	Source	Active Against	Ref.
Sakacins A, M, P, 674, K, and T	*L. sakei*	*L. monocytogenes*	(60–66)
Lactocin 705	*L. casei* CRL705	*S. aureus, L. monocytogenes, S. pyogenes*	(68–69)
Curvacin A	*L. curvatus*	*L. monocytogenes*	(62)
Curvaticin 13	*L. curvatus*	*L. monocytogenes*	(70)
Curvaticin FS47	*L. curvatus*	*L. monocytogenes*	(70)
Plantiricins BN, D, LP84, C19, C, and F	*L. plantarum*	Some Gram-positive and Gram-negative bacteria	(72–74)
Acidophilin 801	*L. acidophilus* IBB 801	Some Gram-positive and Gram-negative bacteria,	(75)
Pediocin AcH	*L. plantarum* ALC 01	*L. monocytogenes*	(76)
Buchnericin LB	*L. buchneri*	*L. monocytogenes, S. aureus, Bacillus* spp, *Pediococcus* spp, *Leuconostoc* spp, *Enterococcus* spp. and *Lactococcus* spp.	(77)
Acidocin LF 221 A and B	*L. gasseri* LF221	Gram-positive bacteria	(78)
Acidocin CH5	*L. acidophilus* CH5	Gram-positive bacteria	(79)
Lactobin A	*L. amylovorus* LMG P-13139	Some Gram-positive and Gram-negative bacteria	(80)
Amylovorin L471	*L. amylovorus* DCE 471	Some Gram-positive and Gram-negative bacteria	(80)
Reuterin	*L. reuteri*	*L. monocytogenes*	(85)

coccus, Pediococcus, Leuconostoc, Enterococcus, and *Lactococcus,* but no inhibitory activity of this bacteriocin was observed on Gram-negative bacteria (77). Another strain of *Lactobacillus, L. gasseri* LF 221, produced two bacteriocins named acidocin LF221 A and LF221 B (78). These two bacteriocins were found to be similar to gassericin A, which is known to have antibacterial activity against a number of Gram-positive food-borne pathogens (29, 78). Recently, a bacteriocin named acidocin CH5 produced by *L. acidophilus* CH5 was characterized and purified by Chumchalova et al. (79), who demonstrated that acidocin CH5 showed strong inhibitory activity on Gram-positive bacteria. Lactobin A and amylovorin L471 are the latest additions to the list of bacteriocins from *Lactobacillus.* Lactobin A and amylovorin L471 are produced by *L. amylovorus* LMG P-13139 and *L. amylovorus* DCE 471, respectively (80). Both bacteriocins have similarities in terms of amino acid sequence; however, the inhibitory spectrum of amylovorin 471 is much wider than that of lactobin A.

The potential application of bacteriocins produced by *Lactobacillus* strains to control pathogens in foods has been evaluated extensively in culture media or food models. Many of these bacteriocins have been or are being used experimentally as food biopreservatives.

13.2.2.3 *Use of* Lactobacilli *Bacteriocins in Food Preservation*

The utilization of bacteriocins from *Lactobacillus* to control food-borne pathogens has been reported by several researchers in recent years (Table 13.4). However, for widespread commercial use of *Lactobacillus* bacteriocins to occur, factors such as spectrum of inhibition, heat stability, solubility, safety, and probiotic activity of the bacteriocins need to be properly characterized to demonstrate their safe use in food products.

One of the most studied types of *Lactobacillus* bacteriocins used in food products is pediocin AcH. The activity spectrum of pediocin AcH is relatively wide, and its action leads to cell lysis (81). Many studies have reported good antimicrobial activity of pediocin AcH. For instance, the growth of *L. monocytogenes* in Muenster cheese was suppressed at day 11 of ripening after application of pediocin AcH on the surface of the cheese (82). A preparation with pediocin AcH bound to its heat-killed producer cells *L. plantarum* WHE 92 was used to control the growth of *L. monocytogenes* and spoilage bacteria in cooked sausage (81). In this study, pediocin AcH preparation reduced the number of *L. monocytogenes* from the initial level of 2.7 log CFU/g sausage to <2 log CFU/g, whereas no inhibitory effect of the preparation was observed on spoilage bacteria. In a recent study, antilisteral activity of pediocin AcH was studied on red smear cheese (81). Near complete inhibition of *L. monocytogenes* was observed during the ripening consortium over a 4-month period when pediocin was used.

The Sakacin group of *Lactobacillus* bacteriocins is also commonly used to control food-borne pathogens, mainly in meat and meat products. Sakacin A, produced by *L. sake*, was found to be highly effective in inhibiting the growth of *L. monocytogenes* in minced meat and raw pork (83). Hugas et al. (65) investigated the inhibitory effect of sakacin K on *Listeria* in dry fermented sausages and found that this bacteriocin possesses a potent activity on *Listeria*. The inhibitory activity of sakacin P on *L. monocytogenes* was investigated in vacuum-packed cold smoked salmon stored at 4°C for 4 weeks (67). Addition of sakacin P on the surface of cold smoked salmon had a strong bacteriostatic effect on *L. monocytogenes* during the entire storage period. In another study, Katla et al. (84) evaluated the potential of sakacin P for the

TABLE 13.4

Application of Lactobacillus Bacteriocins as Food Biopreservatives

Bacteriocin	Example of Food Application	Active Against	Ref.
Pediocin AcH	Muenster cheese, cooked sausage	*L. monocytogenes*	(81, 82)
Sakacin A	Meat, pork	*L. monocytogenes*	(83)
Sakacin K	Dry fermented sausage	*L. monocytogenes*	(67)
Sakacin P	Cold smoked salmon, chicken cold-cuts	*L. monocytogenes*	(84)
Reuterin	Beef sausages	*L. monocytogenes*	(85)
Curvacin	Pork steaks and ground beef	*L. monocytogenes*	(86)
Plantaricin D	Ready-to-eat salad	*L. monocytogenes*	(73)

inhibition of *L. monocytogenes* in vacuum-packed chicken cold cuts. The addition of sakacin P had an inhibitory effect on the growth of *L. monocytogenes* at a concentration of 3.5 µg/g, whereas a lower concentration of 12 ng/g sakacin P slowed the growth rate of *L. monocytogenes* during 4 weeks of storage, but it did not show a complete inhibitory effect. Another type of bacteriocin, reuterin, produced by *L. reuteri*, demonstrated strong inhibitory activity on *L. monocytogenes*, but not on *Salmonella* spp. in beef sausages (85). In their study, Mauriello et al. (86) investigated the possibility of the use of bacteriocin in food packaging to control the growth of *L. monocytogenes* in pork steaks and ground beef. The antimicrobial packaging using curvacin produced by *L. curvatus* 32Y reduced *L. monocytogenes* growth for about 1 log cycle after 24 h of storage. It was concluded that antimicrobial packages containing bacteriocins can be used as biopreservatives to reduce the risk of pathogen development as well as to extend the shelf life of food products.

In recent years, salad bars containing freshly cut vegetables in restaurants and supermarkets have become very popular among consumers. During handling of fresh produce, difficulties of maintaining cold temperature at all times has been a problem that provides an opportunity for pathogens like *L. monocytogenes* to grow. To overcome this problem, researchers evaluated the possibilities of using bacteriocin and bacteriocin-producing cultures in fresh vegetables. This possibility was addressed by Franz et al. (73) who demonstrated that plantaricin D, produced by *L. plantarum* BFE, inhibited the growth of *L. monocytogenes* in ready-to-eat salad. Lactic acid fermentation of vegetables like cabbage is also a common method of preserving fresh vegetables in many countries. Therefore, the use of bacteriocin in combination with lactobacilli starter cultures could be a beneficial method for the prevention of spoilage and pathogenic bacteria in food products, such as fresh produce and salad bars.

13.2.3 Other Bacteriocin Producing Lactic Acid Bacteria

The genera of lactic acid bacteria include *Lactococcus*, *Streptococcus*, *Lactobacillus*, *Pediococcus*, *Leuconostoc*, *Enterococcus*, *Carnobacterium*, and *Propionibacterium*. In the last decade, there has been a significant increase in the use of lactic acid bacteria in foods due to their antibacterial activity against spoilage and pathogenic bacteria. Over the years, extensive research into the antagonistic activity of these bacteria has led to the discovery of bacteriocins, many of which have been or are currently used in food products. The most-studied and well-characterized bacteriocin of lactic acid bacteria is nisin produced by *Lactococcus lactis*. Nisin was discovered by Rogers and Whittier in 1928 (87). It has a broad spectrum of inhibition and kills a wide range of Gram-positive pathogenic bacteria including *L. monocytogenes* and *Clostridium* spp. in cheese products (88). Nisin was approved in 1988 as a food additive in the U.S. (89). The efficiency of nisin in preventing the growth of spoilage bacteria has been proven in a number of food systems (90). A recent review on the use of nisin in food products was also published by Cleveland et al. (91). This review gives ample details on the efficacy and mechanism of action of this bacteriocin against pathogenic and spoilage bacteria.

Pediocin PA-1, produced by by *Pediococcus acidilactici*, is the most promising bacteriocin for meat products. Pediocin PA-1 is active against *L. monocytogenes* and *L. curvatus*, a spoilage organism (92, 93). Pediocin PA-1 successfully controlled the growth of *L. monocytogenes* in raw chicken (94), while the utilization of bacteriocin producing *P. acidilactici* JD1-23 and *P. acidilactici* PAC 1.0 as fermenting agents reduced the number of *L. monocytogenes* by 12 log CFU/g dry sausage (95).

Numerous strains of enterococci associated with food systems, mainly *E. faecalis* and *E. faecium*, produce bacteriocins called enterocins. Enterocins possess inhibitory activity against *L. monocytogenes, S. aureus, Clostridium* spp., and *Vibrio cholerae* (82, 96–99). Numerous studies investigated enterocins and their potential application in dairy and meat products. Aymerich et al. (66) studied the effect of enterocins A and B against *L. innocua* in cooked ham, minced pork meat, deboned chicken breast, pate, and slightly fermented Italian sausages called espetec. Enterocins A and B at a concentration of 4800 Arbitrary Units (AU)/g reduced the number of *L. innocua* by 8 log cycles in cooked ham, by 9 log cycles in pate, by 5 log cycles in deboned chicken breast, by 3 log cycles in minced pork meat, and by 3 log cycles in espetec. In a recent study, antilisteral activity of enterocin 416K1, produced by *E. casseliflavus* 416K1 IM, was investigated in "cacciatore," Italian sausages (100) and found it significantly reduced the number of *L. monocytogenes* in cacciatore during 20 d of storage. The antilisteral activity of enterocins produced by *E. faecalis* and *E. faecium* in cheese was reported by several investigators (98, 101–103). The inhibition of *L. innocua* by enterocin-producing *E. faecium* was evaluated in soft Italian cheese, Taleggio (101). This study showed that enterocin-producing starter culture *E. faecium* inhibited *L. innocua* in the first 55 h of cheese ripening due to the combined synergistic effects of pH decrease and bacteriocin production. The inhibitory effect of enterocin 4 against *L. monocytogenes* strains Ohio and Scott A in Manchego cheese was investigated by Nunez et al. (98). *L. monocytogenes* strain Ohio counts decreased by 3 log cycles after 8 h and by 6 log cycles after 7 d in cheese treated with enterocin 4, whereas no inhibitory effect of enterocin 4 was observed on *L. monocytogenes* strain Scott A in Manchego cheese. The antagonistic effect of enterocin CCM 4231 from *E. faecium* was studied on *L. innocua* Li1 in bryndza, a traditional Slovak cheese where the number of *L. innocua* decreased by day 4, indicating a strong antagonistic effect of this bacteriocin to *L. innocua* (102). In a recent study, conducted by Foulquie et al. (103), two strains of bacteriocin-producing *E. faecium* RZS C5 were effective in extending the shelf life of cheddar cheese. From the multitude of studies above, there is strong evidence that enterococci, either in the form of starter cultures or their bacteriocins, could find potential application as biopreservatives in the processing of meat and dairy products.

An important class of lactic acid bacteria with strong bacteriocin-producing activity is *Carnobacterium* spp., which are able to produce bacteriocins with strong antilisteral activity at low temperatures and high sodium chloride content (104). In addition, *Carnobacterium* spp. can grow in foods with low carbohydrate content such as fish (105). Several bacteriocins produced by *Carnobacterium* spp. have been characterized, including carnobacteriocins BM1 and B2 from *C. piscicola* LV17B, divergicin A produced by *C. divergens* NCIMB 702855, divercin V41 from *C. divergens* V41, divergicin 750 produced by *C. divergens* 750, piscicocin V1a from *C. piscicola* V1, and carnocin CP5 from *C. piscicola* CP5 (105–111). In a recent study, bacteriocinogenic activity of *C. piscicola* CS526 was studied on *L. monocytogenes* in cold smoked salmon (112). *C. piscicola* CS526 suppressed the growth of *L. monocytogenes* in cold smoked salmon within 12 d of storage. Like most *Carnobacterium* bacteriocins, divergicin M35 produced by *C. divergens* M35 displayed high antilisteral activity but no inhibitory effect against *Lactococcus lactis*, *P. acidilactici*, *Streptococcus thermophilus*, and *E. coli* (113). Due to its powerful antilisteral activity, divergicin M35 may have a potential application as biopreservative for lightly processed seafood where *Listeria* spp. can be a serious problem.

Several investigators have examined the antimicrobial and bacteriocin production activity of *Leuconostoc* spp. Leucocin A is a bacteriocin produced by *Leu. gelidum* UAL187 and shows a broad inhibitory spectrum against spoilage lactic acid bacteria, some strains of *L. monocytogenes*, and *Enterococcus faecalis* (114). The strain of *Leu. mesenteroides* UL5 isolated from cheddar cheese was reported to produce a heat-stable bacteriocin called mesenterocin 5, which is bactericidal to a wide range of *Listeria* spp. but not to strains of lactic acid bacteria (115). Leuconocin S produced by an atypical strain of *Leu. paramesenteroides* was studied by Lewus et al. (116), who found that this bacteriocin had bacteriostatic effect on *L. sake* ATCC 15521 over a wide pH range. In a recent study, a bacteriocin (mesentericin y105) closely related to leucocin A and produced by *Leu. mesenteroides* Y105 displayed antagonistic activity against Gram-positive bacteria including *L. monocytogenes* and *Enterococcus faecalis* (117). The bacteriocin, leuconocin S produced by *Leu. paramesenteroides*, is heat stable and has a broad antibacterial spectrum against *C. botulinum*, *L. monocytogenes*, *S. aureus*, and *Yersinia enterocolitica*. Leucocin F10 is another bacteriocin type produced by *Leu. carnosum* strain (118) and possesses strong inhibitory activity on *Listeria* spp., *Leuconostoc* spp., *Enterococcus faecalis*, and *L. sake*. Recently, a strain of *Leu. carnosum* 4010 was reported to produce two bacteriocins named leucocins 4010 (119). Leucocins 4010 are very similar to leucocin A and exhibit strong antilisteral activity in meat products without causing any undesirable effect on the product's flavor. Hornbaek et al. (120) further studied the antilisteral effect of leucocins 4010 alone or in combination with sodium chloride and sodium nitrate in a gelatin system. The presence of sodium nitrate and sodium chloride reduced the antilisteral activity of leucocins 4010 in the gelatin system compared to the use of leucocins 4010 alone at concentrations of 5.3 and 10.6 AU/mL. This study concluded that the application of leucocins

4010 might be more efficient in meat products that contain low sodium chloride and sodium nitrate.

Propionibacterium spp. play an important role in dairy fermentations as starter cultures responsible for the characteristic flavor and eye production in Swiss-type cheeses. Numerous bacteriocins produced by *Propionibacterium* spp. have been isolated and characterized. Among these, propionicin PLG-1, isolated from *P. thoenii* P127, exhibited a broad spectrum of inhibition against closely related species, some Gram-positive and Gram-negative bacteria, as well as some yeasts and molds (121). Strain *P. thoenii* P126 was reported to produce a bacteriocin called jenseniin G (122). Jenseniin G is a heat-stable bacteriocin with strong inhibitory activity on *L. bulgaricus* NCDO 1489, *L. delbrueckii* supsp. *lactis* ATCC 4797, *Lac. cremoris* NCDO 799, and *Lac. lactis* subsp. *lactis* C2. The bacteriocin propionicin T1 isolated from *P. thoenii* 419 is the only *Propionibacterium* bacteriocin that meets the grouping criteria of Klaenhammer bacteriocin classification (123, 124). Another strain of *Propionibacterium*, *P. jensenii* DF1, showed strong antimicrobial activity against closely related species and, therefore, was investigated further for bacteriocin production (125). It was reported that this bacterium produces a specific bacteriocin known as propionicin SM1. Two bacteriocins produced by cutaneous species of *Propionibacterium* have also been described, namely, acnecin isolated from *P. acnes* CN-8 and a bacteriocin-like substance produced by *P. acnes* RTT 108 (126-127). In 2004, a new bacteriocin was added to the list of *Propionibacterium* bacteriocins. Thoeniicin 447, isolated from *P. thoenii* 447, acts as bactericidal against *L. delbrueckii* subsp. *bulgaricus* and as a bacteriostatic against *P. acnes* (128). Although the inhibitory effect of bacteriocins produced by *Propionibacterium* against several bacteria has been documented, there is no evidence of their use in controlling the growth of pathogens and spoilage bacteria in food products.

Several bacteriocins from lactic acid bacteria have already been isolated and characterized from foods (Table 13.5). Many of these bacteriocins are considered safe and beneficial since they have been an important part of the human diet for centuries in the form of microbial activity by-products in foods such as cheese, meat, fish, and fermented food products.

13.3 Conclusions

The effectiveness of bifidobacteria, lactobacilli, and other lactic acid bacteria as biopreservatives is well documented. It is clear that some lactic acid bacteria possess strong antimicrobial activity against many pathogens, and thus they have been or have the potential to be used as biopreservatives in food products. During the last decade, research has focused on the isolation and characterization of antimicrobial compounds from lactic acid bacteria due partly to their beneficial health effects as probiotics. To date, nisin is the

TABLE 13.5

Examples of Bacteriocins Produced by Other Lactic Acid Bacteria and Their Application in Foods

Bacteriocin(s)	Source	Examples of Food Application	Active Against	Ref.
Nisin	*Lc.lactis*	Dairy products	*L. monocytogenes, Clostridium* spp.	(88)
Pediocin PA-1	*P. acidilactici*	Raw chicken	*L. monocytogenes, Lactobacillus curvatus*	(92–93)
Enterocins A and B	*E. faecalis*	Cooked ham, minced pork, deboned chicken breats, pate, and "Espetec"	*L. innocua*	(66)
Enterocin 416K1	*E. casseliflavus* 416K1 1M	"Cacciatore"	*L. monocytogenes*	(100)
Enterocin 4	*E. faecalis*	Manchego cheese	*L. monocytogenes*	(98)
Enterocin CCM 4231	*E. faccium*	"Bryndza"	*L. innocua*	(102)
Divergicin M35	*C. divergens* M35	Fish	*L. monocytogenes*	(113)
Leucocin A	*Leu. gelidum* UAL187	—	Spoilage lactic acid bacteria, *L. monocytogenes,* and *E. faecalis*	(114)
Mesenterocin 5	*Leu. mesenteroides*	Cheddar cheese	*Listeria* spp.	(115)
Mesentericin y105	*Leu. mesenteroides* Y105	—	Some Gram-positive bacteria, *L. monocytogenes,* and *E. faecalis*	(117)
Leucin F10	*Leu. carnosum*	—	*Listeria* spp., *Leuconostoc* spp., *E. faecalis,* and *L. sake*	(118)
Leucocins 4010	*Leu. carnosum* 4010	Meat products	*L. monocytogenes*	(119–120)
Propionicin PLG-1	*P. thoenii* P127	—	Some Gram-positive and Gram-negative bacteria	(121)
Jenseniin G	*P. thoenii* P126	—	*L. bulgaricus, L. delbrueckii* subsp. *lactis, Lac. cremoris,* and *Lac. lactis* subsp. *lactis* C2	(122)
Thoenicin 447	*P. thoenii* 447	—	*L. delbrueckii* subsp. *bulgaricus* and *P. acnes*	(128)

only purified antimicrobial by-product of *Lc. lactis* that has received GRAS status for use in commercial food applications; other bacteriocins, such as pediocin, lactocin, leucocin, and enterocins might have future applications in food systems given the large number of scientific studies showing their efficacy and safety. However, the mechanisms of action of these bacteriocins need to be thoroughly determined in order to ensure safety.

References

1. R Harvenaar, T Ten Brink, JHJ Huis in't Veld. In: R Fuller, Ed., *Probiotics, The Scientific Basis*. Chapman & Hall, London, 1992, pp. 209–224.
2. WEV Lankaputhra, P Shah. Antimutagenic properties of probiotic bacteria and organic acids. *Mutat. Res.*, 397:169–182, 1998.
3. HW Modler, RC McKellar, M Yaguchi. Bifidobacteria and bifidogenic factors. *Can. Inst. Food Sci. Technol. J.*, 23: 29–41, 1991.
4. B Ray. Bacteriocins of starter culture bacteria as food biopreservatives: an overview. In: B Ray, M Daeschel, Eds., *Food Biopreservatives of Microbial Origin*. CRC Press, Boca Raton, FL, 1992, pp. 125–134.
5. ME Sanders, DC Walker, KM Walker, K Aoyama, TR Klaenhammer. Performance of commercial cultures in fluid milk applications. *J. Dairy Sci.*, 79:943–955, 1996.
6. L O'Sullivan, RP Ross, C Hill. Potential of bacteriocin-producing lactic acid bacteria for improvements in food safety and quality. *Biochimie*, 84:59–604, 2002.
7. G Zarate, S Gonzalez, AP Chaia. Assessing survival of dairy propionibacteria in gastrointestinal conditions and adherence to intestinal epithelia. *Methods Mol. Biol.*, 268:423–432, 2004.
8. J Verluyten, F Leroy, L De Vuyst. Influence of complex nutrient source on growth of and curvacin a production by sausage isolate *Lactobacillus curvatus* LTH 1174. *Appl. Environ. Microbiol.*, 70:5081–5088, 2004.
9. R Hartimink, FM Rombouts. Comparison of media for the detection of bifidobacteria, lactobacilli and total anaerobes from faecal samples. *J. Microbiol. Methods*, 36:181–192, 1999.
10. MR Acharya, RK Shah. Selection of human isolates of Bifidobacteria for their use as probiotics. *Appl. Biochem. Biotechnol.*, 102-103:81–98, 2002.
11. R Fekety. Guidelines for the diagnosis and management of *Clostridium difficile*-associated diarrhea and colitis. American College of Gastroenterology, Practice Parameters Committee. *Am. J. Gastroenterol.*, 92:739–750, 1997.
12. JA Vanderhoof, DB Whitney, DL Antonson, TL Hanner, JV Lupo, RJ Young. *Lactobacillus* GG in the prevention of antibiotic-associated diarrhea in children. *J. Pediatr.*, 135:564–568, 1999.
13. GW Elmer, LV McFarland, CM Surawicz, L Danko, RN Greenberg. Behaviour of *Saccharomyces boulardii* in recurrent *Clostridium difficile* disease patients. *Aliment. Pharmacol. Ther.*, 13:1663–1668, 1999.
14. MF Bernet, D Brassart, JR Neeser, AL Servin. *Lactobacillus acidophilus* La1 binds to cultured human intestinal epithelial cell lines and inhibits cell attachment and cell invasion by enterovirulent bacteria. *Gut*, 35:483–489, 1994.

15. E Isolauri, M Juntunen, T Rautanen, P Sillanaukee, T Koivula. A human *Lactobacillus* strain (*Lactobacillus casei* sp. strain GG) promotes recovery from acute diarrhea in children. *Pediatrics*, 88:90–97, 1991.

16. JM Saavedra, NA Bauman, I Oung, JA Perman, RH Yolken. Feeding of *Bifidobacterium bifidum* and *Streptococcus thermophilus* to infants in hospital for prevention of diarrhoea and shedding of rotavirus. *Lancet*, 344:1046–1049, 1994.

17. MH Coconnier, V Lievin, E Hemery, AL Servin. Antagonistic activity against *Helicobacter* infection *in vitro* and *in vivo* by the human *Lactobacillus acidophilus* strain LB. *Appl. Environ. Microbiol.*, 64:4573–4580, 1998.

18. D Haller, C Bode, WP Hammes, AM Pfeifer, EJ Schiffrin, S Blum. Non-pathogenic bacteria elicit a differential cytokine response by intestinal epithelial cell/leucocyte co-cultures. *Gut*, 47:79–87, 2000.

19. P Gionchetti, F Rizzello, A Venturi, P Brigidi, D Matteuzzi, G Bazzocchi, G Poggioli, M Miglioli, M Campieri. Oral bacteriotherapy as maintenance treatment in patients with chronic pouchitis: a double-blind, placebo-controlled trial. *Gastroenterology*, 119:305–309, 2000.

20. M Kalliomaki, S Salminen, H Arvilommi, P Kero, P Koskinen, E Isolauri. Probiotics in primary prevention of atopic disease: a randomised placebo-controlled trial. *Lancet*, 357:1076–1079, 2001.

21. Z Yildirim, MG Johnson. Characterization and antimicrobial spectrum of Bifidocin B, a Bacteriocin produced by Bifidocin B, a Bacteriocin produced by *Bifidbacterium bifidum* NCFB 1454. *J. Food Prot.*, 61: 47–51, 1998.

22. AK Misra, RK Kuila. Antimicrobial substances from *Bifidobacterium bifidum*. *Indian J. Dairy Sci.*, 48:612–614, 1995.

23. J Meghrous, P Euloge, AM Junelles, J Ballongue, H Petitdemange. Screening of *Bifidobacterium* strains for bacteriocin production. *Biotechnol. Lett.*, 12: 575–580, 1990.

24. MH Coconnier, MF Bernet, S Kerneis, G Chauviere, J Fourniat, AL Servin. Inhibition of adhesion of enteroinvasive pathogens to human intestinal Caco-2 cells by *Lactobacillus acidophilus* strain LB decreases bacterial invasion. *FEMS Microbiol. Lett.*, 110:299–305, 1993.

25. PD Midolo, JR Lambert, R Hull, F Luo, ML Grayson. *In vitro* inhibition of *Helicobacter pylori* NCTC 11637 by organic acids and lactic acid bacteria. *J. Appl. Bacteriol.*, 79:475–479, 1995.

26. P Chaveerach, LJ Lipman, F van Knapen. Antagonistic activities of several bacteria on *in vitro* growth of 10 strains of *C. jejuni/coli*. *Int. J. Food Microbiol.*, 90:43–50, 2004.

27. R De Waard, J Garssen, GC Bokken, JG Vos. Antagonistic activity of *Lactobacillus casei* strain shirota against gastrointestinal *Listeria monocytogenes* infection in rats. *Int. J. Food Microbiol.*, 73:93–100, 2002.

28. AC Majhenic, K Venema, GE Allison, BB Matijasic, I Rogelj, TR Klaenhammer. DNA analysis of the genes encoding acidocin LF221 A and acidocin LF221 B, two bacteriocins produced by *Lactobacillus gasseri* LF221. *Appl. Microbiol. Biotechnol.*, 63:705–714, 2004.

29. Y Kawai, R Kemperman, J Kok, T Saito. The circular bacteriocins gassericin A and circularin A. *Curr. Protein Peptide Sci.*, 5:393–398, 2004.

30. PS Mead, L Slutsker, V Dietz, L McCaig, J Bresee, C Shapiro, P Griffin, RV Tauxe. Food related illness and death in the United States". *Emerg. Infect. Dis.*, 5(5): 607–625, 1999.

31. S Orla-Jensen. La Classification des bactéries lactiques. *Lait*, 4:468–474, 1924.

32. V Scardovi. Bifidobacterium. In: *Berger's Manual of Systematic Bacteriology*, Vol. 2, 9th ed., PH Senath, NS Mari, ME Sharpe, JG Holt, Eds., Williams & Wilkins, Baltimore, 1986, p. 1418.

33. SK Anand, RA Srinivasan, LK Rao. Antibacterial activity associated with *Bifidobacterium bifidum*. *Cult. Dairy Prod.*, 6:8, 1985.

34. JC Faure, A Schellenberg, A Bexter, HP Wurzner. Barrier effect of *Bifidobacterium longum* on *Escherichia coli* in germ-free rat. *J. Nutr. Vit. Res.*, 52:225–228, 1982.

35. KH Kang, HJ Shin, YH Park, TS Lee. Studies on the antibacterial substances produced by lactic acid bacteria: purification and some properties of antibacterial substance 'Bifilong' produced by *B. longum*. *Korean J. Dairy Sci.*, 11:204–216, 1989.

36. MF Bernet, D Brassart, JR Neeser, AL Servin. Adhesion of human bifidobacterial strains to cultured human intestinal epithelial cells and inhibition of enteropathogen cell interactions. *Appl. Environ. Microbiol.*, 35:1066–1073, 1993.

37. GR Gibson, X Wang. Regulatory effects of bifidobacteria on the growth of other colonic bacteria. *J. Appl. Bacteriol.*, 77:412–420, 1994.

38. OA Oyarzabal, DE Conner. *In vitro* fructooligosaccharide utilization and inhibition of Salmonella spp. by selected bacteria. *Poult. Sci.*, 74:1418–1425, 1995.

39. K O'Riordan, GF Fitzgerald. Evaluation of bifidobacteria for the production of antimicrobial compounds and assessment of performance in cottage cheese at refrigeration temperature. *J. Appl. Microbiol.*, 85:103–114, 1998.

40. M Tojo, T Oikawa, Y Morikawa, N Yamashita, S Iwata, Y Satoh, J Hanada, R Tanaka. The effects of *Bifidobacterium breve* administration on *Campylobacter enteritis*. *Acta Pediatr. Jpn.*, 29:160–167, 1987.

41. T Tomoda, N Yasua, T Kageyama. Intestinal Candida overgrowth and Candida infection in patients with leukemia: effect of *Bifidobacterium* administration. *Bifidobacteria Microflora*, 7:71, 1988.

42. M Gagnon, EE Kheadr, G Le Blay, I Fliss. *In vitro* inhibition of *Escherichia coli* 157:H7 by bifidobacterial strains of human origin. *Int. J. Food Microbiol.*, 92:69–78, 2004.

43. V Lievin, I Peiffer, S Hudault, F Rochat, D Brassart, JR Neeser, AL Servin. *Bifidobacterium* strains from resident infant human gastrointestinal microflora exert antimicrobial activity. *Gut*, 47:646–652, 2000.

44. Q Shu, HS Gill. A dietary probiotic (*Bifidobacterium lactis* HN019) reduces the severity of *Escherichia coli* O157:H7 infection in mice. *Med. Microbiol. Immunol.*, 189:147–152, 2001.

45. S Fujiwara, H Hashiba, T Hirota, JF Forstner. Proteinaceous factor(s) in culture supernatant fluids of bifidobacteria which prevents the binding of enterotoxigenic *Escherichia coli* to gangliotetraosylceramide. *Appl. Environ. Microbiol.*, 63:506–512, 1997.

46. MM Rinne, M Gueimonde, M Kalliomaki, U Hoppu, SJ Salminen, E Isolauri. Similar bifidogenic effects of prebiotic-supplemented partially hydrolyzed infant formula and breastfeeding on infant gut microbiota. *FEMS Immunol. Med. Microbiol.*, 43:59–65, 2005.

47. CR Kim, JO Hearnsberger, AP Vickery, CH White, DL Marshall. Sodium acetate and bifidobacteria increase shelf-life of refrigerated channel catfish. *J. Food Sci.*, 60:25–27, 1995.

48. MM Al-Dagal, WA Bazaraa. Extension of shelf life of whole and peeled shrimp with organic acid salts and bifidobacteria. *J. Food Prot.*, 62:51–56, 1999.

49. I Al-Sheddy, MM Al-Dagal, WA Bazaraa. Microbial and sensory quality of fresh camel meat treated with organic acid salts and/or bifidobacteria. *J. Food Sci.*, 64:336–339, 1999.

50. I Goktepe, M Ahmedna, H Nasri. Use of Bifidobacteria and organic acid salts to extend the shelf-life of refrigerated chicken. The 2002 Annual Meeting of the American Society of Microbiology, Salt Lake City, UT, 2002, p. 109.

51. C Forestier, C De Champs, C Vatoux, B Joly. Probiotic activities of *Lactobacillus casei rhamnosus*: *in vitro* adherence to intestinal cells and antimicrobial properties. *Res. Microbiol.*, 152:167–173, 2001.

52. LZ Jin, YW Ho, MA Ali, N Abdullah, S Jalaludin. Effect of adherent *Lactobacillus* spp. on *in vitro* adherence of *Salmonellae* to the intestinal epithelial cells of chicken. *J. Appl. Bacteriol.*, 81:201–206, 1996.

53. BV Balasubramanyam, MC Varadaraj. Cultural conditions for the production of bacteriocin by a native isolate of *Lactobacillus delbrueki* ssp. *bulgaricus* CFR 2028 in milk medium. *J. Appl. Microbiol.*, 84:97–102, 1998.

54. M Vescovo, S Torriani, C Orsi, F Macchiarolo, G Scolari. Application of anti-microbial-producing lactic acid bacteria to control pathogens in ready-to-use vegetables. *J. Appl. Bacteriol.*, 81:113–119, 1996.

55. Y Aiba, N Suzuki, AM Kabir, A Takagi, Y Koga. Lactic acid-mediated suppression of *Helicobacter pylori* by the oral administration of *Lactobacillus salivarius* as a probiotic in a gnotobiotic murine model. *Am. J. Gastroenterol.*, 93:2097–2101, 1998

56. MF Bernet-Camard, V Lievin, D Brassart, JR Neeser, AL Servin, S. Hudault. The human *Lactobacillus acidophilus* strain LA1 secretes a nonbacteriocin anti-bacterial substance(s) active *in vitro* and *in vivo*. *Appl. Environ. Microbiol.*, 63:2747–2753, 1997.

57. J Filho-Lima, E Vieira, J Nicoli. Antagonistic effect of *Lactobacillus acidophilus, Saccharomyces boulardii*, and *Escherichia coli* combinations against experimental infections with *Shigella flexneri* and *Salmonella enteritidis* subsp. *typhimurium* in gnotobiotic mice. *J. Appl. Microbiol.*, 88:365–370, 2000.

58. HS Gill, Q Shu, H Lin, KJ Rutherfurd, ML Cross. Protection against translocating *Salmonella typhimurium* infection in mice by feeding the immuno-enhancing probiotic *Lactobacillus rhamnosus* strain HN001. *Med. Microbiol. Immunol.*, 190:97–104, 2001.

59. GC Upreti, RD Hindsdill. Production and mode of action of lactocin 27: bacteriocin from a homofermentative *Lactobacillus*. *Antimicrob. Agents Chemother.*, 7:139–145, 1975.

60. U Schillinger, FK Lucke. Antibacterial activity of *Lactobacillus sake* isolated from meat. *Appl. Environ. Microbiol.*, 55:190–11906, 1989.

61. OJ Sobrino, JM Rodriguez, WL Moreira, LM Cintas, MF Fernandez, B Sanz, PE Hernandez. Sakacin M, a bacteriocin-like substance from *Lactobacillus sake* 148. *Int. J. Food Microbiol.*, 6:215–225, 1992.

62. P Tichaczek, J Nissen-Meyer, I Nes, R Vogel, W Hammes. Characterization of the bacteriocins curvacin A from *Lactobacillus curvatus* LTH1174 and sakacin P from *Lactobacillus sake* LTH673. *Syst. Appl. Microbiol.*, 15:460–468, 1992.

63. A Holck, L Axelsson, K Huhne, L Krockel. Purification and cloning of sacasin 674, a bacteriocin from *Lactobacillus sake* Lb674. *FEMS Microbiol. Lett.*, 115:143–150, 1994.

64. J Samelis, S Roller, J Metaxopoulos. Sakacin B, a bacteriocin produced by *Lactobacillus sake* isolated from Greek dry fermented sausages. *J. Appl. Bacteriol.*, 76:475–486, 1994.

65. M Hugas, M Garriga, MT Aymerich, JM Monfort. Inhibition of Listeria in dry fermented sausages by the bacteriocinogenic *Lactobacillus sake* CTC494. *J. Appl. Bacteriol.*, 79:322–330, 1995.

66. T Aymerich, MG Artigas, M Garriga, JM Monfort, M Hugas. Effect of sausage ingredients and additives on the optimization of enterocin A and B by *Enterococcus faecium* CTC492. Optimization of *in vitro* production and anti-listeral effect in dry fermented sausages. *J. Appl. Microbiol.*, 88:686–694, 2000.

67. T Katla, T Moretro, IM Aasen, A Holck, L Axelsson, K Naterstad. Inhibition of *Listeria monocytogenes* in cold smoked salmon by addition of sakacin P and/or live *Lactobacillus sakei* cultures. *Food Microbiol.*, 18:431–439, 2001.

68. GM Vignolo, F Suriani, A Pesce de Ruiz Holgado, G Oliver. Antibacterial activity of Lactobacillus strains isolated from dry fermented sausages. *J. Appl. Bacteriol.*, 75:344–349, 1993.

69. G Vignolo, S Fadda, MN de Kairuz, AA de Ruiz Holgado, G Oliver. Control of *Listeria monocytogenes* in ground beef by 'Lactocin 705', a bacteriocin produced by *Lactobacillus casei* CRL 705). *Int. J. Food Microbiol.*, 29:397–402, 1996.

70. KI Garver, PM Muriana. Purification and partial amino acid sequence of curvaticin FS47, a heat-stable bacteriocin produced by *Lactobacillus curvatus* FS47. *Appl. Environ. Microbiol.*, 60:2191–5, 1994.

71. A. Axelsson AL, A Holck. The genes involved in production of and immunity to sakacin A, a bacteriocin from *Lactobacillus sake* Lb706. *J. Bacteriol.*, 177:2125–2137, 1995.

72. A Atrih, N Rekhif, M Michel, G Lefebrve. Detection of bacteriocins produced by *Lactobacillus plantarum* strains isolated from different foods. *Microbios*, 75:117–123, 1993.

73. CM Franz, TM Du, NA Olasupo, U Schillinger, WH Holzapfel. Plantaricin D, a bacteriocin produced by *Lactobacillus plantarum* BFE 905 ready-to-eat salad. *Lett. Appl. Microbiol.*, 26:231–235, 1998.

74. CB Lewus, A Kaiser, TJ Montville. Inhibition of food-borne bacterial pathogens by bacteriocins from lactic acid bacteria isolated from meat. *Appl. Environ. Microbiol.*, 57:1683–1688, 1991.

75. M Zamfir, R Callewaert, PC Cornea, L Savu, I Vatafu, L De Vuyst. Purification and characterization of a bacteriocin produced by *Lactobacillus acidophilus* IBB 801. *J. Appl. Microbiol.*, 87:923–931, 1999.

76. J Ennahar, D Aoude-Werner, O Sorokine, A Van Dorsselaer, F Bringel, JC Hubert, C Hasselmann. Production of pediocin AcH by *Lactobacillus plantarum* WHE 92 isolated from cheese. *Appl. Environ. Microbiol.*, **62**:4381–4387, 1996.

77. Z Yildirim, YK Avsar, M Yildirim. Factors affecting the adsorption of buchnericin LB, a bacteriocin produced by Lactobacillus correction of *Lactocobacillus buchneri*. *Microbiol. Res.*, 157:103–107, 2002.

78. WM Zhu, W Liu, DQ Wu. Isolation and characterization of a new bacteriocin from *Lactobacillus gasseri* KT7. *J. Appl. Microbiol.*, 88:877–886, 2000.

79. J Chumchalova, J Stiles, J Josephsen, M Plockova. Characterization and purification of acidocin CH5, a bacteriocin produced by *Lactobacillus acidophilus* CH5. *J. Appl. Microbiol.*, 96:1082–1089, 2004.

80. L De Vuyst, L Avonts, P Neysens, B Hoste, M Vancanneyt, J Swings, R Calle-waert. The lactobin A and amylovorin L471 encoding genes are identical, and their distribution seems to be restricted to the species *Lactobacillus amylovorus* that is of interest for cereal fermentations. *Int. J. Food Microbiol.*, 90:93–106, 2004.

81. M Loessner, S Guenther, S Steffan, S Scherer. A Pediocin-producing *Lactobacillus plantarum* strain inhibvits *Listeria monocytogenes* in a multispecies cheese surface microbial ripening consortium. *Appl. Environ. Microbiol.*, 69:1854–1857, 2003.

82. S Ennahar, D Aoude-Werner, O Assobhei, C Hasselmann. Antilisterial activity of enterocin 81, a bacteriocin produced by *Enterococcus faecium* WHE 81 isolated from cheese. *J. Appl. Bacteriol.*, 85:521–526, 1998.

83. U Schillinger, M Kaya, FK Lucke. Behavior of *Listeria monocytogenes* in meat and its control by a bacteriocin-producing strain of *Lactobacillus sake*. *J. Appl. Bacteriol.*, 70:473–478, 1991.

84. T Katla, T Moretro, I Sveen, IM Aasen, L Axelsson, LM Rorvik, K Naterstad. Inhibition of *Listeria monocytogenes* in chicken cold cuts by addition of sakacin P and sakacin P-producing *Lactobacillus sakei*. *J. Appl. Microbiol.*, 93:191–196, 2002.

85. H Kuleasan, ML Cakmakci. Effect of reuterin produced by *Lactobacillus reuteri* on the surface of sausages to inhibit the growth of *Listeria monocytogenes* and *Salmonella* spp. *Nahrung*, 46:408–410, 2002.

86. G Mauriello, D Ercolini, A La Storia, A Casaburi, F Villani. Development of polyethene films for food packaging activated with an antilisteral bacteriocin from *Lactobacillus curvatus* 32Y. *J. Appl. Microbiol.*, 97:314–322, 2004.

87. LA Rogers, ED Whittier. Limiting factors in lactic fermentation. *J. Bacteriol.* 16:211–229.

88. S Wessels, B Jelle, I Nes. Bacteriocins of Lactic Acid Bacteria. Report of the Danish Toxicology Centre, Denmark, 1998.

89. U.S. FDA. Nisin Preparation: Affirmation of GRAS status as direct human food ingredient. Federal Register, 53, April 6, 1988.

90. J Delves-Broughton, P Blackburn, RJ Evans, J Hugenholtz. Applications of the bacteriocin, nisin. *Antonie van Leeuwenhoek*, 69:193–202, 1996.

91. J Cleveland, TJ Montville, IF Nes, ML Chikindas. Bacterocins: safe, natural antimicrobials for food prservation. *Int. J. Food Microbiol.*, 71:1–20, 2001.

92. G Baccus-Taylor, KA Glass, JB Luchansky, AJ Maurer. Fate of *Listeria monocytogenes* and pediococcal starter cultures during the manufacture of chicken summer sausage. *Poult. Sci.*, 72:1772–1778, 1993.

93. MJ Coventry, K Muirhead, MW Hickey. Partial characterisation of pediocin PO2 and comparison with nisin for biopreservation of meat products. *Int. J. Food Microbiol.*, 26:133–145, 1995.

94. JH Goff, AK Bhunia, MG Johnson. Complete inhibition of low levels of *Listeria monocytogenes* on refrigerated chicken meat with Pediocin AcH bound to heat-killed *Pediococcus acidilactici* cells. *J. Food Prot.*, 59:1187–1192, 1996.

95. ED Berry, RW Hutkins, RW Mandigo. The use of bacteriocin-producing *Pediococcus acidilactici* to control post-processing *Listeria monocytogenes* contamination of frankfurters *J. Food Prot.*, 54:681–686, 1991.

96. G Giraffa. Enterococcal bacteriocins: their potential use as anti-*Listeria* factors in dairy technology. *Food Microbiol.*, 12:551–556, 1995.

97. MAP Franz, U Schillinger, WH Holzapfel. Production and characterisation of enterocin 900, a bacteriocin produced by *Enterococcus faecium* BFE 900 from black olives. *Int. J. Food Microbiol.*, 29:255–270, 1996.

98. M Nunez, JL Rodriguez, E Garcia, P Gaya, M Medina. Inhibition of *Listeria monocytogenes* by enterocin 4 during the manufacture and ripening of Manchego cheese. *J. Appl. Microbiol.*, **83**:671–677, 1997.

99. P Sarantinoupoulos, F Leroy, E Leontopoulou, M Georgalaki, G Kalantzopoulos, E Tsakalidou, L De Vuyst. Bacteriocin production by *Enterococcus faecium* FAIR-E 198 in view of its application as adjunct starter in Greek Feta cheese making. *Int. J. Food Microbiol.*, **72**:125–136, 2002.

100. C Sabia, S de Niederhausern, P Messi, G Manicardi, M Bondi. Bacteriocin-producing *Enterococcus casseliflavus* IM 416K1, a natural antagonist for control of *Listeria monocytogenes* in Italian sausages (Cacciatore). *Int. J. Food Microbiol.*, 87:173–179, 2003.

101. G Giraffa, E Neviani, GT Tarelli. Antilisteral activity by enterococci in a model predicting the temperature evolution of Toleggio, an Italian soft cheese. *J. Dairy Sci.*, 77:1176–1182, 1994.

102. A Laukova, S Czikkova. Antagonistic effect of enterocin CCM 4231 from *Enterococcus faecium* on "bryndza," a traditional Slovak dairy product from sheep milk. *Microbiol. Res.*, 156:31–34, 2001.

103. MMR Foulquie, MC Rea, TM C, L De Vuyst. Applicability of a bacteriocin-producing *Enterococcus faecium* as a co-culture in cheddar cheese manufacture. *Int. J. Food Microbiol.*, 81:73–84, 2003.

104. RL Buchanan, LK Bagi. Microbial competition: effect of culture conditions on the suppression of *Listeria monocytogenes* Scott A by *Carnobacterium piscicola*. *J. Food Prot.*, 60:254–261, 1997.

105. V Stohr, JJ Joffraud, M Cardinal, F Leroi. Spoilage potential and sensory profile associated with bacteria isolated from cold-smoked salmon. *Food Res. Int.*, 34:797–806, 2001.

106. LE Quadri, M Sailer, KL Roy, JC Vederas, ME Stiles. Chemical and genetic characterization of bacteriocins produced by *Carnobacterium piscicola* LV17B. *J. Biol. Chem.*, 269:12204–12211, 1994.

107. RW Worobo, JV Belkum, M Sailer, KL Roy, JC Vederas, ME Stiles. A signal peptide secretion-dependent bacteriocin from *Carnobacterium divergens*. *J. Bacteriol.*, 177:3143–3149, 1995.

108. A Métivier, MF Pilet, X Dousset, O Sorokine, P Anglade, M Zagorec, JC Piard, D Marion, Y Cenatiempo, C Fremaux. Divercin V41, a new bacteriocin with two disulphide bonds produced by *Carnobacterium divergens* V41: primary structure and genomic organization. *Microbiology*, 144:2837–2844, 1998.

109. A Holck, L Axelsson, U Schillinger. Divergicin 750, a novel bacteriocin produced by *Carnobacterium divergens* 750. *FEMS Microbiol. Lett.*, 136:163–168, 1996.

110. P Bhugaloo-Vial, X Dousset, A Métivier, O Sorokine, P Anglade, P Boyaval, D Marion. Purification and amino acid sequences of piscicocins V1a and V1b, two class IIa bacteriocins secreted by *Carnobacterium piscicola* V1 that display significantly different levels of specific inhibitory activity. *Appl. Environ. Microbiol.*, 62:4410–4416, 1996.

111. S Herbin, F Mathieu, F Brule, C Brablant, G Lefebvre, A Lebrihi. Characteristics and genetic determinants of bacteriocin activities produced by *Carnobacterium piscicola* CP5 isolated from cheese. *Curr. Microbiol.*, 35:319–326, 1997.

112. K Yamazaki, M Suzuki, Y Kawai, N Inoue, TJ Montville. Inhibition of *Listeria monocytogenes* in cold-smoked salmon by *Carnobacterium piscicola* isolated from frozen surimi. *J. Food Prot.*, 66:1420–1425, 2003.

113. I Tahiri, M Desbiens, R Benech, E Kheadr, C Lacroix, S Thibault, D Ouellet, I Fliss. Purification, characterization and amino acid sequencing of divergicin M35: a novel class IIa bacteriocin produced by *Carnobacterium divergens* M35. *Int. J. Food Microbiol.*, 97:123–136, 2004.
114. JW Hastings, M Sailer, K Johnson, KL Roy, JC Vederas, ME Stile. Characterization of Leucocin A-UAL 187 and cloning of the bacteriocin gene from *Leuconostoc gelidum. J. Appl. Bacteriol.*, 173:7491–7500, 1991.
115. H Daba, S Pandian, JF Gosselin, RE Simard, J Huang, C Lacroix. Detection and activity of a bacteriocin produced by *Leuconostoc mesenteroides. Appl. Environ. Microbiol.*, 57:3450–3456, 1991.
116. CB Lewus, S Sun, TJ Montville. Production of an amylase-sensitive bacteriocin by an atypical *Leuconostoc paramesenteeroides* strain. *Appl. Environ. Microbiol.*, 58:143–147, 1992.
117. D Morisset, JM Berjeaud, D Marion, C Lacombe, J Frere. Mutational analysis of mesentericin y105, an anti-Listeria bacteriocin, for determination of impact on bactericidal activity, *in vitro* secondary structure, and membrane interaction. *Appl. Environ. Microbiol.*, 70:4672–4680, 2004.
118. E Parente, M Moles, A Ricciardi. Leucocin F10, a bacteriocin from *Leuconostoc carnosum. Int. J. Food Microbiol.*, 33:231–243, 1996.
119. BB Budde, T Hornbaek, T Jacobsen, V Barkholt, AG Koch. *Leuconostoc carnosum* 4010 has the potential as a new protective culture for vacuum-packaged meats: culture isolation, bacteriocin identification, and meat application. *Int. J. Food Microbiol.*, 83:171–184, 2003.
120. T Hornbaek, TF Brocklehurst, BB Budde. The antilisteral effect of *Leuconostoc carnosum* 4010 and leucocins 4010 in the presence of sodium chloride and sodium nitrate examined in a structured gelatin system. *Int. J. Food Microbiol.*, 92:129–140, 2004.
121. WJ Lyon, BA Glatz. Partial purification and characterization of a bacteriocin produced by a strain of *Propionibacterium thoenii. Appl. Environ. Microbiol.*, 57:701–706, 1991.
122. DA Grinstead, SF Barefoot. Jenseniin G, a heat-stable bacteriocin produced by *Propionibacterium jensenii* P126. *Appl. Environ. Microbiol.*, 58:215–220, 1992.
123. TR Klaenhammer. Bacteriocins of lactic acid bacteria. *Biochimie*, 70:337–349, 1988.
124. T Faye, T Langsrud, I Nes, H Holo. Biochemical and genetic characterization of propionicin T1, a new bacteriocin from *Propionibacterium thoenii. Appl. Environ. Microbiol.*, 66:4230–4236, 2000.
125. S Miescher, M Stierli, M Teuber, L Meile. Propionicin SM1, a bacteriocin from *Propionibacterium jensenii* DF1: isolation and characterization of the protein and its gene. *Syst. Appl. Microbiol.*, 23:174–184, 2000.
126. S Fujimara, T Nakamura. Purification and properties of a bacteriocin-like substance (acnecin) of oral *Propionibacterium acnes. Antimicrob. Agents Chemother.*, 14:893–898, 1978.
127. GE Paul, SJ Booth. Properties and characteristics of a bacteriocin-like substance produced by *Propionibacterium acnes* isolated from dental plaque. *Can. J. Microbiol.*, 34:1344–1374, 1988.
128. IR Van der Merwe, R Bauer, TJ Britz, LMT Dicks. Characterization of thoeniicin 447, a bacteriocin isolated from *Propionibacterium thoenii* strain 447. *Int. J. Food Microbiol.*, 92:153–160, 2004.

14

EU Perspectives on Food, Gastrointestinal Tract Functionality, and Human Health

Tiina Mattila-Sandholm, Liisa Lähteenmäki, and Maria Saarela

CONTENTS

14.1 Introduction Development of Efficacious Functional Foods for Intestinal Health

Health is one of the main reasons behind food choices, but still other factors are taken into consideration by the consumer. As pleasantness and sensory quality can be experienced directly, they are known to be the essential factors in repeated choices. Even with foods aimed for health, the taste has to be good for securing repeated choices. Probiotics modifying the gut microbiota have physiological effects that cannot be experienced directly. The credibility of the health-related messages is, therefore, the critical goal aiming to guarantee that the product has a reward value for the consumer. Consumers are selective in their attention so that messages that are relevant to us and which are congruent with our earlier beliefs are easily accepted. Therefore, food-related messages have to correspond with our existing beliefs and motivation to use a product. Preventing a possible stomach upset has little relevance to a healthy person, but can make a vast difference for someone suffering from inflammatory bowel diseases (IBDs) or someone who easily catches traveler's diarrhea.

There is an ever growing demand for developing foods with specific functionalities that increase health and well-being. This need originates from the notion that consumers at critical ages, young and old, are prone to food-related diseases and gastrointestinal (GI) disorders. These can be prevented by consumption of foods with specific functionality for health. Finally, there is an increase in the consumer's awareness of and desire for innovative and healthy foods that meet the strategy of many of the European industries to develop functional foods, which represent the fastest growing market that has already approached a volume of 1.5 billion Euros in Europe alone.

Keeping this in mind, the microbes with health impact are and will remain an important functional ingredient for years to come. New strains will be identified and foods will be developed to fulfill the needs of specific consumer groups. Increased understanding of interactions between gut microbiota, diet, and the host will open up possibilities of producing novel ingredients for nutritionally optimized foods, which promote consumer health through microbial activities in the gut.

The European Commission, through its 5th Framework Programme, is investing a substantial research effort in the intestinal microbiota, its interaction with its host and methods to manipulate its composition, and activity for the improvement of human health. Eight multicenter interdisciplinary research projects currently cover a variety of research topics required for the development of efficacious probiotic foods, ranging from understanding of probiotic mechanisms at a molecular level to developing technologies to ensure delivery of stable products and demonstrating safety and efficacy in specific disorders. This concerted research effort promises to provide an appreciable understanding of the human intestinal microbiota's role in health and disease and new approaches and products to tackle a variety of intestinal afflictions.

The Food, GI-tract Functionality and Human Health (PROEUHEALTH) Cluster brings together eight complementary, multicenter interdisciplinary research projects (Figure 14.1 and Table 14.1). All have the common aim of improving the health and quality of life of European consumers. The collaboration involves 64 different research groups from 16 different European countries and is coordinated by leading scientists. The research results from the cluster are disseminated through annual workshops and through the activities of three different platforms: a science, an industry, and a consumer platform (Figure 14.2 and website: http://proeuhealth.vtt.fi). The cluster started in 2001 and will end in 2005 (1–10).

14.2 Description of Projects

14.2.1 Developing Research Tools

14.2.1.1 Development and Application of High Throughput Molecular Methods for Studying the Human Gut Microbiota in Relation to Diet and Health — MICROBE DIAGNOSTICS

The microbial gut ecosystem has a major impact on the health and well-being of humans. Since the makeup and the activity of the gut microorganisms, collectively known as the microbiota, are highly influenced by diet, the project is designed to better understand its impact on human health and to provide easy-to-use methods for monitoring its composition. Novel methods are developed to understand and exploit the nutrition-driven impact of the human gut microbiota on health. Since the currently used methods for analyzing the intestinal microbiota are time consuming and tedious, high throughput methods for the automated detection of fluorescently labeled cells based on microscopic image analysis, flow cytometry, and DNA arrays

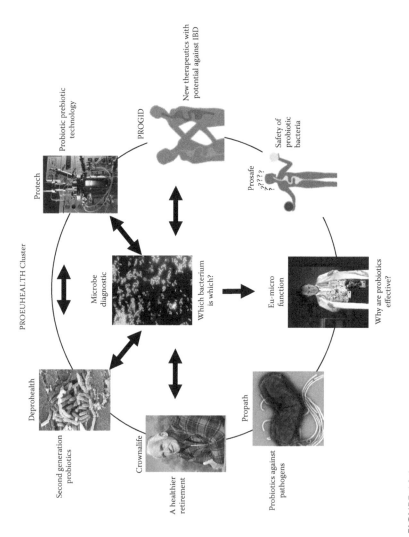

FIGURE 14.1
The layout of the PROEUHEALTH cluster projects.

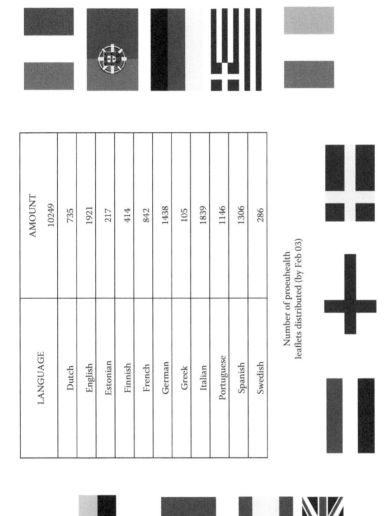

LANGUAGE	AMOUNT
Dutch	10249
English	735
Estonian	1921
Finnish	217
French	414
German	842
Greek	1438
Italian	105
Portuguese	1839
Spanish	1146
Swedish	1306
	286

Number of proeuhealth
leaflets distributed (by Feb 03)

FIGURE 14.2
Consumer platform dissemination efforts.

will be developed. *In situ* detection methods for monitoring bacterial gene expression at the cellular level will be developed to monitor the impact of dietary constituents on the transcription of bacterial genes. Samples from human populations will be analyzed with the developed methods to identify important factors of microbiota development, composition, and activity.

Objectives — The project is aimed at developing, refining, validating, and automating the most advanced molecular methods for monitoring the composition of the gut microbiota and bacterial gene expression in selected human populations in response to diet and lifestyle. The specific objectives of the project are (1) to improve and facilitate gut microbiota monitoring with molecular methods, (2) to understand antagonistic and synergistic interactions of the intestinal microbiota in response to nutrition, and (3) to find links between major dysfunctions and the intestinal microbiota.

Results and Future Targets — The project is expected to provide methods that allow the rapid detection of intestinal bacteria and their activities. The application of the developed methods will provide baseline data on the intestinal microbiota composition in response to origin and lifestyle. The project will also provide fundamental information on the biodiversity and phylogeny of the human gut microbiota. The data obtained will be used to develop a mechanistic concept for diet-induced microbiota development and to define the role of the intestinal microbiota in disease development. The project, thereby, contributes to investigate the role and impact of food on physiological function, the development of foods with particular benefits, and links between diet and chronic diseases.

During the first 2 years of the project, progress has been made in increasing the knowledge on the bacteria present in the human intestinal tract. This was performed by using a variety of growth substrates and growth media and by isolating more than 200 bacterial isolates using different enrichment approaches. Although the majority of these isolates belonged to known species, a considerable number of them represented novel species. Two isolates even led to the description of two new genera (*Dorea* and *Anaerostipes*). In parallel, additional sequences were retrieved by culture-independent methodology. All additional sequences have been implemented into private and public databases. This highly improved database has been used to develop new oligonucleotide probes for the culture-independent detection of intestinal bacteria. The panel of probes has been extended to subgroups of the large bacterial clusters in the gut. Probes have also been designed and validated to detect the new isolates and to determine their numerical importance and their distribution in human individuals.

An important aim in the project is to develop high throughput methods for the detection of fecal bacteria. MICROBE DIAGNOSTICS automated image analysis and flow cytometry-based methods are developed for the rapid detection and enumeration of fluorescent bacteria in fecal samples. Furthermore, the development of a high throughput approach based on the application of DNA arrays for the rapid detection of specific rRNAs of dominant human fecal bacteria has started. To better understand the processes related to the microbiota in the human gut, a first step has been taken in an attempt to visualize gene expression at the cellular level: intracellular PCR amplification of a plasmid-borne gene has been demonstrated (11–23).

MICROBE DIAGNOSTICS: highlights of the 2nd year

- Development of high throughput microarrays with over 100 different phylogenetic probes
- Isolation and characterization of several new GI tract isolates, 20 new species, 10 new genera
- Use of cell sorting
- Molecular methods to monitor functional gene expression
- Development of functional probes for determining *in situ* microbial glycohydrolase activity

14.2.2   Understanding Mechanisms

14.2.2.1   *Probiotic Strains with Designed Health Properties — DEPROHEALTH*

Lactic acid bacteria (LAB) are well known for their extensive use in the preparation of fermented food products. In addition, the potential health benefits they may exert in humans have been intensively investigated during the last century. However, the mechanisms underlying the health-promoting traits attributed to LAB remain vastly unknown, and this has impaired the rational design of probiotic screening methods with accurate predictive value. This project aims at establishing a correlation between *in vitro* tests and mouse models mimicking important human intestinal disorders such as inflammatory bowel disease (IBD), *Helicobacter pylori*, and rotavirus infections. As these diseases correspond to major public health problems, a second generation of probiotic strains with enhanced prophylactic or therapeutic properties will be designed in the project. These designed strains and the isogenic parental ones will be used to unravel mechanisms involved in the immunomodulation capacity of specific *Lactobacillus* strains.

Objectives — The general aim of this project is to acquire knowledge about the molecular factors affecting the immunomodulation and/or immunogenicity of selected probiotic lactobacilli to allow for developing isolates with enhanced protective or therapeutic effect. The objectives of the proposal are, on one hand, to unravel mechanisms and identify key components of the immunomodulation capacity of probiotic lactobacilli, and on the other hand, to design a second generation of probiotic strains with enhanced properties against GI disorders.

Two types of GI diseases are targeted: inflammations such as IBD and infections such as those caused by *H. pylori* and rotavirus. For each of them, a therapeutic or prophylactic (i.e., vaccine) recombinant probiotic strain will be constructed and tested in relevant animal models to evaluate its capacity to induce or modulate the immune response in the proper way. These data will be correlated to *in vitro* testing of the immunomodulation capacity with the attempt to identify and develop screening methods that will allow the identification of efficient probiotic strains. Recombinant DNA technology will also be used to assess the importance of specific bacterial cell wall components and adhesion factors in immunomodulation.

Results and Future Targets — In the project, two types of modified probiotic strains will be constructed: mutant strains affected in their cell wall composition and adhesion proteins as well as recombinant strains with enhanced therapeutic or protective properties, focusing on GI diseases of inflammatory or infectious origin. The final goal is to prove that designed probiotic strains can be used as original therapeutic agents. If successful, they would lead to novel anti-inflammatory treatments or oral vaccines against *H. pylori* and rotavirus. It is also expected that, by the end of the project, major progress will have been achieved toward the rational design of simplified probiotic screening methods with accurate predictive value.

In the project, different *in vitro* tests have been evaluated for their potential to classify probiotic strains according to their immunomodulation profile. This *in vitro* evaluation has been extended by studies conducted in mouse colitis models, which confirmed that given strains exert a beneficial effect on intestinal inflammation. A good correlation seems to exist between the *in vitro* peripheral blood mononuclear cultures (PBMC) stimulation assay and the 2,4,6-trinitrobenzene sulfonic acid-induced (TNBS) mouse model for strains that are characterized by a clear-cut anti- or proinflammatory profile. Mutants of *L. plantarum* NCIMB8826 synthesizing altered lipoteichoic acids and, as a consequence, a cell wall with different biochemical properties, have been obtained. One of the mutants seemed to exhibit a significantly enhanced anti-inflammatory profile compared to its parental strain. Recombinant strains producing a protective antigen from *H. pylori* or rotavirus have been obtained, and immunization experiments have been started in mice. Significant limitation of the *H. pylori* infection was obtained after intragastric immunization of mice with the best constructions (24–32).

DEPROHEALTH: highlights of the 2nd year:

- Protective effect of LAB in a colitis model
- Improved delivery of antigens by specific cell envelope mutations in selected LAB
- Delivery of anti-inflammatory molecule (IL-10) in human (approved genetically modified organism (GMO) trials ongoing)

14.2.2.2 Molecular Analysis and Mechanistic Elucidation of the Functionality of Probiotics and Prebiotics in the Inhibition of Pathogenic Microorganisms to Combat Gastrointestinal Disorders and to Improve Human Health — PROPATH

Recently, a lot of attention has been paid to the health-promoting properties of lactobacilli and bifidobacteria. However, a major problem is that many of these health-promoting properties are still questioned. For instance, the fundamental basis of the inhibition of Gram-negative pathogenic bacteria like the enterovirulent diarrheagenic *Salmonellae* and *H. pylori* causing gastritis and gastric ulcer disease by probiotic LAB has not been elucidated. This project will focus on the identification of the responsible inhibitory compounds. In addition, the mechanism of the inhibition of Gram-negative pathogens by probiotic lactobacilli and bifidobacteria will be studied.

Objectives — The aim is first to obtain a selection of probiotic lactobacilli and bifidobacteria that display a clear inhibition of diarrheagenic Gram-negative pathogenic bacteria and *H. pylori*. Thereafter, the project focuses on identifying the metabolite(s) responsible for the inhibition and/or killing of Gram-negative pathogenic bacteria as well as in elucidating the conditions and kinetics of the production of antimicrobials active toward Gram-negative pathogens and to predict their *in vivo* action. One of the aims will be to establish coculture models (simulated gut fermentations, human cell lines, animal models) and to study the interaction between inhibitory strains of LAB and Gram-negative pathogens causing GI disorders. Finally, the selected probiotic LAB will be tested in clinical studies.

Results and Future Targets — The results of the molecular analysis and mechanistic elucidation of the functionality of probiotics and prebiotics in the inhibition of pathogenic microorganisms will result in a collection of probiotic strains of lactobacilli and bifidobacteria. This will allow us to determine their characteristic and inhibitory spectrum, identify compounds responsible for the inhibition of Gram-negative pathogenic bacteria, determine conditions of antimicrobial production in the gut environment, and establish coculture models showing the inhibitory action by the probiotic strains.

From an initial project, culture collection of more than 850 strains, 15 strains (8 existing and 7 novel strains) inhibiting Gram-negative pathogenic bacteria have been selected. The existing strains either commercial probiotic strains or strains for which results indicating an inhibitory effect toward Gram-negative pathogens (*Escherichia coli*, *Salmonella enterica* serovar Typhimurium, *H. pylori*) were available, or strains were used for which validated literature data were available. The new isolates were derived from fermented foods and human or animal feces. The inhibitory effect of lactobacilli and bifidobacteria toward Gram-negative pathogens was dependent on the culture conditions. Certain conditions allowed discriminating between the pH-lactic acid effect and the effect probably by secreted antimicrobial compounds other than lactic acid.

Bacteriocin preparations from *L. johnsonii* LA1 and *L. amylovorus* DCE 471 displayed a clear killing effect toward *H. pylori.* Both the clinical isolates tested and the reference strains of *H. pylori* were sensitive for these two bacteriocins. The bacteriocin preparation of a culture of *B. longum* BB536 gave a narrow inhibition zone against *E. coli* C1845 only. The other bacteriocin-positive *Bifidobacterium* strains inhibited several Gram-positive indicators. Bacteriocins were isolated from *L. acidophilus* IBB 801, *L. casei* YIT 9029, *L. gasseri* K7, and *L. johnsonii* LA1. The bacteriocins were heat stable, stable in a pH range from 2.2 to 8.0, and sensitive to proteolytic enzymes, but not to other enzymes. They possessed a molecular mass between 3.0 and 6.5 kDa (33–43).

PROPATH: highlights of the 1st year:

- High throughput screening antimicrobial activity against *H. pylori* and *Salmonella*
- Germ-free mice studies of colonization by LAB
- New antimicrobial compounds discovered

14.2.2.3   *Functional Assessment of Interactions between the Human Gut Microbiota and the Host — EU & MICROFUNCTION*

The aim of this project is to identify the effects of dietary modulation on the human GI microbiota. The influence of live microbial food supplements (probiotics), dietary carbohydrates known to have a selective metabolism (prebiotics), and a combination of these two approaches (synbiotics) will be ascertained. The main objectives are to clarify effects on the normal gut microbiota, and on host GI function, as well as determine mechanisms involved in pro-, pre-, and synbiotic functionality. These objectives will be achieved through exploitation of model systems and state-of-the-art technology.

Objectives — The principal objectives are to determine the efficacy and safety of probiotics and prebiotics, determine effective doses and combinations, identify mechanisms of action, and investigate impacts on host function. An important aspect will be the development of new synbiotics and their use in a human trial. The work planned aims to identify the mechanisms of effect and to produce valuable information on the influence of dietary intervention on the activities of human gut microbiota. In addition, it provides essential means of validating probiotics and prebiotics, and will give information on the optimal combinational approach.

There is currently an imperative requirement to identify the realistic health outcomes associated with probiotic and prebiotic intake and, importantly, give rigorous attention toward determining their mechanisms of effect. This project will aim to do so through investigating probiotic and prebiotic influence on host functionality, including microbiological and physiological aspects.

Results and Future Targets — The aim is to identify mechanistic interrelationships through fundamental scientific approaches. The following milestones will be achieved: efficient prebiotics and required dosage, active synbiotics, effects of functional foods on bacterial translocation, effects on host gene interactions, safety of functional foods determined, and effects on selected health indices in humans.

During the first project year, useful prebiotics were identified with batch culture fermentations. The work has continued using complex models of the human colon. Novel prebiotics are also being manufactured and compared. Administration of different probiotics in an acute liver injury and an induced colitis rat model showed different protective effects upon bacterial translocation and disease status. The antagonistic activity (against 8 common gut pathogens) of lactobacilli and bifidobacteria has been determined. Fermentation profiles, antioxidant capacity, amine formation, and inhibitory effects of the probiotics against one another have been determined. In the work on microbe-host interactions, gut bacteria able to utilize mucin have been identified. Microbial diversity of mucin degradation has been identified, including the isolation of a new genus within mucin-degrading bacteria. In the work package dealing with safety issues, a collection of *Bifidobacterium* and *Lactobacillus* strains has been subjected to several safety criteria. No safety issues have so far arisen. In the work package on synbiotic administration in humans, ethical permission for a human trial to test the effects of a synbiotic on selected health parameters has been given and recruitment is proceeding (44–47).

EU & MICROFUNCTION: highlights of the 1ˢᵗ year

- Cross talk between microbes and host gut tissues
- Optimal combination of prebiotics and probiotics determined
- New mucus-degrading bacterium isolated
- Human trial on synbiotics commenced

14.2.3   Investigating Health Effects

14.2.3.1   *Probiotics and Gastrointestinal Disorders: Controlled Trials of European Union Patients — PROGID*

This project will assess the efficacy of two previously selected probiotic microorganisms, administered as dried fermented milk products, in alleviating the effects of IBD–Crohn's disease and ulcerative colitis. Specifically, two distinct long-term (12 month), large-scale, multicentered, randomized, double-blind, placebo-controlled feeding trials will be performed within a subset of the European Union population suffering from these GI disorders.

Objectives — In the assessment of probiotic efficacy, most studies performed so far have been small, uncontrolled, and poorly documented with imprecise definition of the endpoints. In response, the European Commission has funded the PROGID project, which will evaluate two specific probiotic microorganisms in patients with ulcerative colitis or Crohn's disease from diverse geographical locations. The participating centers in these studies will assess the efficacy of *Bifidobacterium infantis* UCC35624 and *Lactobacillus salivarius* UCC118 in 1-year, randomized, double-blind, placebo-controlled trials for maintenance of remission. Both of the selected probiotic strains have a history of safety and efficacy in healthy adults and relapsed IBD patients.

Results and Future Targets — The anticipated results from the PROGID project include: the provision of qualitative and quantitative evidence that the evaluated probiotics may (or may not) have a role in maintaining remission of IBD; confirmation, or otherwise, of the involvement of specific members of the GI microbial flora populations as causative or contributory agents of IBD; generation of linear physiological and immunological data relevant to the disease and remission states of IBD; creation of a greater awareness among the EU population of functional foods and their potential benefits in maintaining healthy lifestyles; and the establishment of a repository of biological samples obtained from across the EU.

During the first 2 years of the PROGID project, partners have collected extensive clinical databases of IBD patients. Identification of a large database of patients is essential for a clinical trial to be successful. By identifying this number of patients as potential study participants, the ambitious target to enroll 360 patients is achievable. From those recruited, about 200 patients have been enrolled (as of January 2003) and are currently either feeding, have completed the study, or have relapsed.

The ulcerative colitis clinical trial has proceeded well, with 120 patients enrolled. Two-thirds of the targeted 180 patients have been enrolled in the ulcerative colitis study, and it is anticipated that this target will be achieved. The Crohn's disease study started later than the ulcerative colitis clinical trial, and patients are currently (as of January 2003) at various stages throughout the study.

Biological samples for microbiology and immunology analyses are continuously being obtained from patients participating in both the Crohn's disease and ulcerative colitis studies. These samples are being appropriately stored and are undergoing analysis.

In addition to conventional disease activity indices, PROGID is also exploring the use of fecal DNA, cytokines, and calprotectin as markers of subclinical disease activity. PROGID partners are also looking at molecular techniques for surveying the gut microbiota in patients with active disease, those in remission, and those taking probiotics and placebo. These results will provide not only data on the efficacy of probiotics, but also data on compliance and the microbiota of the patients over a prolonged period of time.

Due to the nature of the studies (double blinded), no analysis of the results have been undertaken while the study is ongoing; otherwise the blindness of the study may be sacrificed. Hence, definitive results will not be available until the study has been completed and the patients' codes broken (48–58).

PROGID: the highlights of the 2nd year:

- Ulcerative colitis and Crohn's disease patients enrolled for an approved, EU-wide trial
- Gene-based biomarker for mucosal lesions (β-globin)
- Ulcerative colitis and Crohn's disease patients show unstable intestinal microbiota

14.2.3.2 Functional Food, Gut Microflora, and Healthy Aging — CROWNALIFE

Elderly people represent an increasing fraction of the European population. Their higher susceptibility to degenerative and infectious diseases has led to rising public health and social concerns. Appropriate preventive nutrition strategies can be applied to restore and maintain a balanced intestinal microbiota exerting protective functions against the above disorders. The project is based on the application of selected biomarkers in hypothesis-driven human studies. We will identify the structural and functional specificity of the elderly's intestinal microbiota across Europe. Using this baseline information, we will investigate functional food-based preventive nutrition strategies aiming to beneficially affect the functional balance of the elderly's intestinal microbiota. Expected outcomes include nutritional recommendations as well as new concepts and prototype functional foods, specifically adapted for health benefits to the elderly population.

Objectives — The overall objective is to improve the quality of life of the elderly throughout the third age, with emphasis on the preservation of the period of independence recognized as the crown of life. The focus is on preventive nutrition and the application of functional food to derive health benefits for the ever-increasing European elderly population. Based on hypothesis-driven human studies, the specific objectives of the project are to assess structural and functional alterations of the intestinal microbiota with aging and across Europe, and to validate functional food-based preventive nutrition strategies to restore and maintain a healthy intestinal microbiota in the elderly. Implementations include nutritional recommendations as well as new concepts and prototype functional food specifically adapted for health benefits to the elderly population.

Results and Future Targets — The results will establish six steps: (1) the assessment of the gut microbiota diversity and composition in the European elderly, and identification of its alterations with aging (baseline human study); (2) the assessment of modulation of the intestinal microbiota (intervention human study with functional food), and potential health benefits toward degenerative pathologies and infectious diseases; (3) the improved health status of the aging population via specific nutritional recommendations; (4) the nutritional guidelines based on the complete assessment of a synbiotic product; (5) the design and provision of adapted food supplements directed toward intestinal microbial function; and (6) the validation of processes and rationale for the design of a new generation of functional foods to satisfy health benefits for the elderly, based on innovative technologies.

Results obtained so far have been focused on the assessment of specificities of the gut microbiota in the elderly: About 100 bacterial strains have been isolated from the elderly gut microbiota, among which totally novel (yet unknown) strains of bifidobacteria and lactobacilli have been identified. *In vitro* growth inhibition tests using *Campylobacter jejuni*, *Escherichia coli* (both EPEC and VTEC) and *Clostridium difficile* as target organisms indicated that 15 strains could be considered as potential probiotic strains. Among these, a set of 5 strains showed the highest inhibitory activities and have been selected for further analysis. An extreme microbial species diversity was evidenced within the elderly gut microbiota in comparison to former data from younger age groups. Totally new lines of descent (clusters and/or species) were identified within the elderly gut microbiota, with a high prevalence.

The synbiotic mixture of *Bifidobacterium animalis* DN-173 010 and prebiotic (Raftilose Synergy1) to be used in the trial with the elderly has been optimized and produced. Elderly persons and young adults have completed the Baseline Human Study of CROWNALIFE, which took place in Germany, Italy, Sweden, and France. Elderly persons have also completed the Intervention Human Study of CROWNALIFE which took place in Germany (59–73).

CROWNALIFE: highlights of the 2nd year:

- New microbes isolated and characterized from the elderly
- Novel bacterial diversity in the gut microbiota of elderly individuals has been detected based on the large 16S rDNA sequence library
- Base-line studies in the elderly have commenced
- Novel probiotic isolates have been isolated (15 candidate probiotics)
- Twenty recommended probes for assessing the composition of the gut microbiota in the elderly have been generated

14.2.3.3 *Biosafety Evaluation of Probiotic Lactic Acid Bacteria Used for Human Consumption — PROSAFE*

Safety assessment is an essential phase in the development of any new pro- or prebiotic functional food. The safety record of probiotics is good, and lactobacilli and bifidobacteria have a long history of safe use. However, all probiotic strains must be evaluated for their safety before being used in human clinical studies and in functional food products. Conventional toxicology and safety evaluation alone is of limited value in the safety evaluation of probiotic bacteria. Instead, a multidisciplinary approach is necessary involving contributions from pathologists, geneticists, toxicologists, immunologists, gastroenterologists, and microbiologists.

Objectives — The safe use for human consumption of probiotic strains, selected LAB including lactococci, lactobacilli, pediococci and bifidobacteria, and other food-associated microorganisms such as enterococci will be assured by proposing criteria, standards, guidelines, and regulations. Furthermore, procedures and standardized methodologies of premarketing biosafety testing and postmarketing surveillance will be provided. The specific objectives of the project will include five stages: (1) the taxonomic description of probiotics, LAB, and other food-related microorganisms; (2) the detection of resistance and horizontal transfer of antibiotic resistance genes; (3) the careful analysis and definition of the nonpathogenic status of probiotic LAB; (4) the immunological adverse effects of the studied bacteria in the experimental allergic encephalomyelitis (EAE) model, and (5) the survival, colonization, and genetic stability of probiotic strains in the human gut.

Results and Future Targets — Future results will finalize important goals relating to probiotic strains, selected LAB, and other food-associated microorganisms and consists of establishing a culture collection and database; providing standardized methodologies to detect antibiotic resistance; investigating the potential virulence properties, and their association with clinical disease and results obtained in rat endocarditis models; studying potential immunological adverse effects; analyzing the genetic stability and colonization of probiotic strains in the human GI tract; and providing recommendations for biosafety evaluation of probiotic strains.

During the first project year, a strain collection of more than 750 strains was set up. The strains belong mainly to the following genera: *Lactobacillus, Bifidobacterium, Enterococcus, Lactococcus,* and *Pediococcus,* and they were obtained from PROSAFE and PROEUHEALTH partners, from industries, culture collections, and researchers. The strains are of nutritional (probiotic, food), human (mainly fecal and blood isolates), and animal origin.

Characterization of the strains focused on members of the genera *Lactobacillus* and *Bifidobacterium,* which are considered to be taxonomically most complex and for which identification problems were to be expected. For the identification and fingerprinting of *Lactobacillus* and *Bifidobacterium,* an amplified fragment length polymorphism (AFLP) method and an interspersed repetitive sequences (rep)-based PCR method, respectively, were developed. Eighty-six percent of the submitted bifidobacteria and 99% of the submitted lactobacilli were successfully identified to the species level.

During the first project year, over 100 strains of enterococci and almost 300 lactobacilli were tested for their antibiotic susceptibility to 16 antibiotics. Minimal inhibitory concentrations (MICs) were determined for all antibiotic classes including the penicillins, aminoglycosides, glycopeptides, quinupristin/dalfopristin (Q/D), macrolides, and tetracyclines.

For the enterococci, these tests were performed according to the National Committee for Clinical Laboratory Standards (NCCLS) guidelines. For lactobacilli, however, no standardized and validated methods are available, and, therefore, a new procedure for antibiotic susceptibility testing of these organisms was developed. In parallel to the MIC determinations of enterococci and lactobacilli, PCR for detection of a wide range of antibiotic resistance genes was established, including the following genes: *aadE-aphA, aadE, aph2"-aac6', cat pC194, cat piP501, ermA, ermB, ermC2, satA (= vatD), satG (= vatE), vanA, tetK2, tetL, tetM*.

Among the *Enterococcus faecium* strains tested, seven were probiotic and only one possessed resistance (only to erythromycin, but not mediated by *ermA, ermB,* or *ermC*). The staphylococcal transposon *Tn554* with *ermA* was detected for the first time in an *E. faecium* strain; *erm*B-mediated ERY resistance was frequently found in enterococci. In a few *E. faecium* isolates, *satA (vatD)* could be detected. Other resistance genes that were present in the examined enterococci were *aadE-aphA, aadE, aph2"-aac6', cat pC194, cat piP501, vanA, tetL, tetM*. Of these latter-mentioned genes, *aadE, catPC194, ermC, vanA,* and *tetK2* were not found among the 7 *E. faecalis* strains.

Among the lactobacilli examined, resistance frequencies were low except for the glycopeptides, fusidic acid, and cotrimoxazole. The lactobacilli tested used as probiotics (n = 81) and in nutrients (n = 7) were separately evaluated, and these isolates showed similar resistance frequencies as the other lactobacilli of human origin.

In the project, a multiplex PCR system for the rapid detection of four known virulence factors of *E. faecalis* (gelatinase, enterococcal surface protein, cytolysine, and aggregation substance) was also developed. A PCR has also been developed to rapidly detect a potentially new virulence gene, hyaluronidase(*hyl*) in *E. faecium*. (74, 75)

PROSAFE: highlights of the 1st year:

- Six hundred strains obtained from the industry and scientists
- AFLP typing and antibiotic susceptibility testing of most isolates performed
- Database establishment of probiotic safety commenced

14.2.4   Probiotic and Prebiotic Technologies

14.2.4.1   *Nutritional Enhancement of Probiotics and Prebiotics: Technology Aspects on Microbial Viability, Stability, Functionality, and on Prebiotic Function — PROTECH*

Maintaining the functional properties of probiotics during manufacture, formulation, and storage are essential steps in delivering health benefits to consumers from these products. The overall objective of this project is to address and overcome specific scientific and technological hurdles that impact the performance of functional foods based on probiotic–prebiotic interactions. Such hurdles include the lack of a strong knowledge on the primary factors responsible for probiotic viability, stability, and performance. Limited information is available on the impact of processing, storage, and of food matrices or food constituents on probiotic viability, stability, and functionality. Furthermore, sufficient data are missing about the interactions between probiotics and prebiotics in functional foods prior to consumption.

Objectives — This project has three general objectives: to explore effects of processing on the stability and functionality of probiotics and on the performance of prebiotics, to apply selected processing techniques for prebiotic modification to identify and optimize probiotic–prebiotic combinations, and to use the information generated as the basis for new process and product options.

The specific objectives of the project are to consolidate quantitative data of processing effects on the physiology and viability of probiotic organisms, to investigate the influence of media composition and processing conditions on probiotic stability, to optimize probiotic–prebiotic interaction for maximum probiotic performance, to develop new prebiotics tailor-made for the stabilization and optimum performance of probiotics, and to examine the effect of growth conditions and stress treatments on the functionality of probiotic bacteria.

Results and Future Targets — Expected achievements include the establishment of unique data sets that include the identification of critical process parameters for probiotics and prebiotics and results from systematic studies suggesting means to overcome existing process and product limitations. The compilation of protocols for probiotic performance, prebiotic function, and probiotic–prebiotic interactions will also be provided. Furthermore, it is expected to establish probiotic viability models and functionality biomarkers. In addition, it is attempted to achieve optimization of probiotic viability, stability in culture, and real food systems at pilot-plant scale, generation and modification of unique prebiotics, of probiotic interactions, and of environmentally and processing-induced functionality of probiotics. Application of the expected results will lead to new process concepts for probiotics, for prebiotics, and for probiotic–prebiotic combinations. Special emphasis of the development of product concept will be on cereal and dairy-based products and on the development and incorporation of unique plant-based prebiotics for optimum interaction between prebiotics' performance and probiotics' function.

During the first 2 project years, the following results were obtained: a better understanding of storage stability and sensitivity of freeze-dried probiotics. This included development of suitable low-cost food-grade growth media for the strains and data on the effects of cryoprotective agents on probiotic stability. Studies on the response of probiotic strains to stress treatments were conducted, and potential stress markers for monitoring probiotic stability were identified. Furthermore, drying and shelf-life stability studies on stress-treated probiotics were performed, and data on pressure effects on probiotics were compiled. Studies on prebiotics included prebiotic feeding trials on rats. To develop novel prebiotics, controlled enzymatically catalyzed transfer reactions to obtain specific galactooligosaccharides were performed (76–84).

PROTECH: highlights of the 2nd year:

- Database of viability, stability, stress factors in progress
- Detection of strain-specific synbiotics
- Improved survival of probiotics through induced tolerance
- Functional genomics analysis of stress response

14.3 Consumers and Perceived Health Benefits of Probiotics

Science-based knowledge on how probiotic bacteria can promote our well-being is increasing rapidly. The success of probiotic products, however, will be determined by consumers' willingness to buy and eat them. The perceived benefits in these new products are key factors for consumer acceptance. Functional foods, i.e., products promising specified effects on physiological functions represent a new kind of health message for consumers. In traditional nutritional messages, the emphasis is on diet and on avoiding or favoring certain types of foods rather than giving recommendations on any particular products. Thus, it is not surprising that general health interest, which measures people's willingness to comply with nutritional recommendations, has not been strongly linked with willingness to use functional foods. Functionality can give additional value to the product, but still choice decisions within food categories are also based on taste, convenience, price, and familiarity.

Developing probiotic foods also introduces a novelty aspect to the product. In general, consumers tend to be suspicious toward novelty in foods because it means uncertainty and threat. Therefore, the benefits promised by these new food products will be weighed against the perception of possible risks. Highly developed technologies are often involved in manufacturing new probiotic foods. In several European countries, healthiness has been associated with naturalness. The need for highly advanced technology in the production of probiotic foods may lower the perceived naturalness, which may create distrust toward functional foods among some consumers. Consumers seem to find functions that enforce the natural properties of a product more acceptable than those functions that are artificially added or are not in accordance with the earlier image of the product. Therefore, in addition to technological considerations, adding probiotics into food products requires understanding of the existing beliefs consumers have about these products.

The claims attached to probiotics or any functional products need to be based on adequate and sound scientific evidence produced by a set of studies. Results of these studies are often based on probabilities, and translating the likelihood into consumer language is a complex task, since the scientific and everyday thinking differ from each other radically. Consumers favor clear-cut thinking: something either is good or bad for you, but not good in some quantities and bad if eaten too little or too much, because understanding dose responses is hard. Consumers also tend to make assumptions from the data that are easily available for them rather than trying to consider all valid points from different angles as the scientific approach requires. Consumers tend to be wary of linking food with reduction of disease, as this implies associating food that is commonly considered as a source of pleasure with something unpleasant, even if the promised effect would be positive. This conforms with another typical feature in everyday thinking, one which tends to associate any two things that appear at the same time regardless whether they are causally connected or not. Food and health, in general, are both very sensitive topics, and this adds to the challenge of communicating possible health-related claims to consumers in comprehensible, usable, and nonalarming manner.

The aim of the consumer platform in the PROEUHEALTH cluster is to provide consumers with information about the ongoing EU-funded research on probiotics. The core activity for spreading the information is the website (proeuhealth.vtt.fi). The platform has converted the objectives of the eight scientific cluster projects into consumer language, and the one-page leaflet about the projects can be downloaded from the website. The leaflets have been translated into major EU languages, and over 10,000 copies have been distributed around Europe as of February 2003 (Figure 14.3). About half the copies have been downloaded from the website, and the other half have been sent out as paper copies. The website can be freely accessed and provides a channel to send messages to scientists working on the projects. The biggest interest has been in the possibility to alleviate the symptoms of intestinal disorders with probiotics. The progress of the projects is being followed, and the platform generates short pieces of news on each project. In addition to the website, these leaflets are sent out to scientific news services, consumer organizations, and health professionals for further dis-

TABLE 14.1

Probiotics, Prebiotics, and Target Pathogens, Models and Population Groups for Probiotic Function and/or Microbial Analysis in the PROEUHEALTH Cluster Projects

Project	Probiotic Strains	Prebiotics	Target Pathogens, Models and Population Groups for Probiotic Function and/or Microbial Analysis
MICROBE DIAGNOSTICS	—	—	Samples from healthy volunteers and from IBD (inflammatory bowel disease; ulcerative colitis) patients
DEPROHEALTH	Recombinant *Lactobacillus* strains *L. plantarum* NCIMB8226 *L. crispatus* 247 & Mu5 *L. casei* BL23 *L. fermentum* PC1	—	Mouse models on IBD Mouse models on *H. pylori* and rotavirus infections
PROGID	*L. salivarius* UCC118 *B. longum* (*infantis*) UCC35624	—	Clinical trial with IBD (Chrohn's disease and ulcerative colitis) patients
CROWNALIFE	*B. animalis* DN-173010 15 novel probiotic strains (both *Lactobacillus* and *Bifidobacterium* strains)	Fructooligosaccharides/inulins: Synergy1	*In vitro* studies with *C. jejuni*, *E. coli* (EPEC, VTEC), and *C. difficile* Volunteers aged 25–45 and over 65 Clinical trial with a synbiotic pair in elderly volunteers
PROTECH	*L. rhamnosus* E-97800 *L. rhamnosus* GG *L. salivarius* UCC500 *B. animalis* (*lactis*) Bb-12	Fructo-oligosaccharides/inulins: Raftiline HP & HPX, Raftilose P95, Synergy1 Polydextroses Pectins (RS 1400, LA 2560) Lactose derivatives: lactitol, lactulose Novel prebiotics generated in the project	Prebiotic (+probiotic) feeding trials in rats

PROPATH	15 existing and novel strains including, e.g., L. johnsonii La1 L. acidophilus IBB 801 L. amylovorous DCE 471 L. casei Shirota L. rhamnosus GG Bifidobacterium sp. F9 and CA1 B. longum BB536	Fructooligosaccharides	In vitro studies with Gram-negative pathogens, mainly E. coli, S. enterica serovar Typhimurium, H. pylori Cell line and mouse model studies with S. enterica serovar Typhimurium, H. pylori Clinical trials with Helicobacter pylori-positive dyspepsia patients, irritable bowel syndrome (IBS) subjects, and infants with acute diarrhea
EU & MICROFUNCTION	L. plantarum 299v L. fermentum ME-3 L. rhamnosus GG L. casei 8700:2 L. gasseri LG1 L. acidophilus La5 B. animalis (lactis) Bb-12 B. longum B46 - other Lactobacillus and Bifidobacterium strains	Fructooligosaccharides/inulins: Raftilose P95, Actilight, Synergy1, Raftiline HP & ST, inulin Blue Agave Galactooligosaccharides Xylooligosaccharides Isomaltooligosaccharides Polydextrose Maltodextrin Lactose derivatives: lactitol, lactulose Novel prebiotics generated in the project	In vitro studies with gut pathogens such as Salmonella spp., Shigella spp., E. coli, H. pylori and C. difficile Rodent models on translocation (rat models for acute liver injury and colitis) Human translocation studies in colon cancer patiens Clinical trial with a synbiotic pair in volunteers aged 20–50
PROSAFE	- collection of over 700 strains of LAB including, e.g., probiotic strains used in other cluster projects and commercial, food, human and animal LAB isolates	—	—

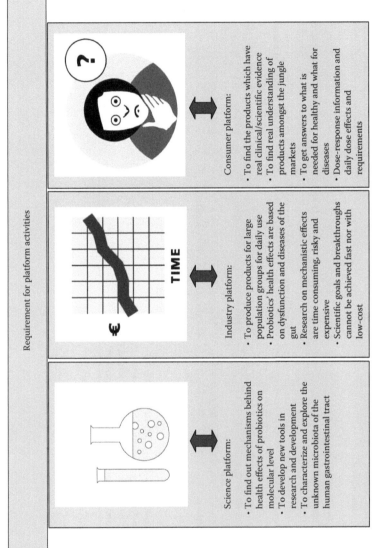

Requirement for platform activities

Science platform:
• To find out mechanisms behind health effects of probiotics on molecular level
• To develop new tools in research and development
• To characterize and explore the unknown microbiota of the human gastrointestinal tract

Industry platform:
• To produce products for large population groups for daily use
• Probiotics' health effects are based on dysfunction and diseases of the gut
• Research on mechanistic effects are time consuming, risky and expensive
• Scientific goals and breakthroughs cannot be achieved fast nor with low-cost

Consumer platform:
• To find the products which have real clinical/scientific evidence
• To find real understanding of products amongst the jungle markets
• To get answers to what is needed for healthy and what for diseases
• Dose-response information and daily dose effects and requirements

Three platforms will disseminate the aims and findings of the cluster to targeted audiences.

FIGURE 14.3
Cluster platforms are designed to meet the needs of different audiences.

semination. To encourage real interaction, the representatives of consumer organizations and relevant patient organizations are invited to present their viewpoints at cluster workshops.

Consumer responses to functional foods and their health-related messages require more research. Communicating the gut health messages effectively to consumers with varying ages and cultural backgrounds remains one of the key issues. Besides this, we need tremendous efforts toward communication and cooperation among medical doctors, nutritionists, and pharmaceutics. As information is a vital factor, trust between all sides is required, so that messages will be given attention. The differences between functional properties of various strains are difficult to explain to consumers, and tools for doing this should be created. For some consumers, food products that work almost like medicines may be hard to approve. In the consumer platform, we have tried to address the above issues so that research and the results gained on probiotics become more familiar and comprehensible for consumers. Probiotic products can improve human health only if they can be trusted and are consumed frequently (85–91).

14.4 Conclusions

Future scientific development will provide the basis for rational development of foods with increased functionality. This notably holds for foods that modulate the GI tract, the primary site of food conversion and uptake where food-borne or systemic disorders are abundant and the innate immune system is stimulated. The body's metabolically most active organ is colonized by a myriad of microbes that contribute to food conversion, communicate with the host, and induce specific responses that contribute to a wide variety of important physiological functions. Hence, the gut represents the site where the host's well-being is affected by foods either directly or by modulating the interplay between microbes and host.

The prime scientific development includes the genomics-related and high throughput technologies that generate new approaches aimed to provide insight into the molecular mechanisms of food functionality, gut health, and microbial function. These are expected to be instrumental in developing new generations of health and other functional foods. In addition, these can be used in human trials in EU patients with various gut-related disorders so that the fruits of the last decade of research can be brought to the realm of evidence-based medicine.

Insight in the basic mechanisms of gut health is also required to further develop biomarkers for microorganisms with specific functions such as probiotics, defined as living microorganisms that beneficially affect the gut balance, as well as prebiotics and other adjunct dietary components that stimulate specific microbial groups in the human gut. Moreover, fundamental knowledge of the consumer's intestinal microbiota will provide targets for developing health foods as well as their production or delivery. This is essential for further innovations in this area where significant scientific progress on health impact has been made that now needs to be backed up by mechanistic explanations.

Heavy use of antimicrobials is leading to serious problems with transferable resistance, and this results in the compelling need to investigate alternative approaches to prevent and treat infections. Evidence is mounting that the use of probiotics may provide a viable alternative adjunct to antimicrobials in antibiotic-associated disorders such as various forms of diarrhea, *Clostridium difficile* infection, or irritable bowel syndrome.

Future research on food and microbes with health impact will continue to develop, specifically aimed at (1) exploring the mechanisms of action of microbes and their health effects in the GI tract, especially in healthy individuals; (2) developing sophisticated diagnostic tools for the gut microbiota and biomarkers for assessing their functionality; (3) examining the effects of food-derived bioactive compounds on GI diseases, GI infections, and allergies; (4) developing new therapeutic and prophylactic treatments for different patient and population groups; (5) realizing molecular understanding of immune modulation by bacteria with health effects; (6) elucidating the role of colon microbiota in the conversion of bioactive compounds; (7) analyzing the effects of the metabolites to the colon epithelium or the effects after absorption; (8) ensuring the stability of microbes with health effects and their bioactive compounds also in new types of food applications by developing feasible technologies; and finally (9) providing safe functional ingredients.

The acceptance of probiotic, prebiotic, and symbiotic products in the future will depend on the solid proof of the health benefits they promise at the moment. Therefore, research that provides scientifically sound evidence to back up health claims is needed. How to make the knowledge produced by science comprehensible to the consumer is the major challenge, as the layman's thinking is based on approximation and black-and-white views, whereas scientific thinking deals with probabilities and degrees of uncertainty. Consumers' trust of the information depends on the source and content of the message. The critical point is that producers of probiotic foods can gradually build and ensure consumer trust.

References

1. AM Kuokka, M Saarela, T Mattila-Sandholm (Eds). VTT Symp. 219. *The Food, GI-tract Functionality and Human Health Cluster PROEUHEALTH, 1st Workshop, Saariselkä, Finland, 1–3 February 2002*, VTT, Espoo, Finland. ISBN 951-38-5729-8, 65 pp.

2. AM Kuokka, M Saarela, T Mattila-Sandholm (Eds). VTT Symposium 226. *The Food, GI-tract Functionality and Human Health Cluster, PROEUHEALTH, 2nd Workshop, Taormina, Italy, 3–5 March 2003*, VTT Espoo, Finland. ISBN 951-38-6276-3, 66 pp.

3. L Lähteenmäki. Consumers and functional foods. In: T Mattila-Sandholm, M Saarela, Eds., *Functional Dairy Products*. Woodhead Publishing, Cambridge, UK, 2003, pp. 346–357.

4. T Mattila-Sandholm, M Blaut, C Daly, L De Vuyst, J Dore, G Gibson, H. Goossens, D Knorr, J Lucas, L Lähteenmaki, A Mercenier, M Saarela, F Shahanan, WM de Vos. The food, GI- tract functionality and human health cluster. *Microb. Ecol. Health Dis.*, 14:65–74, 2002.

5. T Mattila-Sandholm, M Saarela, L Lähteenmaki. The food, GI-tract functionality and human health cluster. In: T Mattila-Sandholm, M Saarela, Eds., *Functional Dairy Products*. Woodhead Publishing, Cambridge, UK, 2003, pp 359–375.

6. T Mattila-Sandholm, M Saarela, L Lähteenmaki. The food, GI- tract functionality and human health cluster. In: I Goktepe, V Juneja, M. Ahmeline, Eds., *Probiotics in Food Safety and Human Health*. Taylor & Francis, Boca Raton, FL, (in press).

7. A Mercenier, T Mattila-Sandholm. The Food, GI-tract Functionality and Human Health European Research Cluster PROEUHEALTH. *Nutr. Metab. Cardiovasc. Dis.*, 11(4):1–5, 2001.

8. R Puupponen-Pimiä, A-M Aura, K-M Oksman-Caldentey, P Myllärinen, M Saarela, T Mattila-Sandholm, K Poutanen. Development of functional ingredients for gut health. *Trends Food Sci. Technol.*, 13:3–11, 2002.

9. M Saarela, L Lähteenmäki, R Crittenden, S Salminen, T Mattila-Sandholm. Gut bacteria and health foods: the European perspective. *Int. J. Food Microbiol.*, 78:99–117, 2002.

10. M Saarela, J Mättö, T Mattila-Sandholm. Safety aspects of Lactobacillus and Bifidobacterium species originating from human oro-gastrointestinal tract or from probiotic products. *Microb. Ecol. Health Dis.*, 14:233–240, 2002.

11. M Blaut. Development and application of high throughput molecular methods for studying the human gut microbiota in relation to diet and health. *The Food, GI-tract Functionality, and Human Health Cluster PROEUHEALTH, 1st Workshop, Saariselkä, Finland, 1–3 February 2002*, VTT Symp. 219, VTT Espoo, Finland, pp. 17–18.

12. M Blaut, MICROBE DIAGNOSTICS: Who's who in the intestinal microbiota. *The Food, GI-tract Functionality and Human Health Cluster PROEUHEALTH, 2nd Workshop, Taormina, Italy. 3–5 March 2003*, VTT Symp., VTT Espoo, Finland, 226, pp 17–18.

13. SM Finegold, M-L Väisänen, DR Molitoris, Y Song, C Liu, MD Collins, PA Lawson. *Cetobacterium somerae* sp. nov., from human faeces and emended description of the genus *Cetobacterium*. *Syst. Appl. Microbiol.*, 26:177–181, 2002.

14. NA Fitzsimons, ADL Akkermans, WM de Vos, EE Vaughan. Bacterial gene expression detected in human faeces by reverse-transcription-PCR, J. Microbiol. Meth. 55:133–140, 2003.

15. S Konstantinov, N Fitzsimons, EE Vaughan, ADL Akkermans. From composition to functionality of the intestinal microbial communities. In: GW Tannock, Ed., *Probiotics and Prebiotics: Where Are We Going?* Horizon Scientific, London, 2002, pp. 59–84.

16. L Rigottier-Gois, A-G Le Bourhis, G Gramet, V Rochet, J Doré. Fluorescent in situ hybridisation combined with flow cytometry and hybridisation of total RNA to analyse the composition of microbial communities in human feces using 16S rRNA probes. *FEMS Microbiol. Ecol.*, 1465:1–9, 2002.

17. RM Satokari, EE Vaughan, M Saarela, J Mättö, WM de Vos. Molecular approaches for the detection and identification of bifidobacteria and lactobacilli in the human gastrointestinal tract, Syst. Appl. Microbiol. 26:572–584, 2003.

18. A Schwiertz, GL Hold, SH Duncan, B Gruhl, MD Collins, PA Lawson, H Flint, M Blaut. *Anaerostipes caccae* gen. nov., sp. nov., a new saccharolytic, acetate-utilising, butyrate-producing bacterium from human faeces. *Syst. Appl. Microbiol.*, 25:46–51, 2002.

19. A Schwiertz, U Lehmann, G Jacobasch, M Blaut. Influence of resistant starch on the SCFA production and cell counts of butyrate-producing *Eubacterium* spp. in the human intestine. *J. Appl. Microbiol.*, 93:157–162, 2002.

20. R Simmering, D Taras, A Schwiertz, G Le Blay, B Gruhl, PA Lawson, MD Collins, M Blaut. *Ruminococcus luti* sp. nov., isolated from a human faecal sample. *Syst. Appl. Microbiol.*, 25:189–193, 2002.

21. T Steer, MD Collins, GR Gibson, H Hippe, PA Lawson. *Clostridium hathewayi* sp nov., from human faeces. *Syst. Appl. Microbiol.*, 24:353–357, 2001.

22. EE Vaughan, MC de Vries, EG Zoetendal, K Ben-Amor, ADL Akkermans, WM de Vos. The intestinal LABs. *Antonie van Leeuwenhoek*, 82:341–352, 2002.

23. M Blaut, MD Collins, GW Welling, J Doré, J van Loo, W de Vos. Molecular biological methods for studying the gut microbiota: the EU human gut flora project. *Br. J. Nutr.*, 87, S2:S203–211, 2002.

24. E Acedo-Félix, G y Pérez-Martínez. Significant differences between *Lactobacillus casei* subsp. *casei* ATCC 393T and a commonly used plasmid-cured derivative revealed by a polyphasic study. *Int. J. Syst. Evol. Microbiol.*, 53:67–75, 2003.

25. A Ciabattini, R Parigi, MR Oggioni, G Pozzi. Oral priming of mice by recombinant spores of Bacillus subtilis. *Vaccine*, 22:4139–4143, 2004.

26. MT Gil, I Perez-Arellano, J Buesa, G y Perez-Martínez. Expression and secretion of rotavirus VP8* in Lactococcus lactis. *FEMS Microbiol. Lett.*, 203:269–274, 2001.

27. H Marcotte, S Ferrari, C Cesena, G Pozzi, L Hammarström, L Morelli, MR Oggioni. The aggregation promoting factor of *Lactobacillus crispatus* M247 and its genetic locus. J. Appl. Microbiol. 97:749–756, 2004.

28. A Mercenier. Probiotic strains with designed health properties. *The Food, GI-tract Functionality and Human Health Cluster PROEUHEALTH, 1st Workshop, Saariselkä, Finland. 1–3 February 2002.* VTT Symposium 219, VTT Espoo, Finland, pp. 23–24.

29. A Mercenier. DEPROHEALTH: Second generation probiotics. *The Food, GI-tract Functionality and Human Health Cluster PROEUHEALTH, 2nd Workshop, Taormina, Italy. 3–5 March 2003.* VTT Symposium 226, VTT Espoo, Finland, 226, p. 19.

30. M Oggioni, A Ciabattini, AM Cuppone, G Pozzi. Bacillus spores for vaccine delivery. Review, *Vaccine*, 21:2/96–2/101, 2003.

31. I Pérez-Arellano, G y Pérez-Martínez. Optimisation of the green fluorescent protein (GFP) expression from a lactose-inducible promoter system in *Lactobacillus casei. FEMS Microbiol. Lett.*, 222:123–127, 2003.

32. L Steidler, S Neirynck, N Huyghebaert, V Snoeck, A Vermeire, B Goddeeris, E Cox, JP Remon, E Remaut. Biological containment of genetically modipred Lactococcus lactis for human interleukin 10. Delivery in the intestine. *Nature Biotechnol.*, 21:785–789, 2003.

33. L Avonts, L De Vuyst. Bacteriocin production by probiotic lactobacilli. Nederlandse Darmendag *Microbiology of the GI- tract*, Lelystad, The Netherlands, October 11, 2002, p. 18.

34. E Avonts, DA Bax, JG Kusters, T Sako, L De Vuyst. Bacteriocin production by probiotic *Lactobacillus* strains may cause the antibacterial activity against Helicobacter pylori. *The Food, GI-tract Functionality and Human Health Cluster PROEUHEALTH, 1st Workshop, Saariselkä, Finland. 1–3 February 2002*. VTT Symp. 219, VTT, Espoo, Finland, pp. 44–45.

35. L De Vuyst. PROPATH: Molecular analysis and mechanistic elucidation of the functionality of probiotics and prebiotics in the inhibition of pathogenic microorganisms to combat gastrointestinal disorders and to improve human health. *The Food, GI-tract Functionality and Human Health Cluster PROEUHEALTH, 1st Workshop, Saariselkä, Finland. 1–3 February 2002*. VTT Symp., VTT Espoo, Finland, 219, pp. 26–27.

36. L De Vuyst, L Avonts, E Makras, H Holo, P Servin, P Managkoudakis, E Tsakalidou, G Kalantzopoulos, D Sgouras, A Mentis, L Savu, S Cappelle, I Nes. Probiotics against pathogenic microbes in the gastrointestinal tract. *The Food, GI-tract Functionality and Human Health Cluster PROEUHEALTH, 2nd Workshop, Taormina, Italy. 3–5 March 2003*. VTT Symp., VTT, Espoo, Finland, 226, pp. 24–25.

37. MD Georgalaki, E Van den Berghe, D Kritikos, B Devreese, J Van Beeumen, G Kalantzopoulos, L De Vuyst, E Tsakalidou. Macedocin: a food grade lantibiotic produced by *Streptococcus macedonicus* ACA-DC 198. *Appl. Environ. Microbiol.*, 68:5891–5903, 2002.

38. PA Maragkoudakis, C Miaris, D Sgouras, A Mentis, G Kalantzopoulos, E Tsakalidou. *In vitro* evaluation of the probiotic potential of *Lactobacillus* strains and their application in yoghurt. *The Food, GI-tract Functionality and Human Health Cluster PROEUHEALTH, 1st Workshop, Saariselkä, Finland. 1–3 February 2002*. VTT Symp., VTT, Espoo, 219, p. 51.

39. P Neysens, B Degeest, W Vansieleghem, B Pot, L De Vuyst. The effect of tyrosine supplementation on the production of *p*-cresol by monocultures of the gut microflora during *in vitro* batch fermentations. In: *Beyond Antimicrobials — the Future of Gut Microbiology. RRI-INRA 2002, June 12–15*. Aberdeen, U.K., 2002, p.101.

40. P Neysens, B Degeest, W Vansieleghem, B Pot, L De Vuyst. The effect of tyrosine supplementation on the production of *p*-cresol by monocultures of the gut microflora during *in vitro* batch fermentations. *Reprod. Nutr. Dev.*, 42(Suppl. 1):67, 2002.

41. AL Servin, M-H Coconnier. Adhesion of probiotic strains to the intestinal mucosa and interaction with pathogens. Best Pract. Res. Clin. Gastroenterol. 17:741–754, 2003.

42. D Sgouras, P Maragkoudakis, K Petraki, B Martinez, S Michopoulos, E Tsakalidou, G Kalantzopoulos, A Mentis. Inhibition of *H. pylori* colonization and associated gastritis in the HpSS1 C57BL/6 mouse model via administration of the probiotic *Lactobacillus paracasei* subsp. *paracasei* ACA-DC 6002. *The Food, GI-tract Functionality and Human Health Cluster PROEUHEALTH, 1st Workshop, Saariselkä, Finland. 1– 3 February 2002*. VTT Symp. 219, VTT Espoo, Finland, 219, p. 60.

43. L De Vuyst. Inhibitory effects of probiotic lactic acid bacteria. *Méte – Szie-étk - Flair-flow Workshop Functional Foods — Probiotics, Budapest*, Hungary, 2003, pp. 19–21.

44. G Gibson. Overview of EU and MICROFUNCTION project. *The Food, GI-tract Functionality and Human Health Cluster PROEUHEALTH, 1st Workshop, Saariselkä, Finland. 1–3 February 2002.* VTT Symp. 219, VTT Espoo, Finland, pp. 30–31.

45. G Gibson. Update on EU and MICROFUNTION project (Interactions between gut microbiota and the host). *The Food, GI-tract Functionality and Human Health Cluster PROEUHEALTH, 2nd Workshop, Taormina, Italy. 3–5 March 2003.* VTT Symp. 226, VTT, Espoo, pp. 26–27.

46. DIA Pereira, GR Gibson. Cholesterol assimilation by lactic acid bacteria and bifidobacteria isolated from the human gut. The Food, GI-tract Functionality and Human Health Cluster PROEUHEALTH, 1st Workshop, Saariselkä, Finland. 1–3 February 2002, VTT Symposium 219, VTT, Espoo, Finland, p.57.

47. T Scantlebury-Manning, C Vernazza, R Rastall, G Gibson. Functional assessment of interactions between the human gut microbiota and the host: WP1 and 2 (prepiotics and synbiotics). *The Food, GI-tract Functionality and Human Health Cluster PROEUHEALTH, 2nd Workshop, Taormina, Italy. 3–5 March 2003.* VTT Symp. 226, VTT, Espoo, Finland, p. 65.

48. C Dunne, F Shanahan. Role of probiotics in the treatment of intestinal infections and inflammation. *Curr. Opin. Gastroenterol.*, 18:40–45, 2002.

49. J McCarthy, L O'Mahony, L O'Callaghan, B Sheil, E Vaughan, N Fitzsimons, J Fitzgibbon, B Kiely, G O'Sullivan, JK Collins, F Shanahan. Double-blind, placebo-controlled trial of two probiotic strains interleukin 10-knockout mice and mechanistic link with cytokine balance. *Gut*, 52:975–980, 2003.

50. P Ryan, MW Bennett, S Aarons, G Lee, JK Collins, GC O'Sullivan, J O'Connell, F Shanahan. PCR detection of *Mycobacterium paratuberculosis* in Crohn's disease granulomas isolated by laser capture microdissection. *Gut*, 51:665–670, 2002.

51. R Ryan, S Aarons, MW Bennett, G Lee, GC O'Sullivan, J O'Connell, F Shanahan. *Mycobacterium paratuberculosis* detected by nested PCR in intestinal granulomas isolated by LCM in cases of Crohn's disease. In: J O'Connell, Ed., *Methods in Molecular Biology. RT-PCR Protocols.* Humana Press, Totowa, NJ, 2002, 93:205–211.

52. F Shanahan. Understanding inflammatory bowel disease. In: R Modigliani, Ed., Research and Clinical Forums 2002;24(1):207–210. IBD and Salicylates – 5. Kent, England, Wells Medical Holdings Ltd.

53. F Shanahan. Implications of pathophysiology for the clinical management of inflammatory bowel disease. In: J Satsangi, L Sutherland, Eds., *Inflammatory Bowel Diseases*, 4th ed., Elsevier Science, 2003.

54. F Shanahan. Crohn's disease. *Lancet*, 359:62–69, 2002.

55. F Shanahan. Inflammatory bowel disease: immuno-diagnostics, immunotherapeutics and ecotherapeutics. *Gastroenterology*, 120:622–635, 2001.

56. F Shanahan. The host-microbe interface within the gut. Best Pract. Res. Clin. Gastroenterol. 16:915–931, 2002.

57. F Shanahan. Probiotics and inflammatory bowel disease: from fads and fantasy to facts and future. *Br. J. Nutr.*, 88(Suppl. 1): 5–9, 2002.

58. S Targan, F Shanahan, L Karp. *Inflammatory Bowel Disease: From Bench to Bedside*, 2nd ed., Kluwer, Dordrecht, 2003, p. 903.

59. J Doré, M Blaut, R Rastall, I Rowland, A Cresci, E Norin, J Van Loo, C Cayuela. Highlights of CROWNALIFE — Functional Foods, Gut Microflora and Healthy Ageing. *The Food, GI-tract Functionality and Human Health Cluster PROEU-HEALTH, 1st Workshop Saariselkä, Finland. 1–3 February 2002.* VTT Symp. 219, VTT, Espoo, Finland, p. 19.

60. J Doré. GI-tract and the elderly. Presented at Probiotics, Prebiotics, New Foods Meeting, Rome, Italy, 2–4 September 2001.

61. C Gill, I Rowland. Diet and Cancer: assessing the risk *Br. J. Nutr.*, 3:73–87, 2002.

62. E Likotrafiti, KM Tuohy, S Silvi, MC Verdenelli, R Casciano, C Orpianesi, A Cresci, CR Gibson, RA Rastall Molecular taxonomy of probiotic strains isolated from the elderly gut microflora. The Food, GI-tract Functionality and Human Health Cluster PROEUHEALTH, 2nd Workshop Taormina, Italy. 3–5 March 2003, VTT symposium 226, VTT, Espoo, Finland, 2003.

63. K Saunier, M Messaoud, KM Tuohy, M Sutren, A Cresci, J DoréCulture inde-pendent molecular analysis of the elderly faecal microflora reveals an extreme complexity. Poster presentation at The Food, GI-tract Functionality and Human Health Cluster PROEUHEALTH, 2nd Workshop Taormina, Italy. 3–5 March 2003. VTT Symposium 226, VTT, Espoo, Finland, 2003, p.64.

64. S Silvi, MC Verdenelli, R Casciano, C Orpianesi, A Cresci. *Bifidobacterium* and *Lactobacillus* strains from faecal samples of elderly subjects for a possible pro-biotic use in functional foods. The Food, GI-tract Functionality and Human Health Cluster PROEUHEALTH, 1st Workshop Saariselkä, Finland. 1–3 Feb-ruary 2002. VTT Sympossium 219, VTT, Espoo, Finland, 2002, p.61.

65. S Silvi, MC Verdenelli, C Orpianesi, A Cresci, I Rowland. Poster presentation EU project Crowna-life: Functional Foods, Gut Microflora and Healthy Ageing. Food & Nutrition for Better Health, Santa Maria Imbaro – Lanciano, Italy, 13–15 June 2001.

66. S Silvi, MC Verdenelli, C Orpianesi, A Cresci, I Rowland. Cibi funzionali, microflora intestinale e invecchiare in salute: il progetto europeo Crownalife,. 3° Convegno FISV, Riva del Garda (TN), Italy. 21–25 September 2001.

67. S Silvi, MC Verdenelli, C Orpianesi, A Cresci, I Rowland. Poster presentation at EU project Crownalife: Functional Foods, Gut Micro-flora and Healthy Age-ing". 6th Karlsruhe Nutrition Symposium, Karlsruhe, Germany. 21–23 October 2001.

68. S Silvi, MC Verdenelli, C Orpianesi, A Cresci, I Rowland. EU project Crownalife: Functional Foods, Gut Micro-flora and Healthy Ageing. 3° Conferenza inter-nazionale di antropologia e storia della salute e delle malattie: Vivere e "curare" la vecchiaia nel mondo, Genova, Italy. 13–16 marzo 2002.

69. C Gill, I Rowland. Diet and cancer. *Br. J. Nutr.*, 88(Suppl):73–87, 2002.

70. K Saunier, J Doré. Gastrointestinal tract and the elderly:functional foods, gut microflora and healthy ageing. *Dig. Liver Dis.*, 3 (Suppl 2):19–24, 2002.

71. S Silvi, MC Verdenelli, C Orpianesi, A Cresci. EU project Crownalife: Functional Foods, Gut Microflora and Healthy Ageing. Isolation and identification of *Lactobacillus* and *Bifidobacterium* strains from faecal samples of elderly subjects use in functional foods. *J. Food Eng.*, 56(2–3):195–200, 2003.

72. S Silvi, MC Verdenelli, C Orpianesi, A Cresci, I Rowland. "Functional Foods, Gut Microflora and Healthy Ageing EU project Crownalife". In: A Guerci, S Consigliere, Eds., *Curare la vecchiaia.* Erga Edizioni, Genova, Switzerland, 2002, pp. 125–131.

73. R Van der Meulen - 'Kinetische analyse van de groei van een probiotische stam van *Bifidobacterium* animalis of inulin-type fructanen / Kinetic analysis of the growth of a probiotic strain of *Bifidobacterium animalis* on inulin type fructans'. Engineering thesis.

74. H Goossens. Biosafety evaluation of probiotic lactic acid bacteria used for human consumption (PROSAFE). *The Food, GI-tract Functionality and Human Health Cluster PROEUHEALTH, 1st Workshop, Saariselkä, Finland. 1–3 February 2002.* VTT Symp. 219, VTT Espoo, Finland, pp. 28–29.

75. H Goossens, V Vankerckhoven, C Vael, T Van Autgaerden, M Vancanneyt, G Huys, R Temerman, J Swings, I Klare, C Konstabel, G Werner, W Witte, M-B Romond, M-F Odou, C Mullie, P Moreilloin, J Knol, A Mensink, E Wiertz. Biosafety evaluation of probiotic lactic acid bacteria used for human consumption (PROSAFE). First Year's Results. *The Food, GI-tract Functionality and Human Health Cluster PROEUHEALTH, 2nd Workshop, Taormina, Italy. 3–5 March 2003.* VTT Symp. 226, VTT, Espoo, Finland, pp 28–29.

76. E Ananta, D Knorr. Pressure induced thermotolerance of *Lactobacillus rhamnosus* GG. *Food Res. Int.*, 36:991–997, 2002.

77. E Ananta, M Ponanti, D Knorr. Improvement of survival rate of probiotic bacteria durino spray-drying using high pressure and heat pre-treatment. Poster presented at IBERDESH 2002 – Symposium Drying: Process, Structure and Functionality, Valencia, Spain. 25–27 September 2002.

78. C Desmond, C Stanton, GF Fitzgerald, K Collins, RP Ross. Environmental adaptation of probiotic lactobacili towards improvement of performance during spray drying. *Int. Dairy J.*, 11:801–808, 2001.

79. A Klijn, E Restle, I Jankovic, F Aymes, R Zink. Determination of the viability kinetics of Bifidobacteria in fermented dairy systems. Poster presented at 7th Symposium on Lactic Acid Bacteria: Genetics, Metabolism, and Application, Egmond aan Zee, NL. 1–5 Sept 2002.

80. A Laitila, M Saarela, L Kirk, M Siika-aho, A Haikara, T Mattila-Sandholm, I Virkajärvi. Malt sprout extract medium for cultivation of *Lactobacillus plantarum* protective clutures. *Lett. Appl. Microbiol.*, 39:336–340, 2004.

81. L Alakemi, J Mättö, I Virkajärvi, M Saarela. Application of microplate scale fluorochrome staining assay for the assessment of viability of probiotic preparations. *J. Microbiol. Meth.* 62:25–35, 2005.

82. I Virkajärvi, J Mättö, A Vaari, M Saarela. Effect of fermentation time and freeze-drying medium on viability and characteristics of probiotics. the Food, GI-tract Functionality and Human Health Cluster PROEUHEALTH, (2nd Workshop) Taormina, Italy. 3–5 March 2003, VTT Symposium 226, VTT, Espoo, Finland, 2003, p.66.

83. I Virkajärvi, J Mättö, A Vaari, M Saarela. Effect of fermentation time and freeze-drying medium on viability and characteristics of probiotics. NFIF2003, New Functional Ingredients and Foods, Copenhagen, Denmark, 9–10 April 2003.

84. I Virkajärvi, A Vaari, H Siren, T Mattila-Sandholm, M Saarela. Development of a fermentation medium for probiotic lactobacilli and bifidobacteria. Poster presented at the Food, GI-tract Functionality and Human Health Cluster PROEUHEALTH, (1st Workshop) Saariselkä, Finland. 1–3 February 2003.

85. T Bech-Larsen, KG Grunert, JB Poulsen. The acceptance of functional foods in Denmark, Finland and the United States. MAPP Working Paper No. 73, The Aarhus School of Business, 2001, p. 73.

86. K Grunert, T Bech-Larsen, L Bredahl. Three issues in consumer quality perception and acceptance of dairy products. *Int. Dairy J.*, 10:575–584, 2000.

87. L Lähteenmäki, A Arvola. Food neophobia and variety seeking consumer fear or demand for new food products. In: LJ Frewer, E Risvik, H Schifferstein, Eds., *Food, People and Society. A European Perspective of Consumer Food Choices.* Springer-Verlag, Berlin, 2001, pp. 161–175.

88. R Lappalainen, J Kearney, MA Gibney. Pan-European survey of consumers attitudes to food, nutrition and health: an overview. *Food Qual. Preference*, 9:467–478, 1998.

89. JB Poulsen. Danish consumers' attitudes towards functional foods. MAPP Working Paper No. 62. The Aarhus School of Business, 1999, p. 62.

90. K Roininen L Lähteenmäki, H Tuorila. Quantification of attitudes to health and hedonic characteristics of foods. *Appetite*, 33:71–88, 1999.

91. N Urala, L Lähteenmäki. Reasons behind consumers' functional food choices. *Nutr. Food Sci.*, 33:148–158, 2003.

15

Modulation of Epithelial Function and Local Immune System by Probiotics: Mechanisms Involved

Sandrine Ménard and Martine Heyman

CONTENTS

15.1 Introduction

Probiotics have long been known as beneficial to intestinal function, and their use, often based on empirical knowledge, has become more frequent in a variety of digestive diseases. Probiotics are bacteria that after ingestion transit along the digestive tract and interact directly with the intestinal content and the intestinal epithelial layer and to a lesser extent with the mucosal immune system. Elie Metchnikoff was the first to stress the importance of lactobacilli in microflora in maintaining health and longevity (1). The term "probiotics," popularized by Fuller (2), has been defined as "a live microbial supplement which beneficially affects the host by improving its intestinal microbial balance". This definition was later extended to include other beneficial effects such as immunomodulation. For a long time, based on empirical practice and later on scientific observations, live microbial supplementation (yogurt, fermented milk, and bacterial lyophilisates) has been proposed to control various digestive or extradigestive diseases. In parallel, nutrient factors defined as prebiotics and favoring the development of a specific flora (such as growth-promoting bifidus factors consisting of various oligosaccharides) have also been used as an additional strategy.

An understanding of the effects of probiotics on intestinal physiology requires a better knowledge of the interactions between probiotic bacteria, intestinal epithelial cells, and the mucosal immune system. The aim of this chapter is to review current knowledge of the mechanisms governing the complex relationship between probiotics and intestinal physiology. As a basis of this analysis, the main forces implicated in intestinal homeostasis will first be summarized.

15.2 Forces Driving Intestinal Homeostasis as Targets of Probiotics

Intestinal homeostasis relies upon the equilibrium between absorption (nutrients, ions) and secretion (ions, IgA, mucus), together with the barrier capacity (to pathogens and macromolecules) of the digestive epithelium. These functions are controlled through multiple interactions between the endocrine, neurocrine, stromal, and immune cells or the resident bacterial microflora that regulate epithelial functions. When this homeostatic control is disturbed, chronic inflammation, diarrhea, and disease may occur. An understanding of how probiotics influence digestive diseases needs to take into account the mechanisms involved in the regulation of intestinal epithelial functions.

First, intestinal homeostasis relies on hydroelectrolytic balance (Figure 15.1). This balance can be dysregulated via a decrease in Na$^+$-coupled nutrient absorption or an abnormal stimulation of ionic secretion driving water losses. Water movements are mainly generated by the Na solutes cotransport systems (sugars, amino acids) or chloride (Cl–) secretion across the apical membrane of intestinal epithelial cells, and such transporters or channels are highly regulated structures. Water movement follows ionic movements across the intestinal epithelium: Na$^+$ absorption is driven in part by the Na-glucose cotransport system expressed on villi and leads to a net water absorption, whereas Cl– secretion mainly drives water secretion from the crypts to the intestinal lumen. Therefore, any luminal or serosal factor affecting these transport systems will also affect electrolyte and water movements. Among the luminal factors, pathogenic bacteria can adhere to the apical membrane of the enterocytes, inducing epithelial dysfunction. Attaching effacing lesions of the brush-border membrane may ensue, or enterotoxins

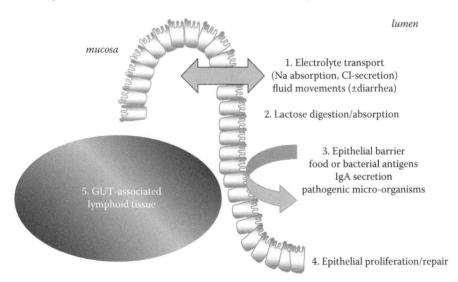

lumen

mucosa

1. Electrolyte transport
(Na absorption, Cl-secretion)
fluid movements (±diarrhea)

2. Lactose digestion/absorption

3. Epithelial barrier
food or bacterial antigens
IgA secretion
pathogenic micro-organisms

5. GUT-associated
lymphoid tissue

4. Epithelial proliferation/repair

FIGURE 15.1
Intestinal homeostasis as targets of probiotics. Intestinal homeostasis mainly relies upon the regulation of fluid and water movements, themselves driven by electrolyte (Na, Cl) transport systems. Probiotics are hypothesized to favor absorption by protecting epithelial integrity and to inhibit exacerbated secretion (for example, toxin-induced chloride secretion). Lactose digestion and epithelial barrier function are also candidates when it comes to explaining the potential beneficial effects of probiotics. Gut-associated lymphoid tissue is also a recognized target as both bacterial determinants or secreted products can influence cytokine secretion directly or indirectly through the activation of intestinal epithelial cells.

can be released, stimulating chloride secretion (and therefore water secretion inducing diarrhea). Besides the specific mechanisms involved in water and electrolyte movements in the gut, osmotic diarrhea can also be induced when a nonabsorbable compound reaches the intestinal lumen and maintains an osmotic gradient between the intestinal lumen and the blood. A typical case of osmotic diarrhea is that induced by lactose malabsorption in cases of lactase deficiency. Serosal factors can also affect the regulation of water movements. Abnormal stimulations of the underlying immune system (mast cells, phagocytes, lymphocytes, dendritic cells) leads to the release of inflammatory mediators capable of altering epithelial function. The enteric nervous system can also be involved through the abnormal release of neuromediators (Met-enkephalin, acetylcholine) known to activate chloride secretion directly. Finally, cytotoxins are also known to disturb intestinal homeostasis by acting on tight junctional complexes and by disrupting epithelial integrity. Thus, any introduced external factor that improves hydroelectrolytic equilibrium and epithelial barrier function may be considered beneficial to the host. Probiotic bacteria may be one of these factors.

15.3   Effect of Probiotics on the Luminal Environment

15.3.1   Probiotic Bacteria: Where Do They Come From?

Bacterial strains used as probiotics have to possess various properties. They must be nonpathogenic, resistant to acid and bile, adhere to intestinal epithelial cells, persist long enough in the digestive tract, produce antimicrobials, modulate the immune response, and be resistant to technological processes (Figure 15.2). Currently used probiotic microorganisms consist mostly of strains of *Lactobacillus*, *Bifidobacterium*, and *Streptococcus*, these being Gram (+) lactic acid bacteria (LAB) which have been used for centuries in the production of fermented dairy products, although some other species (*E. coli*) or other microorganisms (yeast) have also been a matter of interest.

Lactobacilli and bifidobacteria are bacterial strains originating from human microflora. The intestinal microflora is complex and involves more than 400 bacterial species, mainly localized in the colon (10^{14} bacteria). It can be modified by dietary substances favoring the growth of certain bacterial species, for example fructo- or galactooligosaccharides (prebiotics), used to promote the growth of bifidobacteria. Lactobacilli are often part of the intestinal ecosystem, but the numbers present vary widely between different individuals (ranging from 0 to 10^6 CFU/g feces). Bifidobacteria are also part of the human microflora, but the species present differ according to age; for example, newborns are readily colonized by *Bifido-*

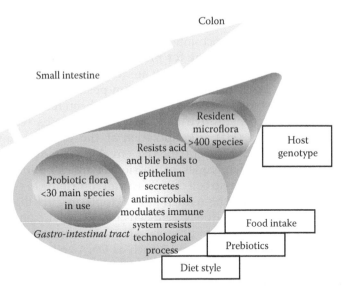

FIGURE 15.2
Relationship between resident microflora and probiotics. Host genotype and food intake influence the establishment and maintenance of the resident microflora. By contrast, ingested bacteria do not persist permanently in the digestive tract and disappear more or less rapidly after ingestion, suggesting that exogenous probiotic bacteria cannot easily colonize the gut.

bacterium breve and *Bifidobacterium infantis* (colonization is favored in breast-fed compared to bottle-fed infants), whereas adults more often host *Bifidobacterium adolescentis*, *Bifidobacterium bifidum*, and *Bifidobacterium longum*. However, ingested bifidobacteria, when administered as probiotics, do not persist permanently in the digestive tract when oral administration has ceased. This may be due to the fact that, on the one hand, the host genotype contributes to the dominant microbial diversity, suggesting specific interactions between microbes and man (3), and on the other hand, microflora establishment and maintenance is highly dependent on food intake and style of diet. In other words, the composition of the indigenous microflora is specific to an individual host, and exogenous bacteria are not easily established.

In addition, studies on probiotics in various animal models suggest an apparent variability in the probiotic efficacy of a given microorganism in a given individual. In humans, where intestinal diseases are heterogeneous and multifactorial, it is likely that the probiotics used will need to be varied according to the disease and/or to its clinical course.

Probiotics transiting through the digestive tract have to deal first with the deleterious effects of the gastric (high acidity) and small intestinal (bile salts) environments. To be a probiotic, a bacterial strain must be resistant to gastric acidity and bile salts.

15.3.2 Survival at Low pH and Resistance to Bile Acids

The acidic environment encountered in the gastrointestinal tract provides a significant survival challenge for probiotic bacteria that need to survive the highly acidic gastric juice if they are to reach the small intestine in a viable state. The mechanisms used by Gram-positive organisms to protect themselves against low-pH environments have been recently reviewed (4). A combination of constitutive and inducible strategies resulting in the removal of protons (H+), alkalinization of the external environment, changes in the composition of the cell envelope, production of general shock proteins and chaperones, expression of transcriptional regulators, and responses to changes in cell density can all contribute to survival. These mechanisms counteract the reduction in cytoplasmic pH that can include loss of activity of the relatively acid-sensitive glycolytic enzymes (which severely affects the ability to produce ATP) and structural damage to the cell membrane and macromolecules such as DNA and proteins. The three main systems involved in LAB pH homeostasis, i.e., the arginine deiminase (ADI) system, the H+-ATPase proton pump, and the GAD (glutamate decarboxylase) system, have been described (4). It has been proposed that amino acid decarboxylases function to control the pH of the bacterial environment by consuming hydrogen ions as part of the decarboxylation reaction. Acid adaptation of *Lactococcus lactis* requires *de novo* protein synthesis (5), as is also the case for *Lactobacillus casei* (all oral LAB) (6). In some strains (e.g., *L. lactis* subsp. *lactis*) acid adaptation is independent of protein synthesis. Heat shock proteins are also induced following acid adaptation of *Lactobacillus delbrueckii* (7).

Bile acids are synthesized in the liver from cholesterol and are secreted from the gallbladder into the duodenum in the conjugated form. These acids then undergo extensive chemical modifications (deconjugation, dehydroxylation, dehydrogenation, and deglucuronidation) in the colon as a result of microbial activity. Both conjugated and deconjugated bile acids exhibit antibacterial activity, inhibiting the growth of strains of *Escherichia coli*, *Klebsiella* sp., and *Enterococcus* sp. *in vitro* (8, 9). However, the deconjugated forms are more inhibitory, and Gram-positive bacteria are found to be more sensitive than Gram-negative bacteria (10). Recent studies have determined the level of bile acid resistance exhibited by several strains of *Lactobacillus* and *Bifidobacterium* (11). These strains exhibited resistance to human bile used at final concentrations of between 0.3 and 7.5%, indicating that they could grow in physiologically relevant concentrations of human bile. It was also reported that *Lactobacillus plantarum* showed high tolerance to consecutive exposure to hydrochloric acid (pH 1.5 to 2.5) and cholic acid (10 m*M*) (12).

The fast adaptation of bacteria to the hostile intestinal environment could be related to their high rate of mutation in the digestive tract. Although initial mutations are favorable to the survival of ingested bacteria, these mutations can turn out to be deleterious to secondary colonization (13), perhaps explaining why the individual host is important in the outcome of a probiotic effect.

15.3.3 Provision of an External Source of Lactase

Lactase insufficiency is characterized by a low concentration of the lactose-cleaving enzyme β-galactosidase (lactase) in the brush-border membrane of small intestinal enterocytes. This hypolactasia causes insufficient digestion of the disaccharide lactose, a phenomenon called lactose malabsorption or maldigestion. There are several forms of lactose malabsorption. Lactase activity is high at birth, decreases in childhood, and remains low in adulthood. This primary hypolactasia is the normal situation for mammals including humans. With the exception of the population of Northern and Central Europe, 70 to 100% of adults worldwide are lactose malabsorbers.

One of the best-documented beneficial effects of yogurt is described in cases of lactose intolerance. In 1984, Kolars et al. (14) and Savaiano et al. (15), showed in lactase-deficient subjects that lactose was absorbed much better from yogurt than from milk, probably due to the intraluminal digestion of lactose by the lactase released from microorganims in yogurt. These results were subsequently confirmed by many other groups (reviewed in Reference 16). In addition, recent work has shown that *Streptococcus thermophilus* is able to produce a β-galactosidase active during its transit in the digestive tract of germ-free mice (17). It is not absolutely proven, however, that the microbial lactase activity, whose maximal activity occurs between pH 6 to 8, persists at the duodenal pH of 5.0. In fact, although β-galactosidase activity in yogurt drops by 80% in the duodenum, one-fifth of the yogurt lactase activity is present in the terminal ileum, suggesting a relative persistence of the protein along the digestive tract. In addition, fresh yogurt is more efficient in facilitating lactose digestion than heated yogurt (18, 19). Thus, bacterial β-galactosidase present in yogurt partly resists luminal hydrolysis and can hydrolyze lactose, at least in the middle and distal parts of the small intestine where the pH is compatible with its enzymatic activity.

15.3.4 Bacterial Interference

Bacterial interference refers to the ability of one microorganism to protect the host against a microbial pathogen by interfering with its adhesion and toxic effects. It has been postulated that signaling molecules produced by commensal or probiotic organisms may activate the host (e.g., by stimulating mucus production) or inhibit the response of the pathogen (e.g., activation of their virulence genes) (20). Until recently, the mechanisms believed to be responsible for bacterial interference were competition for nutrients and epithelial cell-binding sites. However, production of soluble factors that inhibit bacterial toxin activity or alter pathogenic bacterial viability were also recognized as important.

15.3.4.1 Antimicrobial Effect

Several metabolic compounds commonly produced by LAB, including organic acids, fatty acids, hydrogen peroxide, and diacetyl, have antimicrobial effects. Bacteriocins or proteinaceous substances are perhaps the most extensively studied. LAB bacteriocins are biologically active proteins or protein complexes that kill Gram-positive bacteria, usually closely related to the producer strain (21). Among bacteriocins, nisin, which is produced by some *Lactobacillus lactis* subsp. *lactis* strains, is at present the only purified bacteriocin approved for use in products intended for human consumption. The lactococcal bacteriocins are hydrophobic cationic peptides, which form pores in the cytoplasmic membrane of sensitive cells leading to their death. Other antibacterial nonbacteriocins have been described, as exemplified by *L. acidophilus* strain LA1 that produces a protease-resistant antibacterial substance different from lactic acid and having bactericidal activity against a wide range of Gram-negative and Gram-positive pathogens (22). However, it is debatable whether killing the pathogens is the optimal method of interference because in some cases this would exacerbate pathogenicity by triggering the release of more bacterial toxins (23).

15.3.4.2 Antitoxin Effect

A specific antivero cytotoxin (VT) effect has been reported in culture supernatants from *Bifidobacterium longum* (24). In fact VTs are produced by the food-borne pathogen *E. coli* (VTEC) and are associated with hemorrhagic colitis. The B-subunit of VT binds to the gut epithelial receptor globotriaosylceramide (Gb3), whereas the A subunit, after entering the target cell, inhibits protein synthesis. Soluble substances in *B. longum* culture supernatants had an inhibitory effect on Gb3 receptor expression and VT-Gb3 interactions, suggesting their potential beneficial effect.

The yeast *S. boulardii* was shown to secrete a protease displaying a protective effect against *C. difficile*-induced inflammatory diarrhea in humans, possibly through the proteolytic digestion of toxin A and B molecules (25).

15.4 Effect of Probiotics on Intestinal Epithelial Cells

15.4.1 Adhesion to Intestinal Epithelial Cells

The capacity of lactobacilli to adhere to the intestinal epithelium remains controversial. This property is important, because it prevents their rapid

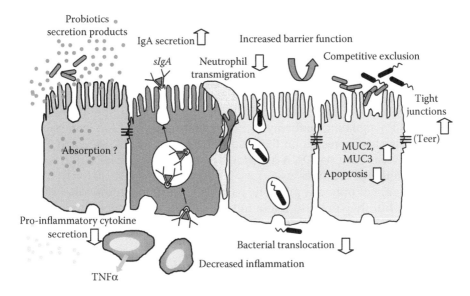

FIGURE 15.3
Possible mechanisms by which probiotics are thought to improve intestinal function: probiotic bacteria can act by direct contact with epithelial cells or through the release of secreted products. Most studies emphasize their properties in increasing the epithelial barrier function by strengthening tight junctions between epithelial cells, increasing mucus protein expression and favoring secretory IgA production. On the other hand, probiotics also reduce inflammatory processes by decreasing pathogenic bacteria translocation and lowering proinflammatory cytokine secretion.

elimination by peristalsis and thus represents an ecologically competitive advantage in the gastrointestinal tract ecosystem.

15.4.1.1 Epithelial Receptors Involved in Bacterial Binding

Despite *in vitro* experiments, with cultured human intestinal cell lines as models of mature enterocytes of the small intestine (26–28), *Lactobacillus* cell surface-associated factors potentially acting as adhesion factors remain to be characterized. Some lactobacilli have the ability to bind to mannose (29), to rat colonic mucin (30, 31), or to glycolipids isolated from rat intestinal mucosa (32). Mechanisms involving proteins or proteinaceous components as mediators of adhesion have been described for some *Lactobacillus* strains, including *Lactobacillus johnsonii* LA1 (27), *L. acidophilus* BG2F04 (33) and LB12 (34), *L. fermentum* 104 (35), and *L. crispatus* (36). Finally, some *Lactobacillus* species may preferentially bind to the follicle-associated epithelium of Peyer's patches as compared with mucus-secreting regions of the small intestine (37).

On the other hand, probiotic bacteria are generally Gram-positive bacteria possessing cell wall constituents (peptidoglycans, lipoteichoic acids) capable of binding to Toll-like receptors such as TLR-2 (38) and able to mediate intracellular signaling and proinflammatory cytokine production. Bacterial

DNA and its immunostimulatory sequences, also known as unmethylated CpG motifs, are also mediators of immune activation involving TLR-9. Intestinal epithelial cells (IECs) have been shown to express receptors for Gram-negative bacteria, such as the lipopolysaccharide (LPS) receptor TLR-4 (39), either on their apical surface (40) or in a cytoplasmic paranuclear distribution (41), although IECs are thought to limit dysregulated LPS signaling by down-regulating expression of MD-2 and TLR-4 while the remainder of the intra-cellular LPS signaling pathway is functionally intact. TLR-5, the flagellin receptor, is also expressed on the basolateral surface of IEC (42), and TLR-9 (43), known as bacterial DNA receptor, has recently been detected in human colonic epithelial cell lines.

Therefore, not only can probiotic bacteria bind to the intestinal epithelial cell layer, but they are also likely to transmit signals via specific receptors on intestinal epithelial cells to the underlying immune system.

15.4.1.2 Receptor-Binding Region of Probiotic Bacteria

Compared with the present knowledge of the adhesive mechanisms of many pathogenic bacteria, only limited information is available about the adhesive surface molecules of probiotics and their tissue receptors. A recent study suggested fibronectin as one of the receptors that mediate the adhesion of lactobacilli to epithelial cells (44). Lectin-like molecules (45, 46) and lipote-ichoic acids (47, 48) have also been shown to function as adhesins in lacto-bacilli. Although proteins have been reported to mediate the adhesion of lactobacilli (33, 50), only a few proteinaceous adhesins have been identified, including a protein component of the bacterial ATP-binding cassette (51), an S-layer protein (52–54), and the cell-surface protein Mub (*Lactobacillus reuteri*) adhering to mucus components (55).

The adhesion of *Lactobacillus* spp. to epithelial cells also depends on bac-terial physiology and physicochemical parameters, which can be modulated by growth conditions (56, 57). The pH has also been reported as of crucial importance in adhesion (58, 59).

15.4.2 Inhibition of Pathogen Binding to Brush-Border Epithelial Cell Membranes

Adhesion to and colonization of mucosal surfaces by probiotic bacteria are possible protective mechanisms against pathogens through competition for binding sites and nutrients, steric hindrance, immune modulation (60), or mucin production (Figure 15.3).

15.4.2.1 Competitive Exclusion

In vitro studies have documented the antagonistic activity of lactobacilli against pathogens as a result of the competitive exclusion of adhesion of pathogenic bacteria to host cells.

Inhibition of the adherence of Gram-negative pathogens to urothelium by bacterial lipoteichoic acid isolated from a *Lactobacillus* strain has been demonstrated *in vitro*, steric hindrance being found to be the major factor in preventing adherence of uropathogens (61). Competitive exclusion was also shown *in vivo* in a rat model of urinary infection. Biosurfactants from *Lactobacillus* isolates were later shown to inhibit the initial adhesion of uropathogenic *Enterococcus faecalis* (62).

Heat-killed *Lactobacillus acidophilus* strain LB that adheres to the intestinal epithelial cell line Caco-2 was shown to inhibit diarrheagenic *Escherichia coli* adhesion in a concentration-dependent manner, by a mechanism that may involve steric hindrance (27). Enteropathogenic *E. coli* K88 colonizes piglet ileum by adhering to the mucosa via K88 fimbriae. Three *Lactobacillus* strains of porcine origin were shown to reduce adhesion of *E. coli* K88 by approximately 50%. Inhibition occurred when mucus was pretreated with spent culture dialysis retentate. The active compound was a fraction of >250,000 molecular weight (63). This factor was further characterized as a 1700-kDa cell wall fragment destroyed by lysosyme treatment and containing glucose, N-acetylglucosamine, and galactose (64).

It is important to underline that inhibiting the binding of enteropathogenic bacteria also probably means inhibiting their deleterious effect on intestinal function, most importantly diarrhea due to active chloride secretion. Enteropathogenic *E. coli* (EPEC) infection causes an increase in short-circuit current (Isc), an index of electrogenic ion transport and water movement. Adding *Lactobacillus plantarum* strain 299v before EPEC infection of Caco-2 cell monolayers reduced the increase in Isc (65). This reduction in the secretory process in response to EPEC infection was possibly due to inhibition of its binding. However, the effect was only produced when the probiotic agent was introduced prior to infection, indicating a preventive rather than therapeutic effect.

15.4.2.2 Mucus Secretion

The epithelial cells lining the intestinal tract provide a physical barrier and have developed inducible innate protective strategies allowing rapid responses to pathogenic challenge. These include secretion of ions and water and elaboration and secretion of antibacterial peptides or more complex molecules such as mucins. Mucins may protect epithelial cells from microbial pathogens by limiting their access through simple steric hindrance and by providing a physicochemical barrier preventing epithelial cell adherence, colonization, toxin delivery, and invasion by pathogenic microorganisms.

In vitro studies led to the hypothesis that the ability of probiotic agents to inhibit adherence of attaching and effacing organisms to intestinal epithelial cells was mediated through their ability to increase expression of MUC2 and MUC3 intestinal mucins (66). It was later shown that selected probiotics, *Lactobacillus* species that have the ability to adhere to intestinal epithelial cells, rapidly induce eukaryotic MUC3 mucin expression. The upregulated

MUC3 mucin gene product was a secreted mucin that had the ability to inhibit enteric pathogen epithelial cell adherence (67). *Lactobacillus casei* GG had also been shown to induce upregulation of MUC-2, probably through binding to specific receptor sites on the enterocyte (68).

15.4.3   Improvement of Epithelial Barrier Function

15.4.3.1   *Transepithelial Electrical Resistance and Intestinal Permeability*

The intestinal barrier function comprises various factors capable of decreasing the absorption of potentially harmful microbial or soluble antigens (Figure 15.3). These factors include the capacity of digestive enzymes to degrade luminal antigens, the presence of an epithelial barrier coated by a mucus layer with entrapped secretory IgA, and epithelial cells firmly connected via tight junctions. The commensal flora is capable of influencing the intestinal barrier function. In germ-free mice, the absence of intestinal microflora is associated with an increase in electrical resistance (an index of barrier integrity) and a decrease in macromolecular transport, suggesting that commensal bacteria influence epithelial physiology (69).

In rat colon, some bacterial strains (*E. coli, K. pneumoniae, S. viridans*) increase small molecule absorption, whereas others, such as *L. brevis*, decrease this permeability (70).

In addition, direct evidence has been obtained *in vivo* that probiotics reinforce the intestinal barrier function. Different studies have suggested that *Lactobacillus casei* GG was able to stabilize intestinal permeability to macromolecules, particularly in cases of acute gastroenteritis in rats (71), and also to reverse the increase in intestinal permeability induced by cow's milk in suckling rats (72). Furthermore, feeding guinea pigs with fermented milk (*Streptococcus thermophilus* and *Bifidobacterium breve)* led to an increase in the transepithelial electrical resistance (TEER) of jejunal fragments mounted in Ussing chambers and to a decrease in the intestinal transport of β-lactoglobulin, as compared to guinea pigs fed nonfermented milk (73). These results indicate a reinforcement of the intestinal barrier in basal conditions.

In vitro evidence that probiotics protect the intestinal barrier has also been provided. The combination of the two probiotics *Streptococcus thermophilus* (ST) and *Lactobacillus acidophilus* (LA) caused a significant increase in TEER of intestinal cell Caco-2 monolayers (74). This effect on TEER was also corroborated by the beneficial effect of ST and LA on the permeability of monolayers infected by enteroinvasive *E. coli*. The effect was demonstrated with the small molecular tracer FITC, but not with the high molecular 10-kDa tracer. The probiotic strains also prevented the disruption of cytoskeletal and tight junctional protein localization and phosphorylation. Spent medium from ST or LA cultures, ST and LA killed with antibiotics, and heat inactivated probiotics failed to induce the same beneficial effects. The alterations in the mucosal barrier function induced by probiotics seem to require biochemical communication between the adherent microorganisms and entero-cytes. The probiotic organism may induce a signal transduction pathway that strengthens tight junctions between enterocytes and thereby reduces paracellular transport of small molecules or larger antigens that cause inflammation. However, soluble compounds released by probiotics are also likely to play a role, as shown by the restoration of normal secretory pro-cesses in epithelial cells infected by enteroinvasive *E. coli* (74) and by *in vitro* studies showing that the epithelial barrier function and resistance to *Salmo-nella* invasion (using T84 intestinal cell line) could be enhanced by exposure to a proteinaceous soluble factor secreted by the probiotic bacteria found in the VSL#3 probiotic mixture (75). VSL#3 (bifidobacteria, lactobacilli, and *Streptococcus salivarius*) was shown to be effective in restoring the barrier function of IL-10 deficient mice and in reducing secretion of TNFα and IFNγ. It was also capable of increasing the electrical resistance of T84 epithelial monolayers by 20% within 6 h (75). In healthy volunteers, indomethacin-induced gastric damage (but not intestinal damage) was prevented by a 5-day administration of live (but not killed) *Lactobacillus* GG (76).

15.4.3.2 Neutrophil Migration and Bacterial Translocation

Neutrophil transmigration in the intestine is associated with a transient barrier dysfunction and elicits a biphasic resistance response representing sequential effects on transcellular and paracellular pathways (77). Neutralization of poly-morphonuclear neutrophil (PMN) transmigration (78) using an anti-CD18 monoclonal antibody that neutralizes binding of PMNs to epithelial cells dra-matically decreases both bacterial invasion and inflammatory destruction of the epithelium. Therefore, conditions that lead to the inhibition of neutrophil migration induced by pathogenic bacteria can be beneficial to the host. Recent studies indicate that *L. plantarum* is beneficial in inhibiting neutrophil migra-tion induced by EPEC, but only when preincubated with host epithelia. Rather than an indirect effect through a secreted substance produced by the probiotic agent, its effect is direct and requires the presence of the bacterium (79). The probiotic bacterium *Lactobacillus* GG also inhibits bacterial translocation of *E. coli C25* in a dose-dependent manner in an *in vitro* cell-culture model (80). This phenomenon was also observed *in vivo*, in a neonatal rabbit model where *Lactobacillus* GG decreased the frequency of *E. coli K1A* bacterial translocation in mesenteric lymph nodes by 46%. However, in a prospective and random-ized study in elective surgical patients, the administration of *Lactobacillus plantarum 299v* was not able to decrease bacterial translocation or septic mor-bidity (81).

15.4.3.3 Antiapoptotic Effect

The deficiency in epithelial barrier function observed in inflammatory diges-tive diseases is predominantly dependent upon increased cytokine production (82) and increased apoptosis of intestinal cells (83). Indeed, a central cytokine in the pathogenesis of inflammatory bowel disease (IBD), tumor necrosis factor alpha (TNFα), regulates both anti- and proapoptotic signaling pathways. The balance between these two signals determines the fate of the cell (84). Recently the effects of *Lactobacillus casei* GG on cytokine-regulated signaling pathways were investigated, and it was shown that this probiotic bacterium prevented cytokine-induced apoptosis in mouse or human intestinal epithelial cells. Culture of *Lactobacillus* GG with colon cells activated antiapoptotic Akt/protein kinase B and inhibited activation of proapoptotic p38/mitogen-activated protein (MAP) kinase by TNF, IL-1α, or IFN-γ (85). Thus, *Lactobacillus* GG promotes intestinal epithelial cell survival through modulation of both anti- and proapoptotic signal transduction pathways. Finally, it should be noted that bacterial DNA containing immunostimulatory sequences (ISS) especially non-methylated CpG motifs, exhibits antiapoptotic properties in colonic epithelial cells in a colitis model in mice (86). It is possible that such ISS sequences form part of the mechanisms by which probiotics ameliorate cell survival and restore mucosal epithelium in inflammatory conditions.

15.4.3.4   *Epithelial Proliferation and Repair*

One may suspect that the beneficial effect of probiotics as adjuvants in the treatment of intestinal inflammation could be due at least in part to their capacity to favor intestinal repair and to induce cell proliferation. Indeed, 7 d of administration of 10^7 CFU of *Lactobacillus casei* or *Clostridium butyricum* to rats fed an elemental diet increased the crypt cell production rate of the jejunum and ileum by 25 to 40%, of the cecum by 70%, and of the distal colon by more than 200% compared with control (87). In another study investigating the influence of *L. rhamnosus* GG on mucosal cell kinetics and morphology in different regions of the intestine of young male rats monoassociated with the bacteria, it was reported that *L. rhamnosus* GG had a compartmentalized effect upon rat intestinal mucosa. After 3 d of monoassociation, the mitotic index for the upper part of the jejunum increased by 14 and 22% as compared to data obtained in germ-free or conventional rats, respectively. These results suggest a mitogenic effect of probiotics on intestinal epithelial cells that may, in pathological situations, favor the reparation of intestinal mucosa.

15.5   Effect of Probiotics on Local Immune Response

15.5.1   Enhancement of Secretory IgA Protection

15.5.1.1   *Secretory IgA Immunity*

When exogenous antigens penetrate via the oral route, especially infectious antigens, a secretory immune response is induced through the release of specific secretory IgM or IgA (Figure 15.3).

On the other hand, colonization of germfree mice with different mixtures of commensal bacteria leads to the rapid appearance of IgA$^+$ plasma cells in the lamina propria (88, 89). It should therefore be expected that giving probiotics via the oral route may be a powerful way to potentiate IgA-dependent secretory immunity. In fact, LAB have been shown to improve the systemic and secretory immune response to luminal antigens. One of the first beneficial effects described with *Lactobacillus rhamnosus* GG was the reinforcement of local immune defense through an enhanced secretion of rotavirus-specific IgA (90, 91). Whether these IgA antibodies have the capacity to counteract virus entry into epithelial cells remains to be established. In healthy volunteers the oral administration of the *L. johnsonii* strain LA1 was shown to increase serum IgA concentrations (92); however, this result was not confirmed by others (93) who showed that the increase in serum IgA was small and that no modification of another Ig was detected. Also, in human volunteers, ingestion of an attenuated *Salmonella typhi* strain, administered in order to mimic an enteropathogenic infection, induced a specific serum IgA response which was fourfold higher in the group supplemented with fermented milk *(Lactobacillus acidophilus* LA1 and bifidobacteria) compared to the control group (92). *Lactobacillus* GG also had an immunostimulating effect on oral rotavirus vaccination in infants. Infants who received *Lactobacillus* GG showed an increased response with regard to rotavirus-specific IgM secreting cells (94).

The enhancement of IgA secretory response against a soluble food antigen was also described in mice fed a whey protein diet with or without *B. longum*. Both total IgA and specific anti-β-lactoglobulin IgA were at significantly higher levels in the small intestine of mice fed the *B. longum*-containing diet than in control mice (95).

15.5.2 Cross Talk between Intestinal Epithelial Cells and Commensal-Probiotic Bacteria: Downregulation of Inflammation?

In normal conditions immunologic tolerance is maintained toward commensal enteric bacteria, which prevents intestinal inflammation. This controlled homeostatic response is lost in susceptible individuals that develop chronic, aggressive immune responses at the intestinal level. Because intestinal microfloras are essential for the development and perpetuation of colitis (96) and seem to be implicated in the pathogenesis of IBD, the rationale for using probiotics has recently emerged (97, 98), and probiotics have been proposed as a therapeutic adjunct in IBD (99–101). Some recent data also support the view that nonpathogenic bacteria may directly alleviate intestinal inflammation through mechanisms altering signal transduction pathways of proinflammatory cytokines, in particular the NFκB transcription factors (102, 103). The mechanisms by which probiotics can downregulate inflammation in digestive diseases may also imply such mechanisms. A recent *in vitro* study suggested that soluble factors produced by probiotics may be involved.

Indeed, conditioned media from *S. thermophilus* and *B. breve* were capable of inhibiting LPS-induced TNFα secretion by human mononuclear cells or the monocytic THP-1 cell line. The mechanism involved was linked to compound(s) of low molecular weight (<3000 Da) which inhibited LPS binding to immune cells and the nuclear translocation of NFkB, both involved in TNFα production (104).

The anti-inflammatory properties of probiotics was further underlined using *Lactobacillus* GG-conditioned media which decreased TNFα production by macrophages *in vitro*, via a contact-independent mechanism (105). In addition, the release of TNFα by inflamed Crohn's disease mucosa was significantly reduced by coculture with *L. casei* or *L. bulgaricus* (104).

Finally, butyrate, produced by bacterial fermentation of unabsorbed carbohydrate in the colon and known to regulate many epithelial cell functions, has been used in the treatment of ulcerative colitis as well as a variety of other diseases characterized by colonic inflammation. Its therapeutic efficacy could be explained by the fact that butyrate is a major source of energy for colonocytes and hence may increase mucosal proliferation, promote cell differentiation, and improve barrier function (107). The cellular signaling events initiated by butyrate enabling it to carry out its protective effect have recently been evaluated. Butyrate was shown to reduce mucosal inflammation and improve epithelial cell integrity and barrier function in experimental colitis in rats. These effects were mediated by inhibition of activation of inducible heat shock protein HSP70, leading to reduced activation of NFκB (108).

Overall, although more research is required to confirm the advantage of using probiotics in IBD, the trend to a beneficial effect of bacterial supplementation as an adjunct to treatment is further supported by these TNFα downregulatory mechanisms.

15.5.2.1   Cytokine Release by Epithelial Cells

IECs located at the interface between commensal and probiotic bacteria and resident mucosal immune cells are involved in the transduction of bacteria-derived signals to the underlying immune system. *In vitro* studies (109) using the polarized intestinal cell line Caco-2 preactivated by coculture with peripheral blood mononuclear cells (PBMC) and challenged apically with *L. sakei* showed stimulation of expression of proinflammatory chemokines (MCP-1, IL-8 mRNA) and cytokines (TNFα and IL-1β mRNA), whereas Caco-2 cells alone remained unresponsive to bacterial challenge. In this study, although *L. johnsonii* showed a reduced ability to induce such proinflammatory cytokines, it increased TGFβ mRNA in PBMC-sensitized Caco-2 cells. In another study, the HT29-19A cell line did not secrete IL-8 in response to lactobacilli and bifidobacteria strains from VSL#3 and *Lactobacillus* GG, whereas *E. coli* Nissle 1917 induced IL-8 secretion in a dose-dependent manner (110). Yet another study using the HT29 epithelial cell line showed that several strains of *Lactobacillus rhamnosus*, *Lactobacillus delbrueckii* and *Lacto-*

bacillus acidophilus suppressed the production of the chemokine RANTES by stimulated HT29 cells. There was also a strain-dependent inhibition of IL-8, TNFα, and TGFβ production (111). Although it is difficult to demonstrate a clear-cut effect by probiotic bacteria on the regulation of cytokine expression by IECs, overall, the results suggest that probiotic-stimulated IECs secrete soluble messengers that may deliver discriminative signals to neighboring immune cells, influencing the immune response to pathogens or ingested antigens.

References

1. E Metchnikoff. *The Prolongation of Life: Optimistic Studies.* Heinemann, London, 1907.
2. R Fuller. Probiotics in man and animals. *J. Appl. Bacteriol.*, 66(5): 365–378, 1989.
3. EE Vaughan,F Schut, HG Heilig, EG Zoetendal, WM de Vos, AD Akkermans. A molecular view of the intestinal ecosystem. *Curr. Issues Intest. Microbiol.*, 1(1): 1–12, 2000.
4. PD Cotter, C Hill. Surviving the acid test: responses of Gram-positive bacteria to low pH. *Microbiol. Mol. Biol. Rev.*, 67(3): 429–453, 2003.
5. F Rallu, A Gruss, E Maguin. *Lactococcus lactis* and stress. *Antonie van Leeuwenhoek*, 70(2–4): 243–251, 1996.
6. Y Ma, TM Curran, RE Marquis. Rapid procedure for acid adaptation of oral lactic-acid bacteria and further characterization of the response. *Can. J. Microbiol.*, 43(2): 143–148, 1997.
7. EM Lim, SD Ehrlich, E Maguin. Identification of stress-inducible proteins in *Lactobacillus delbrueckii* subsp. *bulgaricus. Electrophoresis*, 21(12): 2557–2561, 2000.
8. R Lewis, S Gorbach. Modification of bile acids by intestinal bacteria. *Arch. Intern. Med.*, 130(4): 545–549, 1972.
9. L Stewart, CA Pellegrini, LW Way. Antibacterial activity of bile acids against common biliary tract organisms. *Surg. Forum*, 37: 157–159, 1986.
10. IW Percy-Robb, JG Collee. Bile acids: a pH dependent antibacterial system in the gut? *Br. Med. J.*, 3(830): 813–815, 1972.
11. C Dunne, L O'Mahony, L Murphy, G Thornton, D Morrissey, S O'Halloran, et al. *In vitro* selection criteria for probiotic bacteria of human origin: correlation with *in vivo* findings. *Am. J. Clin Nutr.*, 73(2 Suppl): 386S–392S, 2001.
12. D Haller, H Colbus, MG Ganzle, P Scherenbacher, C Bode, WP Hammes. Metabolic and functional properties of lactic acid bacteria in the gastro-intestinal ecosystem: a comparative *in vitro* study between bacteria of intestinal and fermented food origin. *Syst. Appl. Microbiol.*, 24(2): 218–226, 2001.
13. A Giraud, I Matic, O Tenaillon, A Clara, M Radman, M Fons, et al. Costs and benefits of high mutation rates: adaptive evolution of bacteria in the mouse gut. *Science*, 291(5513): 2606–2608, 2001.
14. JC Kolars, MD Levitt, M Aouji, DA Savaiano. Yoghurt — an autodigesting source of lactose. *N. Engl. J. Med.*, 310: 1–3, 1984.
15. DA Savaiano, A AbouElAnouar, DE Smith, MD Levitt. Lactose malabsorption from yoghurt, pasteurized yoghurt, sweet acidophilus milk, and cultured milk in lactase-deficient individuals. *Am. J. Clin. Nutr.*, 40: 1219–1223, 1984.

16. M de Vrese, A Stegelmann, B Richter, S Fenselau, C Laue, J Schrezenmeir. Probiotics—compensation for lactase insufficiency. *Am. J. Clin. Nutr.*, 73(2 Suppl): 421S–429S, 2001.

17. S Drouault, J Anba, G Corthier. *Streptococcus thermophilus* is able to produce a beta-galactosidase active during its transit in the digestive tract of germ-free mice. *Appl. Environ. Microbiol.*, 68(2): 938–941, 2002.

18. P Marteau, B Flourie, P Pochart, C Chastang, JF Desjeux, JC Rambaud. Effect of the microbial lactase (EC 3.2.1.23) activity in yoghurt on the intestinal absorption of lactose: an *in vivo* study in lactase-deficient humans. *Br. J. Nutr.*, 64: 71–79, 1990.

19. MA Shermak, JM Saavedra, TL Jackson, SS Huang, TM Bayless, JA Perman. Effect of yoghurt on symptoms and kinetics of hydrogen production in lactose-malabsorbing children. *Am. J. Clin. Nutr.*, 62(5): 1003–1006, 1995.

20. G Reid, J Howard, BS Gan. Can bacterial interference prevent infection? *Trends Microbiol.*, 9(9): 424–428, 2001.

21. TR Klaenhammer. Bacteriocins of lactic acid bacteria. *Biochimie*, 70(3): 337–349, 1988.

22. MF Bernet-Camard, V Lievin, D Brassart, JR Neeser, AL Servin, S Hudault. The human *Lactobacillus acidophilus* strain LA1 secretes a nonbacteriocin antibacterial substance(s) active *in vitro* and *in vivo*. *Appl. Environ. Microbiol.*, 63(7): 2747–2753, 1997.

23. S Dundas, WT Todd. Clinical presentation, complications and treatment of infection with verocytotoxin-producing *Escherichia coli*. Challenges for the clinician. *Symp. Ser. Soc. Appl. Microbiol.*, (29): 24S–30S, 2000.

24. SH Kim, SJ Yang, HC Koo, WK Bae, JY Kim, JH Park, et al. Inhibitory activity of *Bifidobacterium longum* HY8001 against Vero cytotoxin of *Escherichia coli* O157:H7. *J. Food Prot.*, 64(11): 1667–1673, 2001.

25. I Castagliuolo, MF Riegler, L Valenick, JT LaMont, C Pothoulakis. *Saccharomyces boulardii* protease inhibits the effects of *Clostridium difficile* toxins A and B in human colonic mucosa. *Infect. Immun.*, 67(1): 302–307, 1999.

26. MF Bernet, D Brassart, JR Neeser, AL Servin. *Lactobacillus acidophilus* LA 1 binds to cultured human intestinal cell lines and inhibits cell attachment and cell invasion by enterovirulent bacteria. *Gut*, 35(4): 483–489, 1994.

27. G Chauvière, MH Coconnier, S Kerneis, A Darfeuille-Michaud, B Joly, AL Servin. Competitive exclusion of diarrheagenic *Escherichia coli* (ETEC) from human enterocyte-like Caco-2 cells by heat- killed *Lactobacillus*. *FEMS Microbiol. Lett.*, 70(3): 213–217, 1992.

28. MH Coconnier, V Lievin, MF Bernet-Camard, S Hudault, AL Servin. Antibacterial effect of the adhering human *Lactobacillus acidophilus* strain LB. *Antimicrob. Agents Chemother.*, 41(5): 1046–1052, 1997.

29. I Adlerberth, S Ahrne, ML Johansson, G Molin, LA Hanson, AE Wold. A mannose-specific adherence mechanism in *Lactobacillus plantarum* conferring binding to the human colonic cell line HT-29. *Appl. Environ. Microbiol.*, 62(7): 2244–2251, 1996.

30. T Mukai, K Arihara, H Itoh. Lectin-like activity of *Lactobacillus acidophilus* strain JCM 1026. *FEMS Microbiol. Lett.*, 77(1–3): 71–74, 1992.

31. N Takahashi, T Saito, S Ohwada, H Ota, H Hashiba, T Itoh. A new screening method for the selection of *Lactobacillus acidophilus* group lactic acid bacteria with high adhesion to human colonic mucosa. *Biosci. Biotechnol. Biochem.*, 60(9): 1434–1438, 1996.

32. K Yamamoto, T Miwa, H Taniguchi, T Nagano, K Shimamura, T Tanaka, et al. Binding specificity of *Lactobacillus* to glycolipids. *Biochem. Biophys. Res. Commun.*, 228(1): 148–152, 1996.

33. MH Coconnier, TR Klaenhammer, S Kerneis, MF Bernet, AL Servin. Protein-mediated adhesion of *Lactobacillus acidophilus* BG2FO4 on human enterocyte and mucus-secreting cell lines in culture. *Appl. Environ. Microbiol.*, 58(6): 2034–2039, 1992.

34. G Chauviere, MH Coconnier, S Kerneis, J Fourniat, AL Servin. Adhesion of human *Lactobacillus acidophilus* strain LB to human enterocyte-like Caco-2 cells. *J. Gen. Microbiol.*, 138 (Pt 8): 1689–1696, 1992.

35. A Henriksson, R Szewzyk, PL Conway. Characteristics of the adhesive determinants of *Lactobacillus fermentum* 104. *Appl. Environ. Microbiol.*, 57(2): 499–502, 1991.

36. M Horie, A Ishiyama, Y Fujihira-Ueki, J Sillanpaa, TK Korhonen, T Toba. Inhibition of the adherence of *Escherichia coli* strains to basement membrane by *Lactobacillus crispatus* expressing an S-layer. *J. Appl. Microbiol.*, 92(3): 396–403, 2002.

37. L Plant, P Conway. Association of *Lactobacillus* spp. with Peyer's patches in mice. Clin Diagn Lab Immunol. 8(2): 320–324, 2001.

38. R Schwandner, R Dziarski, H Wesche, M Rothe, CJ Kirschning. Peptidoglycan-and lipoteichoic acid-induced cell activation is mediated by toll-like receptor 2. *J. Biol. Chem.*, 274(25): 17406–17409, 1999.

39. U Bocker, O Yezerskyy, P Feick, T Manigold, A Panja, U Kalina, et al. Responsiveness of intestinal epithelial cell lines to lipopolysaccharide is correlated with Toll-like receptor 4 but not Toll-like receptor 2 or CD14 expression. *Int. J. Colorectal Dis.*, 18(1): 25–32, 2003.

40. H Imaeda, H Yamamoto, A Takaki, M Fujimiya. *In vivo* response of neutrophils and epithelial cells to lipopolysaccharide injected into the monkey ileum. *Histochem. Cell Biol.*, 118(5): 381–388, 2002.

41. MW Hornef, T Frisan, A Vandewalle, S Normark, A Richter-Dahlfors. Toll-like receptor 4 resides in the Golgi apparatus and colocalizes with internalized lipopolysaccharide in intestinal epithelial cells. *J. Exp. Med.*, 195(5): 559–570, 2002.

42. AT Gewirtz, TA Navas, S Lyons, PJ Godowski, JL Madara. Cutting edge: bacterial flagellin activates basolaterally expressed TLR5 to induce epithelial proinflammatory gene expression. *J. Immunol.*, 167(4): 1882–1885, 2001.

43. M Akhtar, JL Watson, A Nazli, DM McKay. Bacterial DNA evokes epithelial IL-8 production by a MAPK-dependent, NF-kappaB-independent pathway. *FASEB J.*, 17(10): 1319–1321, 2003..

44. Kapczynski DR, Meinersmann RJ, Lee MD. Adherence of *Lactobacillus* to intestinal 407 cells in culture correlates with fibronectin binding. *Curr. Microbiol.*, 2000; 41(2): 136–141.

45. T Mukai, K Arihara, H Itoh. Lectin-like activity of *Lactobacillus acidophilus* strain JCM 1026. *FEMS Microbiol. Lett.*, 77(1–3): 71–74, 1992.

46. VI Morata dA, SN Gonzalez, G Oliver. Study of adhesion of *Lactobacillus casei* CRL 431 to ileal intestinal cells of mice. *J. Food Prot.*, 62(12): 1430–1434, 1999.

47. D Granato, F Perotti, I Masserey, M Rouvet, M Golliard, A Servin, et al. Cell surface-associated lipoteichoic acid acts as an adhesion factor for attachment of *Lactobacillus johnsonii* La1 to human enterocyte-like Caco-2 cells. *Appl. Environ. Microbiol.*, 65(3): 1071–1077, 1999.

48. LA Sherman, DC Savage. Lipoteichoic acids in *Lactobacillus* strains that colonize the mouse gastric epithelium. *Appl. Environ. Microbiol.*, 52(2): 302–304, 1986.

49. PL Conway, S Kjelleberg. Protein-mediated adhesion of *Lactobacillus fermentum* strain 737 to mouse stomach squamous epithelium. *J. Gen. Microbiol.*, 135 (Pt 5): 1175–1186, 1989.

50. G Lorca, MI Torino, V Font d, AA Ljungh. *Lactobacilli* express cell surface proteins which mediate binding of immobilized collagen and fibronectin. *FEMS Microbiol. Lett.*, 206(1): 31–37, 2002.

51. S Roos, P Aleljung, N Robert, B Lee, T Wadstrom, M Lindberg, et al. A collagen binding protein from *Lactobacillus reuteri* is part of an ABC transporter system? *FEMS Microbiol. Lett.*, 144(1): 33–38, 1996.

52. C Schneitz , L Nuotio, K Lounatma. Adhesion of *Lactobacillus acidophilus* to avian intestinal epithelial cells mediated by the crystalline bacterial cell surface layer (S-layer). *J. Appl. Bacteriol.*, 74(3): 290–294, 1993.

53. J Sillanpaa, B Martinez, J Antikainen, T Toba, N Kalkkinen, S Tankka, et al. Characterization of the collagen-binding S-layer protein CbsA of *Lactobacillus crispatus*. *J. Bacteriol.*, 182(22): 6440–6450, 2000.

54. S Avall-Jaaskelainen, A Lindholm, A Palva. Surface display of the receptor-binding region of the *Lactobacillus brev*is S-layer protein in *Lactococcus lactis* provides nonadhesive lactococci with the ability to adhere to intestinal epithelial cells. *Appl. Environ. Microbiol.*, 69(4): 2230–2236, 2003.

55. S Roos, H Jonsson. A high-molecular-mass cell-surface protein from *Lactobacillus reuteri* 1063 adheres to mucus components. *Microbiology*, 148(Pt 2): 433–442, 2002.

56. JA McGrady, WG Butcher, D Beighton, LM Switalski. Specific and charge interactions mediate collagen recognition by oral lactobacilli. *J. Dent. Res.*, 74(2): 649–657, 1995.

57. C Pelletier, C Bouley, C Cayuela, S Bouttier, P Bourlioux, MN Bellon-Fontaine. Cell surface characteristics of *Lactobacillus casei* subsp. casei, *Lactobacillus paracasei* subsp. paracasei, and *Lactobacillus rhamnosus* strains. *Appl. Environ. Microbiol.*, 63(5): 1725–1731, 1997.

58. JD Greene, TR Klaenhammer. Factors involved in adherence of lactobacilli to human Caco-2 cells. *Appl. Environ. Microbiol.*, 60(12): 4487–4494, 1994.

59. A Henriksson, R Szewzyk, PL Conway. Characteristics of the adhesive determinants of *Lactobacillus fermentum* 104. *Appl. Environ. Microbiol.*, 57(2): 499–502, 1991.

60. S Salminen, C Bouley, MC Boutron-Ruaul, JH Cummings, A Franck, GR Gibson, et al. Functional food science and gastrointestinal physiology and function. *Br. J. Nutr.*, 80 Suppl 1: S147–S171, 1998.

61. RC Chan, G Reid, RT Irvin, AW Bruce, JW Costerton. Competitive exclusion of uropathogens from human uroepithelial cells by *Lactobacillus* whole cells and cell wall fragments. *Infect. Immun.*, 47(1): 84–89, 1985.

62. MM Velraeds, HC van der Mei, G Reid, HJ Busscher. Inhibition of initial adhesion of uropathogenic *Enterococcus faecalis* by biosurfactants from Lactobacillus isolates. *Appl. Environ. Microbiol.*, 62(6): 1958–1963, 1996.

63. L Blomberg, A Henriksson, PL Conway. Inhibition of adhesion of *Escherichia coli* K88 to piglet ileal mucus by *Lactobacillus* spp. *Appl. Environ. Microbiol.*, 59(1): 34–39, 1993.

64. AC Ouwehand, PL Conway. Purification and characterization of a component produced by *Lactobacillus fermentum* that inhibits the adhesion of K88 expressing *Escherichia coli* to porcine ileal mucus. *J. Appl. Bacteriol.*, 80(3): 311–318, 1996.

65. S Michai, F Abernathy. *Lactobacillus plantarum* reduces the *in vitro* secretory response of intestinal epithelial cells to enteropathogenic *Escherichia coli* infection. *J. Pediatr. Gastroenterol. Nutr.*, 35(3): 350–355, 2002.

66. DR Mack, S Michail, S Wei, L McDougall, MA Hollingsworth. Probiotics inhibit enteropathogenic *E. coli* adherence *in vitro* by inducing intestinal mucin gene expression. *Am. J. Physiol.*, 276(4 Pt 1): G941–G950, 1999.

67. DR Mack, S Ahrne, L Hyde, S Wei, MA Hollingsworth. Extracellular MUC3 mucin secretion follows adherence of *Lactobacillus* strains to intestinal epithelial cells *in vitro*. *Gut*, 52(6): 827–833, 2003.

68. AF Mattar, DH Teitelbaum, RA Drongowski, F Yongyi, CM Harmon, AG Coran. Probiotics up-regulate MUC-2 mucin gene expression in a Caco-2 cell-culture model. *Pediatr. Surg. Int.*, 18(7): 586–590, 2002.

69. M Heyman, AM Crain-Denoyelle, G Corthier, JL Morgat, JF Desjeux. Postnatal development of protein absorption in conventional and germ-free mice. *Am. J. Physiol.*, 251: G326–G31, 1986.

70. A Garcia-Lafuente, M Antolin, F Guarner, E Crespo, JR Malagelada. Modulation of colonic barrier function by the composition of the commensal flora in the rat. *Gut*, 48(4): 503–507, 2001.

71. E Isolauri, H Majamaa, T Arvola, I Rantala, E Virtanen, H Arvilommi. *Lactobacillus casei* strain GG reverses increased intestinal permeability induced by cow milk in suckling rats. *Gastroenterology*, 105: 1643–1650, 1993.

72. E Isolauri, M Kaila, T Arvola, H Majamaa, I Rantala, E Virtanen, et al. Diet during rotavirus enteritis affects jejunal permeability to macromolecules in suckling rats. *Pediatr. Res.*, 33(6): 548–553, 1993.

73. K Terpend, MA Blaton, C Candalh, JM Wal, P Pochart, M Heyman. Intestinal barrier function and cow's milk sensitization in guinea-pigs fed milk or fermented milk. *J. Pediatr. Gastroenterol. Nutr.*, 28: 191–198, 1998.

74. S Resta-Lenert, KE Barrett. Live probiotics protect intestinal epithelial cells from the effects of infection with enteroinvasive *Escherichia coli* (EIEC). *Gut*, 52(7): 988–997, 2003.

75. K Madsen, A Cornish, P Soper, C McKaigney, H Jijon, C Yachimec, et al. Probiotic bacteria enhance murine and human intestinal epithelial barrier function. *Gastroenterology*, 121(3): 580–591, 2001.

76. M Gotteland, S Cruchet, S Verbeke. Effect of *Lactobacillus* ingestion on the gastrointestinal mucosal barrier alterations induced by indometacin in humans. *Aliment. Pharmacol. Ther.*, 15(1): 11–17, 2001.

77. CA Parkos, SP Colgan, C Delp, MA Arnaout, JL Madara. Neutrophil migration across a cultured epithelial monolayer elicits a biphasic resistance response representing sequential effects on transcellular and paracellular pathways. *J. Cell Biol.*, 117(4): 757–764, 1992.

78. JJ Perdomo, P Gounon, PJ Sansonetti. Polymorphonuclear leukocyte transmigration promotes invasion of colonic epithelial monolayer by Shigella flexneri. *J. Clin. Invest.*, 93(2): 633–643, 1994.

79. S Michai, F Abernathy. *Lactobacillus plantarum* inhibits the intestinal epithelial migration of neutrophils induced by enteropathogenic *Escherichia coli*. *J. Pediatr. Gastroenterol. Nutr.*, 36(3): 385–391, 2003.

80. AF Mattar, RA Drongowski, AG Coran, CM Harmon. Effect of probiotics on enterocyte bacterial translocation *in vitro*. *Pediatr. Surg. Int.*, 17(4): 265–268, 2001.

81. CE McNaught, NP Woodcock, J MacFie, CJ Mitchell. A prospective randomised study of the probiotic *Lactobacillus plantarum* 299V on indices of gut barrier function in elective surgical patients. *Gut*, 51(6): 827–831, 2002.

82. EJ Breese, CA Michie, SW Nicholls, SH Murch, CB Williams, P Domizio, et al. Tumor necrosis factor alpha-producing cells in the intestinal mucosa of children with inflammatory bowel disease. *Gastroenterology*, 106(6): 1455–1466, 1994.

83. M Iwamoto, T Koji, K Makiyama, N Kobayashi, PK Nakane. Apoptosis of crypt epithelial cells in ulcerative colitis. *J. Pathol.*, 180(2): 152–159, 1996.

84. F Yan, SK John, DB Polk. Kinase suppressor of Ras determines survival of intestinal epithelial cells exposed to tumor necrosis factor. *Cancer Res.*, 61(24): 8668–8675, 2001.

85. F Yan, DB Polk. Probiotic bacterium prevents cytokine-induced apoptosis in intestinal epithelial cells. *J. Biol. Chem.*, 277(52): 50959–50965, 2002.

86. D Rachmilewitz, F Karmeli, K Takabayashi, T Hayashi, L Leider-Trejo, J Lee, et al. Immunostimulatory DNA ameliorates experimental and spontaneous murine colitis. *Gastroenterology*, 122(5): 1428–1441, 2002.

87. H Ichikawa, T Kuroiwa, A Inagaki, R Shineha, T Nishihira, S Satomi, et al. Probiotic bacteria stimulate gut epithelial cell proliferation in rat. *Dig. Dis. Sci.*, 44(10): 2119–2123, 1999.

88. PA Crabbe, H Bazin, H Eyssen, JF Heremans. The normal microbial flora as a major stimulus for proliferation of plasma cells synthesizing IgA in the gut. The germ-free intestinal tract. *Int. Arch. Allergy Appl. Immunol.*, 34(4): 362–375, 1968.

89. MC Moreau, R Ducluzeau, D Guy-Grand, MC Muller. Increase in the population of duodenal immunoglobulin A plasmocytes in axenic mice associated with different living or dead bacterial strains of intestinal origin. *Infect. Immun.*, 21(2): 532–539, 1978.

90. M Kaila, E Isolauri, E Soppi, E Virtanen, S Laine, H Arvilommi. Enhancement of the circulating antibody secreting cell response in human diarrhea by a human *Lactobacillus* strain. *Pediatr. Res.*, 32: 141–144, 1992.

91. M Kaila, E Isolauri, M Saxelin, H Arvilommi, T Vesikari. Viable versus inactivated *Lactobacillus* strain GG in acute rotavirus diarrhoea. *Arch. Dis. Child.*, 72(1): 51–53, 1995.

92. H Link Amster, F Rochat, KY Saudan, O Mignot, JM Aeschlimann. Modulation of a specific humoral immune response and changes in intestinal flora mediated through fermented milk intake. *FEMS Immunol. Med. Microbiol.*, 10(1): 55–63, 1994.

93. P Marteau, JP Vaerman, JP Dehennin, S Bord, D Brassart, P Pochart, et al. Effects of intrajejunal perfusion and chronic ingestion of *Lactobacillus johnsonii* strain La1 on serum concentrations and jejunal secretions of immunoglobulins and serum proteins in healthy humans. *Gastroenterol. Clin. Biol.*, 21(4): 293–298, 1997.

94. E Isolauri, J Joensuu, H Suomalainen, M Luomala, T Vesikari. Improved immunogenicity of oral D x RRV reassortant rotavirus vaccine by *Lactobacillus casei* GG. *Vaccine*, 13(3): 310–312, 1995.

95. T Takahashi, E Nakagawa, T Nara, T Yajima, T Kuwata. Effects of orally ingested *Bifidobacterium longum* on the mucosal IgA response of mice to dietary antigens. *Biosci. Biotechnol. Biochem.*, 62(1): 10–15, 1998.

96. C Veltkamp, SL Tonkonogy, YP De Jong, C Albright, WB Grenther, E Balish, et al. Continuous stimulation by normal luminal bacteria is essential for the development and perpetuation of colitis in Tg(epsilon26) mice. *Gastroenterology*, 120(4): 900–913, 2001.

97. L Steidler, W Hans, L Schotte, S Neirynck, F Obermeier, W Falk, et al. Treatment of murine colitis by *Lactococcus lactis* secreting interleukin-10. *Science*, 289(5483): 1352–1355, 2000.

98. KL Madsen, JS Doyle, LD Jewell, MM Tavernini, RN Fedorak. *Lactobacillus* species prevents colitis in interleukin 10 gene-deficient mice. *Gastroenterology*, 116(5): 1107–1114, 1999.

99. M Malin, H Suomalainen, M Saxelin, E Isolauri. Promotion of IgA immune response in patients with Crohn's disease by oral bacteriotherapy with *Lactobacillus* GG. *Ann. Nutr. Metab.*, 40(3): 137–145, 1996.

100. P Gupta, H Andrew, BS Kirschner, S Guandalini. Is *Lactobacillus* GG helpful in children with Crohn's disease? Results of a preliminary, open-label study. *J. Pediatr. Gastroenterol. Nutr.*, 31(4): 453–457, 2000.

101. P Gionchetti, F Rizzello, A Venturi, P Brigidi, D Matteuzzi, G Bazzocchi, et al. Oral bacteriotherapy as maintenance treatment in patients with chronic pouchitis: a double-blind, placebo-controlled trial. *Gastroenterology*, 119(2): 305–309, 2000.

102. AS Neish, AT Gewirtz, H Zeng, AN Young, ME Hobert, V Karmali, et al. Prokaryotic regulation of epithelial responses by inhibition of IkappaB-alpha ubiquitination. *Science*, 289(5484): 1560–1563, 2000.

103. RJ Xavier, DK Podolsky. Microbiology. How to get along—friendly microbes in a hostile world. *Science*, 289(5484): 1483–1484, 2000.

104. S Ménard, C Candalh, J C Bambou, K Terpend, N Cerf-Bensussan, M Heyman. Lactic acid bacteria secrete metabolites retaining anti-inflammatory properties after intestinal transport. *Gut*, 53(6): 821–828, 2004.

105. JA Pena, J Versalovic. *Lactobacillus rhamnosus* GG decreases TNF-alpha production in lipopolysaccharide-activated murine macrophages by a contact- independent mechanism. *Cell Microbiol.*, 5(4): 277–285, 2003.

106. N Borruel, M Carol, F Casellas, M Antolin, F de Lara, E Espin, et al. Increased mucosal tumour necrosis factor alpha production in Crohn's disease can be downregulated *ex vivo* by probiotic bacteria. *Gut*, 51(5): 659–664, 2002.

107. SI Cook, JH Sellin. Review article: short chain fatty acids in health and disease. *Aliment. Pharmacol. Ther.*, 12(6):499–507, 1998.

108. A Venkatraman, BS Ramakrishna, RV Shaji, NS Kumar, A Pulimood, S Patra. Amelioration of dextran sulfate colitis by butyrate: role of heat shock protein 70 and NF-kappaB. *Am. J. Physiol. Gastrointest. Liver Physiol.*, 285(1): G177–G184, 2003.

109. D Haller, C Bode, WP Hammes, AM Pfeifer, EJ Schiffrin, S Blum. Non-pathogenic bacteria elicit a differential cytokine response by intestinal epithelial cell/leucocyte co-cultures. *Gut*, 47(1): 79–87, 2000.

110. KM Lammers, U Helwig, E Swennen, F Rizzello, A Venturi, E Caramelli, et al. Effect of probiotic strains on interleukin 8 production by HT29/19A cells. *Am. J. Gastroenterol.*, 97(5): 1182–1186, 2002.

111. TD Wallace, S Bradley, ND Buckley, JM Green-Johnson. Interactions of lactic acid bacteria with human intestinal epithelial cells: effects on cytokine production. *J. Food Prot.*, 66(3): 466–472, 2003.

16

Probiotics in Cancer Prevention

Kazuhiro Hirayama and Joseph Rafter

CONTENTS

16.1 Introduction

Diet makes an important contribution to cancer, e.g., up to 75% of colorectal cancer cases are thought to be associated with diet, implying that risks of cancer are potentially reducible. Evidence from a wide range of sources supports the view that the colonic microflora is involved in the etiology of cancer. This has led to intense interest in factors such as probiotics that can modulate gut microflora and its metabolism.

Although a myriad of healthful effects have been attributed to the probiotic bacteria (1), perhaps the most interesting and controversial remains that of anticancer activity. Although data from human studies (epidemiology and experimental) for cancer suppression in humans as a result of consumption of probiotic cultures in fermented or unfermented dairy products are limited, there is a wealth of indirect evidence, based largely on laboratory studies, and these will be summarized here. Reports in the literature regarding the anticancer effects of probiotic bacteria fall into the following categories: epidemiological studies, experimental studies in human volunteers, and laboratory animals and *in vitro* studies. Examples of these reports will be given below. The mechanisms by which probiotic bacteria may inhibit colon cancer are still poorly understood. However, several potential mechanisms are being discussed in the literature, and these will also be addressed.

16.2 Epidemiological Studies

As yet, there are few epidemiological studies addressing the association between fermented dairy products and cancer. Consumption of large quantities of dairy products such as yogurt and fermented milk containing *Lactobacillus* or *Bifidobacterium* may be related to a lower incidence of colon cancer (2). An epidemiological study performed in Finland demonstrated that, despite the high fat intake, colon cancer incidence was lower than in other countries because of the high consumption of milk, yogurt, and other dairy products (3, 4).

In two population-based case-control studies of colon cancer, an inverse association was observed for yogurt (5) and cultured milk consumption (6), adjusted for potential confounding variables. In another case-control study, an inverse relationship for yogurt consumption with risk of large colon adenomas in men and women was reported (7). It can also be mentioned that an inverse relationship has been demonstrated between the frequency of consumption of yogurt and other fermented milk products and breast cancer in women (8, 9).

On the other hand, two companion American prospective studies, the 1980–1988 follow-up of the Nurses' Health Study and the 1986–1990 Health Professionals follow-up study, did not provide evidence that intake of dairy products is associated with a decreased risk of colon cancer (10). In a cohort study in The Netherlands, it was shown that the intake of fermented dairy products was not significantly associated with colorectal cancer risk in an elderly population with a relatively wide variation in dairy product consumption, although a weak nonsignificant inverse association with colon cancer was observed (11). In summary, it would appear that the case control studies indicate protective effects, whereas the prospective studies do not.

16.3   Experimental Studies

16.3.1   Studies in Human Volunteers

Consumption of lactobacilli by healthy volunteers has been shown to reduce the mutagenicity of urine and feces associated with the ingestion of carcinogens in cooked meat. When *Lactobacillus acidophilus* was given to healthy volunteers on a fried meat diet known to increase fecal mutagenicity, a lower fecal mutagenic activity was noted on day 3 with *L. acidophilus* compared to day 3 without *L. acidophilus* when fried meat and ordinary fermented milk were given (12). High levels of mutagenicity also appeared in urine on days 2 and 3 of the fried meat and ordinary fermented milk dietary regimen. During *L. acidophilus* administration, the urinary mutagenic activity on days 2 and 3 was significantly lower compared to the ordinary fermented milk period. In most cases, an increase in the number of fecal lactobacilli corresponded to a lower mutagen excretion, particularly in urine. Hayatsu and Hayatsu (13) also demonstrated a marked suppressing effect of orally administered *L. casei* on the urinary mutagenicity arising from ingestion of fried ground beef in the human.

It is possible that the *L. acidophilus* supplements are influencing excretion of mutagens by simply binding them in the intestine. However, lactic acid bacteria have also been shown to affect the host. Mucosal cell proliferative activity in upper colonic crypts of patients with colon adenomas significantly decreased after the administration of *L. acidophilus* and *Bifidobacterium bifidum* cultures (14).

A randomized controlled study in patients with superficial bladder cancer suggested that oral administration of *L. casei* preparation is useful for the prevention of the cancer (15). The 50% recurrence-free interval after transurethral resection of the bladder tumor was prolonged by the treatment to 1.8 times that in the control group. A double-blind trial (16) also demonstrated that oral administration of *L. casei* preparation is safe and effective for preventing recurrence of superficial bladder cancer.

16.3.2   Laboratory Animal Studies

Oral administration of probiotic bacteria has been shown to effectively reduce DNA damage, induced by chemical carcinogens, in gastric and colonic mucosa in rats. Pool-Zobel et al. (17) reported, using the comet assay, that *L. acidophilus*, *L. gasseri*, *L. confusus*, *Streptococcus thermophilus*, *B. breve*, and *B. longum* were antigenotoxic toward N-methyl-N'-nitro-N-nitrosoguanidine (MNNG). These bacteria were also protective toward 1,2-dimethylhydrazine (DMH)-induced genotoxicity. Metabolically active *L. acidophilus* cells, as well as an acetone extract of the culture, prevented MNNG-induced DNA damage, whereas heat-treated *L. acidophilus* was not antigenotoxic. Among different cell fractions from *L. acidophilus*, the peptidoglycan fraction and whole freeze-dried cells were antigenotoxic. Oral administration of probiotics showed strong anticlastogenic action in the chromosome aberration test and the micronucleus test (18). Lactobacilli were effective also after intraperitoneal infection.

Certain strains of lactic acid bacteria have also been found to prevent putative preneoplastic lesions or tumors induced by carcinogens. Goldin et al. (19) showed that a specific strain of *L. casei* subsp. *rhamnosus* designated GG can interfere with the initiation or early promotional stages of DMH-induced intestinal tumorigenesis, and that this effect is most pronounced for animals fed a high-fat diet. Overnight cultures of *L. acidophilus* also inhibited the formation of aberrant crypt foci (ACF), induced by azoxymethane (AOM) (20). ACF are putative preneoplastic lesions from which adenomas and carcinomas may develop. Although *B. adolescentis* culture and its supernatant did not show an inhibitory effect in this study, feeding of bifidobacteria suppressed the ACF formation induced by AOM (21, 22) or DMH (23, 24). Consumption of *B. longum* or inulin was associated with a decrease in AOM-induced colonic small ACF in rats and combined administration significantly decreased the incidence of large ACF (25). In addition, it has been reported that colonization of bacteria with an ability to produce genotoxic compounds and high β-glucuronidase activity enhanced progression of ACF induced by DMH in rats, and that the additional colonization of *B. breve* reduced the number of ACF with four or more crypts or focus and crypt multiplicity, which are reliable predictors of malignancy (26).

Reddy and Rivenson (27) reported that lyophilized cultures of *B. longum* administered in the diet to rats inhibited liver, colon, and mammary tumors induced by the food mutagen 2-amino-3-methyl-3H-imidazo(4,5-*f*)quinoline (IQ). In another study, Kohwi et al. (28) demonstrated the potential of *B. infantis* and *B. adolescentis* injected either subcutaneously or intraperitoneally into BALB/c mice to inhibit 3-methylcholanthrene-induced tumors. Goldin and Gorbach (29) showed that dietary supplements of *L. acidophilus* not only suppressed the incidence of DMH-induced colon carcinogenesis, but also increased the latency period in rats. Feeding of fermented milk increased the survival rate of rats with chemically induced colon cancer (30). Dietary administration of a lyophilized culture of *B. longum* resulted in a significant suppression of colon tumor incidence and tumor multiplicity and also reduced tumor volume induced by AOM in rats (31).

There is additional direct evidence for antitumor activities of probiotic bacteria obtained in studies using preimplanted tumor cells in animal models. It has been demonstrated that feeding of fermented milk or cultures containing lactic acid bacteria inhibited the growth of tumor cells injected into mice (32, 33). Repeated intralesional injection of live or dead *Bifidobacterium* cells inhibited the growth of Meth-A tumor cells transplanted subcutaneously into syngenic BALB/c mice (28). Sekine et al. (34), using whole peptidoglycan isolated from *B. infantis* strain ATCC 15697, reported that a single subcutaneous injection significantly suppressed tumor growth, and five intralesional injections resulted in 70% tumor regression in the mice.

More recently, mindful of the fact that the composition and metabolic activities of the intestinal flora of experimental animals are significantly different from those of humans (35), human flora-associated (HFA) mice were exploited to test the effects of a probiotic mixture on a parameter of relevance for colon carcinogenesis, i.e., DNA adduct formation (36). The probiotic mixture, which contained *S. faecalis* T-110, *Clostridium butyricum* TO-A, and *Bacillus mesentericus* TO-A, had an effect to significantly decrease the DNA adduct formation in the colonic epithelium induced by the food mutagen 2-amino-9H-pyrido[2,3-b]indole (2-amino-alpha-carboline), given by gavage.

Mizutani and Mitsuoka (37, 38) demonstrated that spontaneous liver tumorigenesis in C3H/He male mice is higher in mice harboring intestinal bacteria than in germ-free mice and markedly promoted by association with *Escherichia coli*, *S. faecalis*, and *C. paraputrificum*. This promoting effect was suppressed by addition of certain intestinal bacteria such as *B. longum*, *L. acidophilus*, and *Eubacterium rectale*.

16.3.3 *In Vitro* Studies

L. casei and its derivative peptidoglycan demonstrated *in vitro* cytotoxic activities against various murine (Yac-1, P815, Ehrlich ascites tumor, and mammary carcinoma) and human (K562, KB) tumor cell lines (39). Autolysates of *L. gasseri* improved the repair capacity of cultured human fibroblasts with damaged DNA (40).

Probiotic bacteria were also shown to be effective in mutagenicity assay systems using *Salmonella* strains. Hosono et al. (41) demonstrated the antimutagenic properties of milk cultured with *L. bulgaricus* and *S. thermophilus* on all mutagens tested in an *in vitro* assay system using streptomycin-dependent *Salmonella* strains. They included 2-(2-furyl)-3-(5-nitro-2-furyl) acrylamide, 4-nitroquinoline-N-oxide, and fecal mutagenic extracts from animals as mutagens. Renner and Münzner (18) showed the antimutagenic activities of probiotics in the Ames test with *Salmonella typhimurium* strain TA1538 using beef extract and nitrosated beef extract. Cultured milk using 71 strains of lactic acid bacteria belonging to the genera *Lactobacillus*, *Streptococcus*, *Lactococcus*, and *Bifidobacterium* displayed their characteristic antimutagenic effects on the mutagenicity of MNNG in an *in vitro* assay using *Salmonella typhimurium* strain TA100 as an indicator strain (42). *L. acidophilus* LA106 showed the highest inhibition of 77% among the strains tested.

16.4   Mechanisms by Which Probiotic Bacteria Inhibit Cancer

The precise mechanisms by which probiotic bacteria may inhibit colon cancer are currently unknown. However, such mechanisms might include binding and degrading potential carcinogens; production of antitumorigenic or anti-mutagenic compounds; alteration of the metabolic activities of intestinal microflora; quantitative and/or qualitative alterations in the intestinal micro-flora incriminated in producing putative carcinogen(s) and promoters (e.g., bile acid-metabolizing bacteria); alteration of physicochemical conditions in the colon; enhancing the host's immune response; and effects on the physi-ology of the host. These potential mechanisms are addressed individually below.

16.4.1   Binding and Degrading Potential Carcinogens

Bacterial cells in addition to certain plant cell walls may be an important factor in determining the ratio of bound to free (bioavailable) toxins in the intestine. Mutagenic compounds, commonly found in the Western meat-rich diet, can be bound to the intestinal and lactic acid bacteria *in vitro* and binding correlated well with the reduction in mutagenicity observed after exposure to the bacte-rial strains (43, 44). Morotomi and Mutai (45) investigated the ability of 22 strains of intestinal bacteria to bind the mutagenic pyrolyzates and compared their ability to that of some dietary fibers. 3-amino-1,4-dimethyl-5*H*-pyrido[4,3-*b*]indole (Trp-P-1) and 3-amino-1-methyl-5*H*-pyrido[4,3-*b*]indole (Trp-P-2) were effectively bound to all Gram-positive and some Gram-negative bacterial cells, corn bran, apple pulp, and soybean fiber. The mutagenicity of Trp-P-2 for *Salmonella typhimurium* TA98 in the presence of S9 mix was inhib-ited by the addition of *L. casei* YIT 9018 to the reaction mixture, indicating that bound Trp-P-2 did not cause mutation under the assay conditions. Orrhage et al. (46) have reported on the binding capacity of eight human intestinal or lactic acid bacterial strains for mutagenic heterocyclic amines formed during cooking of protein-rich food. There were only minor differences in the binding capacities of the tested strains, but the mutagenic compounds were bound with markedly different efficiencies. Binding correlated well with the reduc-tion of mutagenicity observed after exposure of the heterocyclic amines to the bacterial strains. The binding was shown to be pH dependent, occurred instan-taneously and was inhibited by the addition of metal salts, indicating a cation-exchange mechanism (45, 46). It has been suggested that cell wall peptidoglycans and polysaccharides are the two most important elements responsible for the binding (47).

Live cells showed higher antimutagenic activity and better efficiency in inhibiting the mutagens than killed bacteria. Live bacterial cells bound or inhibited the mutagens permanently, whereas killed bacteria released mutagens after extraction with dimethylsulfoxide (48). Although little is known about the fate of bound mutagens in the human gastrointestinal system, Zhang and Ohta (49) showed that freeze-dried cells of lactic acid bacteria, intestinal bacteria, and yeast significantly reduced the absorption of Trp-P-1 from the small intestine in rats, and this was accompanied by decreased levels of this food mutagen in blood. Cells of *L. delbrueckii* subsp. *bulgaricus* 2038 and *S. thermophilus* 1131 bind Trp-P-1 and the absorption of Trp-P-1 by the small intestine of F344 rats, investigated using an *in situ* loop technique, were significantly lower in the presence of strain 1131 cells than in the absence of the cells, but the presence of strain 2038 cells had no effect on Trp-P-1 absorption (50). The authors suggested that strain 1131 cells bind to Trp-P-1 at the same pH as that of the small intestine (pH 6 to 7) and thus decrease its absorption. The dietary supplementation of lactic acid bacteria reduced uptake of Trp-P-2 and its metabolites in various tissues of mice (51). In addition, consumption of lactobacilli by human volunteers has been shown to reduce the mutagenicity of urine and feces associated with the ingestion of carcinogens in cooked meat (12, 13).

It has been suggested that the cellular uptake of nitrites by lactic acid bacteria reduce the formation of nitrosamines from nitrates (52). Lactobacilli have also been shown to degrade nitrosamines (53).

16.4.2 Production of Antitumorigenic or Antimutagenic Compounds

Acetone or ethylacetate extracts of skim milk fermented by *S. thermophilus* and/or *L. bulgaricus* showed significant dose response in suppressing mutagenicity of both a direct-acting mutagen and a mutagen requiring S9 activation using *Salmonella typhimurium* strains TA98 and TA100 (54). Samples of reconstituted nonfat dry milk fermented by *L. helveticus* CH65, *L. acidophilus* BG2FO4, *S. salivarius* subsp. *thermophilus* CH3, *L. delbrueckii* subsp. *bulgaricus* 191R, and by a mixture of the latter two organisms were freeze-dried, extracted in acetone, dissolved in dimethylsulfoxide, and assayed for antimutagenicity in the Ames test (*Salmonella typhimurium* TA100) (55). Dose-dependent activity was significant against mutagens tested in all extracts. Compounds that were responsible for activity against the mutagens were less soluble in aqueous solutions than in dimethylsulfoxide.

A soluble compound produced by probiotic bacteria may interact directly with tumor cells in culture and inhibit their growth (56, 57). Lactic acid bacteria significantly reduced the growth and viability of the human colon cancer cell line HT-29 in culture and dipeptidyl peptidase IV, and brush-border enzymes were significantly increased, suggesting that these cells may have entered a differentiation process (58). Milk fermented by *B. infantis, B. bifidum, B. animalis, L. acidophilus,* and *L. paracasei* inhibited the growth of the MCF7 breast cancer cell line, and the antiproliferative effect was not related to the presence of bacteria (59). These findings strongly suggest the presence of an *ex novo* anticarcinogenic compound produced by lactic acid bacteria during milk fermentation or the microbial transformation of some milk components in a biologically active form.

16.4.3 Alteration of the Metabolic Activities of Intestinal Microflora

Many foreign compounds are detoxified by glucuronide formation in the liver before entering the intestine via the bile. The bacterial β-glucuronidase has the ability to hydrolyze many glucuronides due to its wide substrate specificity, and thus may liberate carcinogenic aglycones in the intestinal lumen. Several other bacterial enzymes have also been suggested to be implicated in the carcinogenic process, releasing carcinogens in the intestinal tract. Interestingly, it was these earlier observations that feeding probiotic bacteria supplements in the diets of rodents significantly decreased the activities of some of the above fecal enzyme activities, which focused attention on these bacteria as possible anticancer agents (22, 23, 25, 60). Lactic acid bacteria also reduced the specific activities of fecal enzymes in human volunteer studies (61–64). Goldin and Gorbach (65) studied the effect of feeding *L. acidophilus* strains NCFM and N-2 on the activity of three bacterial enzymes, β-glucuronidase, nitroreductase, and azoreductase, in 21 healthy volunteers. Both strains had similar effects and caused a significant decline in the specific activity of the three enzymes in all subjects after 10 d of feeding. A reversal effect was observed within 10 to 30 d of stopping *Lactobacillus* feeding, indicating that continuous consumption of these bacteria was necessary to maintain the effect. The authors suggested that the observed reduction of these enzymes may explain the earlier reported reduced colon cancer incidence in rats fed viable *L. acidophilus* (29). Administration of the *Lactobacillus*-GG-fermented whey drink to elderly nursing home residents also resulted in a significant decrease in glycocholic acid hydrolase activity and tryptic activity without significant effect on bowel function (66). It is important to mention here that the reports published to date do not always find reductions in the same enzymes, although findings with β-glucuronidase and nitroreductase are most consistently positive. However, we still do not know how or whether a reduction in these enzyme activities affects cancer rates in humans. Indeed, the origin of the carcinogens causing this disease in humans is still to a large extent unknown.

16.4.4 Quantitative and Qualitative Alterations in the Intestinal Microflora

The administration of milk fermented with *L. acidophilus* caused a remarkable decrease in fecal mutagenicity compared with before the administration, and the suppression of mutagenicity appeared to be due to the proliferation of lactobacilli and bifidobacteria in fecal microflora caused by the presence of the administered strain in the human intestine (67). In other studies, consumption of fermented milk containing *L. acidophilus* has been shown to reduce significantly the counts of fecal putrefactive bacteria such as coliforms and increase the levels of lactobacilli in the intestine (2, 61), suggesting that supplemental *L. acidophilus* has a beneficial effect on the intestinal microecology by suppressing the putrefactive organisms that are possibly involved in the production of tumor promoters and putative precarcinogens. However, the mechanisms underlying these effects are still poorly understood.

16.4.5 Alteration of Physicochemical Conditions in the Intestine

Modler et al. (68) have suggested that large bowel cancer could be influenced directly by reducing intestinal pH, thereby preventing the growth of putrefactive bacteria. In rats given inulin-containing diets with or without *B. longum*, an increase in cecal weight and -glucosidase and a decrease in cecal pH were observed (25), though some other studies did not detect a significant change in intestinal pH (23, 69).

Dietary fat has been considered a risk factor for colon cancer, and it has been suggested that this phenomenon may be mediated by increased levels of bile acids (mainly secondary bile acids, produced by the action of bacterial α-dehydroxylase on primary bile acids) in the colon (70). One hypothesis regarding colon carcinogenesis involves a cytotoxic effect on the colonic epithelium exerted by bile acids in the aqueous phase of feces (soluble bile acids), followed by an increased proliferation of cells in the intestine (71). It has been demonstrated that a 6-week administration of *L. acidophilus* fermented milk supplements to colon cancer patients resulted in lower concentrations of soluble bile acids in feces (72). Although the decrease in the concentration of bile acids in this fraction of feces was not significant, it was of interest that decreased levels of soluble bile acids were observed in the colon cancer patients receiving *L. acidophilus* fermented milk supplements. In another study, patients with colonic adenomas participated in a 3-month study, where *L. acidophilus* was administered together with *B. bifidum* (14). During this period, the fecal pH was reduced significantly, and patients having a higher proliferative activity in the upper colonic crypts than that calculated for subjects at low risk for colon cancer showed a significant decrease after therapy with the lactic acid bacteria. In view of the results in the above-mentioned study (72), it is interesting to speculate that this latter effect was in part due to decreased levels of bile acids in the aqueous phase of feces.

16.4.6 Enhancing the Host's Immune Response

One explanation for tumor suppression by lactic acid bacteria may be mediated through an immune response of the host. Sekine et al. (34) suggested that *B. infantis* stimulates the host-mediated response, leading to tumor suppression or regression. The authors further examined the role of the bifidobacterial cell wall preparation (WPG) on the adjuvant activity for *in vivo* immune responses in mice and the induction of antitumor cytokines of mouse peritoneal exudate cells (73, 74). They demonstrated that bifidobacterial WPG may play a role as an immunomodulator.

In addition, there are studies to suggest that probiotic bacteria play an important role and function in the host's immunoprotective system by increasing specific and nonspecific mechanisms to have an antitumor effect (75-77). *L. casei* strain Shirota (LcS) has been shown to have potent antitumor and antimetastatic effects on transplantable tumor cells and to suppress chemically induced carcinogenesis in rodents. Also, intrapleural administration of LcS into tumor-bearing mice has been shown to induce the production of several cytokines, such as interferon-γ, interleukin-1β, and tumor-necrosis factor-α, in the thoracic cavity of mice, resulting in the inhibition of tumor growth and increased survival (78). These findings suggest that treatment with LcS has the potential to ameliorate or prevent tumorigenesis through modulation of the host's immune system, specifically cellular immune responses. An additional study has indicated that oral administration of *L. casei* YIT9018 potentiated systemic immune responses that modified T-cell functions in tumor-bearing mice (79–81) and induced several cytokines which participated in the subsequent immnoresponses (82). *L. casei* YIT9018, viable and/or heat-killed bacteria, inhibited tumor growth and pulmonary metastasis, prolonged the survival period and were effective for regression of established tumors after tumor transplantation (83–85).

It has been also demonstrated that *B. longum* and *B. animalis* promote the induction of inflammatory cytokines (interleukin-6, tumor-necrosis factor-α) in mouse peritoneal cells (86).

16.4.7   Effects on Physiology of the Host

The ileal mucosa as well as the colonic mucosa have the capacity to absorb mutagenic compounds from the intestinal lumen whereafter the compounds are passed into the bloodstream, either unchanged or as metabolites. Lactobacilli are one of the dominant species in the small intestine, and these microorganisms presumably affect metabolic reactions occurring in this part of the gastrointestinal tract. Lactic acid bacteria have also been shown to increase colonic NADPH-cytochrome P-450 reductase activity (17) and glutathione S-transferase levels (21) and to reduce hepatic uridine diphosphoglucuronyl transferase activity (23), enzymes which are involved in the metabolism of carcinogens in rats.

Butyrate, a metabolite of some probiotic strains, promotes differentiation and apoptosis in a variety of colon tumor cell lines. Apoptosis is a central feature in the regulation of cell number and the elimination of nonfunctional, harmful, or abnormal cells in the colon. Therefore, apoptosis induced by butyrate could affect the removal of DNA adducts from the colonic epithelium. Arimochi et al. (20) investigated the effect of several strains of intestinal bacteria on the formation of AOM-induced ACF and DNA adducts in the rat colon. They showed that culture supernatants, not bacterial cells, of *L. acidophilus* inhibited the ACF formation and that the inhibitory effect was due to the enhanced removal of O^6-methylguanine from the colon mucosal DNA. They suggested that a metabolite of *L. acidophilus* in the culture supernatants might induce the O^6-methylguanine-repair enzyme, methylguanine-methyltransferase. Therefore, metabolites of probiotic bacteria might be inducing DNA-adduct-repair enzymes after exposure to mutagens found in the diet.

In addition, it has been demonstrated that dietary administration of lyophilized cultures of *B. longum* strongly suppressed AOM-induced colonic tumor development and that this effect was associated with a decrease in colonic mucosal cell proliferation and colonic mucosal and tumor ornithine decarboxylase, and ras-p21 activities (87).

16.5 Conclusions

The strongest evidence for anticancer effects of probiotics comes from animal studies, and evidence from human studies (epidemiology and experimental) is still limited. An important goal for the future should be carefully designed human clinical trials to corroborate the wealth of experimental studies. Also, as discussed above, there are several possible mechanisms that might explain how probiotic bacteria might protect against tumor development in the colon. However, questions such as what is the major contributing mechanism for a particular bacterial strain or how are the different mechanisms linked are questions to which we currently do not have the answers. It is possible that different strains target different mechanisms. All of the mechanisms have various degrees of support, mainly originating from *in vitro* and animal experiments, and some of them even have some support from human clinical studies.

However, it should be emphasized that great care must be exercised in extrapolating the results of the above *in vitro* and animal studies to the human system. Many of the animal studies exploit specifically bred strains of mice, and whether one can extrapolate antitumor activity in these animals to humans is somewhat unclear. It also must be kept in mind that the composition and metabolic activities of intestinal flora of experimental animals are significantly different from those of humans. Indeed, it has been demonstrated that human intestinal microflora had different effects than mouse microflora concerning DNA adduct formation after exposure to mutagens (88). Exploitation of human flora-associated animals, germ-free animals inoculated with whole human intestinal flora, can be one of the solutions to this problem (89). Results of administering probiotic cultures intravenously, intraperitoneally, and intralesionally (often used in animal studies) may not be compatible with oral consumption in humans. Many of the antitumor activities attributed to lactic cultures have been suggested to involve an enhanced function of the immune response. This effect may not be specific to lactic acid bacteria, and perhaps many microbes administered similarly would produce the same results.

Thus, more work needs to be done to identify the specific strains and strain characteristics responsible for specific antitumor effects and the mechanisms by which these effects are mediated. However, even with the above reservations in mind and mindful of the limited number of human studies available, the use of probiotics for human cancer suppression is interesting, holds promise, and certainly deserves more scrutiny.

Acknowledgments

This work was supported by a grant from the Swedish Cancer Society.

References

1. S Salminen, C Bouley, M-C Boutron-Ruault, JH Cummings, A Franck, GR Gibson, E Isolauri, MC Moreau, M Roberfroid, I Rowland. Functional food science and gastrointestinal physiology and function. *Br. J. Nutr.*, 80(Suppl.): s147–s171, 1998.

2. KM Shahani, AD Ayebo. Role of dietary lactobacilli in gastrointestinal microecology. *Am. J. Clin. Nutr.*, 33: 2448–2457, 1980.

3. SL Malhotra. Dietary factors in a study of colon cancer from cancer registry, with special reference to the role of saliva, milk, and fermented milk products and vegetable fibre. *Med. Hypotheses*, 3: 122–134, 1977.

4. Intestinal Microecology Group, International Agency for Research on Cancer. Dietary fibre, transit time, fecal bacteria, steroids, and colon cancer in two Scandinavian Populations. *Lancet*, 2: 207–211, 1977.

5. RK Peters, MC Pike, D Garabrant, TM Mack. Diet and colon cancer in Los Angeles County, California. *Cancer Causes Control*, 3: 457–473, 1992.

6. TB Young, DA Wolf. Case-control study of proximal and distal colon cancer and diet in Wisconsin. *Int. J. Cancer*, 42: 167–175, 1988.

7. MC Boutron, J Faivre, P Marteau, C Couillault, P Senesse, V Quipourt. Calcium, phosphorous, vitamin D, dairy products and colorectal carcinogenesis: a French case-control study. *Br. J. Cancer*, 74: 145–151, 1996.

8. P van't Veer, JM Dekker, JW Lamers, FJ Kok, EG Schouten, HA Brants, F Sturmans, RJ Hermus. Consumption of fermented milk products and breast cancer: a case-control study in the Netherlands. *Cancer Res.*, 49: 4020–4023, 1989.

9. MG Le, LH Moulton, C Hill, A Kramar. Consumption of dairy produce and alcohol in a case-control study of breast cancer. *J. Natl. Cancer Inst.*, 77: 633–636, 1986.

10. E Kampman, E Giovannucci, P van't Veer, E Rimm, MJ Stampfer, GA Colditz, FJ Kok, WC Willett. Calcium, vitamin D, dairy foods, and the occurrence of colorectal adenomas among men and women in two prospective studies. *Am. J. Epidemiol.*, 139: 16–29, 1994.

11. E Kampman, RA Goldbohm, PA van den Brandt, P van't Veer. Fermented dairy products, calcium, and colorectal cancer in the Netherlands cohort study. *Cancer Res.*, 54: 3186–3190, 1994.

12. A Lidbeck, E Övervik, J Rafter, CE Nord, J-Å Gustafsson. Effect of *Lactobacillus acidophilus* supplements on mutagen excretion in feces and urine in humans. *Microb. Ecol. Health Dis.*, 5: 59–67, 1992.

13. H Hayatsu, T Hayatsu. Suppressing effect of *Lactobacillus casei* administration on the urinary mutagenicity arising from ingestion of fried ground beef in the human. *Cancer Lett.*, 73: 173–179, 1993.

14. G Biasco, GM Paganelli, G Brandi, S Brillanti, F Lami, C Callegari, G Gizzi. Effect of *Lactobacillus acidophilus* and *Bifidobacterium bifidum* on rectal cell kinetics and fecal pH. *Ital. J. Gastroenterol.*, 23: 142, 1991.

15. Y Aso, H Akaza. Prophylactic effect of a *Lactobacillus casei* preparation on the recurrence of superficial bladder cancer. BLP Study Group. *Urol. Int.*, 49: 125–129, 1992.

16. Y Aso, H Akaza, T Kotake, T Tsukamoto, K Imai, S Naito. Preventive effect of a *Lactobacillus casei* preparation on the recurrence of superficial bladder cancer in a double-blind trial. The BLP Study Group. *Eur. Urol.*, 27: 104–109, 1995.

17. BL Pool-Zobel, C Neudecker, I Domizlaff, S Ji, U Schillinger, C Rumney, M Moretti, I Vilarini, R Scassellati-Sforzolini, I Rowland. *Lactobacillus*- and *Bifidobacterium*-mediated antigenotoxicity in the colon of rats. *Nutr. Cancer*, 26: 365–380, 1996.

18. HW Renner, R Münzner. The possible role of probiotics as dietary antimutagens. *Mutat. Res.*, 262: 239–245, 1991.

19. BR Goldin, LJ Gualtieri, RP Moore. The effect of *Lactobacillus* GG on the initiation and promotion of DMH-induced intestinal tumors in the rat. *Nutr. Cancer*, 25: 197–204, 1996.

20. H Arimochi, T Kinouchi, K Kataoka, T Kuwahara, Y Ohnishi. Effect of intestinal bacteria on formation of azoxymethane-induced aberrant crypt foci in the rat colon. *Biochem. Biophys. Res. Commun.*, 238: 753–757, 1997.

21. A Challa, DR Rao, CB Chawan, L Shackelford. *Bifidobacterium longum* and lactulose suppress azoxymethane-induced colonic aberrant crypt foci in rats. *Carcinogenesis*, 18: 517–521, 1997.

22. N Kulkarni, BS Reddy. Inhibitory effect of *Bifidobacterium longum* cultures on the azoxymethane-induced aberrant crypt foci formation and fecal bacterial β-glucuronidase. *Proc. Soc. Exp. Biol. Med.*, 207: 278–283, 1994.

23. H Abdelali, P Cassand, V Soussotte, M Daubeze, C Bouley, JF Narbonne. Effect of dairy products on initiation of precursor lesions of colon cancer in rats. *Nutr. Cancer*, 24: 121–132, 1995.

24. DD Gallaher, WH Stallings, LL Blessing, FF Busta, LJ Brady. Probiotics, cecal microflora, and aberrant crypts in the rat colon. *J. Nutr.*, 126: 1362–1371, 1996.

25. IR Rowland, CJ Rumney, JT Coutts, LC Lievense. Effect of *Bifidobacterium longum* and inulin on gut bacterial metabolism and carcinogen-induced aberrant crypt foci in rats. *Carcinogenesis*, 19: 281–285, 1998.

26. M Onoue, S Kado, Y Sakaitani, K Uchida, M Morotomi. Specific species of intestinal bacteria influence the induction of aberrant crypt foci by 1,2-dimethylhydrazine in rats. *Cancer Lett.*, 113: 179–186, 1997.

27. BS Reddy, A Rivenson. Inhibitory effect of *Bifidobacterium longum* on colon, mammary, and liver carcinogenesis induced by 2-amino-3-methylimidazo[4,5-*f*]quinoline, a food mutagen. *Cancer Res.*, 53: 3914–3918, 1993.

28. T Kohwi, K Imai, A Tamura, Y Hashimoto. Antitumor effect of *Bifidobacterium infantis* in mice. *Gann*, 69: 613–618, 1978.

29. BR Goldin, SL Gorbach. Effect of *Lactobacillus acidophilus* dietary supplements on 1,2-dimethylhydrazine dihydrochloride-induced intestinal cancer in rats. *J. Natl. Cancer Inst.*, 64: 263–265, 1980.

30. LA Shackelford, DR Rao, CB Chawan, SR Pulusani. Effect of feeding fermented milk on the incidence of chemically-induced colon tumors in rats. *Nutr. Cancer*, 5: 159–164, 1983.

31. J Singh, A Rivenson, M Tomita, S Shimamura, N Ishibashi, BS Reddy. *Bifidobacterium longum*, a lactic acid-producing intestinal bacterium inhibits colon cancer and modulates the intermediate biomarkers of colon carcinogenesis. *Carcinogenesis*, 18: 833–841, 1997.

32. BA Friend, RE Farmer, KM Shahani. Effect of feeding and intraperitoneal implantation of yogurt culture cells on Ehrlich ascites tumor. *Milchwissenschaft*, 37: 708–710, 1982.

33. I Kato, S Kobayashi, T Yokokura, M Mutai. Antitumor activity of *Lactobacillus casei* in mice. *Gann*, 72: 517–523, 1981.

34. K Sekine, T Toida, M Saito, M Kuboyama, T Kawashima, Y Hashimoto. A new morphologically characterized cell wall preparation (whole peptidoglycan) from *Bifidobacterium infantis* with a higher efficacy on the regression of an established tumor in mice. *Cancer Res.*, 45: 1300–1307, 1985.

35. K Hirayama, K Itoh, E Takahashi, T Mitsuoka. Comparison of composition of faecal microbiota and metabolism of faecal bacteria among 'human-flora-associated' mice inoculated with faeces from six different human donors. *Microb. Ecol. Health Dis.*, 8: 199–211, 1995.

36. H Horie, M Zeisig, K Hirayama, T Midtvedt, L Möller, J Rafter. Probiotic mixture decreases DNA adduct formation in colonic epithelium induced by the food mutagen 2-amino-9*H*-pyrido(2,3-*b*)indole in a human-flora associated mouse model. *Eur. J. Cancer Prev.*, 12: 101–107, 2003.

37. T Mizutani, T Mitsuoka. Effect of intestinal bacteria on incidence of liver tumors in gnotobiotic C3H/He male mice. *J. Natl. Cancer Inst.*, 63: 1365–1370, 1979.

38. T Mizutani, T Mitsuoka. Inhibitory effect of some intestinal bacteria on liver tumorigenesis in gnotobiotic C3H/He male mice. *Cancer Lett.*, 11: 89–95, 1980.

39. GA Fichera, G Giese. Non-immunologically-mediated cytotoxicity of *Lactobacillus casei* and its derivative peptidoglycan against tumor cell lines. *Cancer Lett.*, 85: 93–103, 1994.

40. H Weirich-Schwaiger, HG Weirich, M Kludas, M Hirsch-Kauffmann. Improvement of the DNA repair capacity of human fibroblasts by autolysates of *Lactobacillus gasseri*. *Arzneim. Forsch.*, 45: 342–344, 1995.

41. A Hosono, T Kashina, T Kada. Antimutagenic properties of lactic acid-cultured milk on chemical and fecal mutagens. *J. Dairy Sci.*, 69: 2237–2242, 1986.

42. M Hosoda, H Hashimoto, H Morita, M Chiba, A Hosono. Antimutagenicity of milk cultured with lactic acid bacteria against *N*-methyl-*N'*-nitro-*N*-nitrosoguanidine. *J. Dairy Sci.*, 75: 976–981, 1992.

43. BL Pool-Zobel, R Münzner, WH Holzapfel. Antigenotoxic properties of lactic acid bacteria in the *S. typhimurium* mutagenicity assay. *Nutr. Cancer*, 20: 261–270, 1993.

44. XB Zhang, Y Ohta. *In vitro* binding of mutagenic pyrolyzates to lactic acid bacterial cells in human gastric juice. *J. Dairy Sci.*, 74: 752–757, 1991.

45. M Morotomi, M Mutai. *In vitro* binding of potent mutagenic pyrolysates to intestinal bacteria. *J. Natl. Cancer Inst.*, 77: 195–201, 1986.

46. K Orrhage, E Sillerström, J-Å Gustafsson, CE Nord, J Rafter. Binding of mutagenic heterocyclic amines by intestinal and lactic acid bacteria, *Mutat. Res.*, 311: 239–248, 1994.

47. XB Zhang, Y Ohta. Binding of mutagens by fractions of the cell wall skeleton of lactic acid bacteria. *J. Dairy Sci.*, 74: 1477–1481, 1991.

48. WEV Lankaputhra, NP Shah. Antimutagenic properties of probiotic bacteria and of organic acids. *Mutat. Res.*, 397: 169–182, 1998.

49. XB Zhang, Y Ohta. Microorganisms in the gastrointestinal tract of the rat prevent absorption of the mutagen-carcinogen 3-amino-1,4-dimethyl-5*H*-pyrido[4,3-*b*]indole. *Can. J. Microbiol.*, 39: 841–845, 1993.

50. M Terahara, S Meguro, T Kaneko. Effects of lactic acid bacteria on binding and absorption of mutagenic heterocyclic amines. *Biosci. Biotech. Biochem.*, 62: 197–200, 1998.

51. K Orrhage, A Annas, CE Nord, EB Brittebo, JJ Rafter. Effects of lactic acid bacteria on the uptake and distribution of the food mutagen Trp-P-2 in mice. *Scand. J. Gastroenterol.*, 37: 215–221, 2002.

52. CF Fernandes, KM Shahani, MA Amer. Therapeutic role of dietary lactobacilli and lactobacillic fermented dairy products. *FEMS Microbiol. Lett.*, 46: 343–356, 1987.

53. IR Rowland, P Grasso. Degradation of *N*-nitrosamines by intestinal bacteria. *Appl. Microbiol.*, 29: 7–12, 1975.

54. AR Bodana, DR Rao. Antimutagenic activity of milk fermented by *Streptococcus thermophilus* and *Lactobacillus bulgaricus*. *J. Dairy Sci.*, 73: 3379–3384, 1990.

55. SR Nadathur, SJ Gould, AT Bakalinsky. Antimutagenicity of fermented milk. *J. Dairy Sci.*, 77: 3287–3295, 1994.

56. GV Reddy, BA Friend, KM Shahani, RE Farmer. Antitumor activity of yogurt components. *J. Food Protect.*, 46: 8–11, 1983.

57. GV Reddy, KM Shahani, MR Benerjee. Inhibitory effect of yogurt on Ehrlich ascites tumor cell proliferation. *J. Natl. Cancer Inst.*, 50: 815–817, 1973.

58. L Baricault, G Denariaz, J-J Houri, C Bouley, C Sapin, G Trugnan. Use of HT-29, a cultured human colon cancer cell line, to study the effect of fermented milks on colon cancer cell growth and differentiation. *Carcinogenesis*, 16: 245–252, 1995.

59. A Biffi, D Coradini, R Larsen, L Riva, G Di Fronzo. Antiproliferative effect of fermented milk on the growth of a human breast cancer cell line. *Nutr. Cancer*, 28: 93–99, 1997.

60. BR Goldin, SL Gorbach. Alterations of the intestinal microflora by diet, oral antibiotics and *Lactobacillus*: decreased production of free amines from aromatic nitro compounds, azo dyes and glucuronides. *J. Natl. Cancer Inst.*, 73: 689–695, 1984.

61. AD Ayebo, IA Angelo, KM Shahani. Effect of ingesting *Lactobacillus acidophilus* milk upon fecal flora and enzyme activity in humans. *Milchwissenschaft*, 35: 730–733, 1980.

62. WH Ling, R Korpela, H Mykkanen, S Salminen, O Hanninen. *Lactobacillus* strain GG supplementation decreases colonic hydrolytic and reductive enzyme activities in healthy female adults. *J. Nutr.*, 124: 18–23, 1994.

63. BR Goldin, L Swenson, J Dwyer, M Sexton, SL Gorbach. Effect of diet and *Lactobacillus acidophilus* supplements on human fecal bacterial enzymes. *J. Natl. Cancer Inst.*, 64: 255–261, 1980.

64. S Spanhaak, R Havenaar, G Schaafsma. The effect of consumption of milk fermented by *Lactobacillus casei* strain Shirota on the intestinal microflora and immune parameters in humans. *Eur. J. Clin. Nutr.*, 52: 899–907, 1998.

65. BR Goldin, SL Gorbach. The effect of milk and lactobacillus feeding on human intestinal bacterial enzyme activity. *Am. J. Clin. Nutr.*, 39: 756–761, 1984.

66. WH Ling, O Hanninen, H Mykkanen, M Heikura, S Salminen, A von Wright. Colonization and fecal enzyme activities after oral *Lactobacillus* GG administration in elderly nursing home residents. *Ann. Nutr. Metabol.*, 36: 162–166, 1992.

67. M Hosoda, H Hashimoto, F He, H Morita, A Hosono. Effect of administration of milk fermented with *Lactobacillus acidophilus* LA-2 on fecal mutagenicity and microflora in the human intestine. *J. Dairy Sci.*, 79: 745–749, 1996.

68. GW Modler, RC McKellar, M Yaguchi. Bifidobacteria and bifidogenic factors. *Can. Inst. Food Sci. Technol. J.*, 23: 29–41, 1990.

69. HP Bartram, W Scheppach, S Gerlach, G Ruckdeschel, E Kelber, H Kasper. Does yogurt enriched with *Bifidobacterium longum* affect colonic microbiology and fecal metabolites in healthy subjects? *Am. J. Clin. Nutr.*, 59: 428–432, 1994.

70. JH Weisburger, EL Wynder. Etiology of colorectal cancer with emphasis on mechanism of action and prevention. In: VT De Vita, S Hellman, SA Rosenberg, Eds., *Important Advances in Oncology.* JB Lippincott, Philadelphia, 1987, pp. 197–220.

71. WR Bruce. Recent hypotheses for the origin of colon cancer. *Cancer Res.*, 47: 4237–4242, 1987.

72. A Lidbeck, U Geltner-Allinger, KM Orrhage, L Ottava, B Brismar, J-Å Gustafsson, JJ Rafter. Impact of *Lactobacillus acidophilus* supplements on the faecal microflora and soluble faecal bile acids in colon cancer patients. *Microb. Ecol. Health Dis.*, 4: 81–88, 1991.

73. K Sekine, E Watanabe-Sekine, T Toida, T Kasashima, T Kataoka, Y Hashimoto. Adjuvant activity of the cell wall of *Bifidobacterium infantis* for *in vivo* immune responses in mice. *Immunopharmacol Immunotoxicol.*, 16: 589–609, 1994.

74. K Sekine, J Ohta, M Onishi, T Tatsuki, Y Shimokawa, T Toida, T Kawashima, Y Hashimoto. Analysis of antitumor properties of effector cells stimulated with a cell wall preparation (WPG) of *Bifidobacterium infantis*. *Biol. Pharm. Bull.*, 18: 148–153, 1995.

75. EJ Schiffrin, F Rochat, H Link-Amster, JM Aeschlimann, A Donnet-Hughes. Immunomodulation of human blood cells following the ingestion of lactic acid bacteria. *J. Dairy Sci.*, 78: 491–496, 1995.

76. C De Simone, R Vesely, B Bianchi Salvadori, E Jirillo. The role of probiotics in modulation of the immune system in man and in animals. *Int. J. Immunother.*, 9: 23–28, 1993.

77. I Kato, T Yokokura, M Mutai. Macrophage activation by *Lactobacillus casei* in mice. *Microbiol. Immunol.*, 27: 611–618, 1983.

78. T Matsuzaki. Immunomodulation by treatment with *Lactobacillus casei* strain Shirota. *Int. J. Food Microbiol.*, 41: 133–140, 1998.
79. I Kato, K Endo, T Yokokura. Effects of oral administration of *Lactobacillus casei* on antitumor responses induced by tumor resection in mice. *Int. J. Immunopharmacol.*, 16: 29–36, 1994.
80. T Matsuzaki, T Yokokura. Inhibition of tumor metastasis of Lewis lung carcinoma in C57BL/6 mice by intrapleural administration of *Lactobacillus casei*. *Cancer Immunol. Immunother.*, 25: 100–104, 1987.
81. T Matsuzaki, T Yokokura, M Mutai. Antitumor effect of intrapleural administration of *Lactobacillus casei* in mice. *Cancer Immunol. Immunother.*, 26: 209–214, 1988.
82. T Matsuzaki, S Hashimoto, T Yokokura. Effects on antitumor activity and cytokine production in the thoracic cavity by intrapleural administration of *Lactobacillus casei* in tumor-bearing mice. *Med. Microbiol. Immunol.*, 185: 157–161, 1996.
83. T Matsuzaki, T Yokokura, I Azuma. Anti-tumor activity of *Lactobacillus casei* on Lewis lung carcinoma and line-10 hepatoma in syngeneic mice and guinea pigs. *Cancer Immunol. Immunother.*, 20: 18–22, 1985.
84. M Asano, E Karasawa, T Takayama. Antitumor activity of *Lactobacillus casei* (LC 9018) against experimental mouse bladder tumor (MBT-2). *J. Urol.*, 136: 719–721, 1986.
85. T Matsuzaki, T Yokokura, I Azuma. Antimetastatic effect of *Lactobacillus casei* YIT9018 (LC 9018) on a highly metastatic variant of B16 melanoma in C57BL/6J mice. *Cancer Immunol. Immunother.*, 24: 99–105, 1987.
86. K Sekine, T Kawashima, Y Hashimoto. Comparison of the TNF -β levels induced by human-derived *Bifidobacterium longum* and rat-derived *Bifidobacterium animalis* in mouse peritoneal cells. *Bifidobacteria Microflora*, 13: 79–89, 1994.
87. BS Reddy. Prevention of colon cancer by pre- and probiotics: evidence from laboratory studies. *Br. J. Nutr.*, 80: S219–S223, 1998.
88. K Hirayama, P Baranczewski, J-E Åkerland, T Midtvedt, L Möller, J Rafter. Effects of human intestinal flora on mutagenicity of and DNA adduct formation from food and environmental mutagens. *Carcinogenesis*, 21: 2105–2111, 2000.
89. CJ Rumney, IR Rowland, TM Coutts, K Randerath, R Reddy, AB Shah, A Ellul, IK O'Neill. Effects of risk-associated human dietary macrocomponents on processes related to carcinogenesis in human-flora-associated (HFA) rats. *Carcinogenesis*, 14: 79–84, 1993.

17

Statistical Considerations for Testing the Efficacy of Probiotics

Mohamed Ahmedna

CONTENTS

17.1 Introduction

Probiotic foods and supplements continue to gain popularity, with a world-wide market estimated to be about $6 billion a year (1). With the expanding functional food market, there is an increase in interest among consumers and researchers in probiotics and their health benefits. Probiotics are defined as "live microorganisms which, when consumed in adequate amounts in food or dietary supplement, confer a health benefit on the host by improving its intestinal microbial balance" (2–3). Potential health benefits of probiotics include increased resistance to infectious diseases (e.g., diarrhea), lowering of blood pressure and serum cholesterol concentration, reduction of allergy, and modulation of cytokine gene expression. In fact, probiotics have been used for treatment of gastrointestinal (GI)-diseases, hypercholesterolemia, lactose intolerance, suppression of procarcinogenic enzymes, and immuno-modulation (4–5). In addition many probiotics have significance in food safety because they produce antibacterial compounds such as bacteriocins and organic acids. To achieve these potential health and safety benefits, desirable probiotic strains must have properties such as the ability to survive transit through digestive tract, adherence to epithelial cells and intestinal mucosa, colonization potential in the human intestinal tract, production of antimicrobial substances against pathogens, safety to humans, and stability during normal storage (1). A number of probiotics such as lactobacilli and bifidobacteria have been shown to strongly adhere to human intestinal cell lines. These probiotics confer health benefits to the host including prevention of food-borne diseases through interference with adherence by pathogenic bacteria, competition for nutrients, release of antibacterial substances, and stimulation of mucosal and systemic host immunity (6).

Despite these potential health benefits, clinical demonstration of the efficacy of probiotics using established medical protocols for drug efficacy remains a challenging task. Unlike drugs, the effect of probiotics tends to be relatively weak, diffuse and often difficult to isolate from other confounding factors related to the diet and lifestyle. Furthermore, health benefits typically associated with probiotics are directly or indirectly affected by a multitude of other factors that are difficult to isolate or control. For instance, traveler's diarrhea is caused by a diverse and changing range of microbial pathogens such as *E. coli, Salmonella, Campylobacter, and Shigella* strains as well as viruses (7). Similar constraints are encountered when testing the cholesterol-lowering and immune-boosting effects of probiotics. Because the effect of the size of treatments with probiotics tends to be modest and subject to multiple potential confounders leading to substantial statistical variance, hypothesis-testing experiments must be well controlled as to minimize unsystematic variability and maximize systematic variance. This goal is typically achieved through appropriate design and controls such as randomization, double-blind, and placebo-controlled studies. This chapter is designed to provide a general overview of some statistical considerations relevant to testing the efficacy of probiotics. The chapter will discuss issues pertaining to experimental design and data analysis of studies intended to demonstrate the efficacy of probiotics. These topics include general considerations on how the types of experimental design, the nature of outcome (dependent) and treatment (independent) variables determine the appropriate statistical analysis and affect the validity and generalizability of inferences. Specific considerations unique to priobiotics that may impact study reliability are also included. The chapter also discusses common designs that may be used for testing the efficacy of probiotics *in vitro* and *in vivo*, issues of validity, power, and sample size, and provides a summary of major designs involving human subjects, with emphasis on randomized, controlled clinical trials.

17.2 Approach to Testing the Efficacy of Probiotics

As in any scientific study, testing of the efficacy of probiotics requires well-defined and specific objectives along with underlying research hypotheses. The methods of measurement of probiotic efficacy should also possess the desirable features of accuracy and precision. Furthermore, researchers must reflect on specific issues with relevance to probitics such as selection of the following:

- Specific, desired "beneficial effects" that one wants to produce by using the selected strains of probiotics
- Known factors that may influence the beneficial effect(s) being investigated (negatively or positively)
- Appropriate "probiotic" strains that can survive passage through the digestive tract and colonize the digestive tract
- Effective dose and colonization of the GI tract by probiotic species, especially because this depends on (i) resistance to gastric acid and bile acids, (ii) growth characteristics of GI-resistant species, and (iii) expression of immune markers by human peripheral blood cells (8)
- Optimal dose range, frequency of administration, and appropriate vehicle for delivery of probiotic strain(s)
- Known and quantifiable biological markers and/or physiological changes that indicate the beneficial effects of probiotics being tested
- Possible mechanisms of anticipated probiotic effect as an explanation of the expected or observed beneficial health effects
- Effective design that isolates probiotic effects from other potential confounding variables.
- Scope of research and adequacy of resources available for the investigation

In addition to the above considerations, the choice of the right outcome or dependent variables, identification of key predictor (independent) variables and selection of appropriate design are critical to ensuring accuracy and precision, two critical features of rigorous scientific methods. Thus, one needs to identify all dependent (outcome) variables of interest and their methods (scales) of measurement. For all the dependent variables, an exhaustive list of independent (predictor) variables and known systematic sources of controllable sources of variance should be made. Known sources of variation that are not under study should be controlled for through appropriate use of blocking designs such as randomized block designs and/or covariate analysis. However, because independent and blocking factors impose stratification that may lead to practical difficulties pertaining to sample size, study participants should be stratified on the fewest possible and most critical independent variables without sacrificing control for extraneous factors or sample size.

The next step is to evaluate available resources for the study and decide on what objectives and hypotheses are achievable and what experimental design would lead to the most efficient use of such resources. The best design is one that has well-defined objectives, estimates error, eliminates extraneous sources of variation, and isolates the effect of experimental treatment(s) with sufficient precision. The appropriate design and statistical analysis are typically chosen based on the type of research questions and hypotheses to be investigated, the nature of dependent and independent variables, and the resources available for the study.

As shown in Figure 17.1, experimental design is critical to reducing unsystematic error variance, accounting for the effect of known extraneous factors, and unmasking the treatment effect, if any exists. The appropriate experimental design also enhances power (the probability of detecting difference if any exists) and enables efficient management of available resources. In terms of appropriate statistical analysis, Table 17.1 illustrates how the nature of dependent variables influences the choice of statistical analysis. Overall, both the nature of dependent and independent variables and the type of design have to be considered together to determine what statistical analysis is most appropriate to use and whether valid statistical inferences can be made regarding the cause-and-effect relationships between independent and dependent variables. Table 17.2 provides a summary of statistical analysis and inferences as affected by study design and the types of dependent and independent variables. The Table also comments on inferences and their generalizability for each type of analysis. There are numerous published textbooks on experimental design and statistical analysis that can provide interested readers with more details (9–13).

17.3 Common Designs for Testing the Efficacy of Probiotics

Health effects of probiotics have been investigated through *in vitro* and *in vivo* (animal and human) studies. In such studies, experimental design plays a major role in study validity and generalizability, typically through randomization and control. For brevity, Table 17.3 summarizes some of the advantages and limitations of randomized and nonrandomized designs suitable to investigate the benefits of probiotics *in vitro* and *in vivo*.

17.3.1 *In Vitro* and *In Vivo* Animal Studies

Considering the fact that probiotic activity is likely to be multifunctional and strain specific and that the health conditions typically treated with probiotics

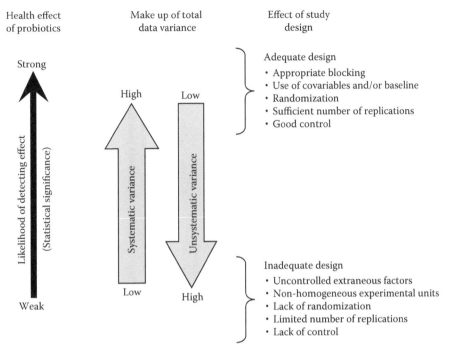

FIGURE 17.1
Importance of appropriate design in unmasking weak effect of probiotics.

are of multifactorial and complex etiology, there is a need for studies that build on data from *in vitro* and animal experiments with specific probiotic strains showing particular modes of actions against defined pathological targets. Properties of probiotics that can be easily studied *in vitro* include resistance to simulate gastric juice and bile acid, growth characteristics of resuscitated cultures, and expression of immune markers by human peripheral blood cells (8). Measurement of *in vivo* activity using animal models is critical in the development of efficacious probiotic strains against particular diseases (7). *In vitro* and animal studies are typically suitable for randomized and controlled experimental designs. *In vitro* and animal studies are excellent and cost-effective ways to screen probiotics with the most potential to benefit the human host and understand their growth and survival patterns as well as their mode of action. In contrast to studies involving human subjects, these studies give researchers the flexibility to use a variety of efficient experimental designs and lead to generalizable inferences about research hypotheses pertaining to potential health effects of probiotics and their mechanisms of action. Given the high cost of clinical trials that typically require consideration of "one strain vs one placebo" experiments to prove efficacy, *in vitro* and *in vivo* studies using animals account for the bulk of the scientific literature on the health benefits of probiotics (14–25). However, despite the many positive findings pertaining to probiotics benefits from *in*

TABLE 17.1

Data Analysis Based on the Nature of Dependent and Independent Variables

Nature of Dependent Variable	Number of Groups (Factors/ Independent Variables)	Nature of Variation	Type of Appropriate Statistical Analysis	Comments
Quantitative (interval or ratio)	One	Within	Single-group *t*-test	Analysis Assumes: Random assignment, independence, normality, and equal variance.
	Two (one factor with two levels)	Between	*t*-test for independent groups	If the factors or independent variables are quantitative, then the appropriate analysis is simple linear regression (for one independent variable) or multiple regression (for two or more quantitative independent variables (see Table 17.2)
		Within	Paired *t*-test for correlated groups	
	More than two (one factor with more than two levels)	Between	Analysis of variance	
		Within	Repeated measure analysis of variance	
	More than two (two or more factors with two or more levels each)	Between	Multiple factor Analysis of Variance	
		Within	Factorial repeated measure analysis of variance	
Ordinal (rank data)	One	Between	One-sample test	Assume random assignment or selection.
	Two	Between (independent)	Mann-Whitney U test	If factors or independent variables are quantitative, then appropriate data analysis would be logistic regression and weighted least square (see Table 17.2).
		Within (matched)	Wilcoxon Rank test	
	Three or more	Between	Kruskal-Wallis ANOVA	
Nominal (qualitative)	One	N/A	Goodness of fit	
	Two or more	N/A	Chi-square	

TABLE 17.2

Statistical Analysis and Inferences as Affected by Study Design and Nature of Dependent and Independent Variables

Design Type	Nature of Independent Variables	Nature of Dependent Variables	Appropriate Statistical Analysis	Parameters	Outcomes
Experimental/ Randomized Designs (e.g. Completely randomized design, Factorial and fractional factorial designs, Randomized block and Latin Square designs, and Cross-Over designs) **These include Randomized Clinical Trials with and without controls**	Interval (Quantitative) Example: dose of probiotics administered, level of gut colonization, and frequency or duration of probiotic treatment.	Interval (Quantitative) Examples: concentration of serum cholesterol, counts of pathogenic bacteria shed by patients, levels of monocytes, IgA, inflammatory markers, cytokine, or bacteriocins.	[a]Simple Regression Or multiple regression if more than two independent variables.	Estimates of regression coefficients for population Coefficient of correlation.	**Test statistic for significance of parameter estimates under null hypothesis; powerful and generalizable inferences, predictive models.**
	Categorical (Qualitative) Examples: probiotic strains, treatment levels (e.g treatment and placebo).	Interval (Quantitative) Examples: see variables listed in cell immediately above.	[a]ANOVA (multiple ANOVA if more than one dependent variables), repeated measure ANOVA or t-test it only two treatments or groups are compared.	Estimates of population means and variance.	**Test statistic for equality of means under null hypothesis, powerful and generalizable inferences; magnitude and direction of difference between treatment means.**
	Interval (Quantitative) Examples: same as examples listed in first cell of this column.	[b]Categorical (Qualitative) Examples: status of chronic condition (presence or absence), relapse rate; descriptors of illness severity; improvement status.	Logistic Regression, Weighted least square	Estimates of regression coefficients for population	**Test statistic for significance of parameter estimates under null hypothesis; powerful and generalizable inferences, predictive models**
	Categorical (Qualitative) See examples listed in second cell of this column.	Categorical (qualitative) See Examples listed in cell immediately above	Categorical Data Analysis (Chi-square & Log linear models)	Estimates of population parameters and odds ratios	**Population parameters under null hypothesis; odds ratios and relative risk; generalizable inferences**
Quasi-experimental Designs (One-group posttest, One-group Pretest-Posttests, Removed and Repeated-treatment designs)	All types listed above	All types listed above	[c]All methods listed above, Time series	Cannot estimate population parameters (unless assumptions are met) but give group characteristics and indication of possible causal relationships	**Yield useful data if appropriate controls are used, useful for generation of hypotheses about cause-effect but give weak indication of cause and effect, sample means and frequencies**
Non-Experimental Designs (e.g. correlational and causal comparative designs)	All types listed above	All types listed above	Limited to analyses such as correlation for quantitative variables and log linear model for qualitative variables; limited statistical tests available	No population parameters only measure of covariation (e.g., coefficient of correlation); Sample parameters for causal comparative	**Measure of strength and direction of covariation or association between variables; limited inferential statistics these designs; inferences are weakened by inability to account for potential confounds**

[a]Assumptions: observations are random, independent, normally distributed, and with equal variance

[b]In cases where dummy variable coding of categorical and interval data are mixed, multiple regression analysis is appropriate

[c]Method validity (i.e., assumptions) needs to be checked. If assumptions are violated, corrective measures should be taken. Otherwise, non-parametric methods are more appropriate.

TABLE 17.3

Comparison of Experimental Designs Applicable for Testing the Health Benefits of Probiotics

Design Type	Overall Advantages & Limitations	Advantages/Limitations by Study Type		
		In vitro	*In vivo* (Animals)	*In vivo* (Humans)
Experimental randomized designs	Yield powerful and generalizable inferences	Less costly than *in vivo* study designs	Typically less costly than human trials	May be very costly
Examples:	Filter out effect of extraneous factors/reduce unsystematic variability	Enable exploration of hypotheses and understanding of possible action mechanisms	Enable testing of hypotheses	Provide ultimate demonstration of probiotic efficacy
Completely randomized design including factorial and fractional factorial designs; Randomized block and Latin square design; and Cross-over design.	Likely to show significant effect if any exists		Explore key physiological and biochemical processes induced by probiotics	Require proof of safety prior to testing
	Provide strong cause and effect indication	Can be tightly controlled		Scope of testing is limited due to ethical issues
(This group of design also include *Randomized Clinical Trials*, the standard for efficacy testing of drugs and probiotics)	May become cumbersome with multi-factors or over-stratification	Relatively straightforward	Gives indication of potential efficacy of probiotics in humans	Strict regulatory requirements
	May impose stratification constraints that affect recruitment/sample size	Limited in scope		Suffers from effect of confounding factors that cannot be eliminated due to practical or ethical reasons
		Least applicable to probiotic effect in humans	Data generated are closely related but not directly applicable to effects in humans	
Quasi-experimental designs and causal/correlational designs	Commonly used with human subjects; typically inexpensive; provide valuable descriptive data and information for generation of hypotheses; useful when randomization is not feasible for practical or ethical reasons and/or when independent variable cannot be manipulated. Limited statistical analysis; inferences are not generalizable; Yield weak indication of cause and effect. Quasi-experimental yield better controls and have more options for statistical analysis/ inferences than causal and correlational designs.			
Examples: Causal comparative & correlational designs				

vitro and animal studies, such data cannot be used to directly predict the efficacy of probiotics in humans. Efficacy in the latter is typically demonstrated through randomized controlled clinical trials (RCTs), which remain the gold standard for proof of efficacy in humans. Thus, the efficacy of probiotics needs to be demonstrated through randomized clinical trials for probiotics to enjoy widespread use as therapeutic agents against specific health conditions and be readily prescribed and recommended by health professionals.

17.3.2 Efficacy Studies Involving Human Subjects

17.3.2.1 Issues of Validity and Power

Unlike *in vitro* and *in vivo* studies using animals, trials involving humans may be difficult to randomize and control for practical, economical, and/or ethical reasons. Thus, experiments involving human subjects require well-thought experimental designs tailored to maximize precision (minimized unsystematic variation), taking into account financial, practical, and ethical constraints. Randomized designs are the most unbiased designs and lead to powerful and generalizable inferences. In cases where randomization is unethical or impractical, quasi-experimental designs are used as the next best designs. Both randomized and quasi-experimental designs manipulate treatment to force it to occur before the effect of treatment in an attempt to show causality. Both types of design can be enhanced by the use of appropriate controls. In situations where no treatment is manipulated, causal comparative or correlational studies offer limited information on possible relationships between independent and dependent variables. Table 17.4 gives a brief comparison of these three groups of experimental designs. Among these three major groups of designs, only randomized designs provide a valid basis for generalizable inferences about cause and effect relationships between treatment (independent variables) and dependent variables. However, regardless of the type of randomized design used, the issues of internal and external validity and statistical power are important to consider.

Internal and External Validity

Most experimental studies are designed or intended to use a finite sample (limited number of data points) to make inferences pertaining to what would happen in the larger population. The validity and generalizability of these inferences depend on the degree of internal and external validities. Internal validity refers to the validity of inferences about whether observed covariation between a treatment and outcome variables reflects a causal relationship between the two variables as the treatment variable is manipulated. On the other hand, external validity is the validity of inferences about whether the cause-effect relationship holds over variation in person, settings, treat-

TABLE 17.4

Summary of Advantages and Limitations of Major Design Types

Design Type	Examples	Situation	Advantages	Limitations
Randomized experimental designs	Completely randomized design (CRD)	Homogeneous subjects with no known source of extraneous variation	Easy to use, contains least stratification of subjects, which helps in sample size/recruitment	Does not allow for control of extraneous factors, but precision could be enhanced with the use of a control group
	Factorial designs	Same as CRD but with two or more treatment combinations	Allows evaluation of multiple treatments and treatment levels and their interaction, can be used within designs that control for extraneous factors/variation	Can be cumbersome and expensive due to large sample size, does not allow for control of extraneous factors unless used within block designs or use covariates
	Block design (complete block and Latin Square designs)	Heterogeneous subjects with one and two known extraneous source of variation, respectively	Eliminates effect of extraneous factor, enhances power	Sources of extraneous variability need to be qualitative, imposes stratification that may affect sample size, requires that extraneous factors be qualitative
	Covariate	Known sources of extraneous variation that are quantitative	Reduces error and enhances power	Sources of extraneous variability need to be quantitative, complex analysis
	Crossover designs	Limited number of subjects due to few subjects available for study or high cost per subject	Minimizes within-subject variability, easy to reach desired sample size, cost effective	Carryover effect, requires correction for repeated measure correlation
Quasi-experimental designs[a] (nonrandomized)	One-group posttest design	Randomization is not feasible, no pretest or control used	Simple and practical in cases where there is good knowledge of how dependent variable behaves	Weak design, absence of control group and pretest makes it difficult to know what would have happened without treatment
	One-group pretest-posttests	Randomization is not feasible, control group available for use	Simple, improved over one-group posttest, can be improved using other design elements such as double pretest and nonequivalent dependent variable	Weak design, absence of control makes it difficult to exclude effect of other factors unrelated to treatment

TABLE 17.4

Summary of Advantages and Limitations of Major Design Types (continued)

Design Type	Examples	Situation	Advantages	Limitations
	Removed treatment design	Randomization is not feasible, treatment can be applied and removed without lingering effect	Simple, gives better (but still weak) indication of possible cause and effect relationship between treatment and dependent variable than the above two examples of quasi-experimental designs	Affected by outliers, thus requiring large sample, not appropriate if removal of treatment is unethical, could lead to unintended effect on dependent variable, suffers from uncertainty about cause–effect
	Repeated treatment	Randomization is not feasible, treatment can be applied cyclically without carryover or subject sensitization	Strong internal validity, better indication of cause and effect	Affected by cyclical events, weak external validity, vulnerable to threats to validity such as subject adaptation to cyclic treatment regimen
	Case control design	When it is not feasible or ethical to experiment, typically used for rare outcomes	Easy and inexpensive, very useful for generating hypotheses about causal relationships between dependent and independent variables	Difficult to choose control cases
	Untreated control with dependent pretest posttest samples	Randomization is not feasible, control group available, post- and pretests can be used on control treatment groups	Best controlled among all quasi-experimental designs; most widely used; comparison group and pre- and posttests facilitate causal plausible inferences; can be further improved by use of double pretest or reversed treatment control groups	Uncertainty about cause and effect remains due to lack of randomization, design is subject to threats to validity such as maturation, history, and selection

Nonexperimental	Causal comparative and correlational	Where independent variable or factor cannot be manipulated and randomization is not used	Least costly, relatively easy to conduct, yields useful information for generation of hypotheses about possible cause and effect	Does not lead to generalizable inferences, yields weak (if any) indication of cause and effect relationships between dependent and independent variables

[a] [Reference 26 includes more details about several other types of quasi-experimental designs. Many quasi-experimental designs combine pretest, posttest, and control for better ability to infer about causal relationships between independent and dependent variables.

ment variables, and measurement variables (26). Randomization enhances internal validity but may not help in external validity. Internal and external validity and threats that affect them are discussed in detail in many research methods books (26–28). However, even for designs selected to meet the requirements of internal and external validity, statistical tests used as the basis for inferences need to have sufficient power to ensure reliability.

Power and sample size

Power is the sensitivity of the statistical procedure used to the differences being sought (probability of detecting difference if any exists) or ability to reduce Type II error, failing to reject the null hypothesis when a difference exists. Power depends on the precision of the research design (control of extraneous sources of variation) and method of measurement. Ways to increase statistical power include (9, 28) the following actions:

1. Use matching, stratification, and blocking.
2. Measure and correct for covariates to reduce extraneous variation in dependent variable.
3. Use homogeneous experimental units or subjects that are responsive to the treatment.
4. Use a within-participant design.
5. Use large sample size.
6. Relax (increase) Type I error.
7. Use other methods such as improving measurement, increasing strength of treatment, and increasing variability between treatment levels.

Among these seven methods to increase power, Methods 1 through 4 can be achieved by a variety of experimental designs. Section 17.3.2.2 provides a summary overview of major randomized designs that can be used to enhance power and could be useful to consider in experiments intended to demonstrate the efficacy of probiotics. Power enhancement Method 5 (use large sample size) is affected by the inherent variability of the process being investigated and the resources available for the study. Method or measurement variability can be used to estimate a sample size that yields a desired power level. It is noteworthy that the computation of power for designs involving three or more groups and multivariate analysis are complex. The required sample size is influenced by variance (σ^2) of the data, the size of the difference between two means (λ), significance level or Type I error (α), the desired power of test ($1-\beta$) or the probability of detecting a difference (λ) between two means at a given probability of Type II error (β) (9). Using these factors, the required number of replications (sample size) to yield a given power level can be estimated using the formula below.

$$\text{Minimum sample size} = 2 *(z_{/2} + z)^2 * (\sigma/\lambda)^2$$

where $z_{/2}$ and z are the standard normal variate values exceeded with probability $\alpha/2$ and β, respectively. Both values can be obtained from standard normal distribution tables available in most statistical books. In preliminary tests for sample size estimation, the coefficient of variation (%CV) is substituted for σ and (%Diff) is substituted for λ, where %CV = 100 * (standard deviation/mean) and %Diff = 100 * (difference between means/mean). For simplicity, %CV and the estimated or expected difference between two means (%Diff), obtained either from preliminary tests or the literature, are substituted for σ and λ, respectively, in the above equation. Table 17.5 provides examples of a sample size calculation at various significance and power levels for a given mean difference (%Diff) and data variability (%CV). As shown in Table 17.5, the required number of replications (sample size) increases when:

1. The coefficient of variation (%CV) or σ increases
2. The size of the difference between means (%Diff) decreases
3. The significance level (α) is decreased
4. The set value of power of test (1–β) increases

Several statistical packages such as nQueryAdvisor, S-Plus (Power aSypower), SAS (Power SAS), SPSS (SamplePower), NCSS (Pass), and Biostat (Power and Precision) have algorithms that generate statistical power curves based on which experimenters can decide on the appropriate sample size depending on their study requirements or constraints. Free power analysis programs such as G*Power and UnifyPow are also available online. In addition, numerous experimental design textbooks (9) also provide statistical power curves for specific tests such as the F-test.

17.3.2.2   *Randomized Experimental Studies*

Random assignment reduces the plausibility of an alternative explanation of an observed effect. Random assignment consists of using a procedure (e.g., random numbers, coin or die toss) that assigns individual experimental units to any given treatment or treatment levels by chance alone. They facilitate causal inferences by making alternative explanations implausible, allowing a valid estimate of the error terms (26).

Randomized controlled human trials (RTCs) to test the efficacy of active ingredients in drugs are the ultimate measures of a drug's potential usefulness. However, most of these designs are expensive, time demanding, and are subject to many constraints to randomization compared to *in vitro* and *in vivo* animal studies. RCTs require randomization in which usually two treatments (e.g., drugs, probiotics) are given to the participants — one is the

TABLE 17.5

Examples of Samples Sizes as Affected by Type I Error, Power Level, and Data Characteristics

Values Set by Experimenter		Parameters Obtained from Preliminary Tests or Literature[a]		Values Estimated by Calculation
Type I error	Desired Power (1-β)	Difference between Means (λ or %Diff)	Data Variability (σ or %CV)	Minimum Required Sample Size
5% [=0.05; two-tailed; z=1.96]	95 (β=0.05; z=1.645)	6	5	18
			10	72
		12	5	5
			10	18
	85 (β=0.15; z=1.036)	6	5	12
			10	50
		12	5	3
			10	12
10% [=0.05; two-tailed; z=1.645]	95 (β=0.05; z=1.645)	6	5	15
			10	60
		12	5	4
			10	15
	85 (β=0.15; z=1.036)	6	5	10
			10	40
		12	5	2
			10	10

[a] A probiotic research example would be evaluation of the effect of the consumption of a probiotic strain or probiotic mixture on a health condition (e.g., serum cholesterol, immune boosting as measured by IgA, cytokine, concentration of diarrheal viruses or pathogens shed, etc.) of subjects participating in a randomized controlled clinical trial where subjects are randomly assigned to the probiotic treatment or a placebo control. Here, for illustration purposes, we assume that preliminary data or the literature suggests that the difference between mean value for treatment and control groups is 6 (small difference) or 12 (larger difference) with a coefficient of variation of 5 (small variability) or 10 (larger variability). Note that the actual unit (scale of measurement) of the dependent and outcome variable would not matter since both mean difference (%Diff) and coefficient of variation (%CV) are expressed in a percentage.

active drug or probiotic preparation that is believed to hold some promise as an effective treatment of a given health condition and the other is a "placebo". The latter is a drug or probiotic preparation which has no expected treatment value and is indistinguishable from the active drug or probiotic preparation. RCTs are conducted in single- or more often double-blind design which ensures that neither the patient nor the researchers are aware of the nature of the treatment they receive or give. This is necessary to minimize experimental bias. Experimenter effects arise from the experimenter knowing (i) the hypothesis being tested, (ii) the nature of experimental and control conditions, and (iii) the condition to which participant is assigned. Subject bias is also another concern if treatment assigned is known to the subject.

In the single-blind design, the individual who administers the treatment does not know the condition to which a participant is assigned. In the more powerful double-blind protocol, the researcher administering treatment is blind to the assignment of each participant, and the participants are blind to their assignment. The blind assignment is enhanced with the use of a control group that receives a placebo (28). The randomized, double-blind, placebo-controlled design is the most powerful of all the RCTs. Table 17.6 provides a description of RCTs. Numerous clinical trials have been used to test the efficacy of probiotics using randomized; randomized, double-blind, placebo-controlled; randomized controlled; randomized placebo-controlled; and randomized blind placebo-controlled trials (29–45). In cases where standard RCTs are not suitable (e.g., impractical or costly), comparison of responses to a range of doses of the test drug generates valuable dose-response data in addition to supporting drug efficacy. Balanced and randomized crossover designs represent another attractive alternative because they factor out the largest source of experimental variance, interindividual variability. A crossover design might be appropriate for probiotic efficacy research, especially if the study is dealing with chronic stable conditions, where within-subject variability is much less than between-subject variability.

17.3.2.3   Quasi-experimental Studies

Quasi-experiments are similar to randomized experiments in terms of purpose and structural attributes, but lack random assignment of subjects for the condition or treatment being investigated.

Causal inferences from any quasi-experiment must meet the basic requirements for all causal relationships that precede the effect, that cause or vary with the effect, and for which alternative explanations for the causal relationship are implausible. Because quasi-experiments do not use random assignment, they rely on other principles to show alternative explanations that are implausible. These include identification and evaluation of threat to internal validity, primacy of control by design (e.g., use of control pretest or posttest to evaluate plausibility of threats to validity), and coherent pattern

TABLE 17.6

Examples of Randomized Clinical Trials (RCTs)

Randomized Clinical Trial	Description	Comments
Randomized	Subjects are randomly assigned to treatment(s) (e.g., probiotic regimen).	Randomization ensures generalizability of inferences, but design precision is further enhanced by the use of appropriate control; experimenter bias is of concern; less elaborate than blind designs.
Randomized controlled	Subjects are assigned randomly either to treatment (s) (e.g., probiotics in a food carrier) or to control group (food carrier with no probiotics).	
Randomized single-blind	Subjects are assigned randomly to treatment(s), but individual who administers the treatment does not know the condition to which a participant is assigned	Powerful and generalizable results/inferences; precision is enhanced with the use of control and blinding of individual administering treatment to reduce experimenter bias. However, double blinding is best in removing experimental bias.
Randomized single-blind controlled or placebo controlled	Subjects are assigned randomly to either treatment(s) or control (placebo) groups, but individual who administers the treatment does not know the group to which a participant is assigned.	
Randomized double blind	The researcher administering treatment(s) is blind to the assignment of each participant, and the participants are blind to their assignment.	Most powerful RCTs, especially the randomized double-blind placebo-controlled design; powerful and generalizable inferences; more elaborate experimental protocol; may be costly but would be top choice if available resources permit.
Randomized double-blind (placebo) controlled	Subjects are randomly assigned to either the treatment(s) or placebo, but the researcher administering treatment(s) is blind to the assignment of each participant and the participants are blind to their assignment.	

matching (26). Quasi-experimental designs identify and reduce the plausi-bility of alternatives to causal explanations of the observed effect but cannot provide the degree of certainty associated with random designs. Quasi-experimental designs are, however, very useful in situations where random-ization is impractical or unethical. In these instances, these designs provide the best indication of causal relationships without the degree of certainty that randomized designs provide. Quasi-experimental designs also yield valuable insights regarding possible cause and effect relationships and enable researchers to formulate research hypotheses. The majority of quasi-experimental designs are used in social science research and are discussed in details in numerous research method books (26–28, 46).

17.3.2.4 Causal Comparative and Correlational Studies

In contrast with experimental and quasi-experimental studies, causal com-parative and correlational studies do not involve manipulation of indepen-dent variables or factors, or experimental design or controls. They are used in situations where manipulation of independent variables is impractical, impossible, or inappropriate.

Causal comparative designs compare groups of subjects drawn from the same population that are different in a critical variable (e.g., probiotic strains in the gut), but otherwise identical. Causal comparative design is used to explore the causal relationship between two variables in instances where it is not possible to manipulate an independent variable or is pro-hibitive to assess changes due to ethical or practical reasons (27). These designs are useful only in suggesting a possible cause and effect relation-ship between variables. However, such an indication of cause and effect is weak because it is difficult to assert that a particular variable is a cause or the result of a condition being investigated. Furthermore, many potential confounders could account for some or all of the observed effects. Data from causal comparative designs can be statistically analyzed using para-metric (e.g., t-test and ANOVA) and nonparametric tests. Table 17.1 pro-vides a systematic way to determine the appropriate statistical analysis based on the type of independent and dependent variables and the type subject variations.

Correlational studies are similar to causal comparative designs, but their main purpose is to explore relationships between or among variables. Cor-relational studies can yield a measure of the direction and strength of the variation or covariation between or among independent and dependent variables and provide a predictive relationship between these variables. However, they cannot be used to make inferences about the cause and effect relationships between dependent and independent variables because the absence of randomization and controls makes it difficult to exclude potential confounding variables that may explain the observed correlation. Correla-tional studies are, however, useful as exploratory and hypothesis-generating studies. Correlational data are analyzed by a variety of methods depending on the nature and number of variables being correlated. If two variables measured on an interval or ratio scale are being studied, a Pearson product moment correlation is used for data analysis, whereas a Spearman rank-order correlation is used if the two variables are ordinal. For multivariate relationships, multiple correlation and canonical correlation are appropriate for data analysis. Most statistical packages (e.g., SAS, SPSS) have algorithms that generate correlation coefficients and test their significance.

17.4 Concluding Remarks

As in any rigorous scientific method, the use of randomization, appropriate experimental design to eliminate the effect of extraneous factors, appropriate controls, and sufficient sample size is key to obtaining powerful and generalizable statistical inferences regarding the health effects of probiotics. To this end, randomized placebo-controlled clinical trials are the best, if cost effective. A variety of other designs such as a crossover design can also yield reliable inferences with less cost (e.g., fewer subjects). In addition to experimental design issues, pinpointing cause and effect relationships between probiotics and any desirable health effect will also require addressing issues specific to probiotic research within the framework of large clinical trials in a manner that clearly demonstrates health effects of probiotics and removes source of confounds. These issues, reported by many researchers (1, 2, 6, 7, 24, 46, 47, 48), include (A) rigorous strain-to-strain comparisons of probiotic performances against specific clinical disorders in specific population groups given the fact that single probiotic strains are often proposed to contribute to a multitude of benefits across many individuals in a test population, (B) determination of dose, duration, and frequency of probiotic treatment for different strains with special emphasis on the minimum therapeutic dose of the probiotic microorganism, (C) evaluation of individual variations in composition and metabolism of consumed probiotic and host flora, (D) identification of microbial determinants of a specific probiotic effect as influenced by the environment such as food matrices because a strain present at high numbers in two food matrices can be probiotically active in one food but not in the other, and (E) evaluation of probiotics' ability to permanently colonize the intestinal tract. Finally, new molecular techniques should enable an accurate assessment of the flora composition and give improved strategies for identifying specific biochemical mechanisms of probiotic effects. Such biotechnological advances should give researchers new tools to address the above issues.

References

1. CE Hoesl, JE Altwein. The probiotic approach: An alternative treatment option in urology. *Eur. Urol.*, 47: 288–296, 2005.
2. F Shanahan. Probiotics in inflammatory bowl disease—therapeutic rationale and role. *Adv. Drug Deliv. Rev.*, 56: 809–818, 2004.
3. T Matilla-Sandholm, S Blum, JK Collins, R Crittenden, W deVos. Probiotics: towards demonstrating efficacy. *Food Sci. Technol.*, 10: 393–399, 1999.
4. P Brigidi, B Vitali, E Swennen, G Bazzocchi, D Matteuzzi. Effects of probiotic administration upon the composition and enzymatic activity of human fecal microflora in patients with irritable bowl syndrome of functional diarrhea. *Res. Microbiol.*, 152: 735–741, 2001.
5. B Mombelli, MR Gismondo. The use of probiotics in medical practice. *Antimicrob. Agents*, 16: 531–536, 2000.
6. CE McNaught, J McFie. Probiotics in clinical practice: a critical review of the evidence. *Nutr. Res.*, 21: 343–353, 2001.
7. KM Tuohy, HM Probert, CW Smejkal, GR Gibson. Using probiotics to improve gut health. *Drug Discov. Today*, 8: 692–700, 2004.
8. WP Hammes, C Hartel. Research approaches for pre- and probiotics: challenges and outlook. *Food Res. Int.*, 35: 165–170, 2002.
9. RO Kuehl. *Statistical Principles of Research Design and Analysis.* 2nd ed., Duxbury Press, Belmont, CA, 1999, 1–562.
10. RJ Freund, WJ Wilson. *Statistical Methods.* 2nd ed., Academic Press, Boston, 2002, 117–580.
11. J Neter, W Wasserman, MH Kutner, CJ Nachtsheim, M Kutner. *Applied Linear Statistical Models.* 4th ed., McGraw Hill, 1996, 3–631.
12. A. Agresti. *Categorical Data Analysis.* 2nd ed., John Wiley & Sons, New York, 2002, 9–625.
13. RA Johnson, DW Wichern. *Applied Multivariate Statistical Analysis.* 5th ed., Prentice Hall, New York, 2001, 149–668.
14. L Pelto, E Isolauri, E LiliusNuutila, S Salminen. Probiotic bacteria down-regulate the milk-induced inflammatory response in milk-hypersensitive subjects but have an immunomodulatory effect in healthy subjects. *Clin. Exp. Allerg.*, 28: 1474–1479, 1998.
15. H Yasui, N Nagaota, A Mike, K Hayakawa, M Ohwaki. Detection of *Bifidobacterium* strains that induce large quantities of IgA. *Microb. Ecol. Health Dis.*, 5: 155–162, 1992.
16. T Matsuzaki. Immunomodulation by treatment with *Lactobacillus casei* strain Shirota. *Int. J. Food Microbiol.*, 41: 133–140, 1998.
17. K Tomita, H Akaza, K Nomoto, K Nomoto, T Yokokura, H Mitsushima, Y Homma, Y Aso. Influence of *Lactobacillus casei* on rat bladder carcinogenesis. *Jpn. J. Urol.*, 85: 655–663, 1994.
18. LA Shackelford, DR Rao, CB Chawan, SR Pulusani. Effect of feeding fermented milk on the incidence of chemically induced colon tumors in rats. *Nutr. Cancer*, 5: 159–164, 1983.
19. BS Reddy. Possible mechanisms by which pro-and prebiotics influence colon carcinogenesis and tumors growth. *J. Nutr.*, 129: 147S–182, 1999.

20. M du Toit, CM Franz, LM Dicks. Characterization and selection of probiotic lactobacilli for a preliminary minipig feeding trial and their effect on serum cholesterol levels, faeces pH and faeces moisture content. *Int. J. Food Microbiol.*, 40: 93–104, 1998.

21. M Fukushima, A Yamada, T Endo, M Nakano. Effects of a mixture of organisms, *Lactobacillus acidophilus* or *Streptococcus faecalis* on delta6-desaturase activity in the livers of rats fed a fat-and cholesterol-enriched diet. *Nutrition*, 15: 373–378, 1999.

22. MP Taranto, M Medici, G Perdigon, AP Ruiz Holgado, GF Vadez. Evidence for hypocholesterolemic effect of *lactobacillus reuteri* in hypercholesterolemic mice. *J. Dairy Sci.*, 81: 2336–2340, 1998.

23. E Isolauri, H Majamaa, T Arvola. *Lactobacillus casei* GG reverses increased intestinal permeability induced by cow milk in suckling rates. *Gastroenterology*, 105: 1643–1650, 1993.

24. K Madsen, A Cornish, P Soper. Probiotic bacteria enhance murine and human intestinal epithelial barrier function. *Gastroenterology*, 121: 580–591, 2001.

25. TR Klaenhammer, MJ Kullen. Selection and design of probiotics. *Int. J. Food Microbiol.*, 50: 45–57, 1999.

26. WR Shadish, TD Cook, DT Campbell. *Experimental and Quasi-Experimental Designs for Generalized Causal Inference.* Houghton Mifflin, Boston, 2002, 13–311.

27. RC Martella, R Nelson, NE Marchand-Martella. *Research Methods: Learning to Become a Critical Research Consumer.* Allyn and Bacon, Boston, 1999, 48–246.

28. AM Graziano, ML Raulin. *Research Methods: A Process of Inquiry*, 4th ed., Allyn and Bacon, Boston, 1999, 150–285.

29. P Gionchetti, F Rizello, A Venturi, P Brigidi, G. Bazzocchi. Oral bacteriotherapy as maintenance treatment in patients with chronic pouchitis: a double-blind, placebo-controlled trial. *Gastroenterology*, 19: 305–309, 2000.

30. M Kalliomaki, S. Salminen, T Poussa, H Arvillommi, E Isolauri. Probitics and prevention of atopic diseases: 4-year follow-up of a randomized placebo-controlled trial. *Lancet*, 361: 1869–1871, 2003.

31. V Gotz, JA Romankiewicz, J Moss, HW Murray. Prophylaxis against ampicillin associated diarrhea with *Lactobacillus* preparation. *Am. J. Hosp. Pharm.*, 36: 754–757, 1979.

32. Y Aso, H Akaza, T Kotake, T Tsukamoto, K Imai, S. Naito. Preventive effect of a *Lactobacillus casei* preparation on the recurrence of superficial bladder cancer in a double-blind trial. The BLP Study Group. *Eur. Urol.*, 27: 104–109, 1995.

33. CM Surawicz, GW Elmer, P Speelman, LV McFarland, J Chinn, G Van Belle. Prevention of antibiotic associated diarrhea by *Saccharomyces boulardii*. *Gastroenterology*, 96: 981–988, 1989.

34. RM Tankanow, MB Ross, IJ Ertel, DG Dickinson, LS McCormick, JF Garfinkel. Double-blind placebo-controlled study of the efficacy of lactinex in the prophylaxis of amoxicillin-induced diarrhea. DICP. *Ann. Pharm.*, 24: 382–384, 1990.

35. PF Wunderlich, L Braun, I Fumagalli, V D'Apuzzo, F Hein, M Karly, R Lodi, G Polittia, F von bank, L Zettner. Double-blind report on the efficacy of lactic acid-producing Enterococcus SF68 in the prevention of antibiotic-associated diarrhea and in the treatment of acute diarrhea. *J. Int. Med. Res.*, 17: 333–338, 1989.

36. M Guslandi, G Mezzi, M Sorghi, PA Testoni *Saccharomyces boulardii* in maintenance treatment of Crohn's disease. *Dig. Dis. Sci.*, 45: 1462–1464, 2000.

37. JK Collins, C Dunne, L Murphy, D Morrissey, L O'Mahony, E O"Sullivan, G. Fitzgerald, B Kiely, GC O'Sullivan, C Daly, P Marteau, F Shanahan. A randomized controlled trial of a probiotic *Lactobacillus* strain in healthy adults: assessment of its delivery, transit, and influence on microbial flora and enteric immunity. *Microb. Ecol. Health Dis.*, 14: 81–89, 2002.

38. W Kruis, E Schutz, P Fric, B Fixa, G Judmaiser, M Stolte. Double blind comparison of an oral Escherichia coli preparation and mesalamine in maintaining remission of ulcerative colitis. *Aliment. Pharmacol. Ther.*, 11: 853–858, 1997.

39. H Majamaa, I Isolauri, M Saxelin, T Vesikaru. Lactic acid bacteria in the treatment of acute rotavirus gastroenteritis. *J. Pediatr. Gastroenterol. Nutr.*, 20: 333–338, 1995.

40. PH Katelaris, I Salam. Lactobacilli to prevent traveller's diarrhoea? *N. Engl. J. Med.*, 333: 1360–1361, 1995.

41. AV Shornikova, IA Casas, E Isolauri, H Mykkanen, T Vesikari. Lactobacillus reuteri as a therapeutic agent in acute infectious diarrhoea in young children. *J. Pediatr. Gastroenterol. Nutr.*, 24: 399–404, 1997.

42. RA Oberhelman, RH Gilman, P Sheen, DN Taylor, RE Black, L Cabera, AG Lescano, R Meza, G Madico. A placebo-controlled trial *Lactobacillus* GG to prevent diarrhoea in undernourished Peruvian children. *J. Pediatr.*, 134: 15–20, 1999.

43. E Isolauri, M Kaila, H Maykkane. Oral bacteriotherapy for viral gastroenteritis. *Dig. Dis. Sci.*, 39: 2595–2600, 1994.

44. JD Pozo-Olano, JH Warren, RG Gomez, MG Cavazos. Effect of a *Lactobacillus* preparation on traveler's diarrhoea. A randomized , double-blind clinical trial. *Gastroenterology*, 74: 829–830, 1978.

45. E Hinton, P Kalakowski, C Singer, M. Smith. Efficacy of *Lactobacillus* GG as a diarrheal preventive in travelers. *J. Trav. Med.*, 4: 41–43, 1997.

46. DT Campbell, JC Stanley. Experimental and Quasi-Experimental designs for Research. Rand McNally, Chicago, 1966, pp 34–70.

47. HS Gill. Probiotics to enhance anti-infective defences in the gastrointestinal tract. *Best Pract. Res. Clin. Gastroenterol.*, 17: 755–773, 2003.

48. CP Tamboli. Probiotics in inflammatory bowel disease: a critical review. *Best Pract. Res. Clin. Gastroenterol.*, 17: 805–820, 2003.

18

The Safety of Probiotics in Foods in Europe and Its Legislation

Arthur C. Ouwehand, Vanessa Vankerckhoven, Herman Goossens,
Geert Huys, Jean Swings, Marc Vancanneyt, and Anu Lähteenmäki

CONTENTS

18.1   Introduction

Fermented foods have been consumed for thousands of years (1). Traditionally one of the main reasons to consume fermented foods has been their extended shelf life, and thus improved safety, compared to the unfermented starting material. Lactic acid bacteria (LAB) are the organisms found most commonly in fermented foods (2), in particular members of the genera *Lactobacillus* and *Lactococcus*. In fact, in Europe the annual consumption of pure LAB biomass, from fermented dairy products alone is estimated to be 3400 t (3). Furthermore, LAB are common inhabitants of the gastrointestinal (GI) tract of many animal species, including humans, where they are thought to fulfill a beneficial role. It is because of this long history of safe and widespread use and their general presence in the GI tract that LAB are considered safe for human consumption.

Starter cultures often have a long history of safe use. However, probiotics often do not. Some strains of *Lactobacillus acidophilus* and *Lactobacillus casei* Shirota have a relatively long history of use (4), but most probiotics have been on the market for only one or two decades. Another difference between starters and probiotics is their origin. Starters are isolated from milk, fermented dairy, and vegetable or meat products, whereas probiotics usually are of intestinal origin. Starter strains will often not survive passage through the GI tract; probiotics, on the other hand, are selected to survive this passage and persist in the GI tract. Thus the apparent safety of lactic starters cannot directly be applied to probiotics.

Although probiotics have a very good safety record, it is nevertheless important to ascertain the safety of new and existing probiotic strains before conducting clinical trials and certainly before marketing. After all, the producer will remain responsible for the safety of the probiotic product.

The current chapter will outline some of the safety considerations to be taken into account before use and marketing of probiotic strains.

18.2   Infections Where Lactic Acid Bacteria Were Involved

18.2.1   Relevance for Probiotics

LAB, and in particular lactobacilli and bifidobacteria, are widely used in the food industry, and this use has, in general, not been associated with disease. Starters and other cultures used in food fermentation have contributed to an increased food safety and health of the consumer. Nevertheless, some rare cases of *Lactobacillus* and even *Bifidobacterium* infections have been reported over the years, and the intestine is thought to be the source of most of these infections (5). Such strains should be stored for future reference of postmarketing surveillance, which is the ultimate safety assessment for probiotics and starters.

18.2.2   Underlying Conditions

Certain underlying conditions have been suggested to enhance the risk of infections with lactobacilli and bifidobacteria. These conditions are mainly related to reduced immune function, although dental manipulation has also been associated with increased risk of infection with these organisms. It is, however, important to put these potential, albeit small, risks in relation to the potential benefit.

18.3   Safety Assessment of Probiotic Microbes

18.3.1   Acute Toxicity

The toxicity of LAB strains to the GI tract mucosa is considered an important aspect of safety, as most probiotics are intended for oral consumption; it is through colonization of the GI tract that probiotics are thought to exert their health-enhancing effects (6). Although acute toxicity tests were originally designed for chemicals, they also give an indication of harmful effects associated with bacteria (7).

Balb-c, Swiss mice or Lewis rats are commonly used to assess acute toxicity. Bacteria (10^7–10^{10} CFU/ml) can be added to the drinking water for 8 to 10 d, 50 µl of bacterial suspension can be orally fed for 28 d, or 1 ml bacterial suspension can be given by gavage for 7 d. During the experiments possible changes in activity, behavior, and general health are observed daily. Water intake, feed intake, and body weight are measured weekly. The animals are euthanized after the experiments, and the organs are examined for gross pathological changes (8–12). The contents of the distal part of the ileum, cecum, and colon can be dissolved and homogenized, after which mucin degradation can be determined colorimetrically (10). Bacterial translocation can be measured by plating blood samples taken by cardiac puncture and by plating homogenized tissue samples of the mesenteric lymph nodes, spleen, and liver. The plates are incubated at 37°C for 72 h. The organisms are identified by randomly amplified polymorphic DNA (RAPD) polymerase chain reaction (PCR) (11, 12). For histological examination, part of the stomach, distal part of the ileum, cecum, and colon are removed, fixed, and embedded in paraffin wax. Sections (5 to 6 µm) are stained with hematoxylin and eosin or Alcian blue to reveal villus architecture, epithelial cell height, and mucosal thickness (10–12). Examples of acute toxicity of LAB are given in Table 18.1.

18.3.2 Endocarditis Model

L. casei and *L. rhamnosus* are the organisms most commonly associated with infective endocarditis (IE), although the incidence of *Lactobacillus* endocarditis is very low (13). Nevertheless, a European Union-sponsored workshop organized by the Lactic Acid Bacteria Industrial Platform proposed that the safety of each strain should be assessed (14). Several models of IE have been developed, such as a rabbit model using specific-pathogen-free male Japanese white rabbits (15). In this model, a polyethylene catheter is passed down the artery and positioned in the left ventricle. During 7 to 10 d, small vegetations of platelets and fibrin can form on the tricuspid valve or the endocardium at points of contact with the catheter. After this period, 1 ml of bacterial suspension, grown under appropriate conditions, washed three times with phosphate-buffered saline (PBS), and resuspended to an optical density of 0.9 ± 0.1 at 600 nm (ca. 10^9 CFU/ml), is injected into the marginal ear vein. Blood samples are taken at intervals from the opposite marginal ear vein. The animals are killed, and small pieces of liver, spleen, and vegetations at the heart valve are removed, weighed, and homogenized in 1 ml of saline. The blood samples and organ samples at 0.1 ml are plated, and colony counts are performed.

TABLE 18.1

Acute Toxicity of Probiotic Lactic Acid Bacteria in Mice

Species	Strain	LD_{50} (g/kg body weight)
Bifidobacterium lactis	HN019	>50
Bifidobacterium longum	a	25
Bifidobacterium longum	BB-536	>50 (0.52 via intraperitoneal route)
Enterococcus faecium	AD1050 [b]	>6.60
Lactobacillus acidophilus	HN017	>50
Lactobacillus casei	Shirota	>2.00
Lactobacillus delbrueckii subsp. bulgaricus	a	>6.00
Lactobacillus fermentum	AD0002[b]	>6.62
Lactobacillus helveticus	a	>6.00
Lactobacillus rhamnosus	GG (ATCC 53103)	>6.00
Lactobacillus rhamnosus	HN001	>50
Lactobacillus salivarius	AD0001 [b]	>6.47
Streptococcus equinus	a,b	>6.39

[a] Strain not known.
[b] Nonviable, heat-inactivated, preparations.

Source: Modified after Reference 104.

18.3.3   Effect on the Immune System

Several models have been developed to study the effect of LAB on the immune system. In mice, strain-dependent cytokine profiles are induced after oral administration of *Lactobacillus* (16). It is likely that local cytokine profiles may favor either T helper cell type 1 (Th1) immunogenic responses, which may be the predominant response in autoimmune diseases, or T helper cell type 2 (Th2) tolerogenic responses, which may contribute to diseases such as atopic allergy (17–20). The proinflammatory cytokines secreted by the epithelium, such as TNF-α, IL-1, IL-6, IL-8, and IL-12, are the hallmarks of the inflammatory responses of the intestine (21). Insight into the proinflammatory cytokine profile of human intestinal cells after their exposure to LAB will help in understanding the interaction between human enterocytes and LAB used as probiotics (21, 22).

In SJL/J mice, lactobacilli are able to enhance or inhibit the development of disease after induction of experimental autoimmune encephalomyelitis (EAE) (16). Bacteria are grown under appropriate conditions. Female SJL/J mice 6 to 12 weeks old, which have a low microbiological burden and low IgE levels, or BALB/c mice, which are Th2 biased, are used. The mice receive a 10^{10}-CFU/ml bacterial suspension intragastrically on 5 alternating days to create a continuous high level of lactobacilli. On day 0, the mice are subcutaneously immunized with 50 to 300 µg proteolipid protein (PLP 139-151) in Freund's complete adjuvant to induce acute EAE. On days 0 to 1 and days 2 to 3 the mice are intravenously injected with 10^{10} CFU/ml of *Bordetella pertussis* to affect the integrity of the blood-brain barrier. The severity of the EAE can be scored using the disability scale: 0 = no clinical signs, 1 = tail weakness, 2 = mild paralysis and ataxia of the hind legs, 3 = severe paralysis or ataxia of the hind legs, 4 = moribund, and 5 = death due to EAE. EAE develops at days 12 to 16 after immunization with a highest score of 3. Blood samples are taken by tail vein puncture on different days. Enzyme-linked immunosorbent assay (ELISA) is used to detect anti-PLP antibodies. Immunohistochemical analysis is performed on cryosections (8 µm) from snap-frozen material. For the detection of IL-1α, IL-1β, IL-2, IL-4, IL-10, IFN-γ, and TNF-α, monoclonal antibodies with an anti-mouse/rat Ig-biotin (Dako) and an HRP-labeled avidin-biotin complex (Dako) are used (16, 23–27).

18.3.4   Metabolism

Lactobacilli are capable of transforming conjugated primary bile salts into free deconjugated and secondary dehydroxylated bile salts in the small bowel (28). The free bile salts may induce diarrhea and intestinal lesions (29), whereas secondary bile salts may exhibit carcinogenicity (30). Excessive deconjugation or dehydroxylation of bile salts by probiotics, through production of the enzyme bile salt hydrolase, could therefore be a potential risk (28).

To study transformation of bile salts, pure cultures of bacteria grown under appropriate conditions are incubated in fresh broth with conjugated primary bile salts (taurocholic acid, glycocholic acid, taurochenodeoxycholic acid, glycochenodeoxycholic acid). Relative amounts of deconjugation activity can be determined by appropriate dilutions of the broth cultures (31), or the amounts of bile salts deconjugated can be determined using high-performance liquid chromatography (HPLC). Samples can be taken before introduction of bile salt into the bacterial suspension, then every 60 min for 6 h and after 24 h. Methanol-acetate buffer is used as the mobile phase and is passed through a 0.45-μm nylon filter to remove undissolved material. For quantification of conjugated bile salts, 200 μl of sample is mixed with 50 μl of internal standard (dexamethasone); 20 μl of the mixture is then passed through a 0.45-μm nylon filter and injected into the HPLC. For quantification of free bile salts, 500 μl is taken for cholic acid derivatization. Of the final free bile salt derivative, 200 μl is filtered through a 0.45-μm polysulfone filter and mixed with 50 μl internal standard (testosterone). Of this mixture, 20 μl is then injected into the HPLC. For quantification of deconjugated bile salts, 2 ml of sample is resuspended in 8 ml 0.9% NaCl in 0.1 M NaOH and 6 ml of mobile phase. The bile salts are recovered by passage through a Sep-Pac cartridge. Dexamethasone is added to the mobile phase as an internal standard. The amount of each bile salt deconjugated is based on the disappearance of each conjugated bile salt (28, 31–33).

18.3.5   Hemolysis

Hemolysis is a common virulence factor among pathogens that serves mainly to make iron available to the microbe and causes anemia and edema in the host.

   Hemolysis can be determined as described by Baumgartner and coworkers (34). Strains to be tested are grown on suitable solid media containing 5% human blood and incubated under appropriate conditions. *Staphylococcus aureus* and *Bacillus cereus* should be included as positive controls for α- and β-hemolysis, respectively.

18.3.6   Resistance to Human Serum-Mediated Killing

The complement system in blood can opsonize bacteria and facilitate their phagocytosis by leukocytes. The activated complement system can also form a complex that kills bacteria (35). Evasion of this will therefore enhance the survival of the bacteria after translocation.

   Serum resistance can be assessed as described by Burns and Hull (36). Blood is collected from at least 10 healthy adult donors and allowed to clot, and the serum is pooled in equal amounts. Part of the serum has to be heated to 56°C for 20 min to inactivate the complement system. Aliquots are frozen at −70°C until use. Bacteria are grown under appropriate conditions, washed twice with PBS (pH 7.2, 10 mM phosphate), and the absorbance (600 nm) is adjusted to 0.5 ± 0.01 in order to standardize the number of bacteria (10^7 to

10^8 CFU/ml). The bacteria were mixed with serum, heat-inactivated serum or PBS, to a final serum concentration of 80% and incubated aerobically 90 min at 37°C. The reaction is stopped by incubating the sample 10 min on ice and making serial dilutions in PBS. Dilutions are plated on appropriate media and incubated. Alternatively, viability can be assessed using flow cytometry and LIVE/DEAD BacLight viability kit (Molecular Probes, Eugene, OR) (37).

18.3.7 Induction of Respiratory Burst

Upon phagocytosis, peripheral blood mononucleocytes (PMNs) produce a burst of reactive oxygen species; this will kill and digest the phagocytosed particles (38). The ability to avoid the induction of such a respiratory burst may enhance the survival of a translocated microbe and hence the risk for infection.

Bacteria are grown under appropriate conditions, washed twice with HEPES-buffered Hank's balanced salt solution (HH; 10 mM HEPES; pH 7.4), and the absorbance (600 nm) is adjusted to 0.5 ± 0.01. PMNs were collected by lysing erythrocytes in freshly collected human blood with 0.8% NH_4Cl. PMNs are washed and resuspended in Hank's balanced salt solution. The concentration of PMNs is determined by flow cytometry or Bürker counting chamber. The measurement of the respiratory burst is then performed as described by Lilius and Marnila (38). To gelatin-coated microtiter plate wells (Cliniplate; Labsystems, Helsinki, Finland), 25 ml Hank's balanced salt solution containing 0.1% gelatin, 20 ml 5-amino-2,3-dihydro-1,4-phthalazine-dione (luminol; 10 mM), 40 ml bacterial suspension were added and incubated 30 min at 37°C. Subsequently, 40 ml PMN suspension is added. For background measurements, PMNs are incubated without bacteria. The plates are incubated at 37°C, and luminescence is measured for at least 2 h with 3-min intervals, e.g., with a Victor2 multilable reader (PerkinElmer, Turku, Finland). Results are presented as the maximum signal (mV/100,000 PMNs) after subtraction of the background and as the average of at least three independent observations with PMNs from different donors.

18.3.8 Platelet Aggregation

The spontaneous aggregation of platelets leads to the formation of thrombi and edema and has been implicated in infective endocarditis (39).

Platelet-rich plasma (PRP) is obtained by centrifuging fresh citrated blood 10 min at 100 × g. Platelet-poor plasma (PPP) is obtained by further centrifuging 10 min at 2350 × g. The PRP is adjusted with PPP to an absorbance (660 nm) of 0.5 ± 0.01 to give a platelet concentration of approximately 2.27 × 108/ml. Platelet aggregation is carried out in an aggregometer, where light transmission through PPP represents 100% aggregation and that through PRP 0% aggregation. The ability to induce platelet aggregation by the test strains was determined by adding 25 µl bacteria (2×10^9 CFU/µl) to 250 µ PRP or PPP, preincubated at 37°C for 5 min. A lag phase of more than 25 min can be assumed to be a negative aggregation (40).

18.3.9   Adhesion to Extracellular Matrix Proteins and Intestinal Mucus

Collagens are the major proteins of the extracellular matrix and may be exposed in injured tissue. Pathogens often have affinity for these proteins as this affinity will give them access to host tissues (41). Fibrinogen is present in high concentrations in plasma and forms the structure of the blood clot. It also coats the outer surface of implanted biomaterials (42). Its presence in wounds and on foreign bodies makes fibrinogen an important substratum for microbial adhesion.

Adhesion to mucus and human intestinal epithelium has been proposed as one of the main criteria for selecting potentially promising probiotics for use in humans (43). It is, however, also the first step in pathogenesis (44) and deserves close attention.

Human collagen type IV and fibrinogen can be purchased from, e.g., Sigma (St. Louis, MO); intestinal mucus can be isolated from feces of healthy adult volunteers by aqueous extraction and dual ethanol precipitation (45). The substrata are passively immobilized (0.5 mg/ml) on microtiter plate wells by overnight incubation at 4°°C. Excess material is washed away twice with HH. Bacteria are grown under appropriate conditions, to the culture media 10 µl/ml of tritiated thymidine (methyl-1,2-[^{3}H]thymidine, 120 Ci/mmol) is added to metabolically radiolabel the bacteria. After growth, the bacteria are washed twice with PBS, and the absorbance (600 nm) is adjusted to 0.25 ± 0.01. The bacteria are added to the wells and incubated for 1 h at 37°C. Nonbound bacteria are removed by washing twice with HH. Bound bacteria are released and lysed with 1% sodium dodecyl sulfate (SDS) in 0.1 M NaOH for 1 h at 60°C. Radioactivity is determined by liquid scintillation and the adhesion expressed as the percentage of radioactivity recovered after adhesion, relative to the radioactivity in the bacterial suspensions added to the immobilized collagen, fibrinogen, or mucus (45).

18.3.10 Adhesion to Tissue Culture Cells

18.3.10.1 *Culturing of Intestinal Cells*

It is cumbersome to use an *in vivo* assay to screen a large number of potential probiotic candidate strains for their adhesion (46). Therefore, *in vitro* models involving human intestinal epithelial cell lines (derived from colonic adenocarcinomas) have been used extensively to assess the adhesive properties of probiotics: HT-29 (ATCC 38-HTB) and Caco-2 (ATCC 2102-CRL) cells (47). In culture, these cell lines spontaneously develop the characteristics of mature enterocytes, including polarization, a functional brush border and apical intestinal hydrolases (48, 49). The HT-29 cells can differentiate into columnar absorptive and mucus-secreting cells by growth adaptation to methotrexate: HT-29MTX cells. The cells are cultured in Dulbecco's modified Eagle's minimal essential medium (DMEM) supplemented with 10 to 20% heat-inactivated (56°C for 30 min) fetal calf or bovine serum (Life Technologies GIBCO, BRL, Rockville, MD) in 10% CO_2/90% O_2 air at 37°C (22, 46, 50–62). The use of 100 µg/ml streptomycin and 100 U/ml penicillin is recommended. The medium can also be supplemented with 1% nonessential amino acids. For the HT-29MTX cell line, DMEM with glutamax is recommended. For maintenance purposes, the cells are passaged weekly using trypsin/EDTA and seeded in 6-well plates. The medium is changed every 2 to 3 d and replaced by DMEM without any supplements at least 1 h prior to use for adhesion experiments. The cells are used at 2 weeks post confluence for the adhesion assays to allow full morphological and functional differentiation (22, 46, 50–62).

18.3.10.2 Adhesion to Intestinal Cells

Bacteria are grown under appropriate conditions. Prior to the adhesion assay all bacterial strains are washed twice with PBS and resuspended in DMEM to a final concentration of 10^8 CFU/ml. Before the adhesion experiment the cells are washed three times with PBS. Different methods can be used to test the adhesion of bacteria to intestinal cells.

For light microscopy, two-chamber slides or 6-well tissue culture plates containing glass coverslips are used on which the cells are seeded. Bacterial suspensions of 0.5 to 3 ml are added to each of the wells and incubated for 1 to 3 h at 37°C in 10% CO_2/90% air. The cells are then washed three to five times with PBS to remove unbound bacteria, fixed with acetone, and stained with Giemsa or fixed with methanol and stained with Gram. The number of adherent bacteria are counted in 20 random microscopic areas using oil immersion. The assay is performed in duplicate or triplicate over several successive passages. Adhesion can be expressed as the number of bacteria adhering to 100 cells (46, 50–56, 63].

For scanning electron microscopy, the adhesion experiment is performed as for light microscopy, but the cells are fixed with 2.5% glutaraldehyde in 0.1 M phosphate buffer for 1 h at room temperature. After two washes with phosphate buffer, cells are postfixed for 30 min at room temperature with 2% osmium tetraoxide in phosphate buffer and washed three times with the same buffer. The cells are dehydrated in graded series of ethanol and amyl acetate. Finally, the cells are dried in a critical point dryer and coated with gold (46, 50, 51, 57, 63).

For radiolabeling, the bacteria can be labeled by addition of ^{14}C-uracil and ^{14}C-adenine or ^{14}C-acetic acid or methyl-1,2-[^{3}H] thymidine (Amersham International, UK). To remove excess radioactivity, the bacteria are washed with PBS + 0.1% sodium azide. The cells are prepared on glass coverslips in

6-well tissue culture plates. To each well, 50 μl to 2 ml of radiolabeled bacteria are added and incubated for 1.5 h at 37°C in 10% CO_2/90% air. The mono-layers are washed three times with PBS to remove unattached bacteria. The cell-associated bacteria and intestinal cells are dissolved in 0.1% SDS/0.1 to 0.9 M NaOH. Each assay has to be conducted in triplicate over several successive passages. Radioactivity is determined by liquid scintillation and the adhesion expressed as the percentage of radioactivity recovered after adhesion, relative to the radioactivity in the bacterial suspensions added to the cells (46, 57–61, 63).

The number of attached bacteria can also be enumerated by plate counting. After 1-h incubation, the monolayers are washed three times with PBS to remove nonadherent bacteria, and the monolayers are lysed with sterilized distilled water. Serial dilutions are then plated to count the number of viable bacteria. The assay is performed in triplicate over several successive pas-sages. The adhesion ratio can be calculated by dividing the number of adher-ent bacteria by the number of added bacteria (22, 54, 62, 64).

For the interpretation of the results, the adhesion index is strongly depen-dent on the *in vitro* conditions used. The percentage of adhesion does not constitute an absolute value and has to be seen in relation to the nonadhesive control strain (65).

18.3.11 Phosphatidylinositol-Specific Phospholipase C (PI-PLC) Activity

PI-PLC activity will damage the host cell membrane and is thought to be important for translocation by some microbes (66).

Strains to be tested are spot inoculated on appropriate solid media and incubated until colonies are visible. Thereafter, PI-PLC activity can be deter-mined by applying an overlay with 20 mg/l L-a-phosphatidylinositol in 1.4% (w/v) agarose. PI-PLC activity is visible as a turbid halo of insoluble dia-cylglycerol around the colony. *B. cereus* is included as a positive control (67).

18.3.12 Mucinolytic Activity

An ability to degrade mucus or its glycoproteins is considered one of the valuable indicators of potential pathogenicity and local toxicity of lumen bacteria (7, 68). Both *in vitro* and *in vivo* assays can be used to evaluate toxicity.

Bacteria are grown under appropriate conditions. For the *in vitro* assay, mucin degradation can be evaluated in liquid medium or in a petri dish. For the first method, 200 μl of bacterial suspension (10^7 CFU/ml) is incubated at 37°C for 2 to 5 d with 10 to 25 ml of PBS, basal medium (10) with or without 1 to 3% glucose, basal medium containing 0.3% hog gastric mucin (HGM) with or without 1 to 3% glucose, or basal medium containing 0.3% human intestinal glycoproteins (HIG). After incubation the supernatants are collected and heat treated to inactivate mucinolytic enzymes. The mucin pellets are precipitated by the addition of chilled ethanol. The supernatants are aspirated and stored while the pellets are washed and resuspended in 0.5 ml 10 mM Tris-HCl buffer. The decrease of total carbohydrates and protein content in the mucin residues is measured by SDS-PAGE (polyacrylamide gel electrophoresis) as this is defined as mucin degradation. The assay is performed in triplicate. Autoclaved samples are used as negative controls, and the degradation ratio can be calculated as [1-concentration in test samples/concentration in control sample] × 100% (6, 10).

For the second method, 10 μl of bacterial suspension is inoculated onto 0.5% HGM and 1.5% agarose incorporated into basal medium. The plates are incubated at 37°C for 72 h and subsequently stained with 0.1% Amido Black in 3.5 M acetic acid for 30 min. They are then washed with 1.2 M acetic acid until a discolored halo around the colony appears. The mucin degradation activity can be defined as the size of the mucin lysis zone.

18.4 Antibiotic Resistance

Resistance to antibiotics can be intrinsic or acquired via specific genetic elements. In the former, the target of the antibiotic does not exist or is very different, and hence the antibiotic is ineffective. In the latter form, a genetically encoded mechanism exists that interferes with the antibiotic's action, such as, e.g., β-lactamases, active efflux of the antibiotic, or modification of the target. This type of antibiotic resistance becomes particularly important when it is transferable.

Antibiotic resistance can be assessed using standard disk diffusion tests on appropriate solid media or antibiotic dilution in broth culture. Intrinsic resistance is usually species dependent (69) and will be present in the majority of strains from a given species. Acquired resistance, on the other hand, will usually be strain dependent. Because transferable antibiotic resistance is a serious safety concern, it needs to be distinguished from intrinsic resistance, as outlined by the Scientific Commission on Animal Nutrition (70), Table 18.2.

18.5 The Importance of Taxonomy

The accuracy and reliability of the species identification of probiotic strains and the label correctness of the probiotic products have long been ignored as true safety aspects. Triggered by the exponential growth of the functional food market in general, it is only since the mid-1990s that commercial probiotic strains have been included in taxonomic studies to obtain an identification up to the species level. Regardless of the country in which studies

TABLE 18.2

Suggested Experiments to Distinguish between Intrinsic and Acquired Antibiotic Resistance, As Suggested by the Scientific Commission on Animal Nutrition (70)

Test	Intrinsic	Acquired
Resistance present in most strains of a given species	+	-
In vitro transfer of resistance	-	+
Known resistance genes	-	-/+
Isolation and sequencing of the resistance gene	+	+
Localization of chromosome	-/+	-/+
House keeping genes flanking the resistance gene	+	-/+
Insertion sequences flanking the resistance gene	-	+

were performed, it is a common observation that bacterial strains with claimed probiotic properties are often misidentified, and current reports indicate that this situation has not significantly improved in recent years.

Only a few taxonomic studies have attempted to directly contact the product manufacturer or strain distributor to obtain pure cultures of strains marketed as probiotic organisms. In fact, the majority of studies that have assessed the identity of probiotic strains were performed on isolates recovered from commercial probiotic products. Recovery and purification of probiotic organisms can be complicated by several factors including the product's shelf life or expiry date, the use of suboptimal cultivation conditions (incubation temperature and atmosphere, culture medium, etc.), and nonrepresentational colony selection (71). As discussed below, these potential shortcomings can be overcome by using a culture-independent identification as an alternative to or in combination with conventional cultivation methods. Most taxonomic studies have so far concentrated on the investigation of yogurts and other fermented milk products, dried food supplements, and fruit drinks containing LAB (bifidobacteria, enterococci, lactobacilli, lactococci, pediococci, and streptococci), although the composition of probiotic products containing *Bacillus* or *Saccharomyces* has also been studied.

The speciation of a pure culture can be considered as a very straightforward process. A wide range of different techniques has been used to obtain a correct identification of probiotic strains (Table 18.3, for more details see Chapter 9 by Temmerman and coworkers in this book). It is common knowledge that phenotypic methods display limited taxonomic resolution for the speciation of probiotic lactobacilli and bifidobacteria, and consequently, it is not surprising that many cases of probiotic product mislabeling can be attributed to the use of only phenotypic methods for taxonomic characterization. Still, carbohydrate fermentation patterning and miniaturized biochemical identification systems remain very popular methods because they require little expertise and because a wide range of phenotypic test systems are commercially available such as API and BIOLOG. Molecular phenotypic (chemotaxonomic) methods such as SDS-PAGE of whole-cell proteins have

TABLE 18.3

Overview of Studies Reporting on Species Identification of Probiotic Strains or Label Correctness of Probiotic Products

Probiotic Strains or Products Tested	Identification Technique(S) Used	Reported Discrepancies with Original Strain or Product Label Information [a]	Ref.
Commercial probiotic strains	16S rDNA sequencing, fatty acid analysis, carbohydrate fermentation	14/29 strains displayed nonmatching species designations	76
Probiotic (fermented milk) products	16S rDNA sequencing, fatty acid analysis, carbohydrate fermentation	6 products contained other species than those stated	76
Food supplements and fermented products	API	13/29 food supplements lacked one or more stated species; 2/5 fermented products contained one additional species	93
Dried *Bacillus* spore probiotics	16S rDNA sequencing, API 50 CH	4/5 products were mislabeled	81
Mild and probiotic yogurts	DNA-DNA hybridization, carbohydrate fermentation	9/15 yogurts contained other species than those stated	77
Bifidobacterium dairy products	DNA-DNA hybridization, carbohydrate fermentation	3/6 products contained other species than those stated	78
Novel-type yogurts	DNA-DNA hybridization	4/16 yogurts contained other species than those stated	79
Dairy products and food supplements	Whole-cell protein profiling	14/30 food supplements were mislabeled; 10/25 dairy products were mislabeled	71
Dairy products and food supplements	16S rDNA-DGGE	2/4 dairy products were mislabeled; 3/5 food supplements were mislabeled	84
Yogurts and lyophilized products	16S rDNA-DGGE, species-specific PCR	1/5 yogurts contained other species than those stated; 6/7 lyophilized products lacked one or more species stated	85

[a] Discrepancies were only mentioned for those products that contained sufficient information on their bacterial species composition.

proven to be much more powerful to distinguish among probiotic *Lactoba-cillus* and *Bifidobacterium* species (71, 72), although a limited resolution has been reported, e.g., for the differentiation of species in the *L. acidophilus* group (73, 74), the *L. plantarum* group (75), and among bifidobacteria. Other chemo-taxonomic techniques such as gas–liquid chromatographic analysis of fatty acid methyl esters (GLC-FAME) have not been found useful for speciation of probiotic LAB due to the limited number of entries in the identification library and the inconsistent extraction of FAME (76). A few studies have performed (microscale) DNA-DNA hybridizations to assess the taxonomic identity of probiotics (77–79). Although this approach is still considered the "gold standard" to define and identify bacterial species, it is generally not recommended for routine identification of probiotic strains because data from this technique are not cumulative and thus do not allow construction of identification libraries (80). Sequencing analysis of the 16S rDNA gene is now widely used as the first choice technique for the phylogenetic position-ing of unknown bacterial strains, and as such it has also been used for the identification of commercial probiotic strains or isolates recovered from pro-biotic products (76, 81). Yeung et al. and co-workers (76) concluded that most probiotic *Lactobacillus* species can be reliably identified using the first 500 bp of the 16S rDNA gene except for *L. gallinarum* and *L. amylovorus* that display high homology in this region of the gene. On the other hand, the authors indicated that partial 16S rDNA sequencing is not adequate for speciation of probiotic bifidobacteria. At present, genomic fingerprinting techniques such as amplified fragment length polymorphism analysis (74, 75) and rep-PCR of repetitive genomic elements such as (GTG)5 (82) and BOX (83) may be the most promising tools currently available to speciate probiotics. Pro-vided that a taxonomic framework of well-documented reference strains is available, these fingerprinting techniques have the potential to provide the species identity and to reveal subspecific relationships of probiotic strains. Finally, there is a current trend to determine the taxonomic identity of com-mercial probiotic strains directly in the product matrix using culture-inde-pendent denaturing gradient gel electrophoresis (DGGE) analysis of 16S rDNA gene regions (84, 85). This approach circumvents the frequently encountered problem of poor strain recovery under laboratory conditions and avoids time-consuming analysis of multiple isolates obtained from a given product.

The importance of taxonomy in relation to the safety of probiotics has been particularly well documented for some groups of the LAB. The identity of probiotic enterococci, in most cases *E. faecium* or *E. faecalis*, can be relatively easily verified by phenotypic or genotypic analyses (86, 87). In contrast, straightforward identification of probiotic lactobacilli and bifidobacteria is more complicated due to their complex taxonomic structure. Among the lactobacilli, the majority of probiotic species either belong to the *Lactobacillus acidophilus* group, the *L. casei* group, or the *L. reuteri* group (73). In the first group, the taxonomic recognition of probiotic strains belonging to *L. acido-philus sensu stricto* is sometimes hampered by its phenotypic resemblance to other species of this group, i.e., *L. amylovorus*, *L. gallinarum*, *L. johnsonii*, *L. gasseri*, and *L. crispatus*, of which the latter three taxa are also used as probiotics. Therefore, the use of DNA-based techniques is recommended to discriminate between representatives of the *L. acidophilus* group (74). Likewise, there are several reported cases of confusion in the *L. casei* group as a result of the unclear taxonomic status of the species *L. casei* and *L. paracasei*, and their affiliation to the other members of *L. rhamnosus* and *L. zeae*. Depending on the taxonomic proposals to reject (88) or to retain (89) the name *L. paracasei*, it is very possible that the same probiotic strain is cited as *L. casei* or as *L. paracasei*. Furthermore, probiotic strains belonging to *L. rhamnosus* are often referred to as *L. casei* although the nomenclatural proposal to change *L. casei* subsp. *rhamnosus* to *L. rhamnosus* already dates from 1989 (90). Strains of the species *L. reuteri* are also often used as probiotics and are phylogenetically situated in the *L. reuteri* group. Because of their close phenotypic relationship to *L. fermentum*, the identity of probiotic representatives of the latter two species is sometimes confounded, but molecular techniques offer a clear differentiation between the two taxa.

In bifidobacteria, the majority of probiotic strains belong to *B. animalis* or *B. lactis*, and *B. longum*, and to a lesser extent to *B. infantis*, *B. breve*, and *B. bifidum*. Whereas the separation of the latter three taxa is usually possible using phenotypic characteristics, unambiguous differentiation between probiotic *B. animalis* or *B. lactis*, and *B. longum* requires molecular phenotypic or genotypic techniques. However, the most important problem in this genus relates to the labeling of many probiotic *B. animalis* strains as *B. lactis*, although DNA-DNA hybridization and genotypic data seem to indicate that the latter species should be rejected on the basis of junior synonymy with *B. animalis* (91, 92).

In order to allow comparison of a product's species content claimed on its label with the identification result obtained by the researcher, it is important to emphasize that the label of the product under investigation should contain sufficiently detailed information on its taxonomic composition. As summarized in Table 18.3, the reported discrepancies can be categorized in different scenarios: (i) the product contains the correct number of species claimed on the label but the identity of the detected species does not match those claimed, (ii) the product contains fewer species than indicated on the label and the identity of the detected species does not necessarily match the claimed species designations, and (iii) the product contains more species than indicated on the label and the identity of the detected species does not necessarily match the claimed species designations. Many of these discrepancies can be linked to the confusing taxonomic position of some probiotic taxa or to the use of instantly recognizable species designations on product labels with no taxonomic standing, or simply result from the fact that classification methods with limited species resolution were used. In addition, it has been speculated that contamination during the production process may also account for some of the reported discrepancies, such as the unlabeled presence of *E. faecium* in probiotic food supplements (71, 93).

18.6 Postmarketing Surveillance

The manufacturer of a probiotic product has the final responsibility for its safety. Once a probiotic product has been introduced into the market, surveillance for any cases of disease related to the probiotics is important to be able to continue to ascertain the safety of the product. Any *Lactobacillus*, *Bifidobacterium* or other LAB isolate associated with disease should be reported, stored, and eventually compared to probiotic strains of the same species (94). This also illustrates that accurate taxonomic identification of both probiotic and clinical isolates is essential; unfortunately, this is not always the case. Genotyping should reveal clonal relatedness of the clinical isolate and probiotic of LAB.

For future reference and research, it is important that the clinical LAB isolates are minimally subcultured to avoid adaptation to laboratory culture conditions and are stored properly. This makes it possible to investigate the characteristics of these isolates and obtain information on potential risk factors (94).

18.7 Next-Generation Probiotics

New probiotic microorganisms and new applications are likely to be introduced into the market. These should require different safety assessment procedures from the "traditional" probiotic applications.

Most probiotics currently on the market belong to the genera *Lactobacillus* and *Bifidobacterium*, although species from other genera are also used as probiotics (Table 18.4). The use of microbes outside the traditional field, such as, e.g., *Oxalobacter formigenes* to reduce the risk for kidney stones (95), will require new standards for safety assessment.

Although genetically modified organisms (GMOs) are not particularly appreciated by the European consumer, this may be different in the case of clear medical applications. GMO probiotics expressing antigens from pathogens would be safe vaccines and would be preferable over attenuated pathogens (96). Probiotics producing anti-inflammatory cytokines (interleukin-10) *in situ* in the intestine may provide novel strategies for the treatment of inflammatory bowel disease (97). Legislation for the assessment of the safety of GMOs is very strict and falls outside the scope of this review.

Probiotics are commonly defined as live microorganisms. However, there is some indication that under certain circumstances nonviable probiotics may also have beneficial health effects (98). Nonviable probiotics would have clear practical and economical benefits: longer shelf life, no need for cooled transportation and storage, etc. There may also be a safety advantage.

TABLE 18.4

Microbial Genera Commonly Used as Probiotics

Genus (species)	Remark
Bifidobacterium	*B. dentium* has been found to be associated with dental caries.
Bacillus	Some strains with a history of safe use are known.
Carnobacterium	Some strains are pathogenic for fish.
Enterococcus	Some strains with a history of safe use are known.
Escherichia coli	Some strains with a history of safe use are known.
Lactobacillus	
Lactococcus	Some strains are pathogenic for fish.
Propionibacterium	Only species of the classical or dairy propionibacteria, cutaneous propionibacteria have been associated with disease.
Saccharomyces cerevisiae	*S. boulardii* is likely to be a variant of *S. cerevisiae*.
Streptococcus	Only *S. thermophilus* has a history of safe use.

Although the risk for infection due to the probiotics lactobacilli and bifidobacteria is extremely small, nonviable probiotics would not pose this risk at all. However, a recent study on nonviable probiotics in infants was associated with adverse gastrointestinal symptoms and diarrhea (99), although this is more likely related to the higher level of *Bacteroides* and *Clostridium* in the group receiving the nonviable probiotics than to the preparation as such.

Most current applications of probiotics are aimed at oral use. Although the probiotics principle can be expected to work in any part of the body where there is a normal microbiota, nonintestinal applications are limited to urogenital tract infections (100). Recent publications indicate that intranasal application of probiotics (α-hemolytic streptococci) may be a feasible option to prevent recurrent otitis media (101). It is obvious that such applications, which are very different from traditional probiotic use, require different safety assessment procedures.

18.8 Legislation Evaluation of Lactic Acid Bacteria

LAB are difficult to classify and regulate. Food supplements containing LAB could be classified as foods or food additives. They could also be classified as drugs, particularly if one makes health claims or medicinal claims concerning their use.

Regular foods or food supplements containing LAB could fall under European novel food regulation, in case the bacterium in question has not been used before in the food in question. Novel foods by legal definition are foods and food ingredients that have not been used for human consumption to a significant degree within the Community before 15 May 1997. Regulation EC 258/97 (102) of 27 January 1997 of the European Parliament and the

TABLE 18.5

Suggested Steps for the Assessment of the Safety of Probiotics and Starter Cultures

1.	Strains with a long history of safe use by humans can be used safely as probiotics.
2.	Strains that belong to species for which no pathogenic strains are known but which do not have a history of safe use are likely to be safe as probiotics but are nevertheless novel foods.
3.	Strains that belong to species for which pathogenic strains are known should be treated as novel foods.
4.	Strains that are genetically modified are novel foods and should be treated as such.
5.	Characterization of the intrinsic properties of new strains. Strains carrying transferable antibiotic resistance genes should not be marketed.
6.	Strains that are not properly taxonomically described (DNA-DNA hybridization, rRNA sequence determination) should not be marketed.
7.	Strains for which the origin is not known should not be marketed.
8.	Assessment of the acute and subacute toxicity of ingestion of extremely large amounts of the probiotics should be carried out.
9.	Estimation of the infective properties *in vitro*, using human cell lines and intestinal mucus, and *in vivo* using sublethally irradiated or immune-compromised animals should be carried out.
10.	Assessment of the effects of the strains on the composition and activity of the human intestinal microbiota should be made.
11.	Epidemiology and postmarketing surveillance should be carried out.

Source: Modified after References 105 and 106.

Council lays out detailed rules for the authorization of novel foods and novel food ingredients.

Companies that want to place a novel food on the EU market need to submit their application in accordance with Commission Recommendation 97/618/EC (103) that concerns the scientific information and the safety assessment report required. Novel foods or novel food ingredients may follow a simplified procedure, only requiring notifications from the company, when they are considered by a national food assessment body as "substantially equivalent" to existing foods or food ingredients.

18.9 Conclusions

Probiotics and especially LAB starters have a long history of safe use. It is, nevertheless, important that existing and particularly new strains for food or feed use are assessed for their absence of potential risk factors. However, absolute safety cannot be guaranteed, and their safety should be viewed in relation to their benefits, which for probiotics and starters is very favorable. Methods to assess the safety of probiotics and starters have been discussed in this chapter and are summarized in Table 18.5. They will provide a scientific rationale for the safe use of probiotic LAB in humans.

References

1. CS Pederson. *Microbiology of Food Fermentations*. 1st ed., AVI Publishing, Westport, CT, 1971, pp. 283.
2. G Mogensen et al. Inventory of microorganisms with a documented history of use in food. *Bull. IDF*, 377: 10–18, 2002.
3. G Mogensen et al. Food microorganisms — Health benefits, safety evaluation and strains with documented history of use in foods. *Bull. IDF*, 377: 4–9, 2002.
4. C Shortt. The probiotic century; historical and current perspectives. *Trends Food Sci. Technol.*, 10: 411–417, 1999.
5. MK Salminen et al. *Lactobacillus* bacteremia during a rapid increase in probiotic use of *Lactobacillus rhamnosus* GG in Finland. *Clin. Infect. Dis.*, 35: 1155–1160, 2002.
6. JS Zhou, PK Gopal, HS Gill. Potential probiotic lactic acid bacteria *Lactobacillus rhamnosus* (HN001), *Lactobacillus acidophilus* (HN017) and *Bifidobacterium lactis* (HN019) do not degrade gastric mucin *in vitro*. *Int. J. Food Microbiol.*, 63: 81–90, 2001.
7. DC Donohue, S Salminen. Safety of probiotic bacteria. *Asian Pac. J. Clin. Nutr.*, 5: 25–28, 1996.
8. M Bernardeau, JP Vernoux, M Gueguen. Safety and efficacy of probiotic lactobacilli in promoting growth in post-weaning Swiss mice. *Int. J. Food Microbiol.*, 77: 19–27, 2002.
9. D Donohue, M Deighton, JT Ahokas, S Salminen. Toxicity of lactic acid bacteria. In: S Salminen, A von Wright, Eds. *Lactic Acid Bacteria*. Marcel Dekker, New York, 1993, pp. 307–313.
10. JG Ruseler-van Embden, LM van Lieshout, MJ Gosselink, P Marteau. Inability of *Lactobacillus casei* strain GG, *L. acidophilus*, and *Bifidobacterium bifidum* to degrade intestinal mucus glycoproteins. *Scand. J. Gastroenterol.*, 30: 675–680, 1995.
11. JS Zhou, Q Shu, KJ Rutherfurd, J Prasad, MJ Birtles, PK Gopal, HS Gill. Safety assessment of potential probiotic lactic acid bacterial strains *Lactobacillus rhamnosus* HN001, *Lb. acidophilus* HN017, and *Bifidobacterium lactis* HN019 in BALB/c mice. *Int. J. Food Microbiol.*, 56: 87–96, 2000.
12. JS Zhou, Q Shu, KJ Rutherfurd, J Prasad, PK Gopal, HS Gill. Acute oral toxicity and bacterial translocation studies on potentially probiotic strains of lactic acid bacteria. *Food Chem. Toxicol.*, 38: 153–161, 2000.
13. F Gasser. Safety of lactic acid bacteria and their occurrence in human clinical infections. *Bull. Inst. Pasteur*, 92: 45–67, 1994.
14. Organizing Committee of the Lactic Acid Bacteria Industrial Platform Workshop. On the Safety of Lactic Acid Bacteria. Report on the Workshop Safety of Lactic Acid Bacteria, Unilever Research Laboratorium, Vlaardingen, The Netherlands, 1994.
15. T Asahara, M Takahashi, K Nomoto, H Takayama, M Onoue, M Morotomi, R Tanaka, T Yokokura, N Yamashita. Assessment of safety of *Lactobacillus* strains based on resistance to host innate defence mechanisms. *Clin. Diagn. Lab. Immunol.*, 10: 169–173, 2003.

16. CB Maassen, JC van Holten, F Balk, MJ Heijne den Bak-Glashouwer, R Leer, JD Laman, WJ Boersma, E Claassen. Orally administered *Lactobacillus* strains differentially affect the direction and efficacy of the immune response. *Vet. Q.*, 20 (Suppl. 3): S81–S83, 1998.

17. PA Gonnella, Y Chen, J Inobe, Y Komagata, M Quartulli, HL Weiner. *In situ* immune response in gut-associated lymphoid tissue (GALT) following oral antigen in TCR-transgenic mice. *J. Immunol.*, 160: 4708–4718, 1998.

18. JD Laman, EJ Thompson, L Kappos. Balancing the Th1/Th2 concept in multiple sclerosis. *Immunol. Today*, 19: 489–490, 1998.

19. A O'Garra. Cytokines induce the development of functionally heterogeneous T helper cell subsets. *Immunity*, 8: 275–283, 1998.

20. S Romagnani. T-cell subsets (Th1 versus Th2). *Ann. Allerg. Asthma Immunol.*, 85: 9–18, 2000.

21. E Isolauri. Probiotics and gut inflammation. *Curr. Opin. Gastroenterol.*, 15: 534–537, 1999.

22. H Morita, F He, T Fuse, AC Ouwehand, H Hashimoto, M Hosoda, K Mizumachi, J Kurisaki. Adhesion of lactic acid bacteria to caco-2 cells and their effect on cytokine secretion. *Microbiol. Immunol.*, 46: 293–297, 2002.

23. K Gerritse, JD Laman, RJ Noelle, A Aruffo, JA Ledbetter, WJ Boersma, E Claassen. CD40-CD40 ligand interactions in experimental allergic encephalomyelitis and multiple sclerosis. *Proc. Natl. Acad. Sci. U.S.A.*, 93: 2499–2504, 1996.

24. JD Laman, CB Maassen, MM Schellekens, L Visser, M Kap, E de Jong, M van Puijenbroek, MJ van Stipdonk, M van Meurs, C Schwarzler, U Gunthert. Therapy with antibodies against CD40L (CD154) and CD44-variant isoforms reduces experimental autoimmune encephalomyelitis induced by a proteolipid protein peptide. *Mult. Scler.*, 4: 147–153, 1998.

25. CB Maassen, C Holten-Neelen, F Balk, MJ Bak-Glashouwer, RJ Leer, JD Laman, WJ Boersma, E Claassen. Strain-dependent induction of cytokine profiles in the gut by orally administered *Lactobacillus* strains. *Vaccine*, 18: 2613–2623, 2000.

26. CB Maassen, WJ Boersma, C Holten-Neelen, E Claassen, JD Laman. Growth phase of orally administered *Lactobacillus* strains differentially affects IgG1/IgG2a ratio for soluble antigens: implications for vaccine development. *Vaccine*, 21: 2751–2757, 2003.

27. JD Laman, L Visser, CBM Maassen, CJA de Groot, LA de Jong, BA 't Hart, M van Meurs, MM Schellekens. Novel monoclonal antibodies against proteolipid protein peptide 139–151 demonstrate demyelination and myelin uptake by macrophages in MS and marmoset EAE lesions. *J. Neuroimmunol.*, 119: 124–130, 2001.

28. P Marteau, MF Gerhardt, A Myara, E Bouvier, F Trivin, JC Rambaud. Metabolism of bile salts by alimentary bacteria during transit in the human small intestine. *Microb. Ecol. Health Dis.*, 8: 151–157, 1995.

29. DC Donohue, S Salminen, P Marteau. Safety of probiotic bacteria. In: S Salminen, A von Wright, Eds., *Lactic Acid Bacteria — Microbiology and Functional Aspects.* Marcel Dekker, New York, 1998, pp. 369–383.

30. PY Cheah. Hypotheses for the etiology of colorectal cancer--an overview. *Nutr. Cancer*, 14: 5–13, 1990.

31. DK Walker, SE Gilliland. Relationship among bile tolerance, bile salt deconjugation, and assimilation of cholesterol by *Lactobacillus acidophilus*. *J. Dairy Sci.*, 76: 956–961, 1993.

32. MM Brashears, SE Gilliland, LM Buck. Bile salt deconjugation and cholesterol removal from media by *Lactobacillus casei*. *J. Dairy Sci.*, 81: 2103–2110, 1998.

33. G Corzo, SE Gilliland. Measurement of bile salt hydrolase activity from *Lactobacillus acidophilus* based on disappearance of conjugated bile salts. *J. Dairy Sci.*, 82: 466–471, 1999.

34. A Baumgartner, M Kueffer, A Simmen, M Grand. Relatedness of *Lactobacillus rhamnosus* strains isolated from clinical specimens and such from food-stuffs, humans and technology. *Lebensm. Wiss. Technol.*, 31: 489–494, 1998.

35. MO Ogundele. Role and significance of the complement system in mucosal immunity: particular reference to the human breast milk complement. *Immunol. Cell Biol.*, 79: 1–10, 2001.

36. SM Burns, SI Hull. Comparison of loss of serum resistance by defined lipopolysaccharide mutants and an acapsular mutant of uropathogenic *Escherichia coli* O75:K5. *Infect. Immun.*, 66: 4244–4253, 1998.

37. M Virta, et al. Determination of complement-mediated killing of bacteria by viability staining and bioluminescence. *Appl. Environ. Microbiol.*, 64: 515–519, 1998.

38. E-M Lilius, P Marnila. Photon emission of phagocytes in relation to stress and disease. *Experientia*, 48: 1082–1091, 1992.

39. CMAP Franz, ME Stiles, KH Schleifer, WH Holzapfel. Enterococci in foods — a conundrum for food safety. *Int. J. Food Microbiol.*, 88: 105–122, 2003.

40. DWS Harty, M Patrikakis, EBH Hume, HJ Oakey, KW Knox. The aggregation of human platelets by *Lactobacillus* species. *J. Gen. Microbiol.*, 139: 2945–2951, 1993.

41. JM Patti, BL Allen, MJ McGavin, M Höök. MCRAMM-mediated adherence of microorganisms to host tissues. *Annu. Rev. Microbiol.*, 48: 585–617, 1994.

42. W-J Hu, JW Eaton, L Tang. Molecular basis of biomaterial-mediated foreign body reaction. *Blood*, 98: 1231–1238, 2001.

43. ME Sanders. Probiotics: considerations for human health. *Nutr. Rev.*, 61: 91–99, 2003.

44. JM Fleckenstein, DJ Kopecko. Breaching the mucosal barrier by stealth: an emerging pathogenic mechanism for enteroadherent bacterial pathogens. *J. Clin. Invest.*, 107: 27–30, 2001.

45. AC Ouwehand, EM Tuomola, YK Lee, S Salminen. Microbial interactions to intestinal mucosal models. *Methods Enzymol.*, 337: 200–212, 2001.

46. PK Gopal, J Prasad, J Smart, HS Gill. *In vitro* adherence properties of *Lactobacillus rhamnosus* DR20 and *Bifidobacterium lactis* DR10 strains and their antagonistic activity against an enterotoxigenic *Escherichia coli*. *Int. J. Food Microbiol.*, 67: 207–216, 2001.

47. S Blum, R Reniero. Adhesion of selected *Lactobacillus* strains to enterocyte-like Caco-2 cells *in vitro*: a critical evaluation of reliability of *in vitro* adhesion assays. 4th Workshop, Demonstration of Nutritional Functionality of Probiotic Foods, Rovaniemi, Finland, 2000.

48. M Pinto, S Appay, P Simon-Assmann, G Chevalier, N Dracopoli, J Fogh, A Zweibaum. Enterocytic differentiation of cultured human colon cancer cells by replacement of glucose by galactose in the medium. *Biol. Cell*, 44: 193–196, 1982.

49. M Pinto, S Robine-Leon, MD Appay, M Kedinger, N Tiadou, E Dussaulx, B Lacroix, P Simon-Assmann, K Haffen, J Fogh, A Zweibaum. Enterocyte like differentiation and polarisation of the human colon carcinoma cell line Caco-2 in culture. *Biol. Cell*, 47: 323–330, 1983.

50. Y Azuma, M Sato. *Lactobacillus casein* NY1301 increases the adhesion of *Lactobacillus gasseri* NY0509 to human intestinal caco-2 cells. *Biosci. Biotechnol. Biochem.*, 65: 2326–2329, 2001.

51. MF Bernet, D Brassart, JR Neeser, AL Ervin. Adhesion of human bifidobacterial strains to cultured human intestinal epithelial cells and inhibition of enteropathogen-cell interactions. *Appl. Environ. Microbiol.*, 59: 4121–4128, 1993.

52. MF Bernet, D Brassart, JR Neeser, A L Servin. *Lactobacillus acidophilus* LA 1 binds to cultured human intestinal cell lines and inhibits cell attachment and cell invasion by enterovirulent bacteria. *Gut*, 35: 483–489, 1994.

53. G Chauviere, MH Coconnier, S Kerneis, J Fourniat, A L Servin. Adhesion of human *Lactobacillus acidophilus* strain LB to human enterocyte-like Caco-2 cells. *J. Gen. Microbiol.*, 138 (Pt 8): 1689–1696, 1992.

54. MH Coconnier, TR Klaenhammer, S Kerneis, MF Bernet, AL Servin. Protein-mediated adhesion of *Lactobacillus acidophilus* BG2FO4 on human enterocyte and mucus-secreting cell lines in culture. *Appl. Environ. Microbiol.*, 58: 2034–2039, 1992.

55. MH Coconnier, MF Bernet, G Chauviere, AL Servin. Adhering heat-killed human *Lactobacillus acidophilus*, strain LB, inhibits the process of pathogenicity of diarrhoeagenic bacteria in cultured human intestinal cells. *J. Diarrhoeal Dis. Res.*, 11: 235–242, 1993.

56. MH Coconnier, MF Bernet, S Kerneis, G Chauviere, J Fourniat, AL Servin. Inhibition of adhesion of enteroinvasive pathogens to human intestinal Caco-2 cells by *Lactobacillus acidophilus* strain LB decreases bacterial invasion. *FEMS Microbiol. Lett.*, 110: 299–305, 1993.

57. A Darfeuille-Michaud, D Aubel, G Chauviere, C Rich, M Bourges, A Servin, B Joly. Adhesion of enterotoxigenic *Escherichia coli* to the human colon carcinoma cell line Caco-2 in culture. *Infect. Immun.*, 58: 893–902, 1990.

58. B Del Re, B Sgorbati, M Miglioli, D Palenzona. Adhesion, autoaggregation and hydrophobicity of 13 strains of *Bifidobacterium longum*. *Lett. Appl. Microbiol.*, 31: 438–442, 2000.

59. C Forestier, C De Champs, C Vatoux, B Joly. Probiotic activities of *Lactobacillus casein rhamnosus*: *in vitro* adherence to intestinal cells and antimicrobial properties. *Res. Microbiol.*, 152: 167–173, 2001.

60. EM Lehto, SJ Salminen. Inhibition of *Salmonella typhimurium* adhesion to Caco-2 cell cultures by *Lactobacillus* strain GG spent culture supernate: only a pH effect? *FEMS Immunol. Med. Microbiol.*, 18: 125–132, 1997.

61. EM Lehto, S Salminen. Adhesion of two *Lactobacillus* strains, one *Lactococcus* and one *Propionibacterium* strain to cultured human intestinal Caco-2 cell line. *Biosci. Microflora*, 16: 13–17, 1997.

62. K Todoriki, T Mukai, S Sato, T Toba. Inhibition of adhesion of food-borne pathogens to Caco-2 cells by *Lactobacillus* strains. *J. Appl. Microbiol.*, 91: 154–159, 2001.

63. T Lesuffleur, A Barbat, E Dussaulx, A Zweibaum. Growth adaptation to methotrexate of HT-29 human colon carcinoma cells is associated with their ability to differentiate into columnar absorptive and mucus-secreting cells. *Cancer Res.*, 50: 6334–6343, 1990.

64. S Elo, M Saxelin, S Salminen. Attachment of *Lactobacillus casei* strain GG to human colon carcinoma cell line caco-2: comparison with other dairy strains. *Lett. Appl. Microbiol.*, 13: 154–156, 1991.

65. YK Lee, CY Lim, WL Teng, AC Ouwehand, EM Tuomola, S Salminen. Quantitative approach in the study of adhesion of lactic acid bacteria to intestinal cells and their competition with enterobacteria. *Appl. Environ. Microbiol.*, 66: 3692–3697, 2000.

66. JG Songer. Bacterial phospholipases and their role in virulence. *Trends Microbiol.*, 5: 156–161, 1997.

67. AV Rodriguez, MD Baigori, S Alvarez, GR Castro, G Oliver. Phosphatidylinositol-specific phospholipase C activity in *Lactobacillus rhamnosus* with capacity to translocate. *FEMS Microbiol. Lett.*, 204: 33–38, 2001.

68. S Salminen, A von Wright, L Morelli, P Marteau, D Brassart, WM De Vos, R Fonden, M Saxelin, K Collins, G Mogensen, SE Birkeland, T Mattila-Sandholm. Demonstration of safety of probiotics — a review. *Int. J. Food Microbiol.*, 44: 93–106, 1998.

69. M Danielsen, A Wind. Susceptibility of *Lactobacillus* spp. to antimicrobial agents. *Int. J. Food Microbiol.*, 82: 1–11, 2003.

70. European Commission. Opinion of the Scientific Committee on Animal Nutrition on the Criteria for Assessing the Safety of Microorganisms Resistant to Antibiotics of Human Clinical and Veterinary Importance. 2002. (http://www.europa.eu.int/comm/food/fs/sc/scan/out108_en.pdf).

71. R Temmerman, B Pot, G Huys, J Swings. Identification and antibiotic susceptibility of bacterial isolates from probiotic products. *Int. J. Food Microbiol.*, 81: 1–10, 2003a.

72. B Pot, T Coenye, K Kersters. The taxonomy of microorganisms used as probiotics with special focus on enterococci, lactococci, and lactobacilli. *Microecol. Ther.*, 26: 11–25, 1997.

73. G Klein, A Pack, C Bonaparte, G Reuter. Taxonomy and physiology of probiotic lactic acid bacteria. *Int. J. Food Microbiol.*, 41: 103–125, 1998.

74. A Gancheva, B Pot, K Vanhonacker, B Hoste, K Kersters. A polyphasic approach towards the identification of strains belonging to *Lactobacillus acidophilus* and related species. *Syst. Appl. Microbiol.*, 22: 573–585, 1999.

75. S Torriani, F Clementi, M Vancanneyt, B Hoste, F Dellaglio, K Kersters. Differentiation of *Lactobacillus plantarum*, *L. pentosus* and *L. paraplantarum* species by RAPD-PCR and AFLP. *Syst. Appl. Microbiol.*, 24: 554–560, 2001.

76. PSM Yeung, ME Sanders, CL Kitts, R Cano, PS Tong. Species-specific identification of commercial probiotic strains. *J. Dairy Sci.*, 85: 1039–1051, 2002.

77. U Schillinger. Isolation and identification of lactobacilli from novel-type probiotic and mild yoghurts and their stability during refrigerated storage. *Int. J. Food Microbiol.*, 47: 79–87, 1999.

78. T Yaeshima, S Takahashi, N Ishibashi, S Shimamura. Identification of bifidobacteria from dairy products and evaluation of a microplate hybridization method. *Int. J. Food Microbiol.*, 30: 303–313,1996.

79. W Holzapfel, P Haberer, J Snel, U Schillinger, JHJ Huis in't Veld. Overview of gut flora and probiotics. *Int. J. Food Microbiol.*, 41: 85–101, 1998.

80. P Vandamme, B Pot, M Gillis, P de Vos, K Kersters, J Swings. Polyphasic taxonomy, a consensus approach to bacterial systematic. *Microbiol. Rev.*, 60: 407–438, 1996.

81. NT Hoa, L Baccigalupi, A Huxham, A Smertenko, PH Van, S Ammendola, E Ricca, SM Cutting. Characterization of *Bacillus* species used for oral bacteriotherapy and bacterioprophylaxis of gastrointestinal disorders. *Appl. Environ. Microbiol.*, 66: 5241–5247, 2000.

82. D Gevers, G Huys, J Swings. Applicability of rep-PCR fingerprinting for identification of *Lactobacillus* species. *FEMS Microbiol. Lett.*, 205: 31–36, 2001.

83. L Masco, G Huys, D Gevers, L Verbrugghen, J Swings. Identification of *Bifidobacterium* species using rep-PCR fingerprinting. *Syst. Appl. Microbiol.*, 26: 557–563, 2003.

84. R Temmerman, I Scheirlinck, G Huys, J Swings. Culture-independent analysis of probiotic products by denaturing gradient gel electrophoresis. *Appl. Environ. Microbiol.*, 69: 220–226, 2003b.

85. S Fasoli, M Marzotto, L Rizzotti, F Rossi, F Dellaglio, S Torriani. Bacterial composition of commercial probiotic products as evaluated by PCR-DGGE analysis. *Int. J. Food Microbiol.*, 82: 59–70, 2003.

86. LA Devriese, B Pot, P Vandamme, K Kersters, F Haesebrouck. Identification of enterococci species isolated from foods of animal origin. *Int. J. Food Microbiol.*, 26: 187–197, 1995.

87. P Descheemaeker, C Lammens, B Pot, P Vandamme, H Goossens. Evaluation of arbitrarily primed PCR analysis and pulsed-field gel electrophoresis of large genomic DNA fragments for identification of enterococci important in human medicine. *Int. J. Syst. Bacteriol.*, 47: 555–561, 1997

88. LMT Dicks, EM Du Plessis, F Dellaglio, F Lauer. Reclassification of *Lactobacillus rhamnosus* ATCC 15820 as *Lactobacillus zee* nom. rev., designation of ATCC 334 as ecotype of *L. casein* subsp. *casein*, and rejection of the name *Lactobacillus paracasei*. *Int. J. Syst. Bacteriol.*, 46: 337–340, 1996.

89. F Dellaglio, GE Felis, S Torriani. The status of the species *Lactobacillus casei* (Orla-Jensen 1916) Hansen and Lessel 1971 and *Lactobacillus paracasei* Collins et al. 1989. Request for an Opinion. *Int. J. Syst. Bacteriol.*, 52: 285–287, 2002

90. MD Collins, BA Phillips, P Zanoni. Deoxyribonucleic acid homology studies of *Lactobacillus casei*, *Lactobacillus paracasei* sp. nov., subsp. *paracasei* and *tolerans*, and *Lactobacillus rhamnosus* sp. nov., comb. nov. *Int. J. Syst. Bacteriol.*, 39: 105–108, 1989.

91. Y Cai, M Matsumoto, Y Benno. *Bifidobacterium lactis* Meile et al. 1997 is a subjective synonym of *Bifidobacterium animalis* (Mitsuoka 1969) Scardovi and Trovatelli 1974. *Microbiol. Immunol.*, 44: 815–820, 2000.

92. M Ventura, R Zink. Rapid identification, differentiation, and proposed new taxonomic classification of *Bifidobacterium lactis*. *Appl. Environ. Microbiol.*, 68: 6429–6434, 2002.

93. JMT Hamilton-Miller, S Shah, JT Winkler. Public health issues arising from microbiological and labelling quality of foods and supplements containing probiotic microorganisms. *Public Health Nutr.*, 2: 223–229, 1999.

94. E Apostolou et al. Good adhesion properties of probiotics: a potential risk for bacteremia? *FEMS Immunol. Med. Microbiol.*, 31: 35–39, 2001.

95. R Kumar, M Mukherjee, M Bhandari, H Kumar, H Sidhu, RD Mittal. Role of *Oxalobacter formigenes* in calcium oxalate stone disease: a study in North India. *Eur. Urol.* 41: 318–322, 2002.

96. C Grangette, H Müller-Alouf, M-C Geoffroy, D Goudercourt, M Turneer, A Mercenier. Protection against tetanus toxin after intragastric administration of two recombinant lactic acid bacteria: impact of strain viability and *in vivo* persistence. *Vaccine*, 20: 3304–3309, 2002.

97. L Steidler. In situ delivery of cytokines by genetically engineered *Lactococcus lactis*. *Antoine van Leeuwenhoek*, 82: 323–331, 2002.

98. AC Ouwehand, SJ Salminen. The health effects of viable and non-viable cultured milks. *Int. Dairy J.*, 8, 749–758, 1998.
99. PV Kirjavainen, SJ Salminen, E Isolauri. Probiotic bacteria in the management of atopic disease: underscoring the importance of viability. *J. Pediatr. Gastroenterol. Nutr.*, 36: 223–227, 2003.
100. G Reid. The potential role of probiotics in pediatric urology. *J. Urol.*, 168: 1512–1517, 2002.
101. K Tano, E Grahn Håkansson, S E Holm, S Hellström. A nasal spray with alpha-haemolytic streptococci as long term prophylaxis against recurrent otitis media. *Int. J. Pediatr. Otorhinolaryngol.*, 62: 17–23, 2002.
102. European Commission. Recommendation 97/618/EC (http://europa.eu.int/smartapi/cgi/sga_doc?smartapi!celexapi!prod!CELEXnumdoc&lg=EN&numdoc=31997H0618&model=guichett)
103. European Commission. Regulation EC N° 258/97 (http://europa.eu.int/smartapi/cgi/sga_doc?smartapi!celexapi!prod!CELEXnumdoc&lg=EN&numdoc=31997R0258&model=guichett)
104. AC Ouwehand, S Salminen. Safety evaluation of probiotics. In: *Functional Dairy Products*. T Mattila-Sandholm, M Saarela, Eds. Woodhead Publishing, Cambridge, UK, pp. 316–336.
105. S Salminen, A von Wright. Current probiotics — Safety assured? *Microl. Ecol. Health Dis.*, 10: 68–77, 1998.
106. S Salminen et al. Demonstration of safety of probiotics — a review. *Int. J. Food Microbiol.*, 44: 93–106, 1998.

19

Probiotics in Food Safety and Human Health: Current Status of Regulations on the Use of Probiotics in Foods in Japan

Yoichi Fukushima and Hisakazu Iino

CONTENTS

19.1 Introduction

Food and dietary habits are influenced by the culture and history. Japanese society has been influenced by Chinese culture for thousands of years. For instance, Japanese are accustomed to the Chinese concept of "Food and Medicine are isogenic (I-syoku-dou-gen)," and have used naturally accepted foods with certain health benefits throughout history. The concept of "Functional Foods" started in Japan where the world's first health-claim approval system, Foods for Specific Health Uses (FOSHU), and the market for functional foods were developed in the last decade. "Probiotics" are defined as live microorganisms which when administered in adequate amounts confer a health benefit on the host (1, 2). The term probiotics has recently recently become a common word, even in the general public in Japan after fermented milk products named "probiotics" were introduced by the industry. The food industry actively informed consumers on the physiological functions and health benefits associated with "probiotics." Hence, Japanese are likely to perceive probiotics as a high-grade yogurt or lactic acid bacteria (LAB) drink with specific health benefits such as immune reinforcement, anti-*Helicobacter pylori* effect, prevention of allergy, cancer, and food poisoning. Japanese traditionally consume fermented foods with viable microorganisms such as "natto" (steamed soybean fermented with bacilli), "miso" (soy paste fermented with yeast and fungi), and sour vegetables fermented by lactobacilli. Although these products are legally considered regular foods and are subject to special regulations, they contribute to the "good and healthy" perception of fermented foods and probiotics.

In terms of dairy probiotics, "Fermented Milks" and "Lactic Acid Bacteria (LAB) Drinks" but not "Yogurt" are the official terms used in the Japanese regulations. The total production of fermented milks and LAB drinks in Japan reached 916,000 and 550,000 KL per family in 2003, respectively. A typical Japanese family spends approximately $100 per year on average for both product categories. In terms of monetary value, Japan has the largest fermented milk market in the world. The Japanese industry initially introduced LAB drinks and developed the market for this food category. Fermented milks, the equivalent of European yogurt, were introduced later. Following the introduction of fermented milks and LAB drinks the regulation of probiotics in Japan was passed to ensure standards and protect consumers against false claims. In this chapter, we describe the current status of probiotics regulations in Japan with focus on the LAB in fermented milks and LAB drinks and related health claims including the FOSHU system, while covering the history as well as future expectations of the probiotic market.

19.2   History and Regulatory Framework of Probiotics in Japan

19.2.1   History of Fermented Milk in Japan

Fermented milk has been consumed by humans for at least 4000 years and throughout civilizations such as the Egyptians, Mesopotamians, Aryans in central Asia, Greeks, and Romans. In Japan, the culture of dairy product intake started about 1350 years ago. Medical literature from the year 984 indicates that milk products were used to cure total debilitation and constipation and improve skin. Long before Metchnikoff discovered in 1902 that putrefaction in the intestines causes early aging and death in humans and that this could be prevented by consumption of sour milk (3), ancient Japanese had already discovered the medicinal benefits of fermented milk and used it as medicine as well as food. The dairy culture disappeared for a time in Japan under physical or political limitation of contact with Western culture, and fermented milk was reintroduced as a medicine for improving gut health and helping with diabetes after the Meiji Restoration in 1868. Japanese pioneers developed unique LAB drinks, one of which is a concentrated type of sterilized fermented milk drink called "CALPIS," that was introduced in 1919. Another is the LAB drink or "YAKULT" which was introduced in 1935. After a sweetened fermented milk hardened by agar and named yogurt was industrialized in 1950, a "yogurt" product corresponding to the European definition was at last introduced in 1971 to Japanese consumers. In the medical field, tablet and powdered forms of viable LAB were also introduced in 1971 as a drug for curing intestinal illnesses. The regulatory system for probiotics in Japan was influenced by the history of the introduction of these products.

19.2.2   Legal Definition and Regulation of Fermented Milk in Japan

Food products manufactured, imported, or sold in Japan are legally regulated by specific laws and standards issued by government ministries according to their respective policies and guidelines. The Ministry of Health, Labor and Welfare (MHLW) passes and enforces the Food Sanitation Laws and the Health Promotion Laws. The latter were established in May 2003 based on the Nutrition Improvement Law and the Pharmaceutical Affairs Law pertaining to proof of efficacy, safety, and hygiene. For dairy products, the Ministerial Ordinance on the Compositional Standards for Milk and Milk Products provides a definition for this category as well as the quality standards for milk products in terms of raw materials, manufacturing procedures, product specifications, methods of inspection, packaging materials, labeling, and methods for storage under the MHLW. The food safety basic law was revised in 2003 and a food safety commission was established to oversee the safety of foods. The Ministry of Agriculture, Forestry and Fisheries (MAFF) has jurisdiction over laws concerning Standardization and Proper Labeling of Agricultural and Forestry Products, while Japan Agricultural Standards (JAS) Law deals with quality assurance and consumers' choice of foods. The JAS Law, revised in 1999, requires that all food products have a quality label according to the specific Quality Labeling Standard of JAS. The standards regulate the necessary labeling items and how to display them on product package. The standards also regulate items prohibited from inclusion on the package label, including words, phrases, or drawings that may mislead consumers. To ensure fair trade and protect consumers, the Fair Trade Commission has specific rules against Unjustifiable Premiums and Misleading Representations. The Ministry of Economy, Trade and Industry has jurisdiction over laws pertaining to Measurement and Product Reliability. Local governments also have district regulations that impact food claims and label requirement.

The Ministerial Ordinance for Milk Products defines fermented milk as "a product obtained by fermenting milk, or milk containing an equal or greater amount of nonfat milk solids, with lactic acid bacteria or yeasts, and forming a paste or liquid, or the frozen product of the same (4)." In the category of fermented milks, there are five subgroups including "plain yogurt" without sweetener or flavor; "hard yogurt" with sweetener or flavor and jellied by adding agar or gratins; "drink yogurt" homogenized with added sweetener and fruit juice; "soft yogurt" with fruit preparation; and "frozen yogurt" deaerated and frozen. The Ministerial Ordinance for Milk Products also defines three other categories of foods with LAB, namely Milk Product-LAB Drinks, Milk Product-LAB Drinks (sterile), and LAB Drinks. YAKULT and CALPIS described above are also categorized in the Milk Product-LAB drinks and Milk Product-LAB Drinks (sterile), respectively.

The type of LAB used for fermentation is not restricted by Japanese regulations Major products and microorganisms used in the production of fermented milks and the LAB drinks in Japan are listed in Table 19.1 *Lactobacillus acidophilus, L. delbrueckii* subsp. *bulgaricus, L. gasseri, L. johnsonii, L. casei, L. rhamnosus, L. helveticus, S. thermophilus*, bifidobacteria, and yeast are used in Japanese fermented milk products. The Fair Competition Rules on the Label of Fermented Milks and LAB Drinks states that bifidobacteria can be categorized as LAB because it produces lactic acid as well as acetic acid (5). In this chapter, we will refer to LAB as the above bacteria along with lactobacilli and bifidobacteria.

The Codex Alimentarius Commission (CAC) issued a Draft Revised Standards for Fermented Milks (6), and the draft standards became effective as of July 2003. Fermented milks are now defined as "a milk product obtained by fermentation of milk in which milk may have been manufactured from products obtained from milk with or without compositional modification, by the action of suitable microorganisms and resulting in reduction of pH with sufficient microbial abundance in the product up to the date of minimum durability." Yogurt previously defined in Codex STAN A11(a)-1975 as

TABLE 19.1

Fermented Milks and Lactic Acid Bacteria (LAB) Drinks in Japan

Product name	Manufacturer	Volume	LAB	Additives or enrichment *	FOSHU health claim**
Fermented Milks					
Bulgaria Yogurt LB81	Meiji Milk	120 g, 500 g	*L. delbrueckii* ssp. *bulgaricus, S. salivarius* ssp. *thermophilus*		1
Bifidus Yogurt	Morinaga Milk	300 g, 500 g	*B. longum, L. bulgaricus, S. thermophilus*		1
Nature Pro GB	Nippon Milk Community	200, 500 g	*L. gasseri, B. longum, L. bulgaricus*		1
Probio Yogurt LG21	Meiji Milk	120 g, 100 ml	*L. gasseri, S. thermophilus*		–
LC1 Yogurt	Nestlé Snow	120 g, 90 g	*L. johnsonii, S. thermophilus*		1
Lactoferrin Yogurt	Morinaga Milk	120 g	*L. bulgaricus, S. thermophilus*	a	–
Dannon BIO	Calpis Ajinomoto Danone	120 g	*B. animalis, L. bulgaricus, S. thermophilus*		–
Aloe Yogurt	Morinaga Milk	80 g, 130 g	*L. bulgaricus, S. thermophilus*		1
Sofuhl Plain	Yakult	100 g	*L. casei*		1
Yogurt Onaka-e-GG!	Takanashi Milk	100 g	*L. rhamnosus*		1
Fermented Milks (drink type)					
Yogurt Onaka-e-GG!	Takanashi Milk	100 ml	*L. rhamnosus*		1
Bulgaria Nomu Yogurt LB81	Meiji Milk	100 ml, 240 ml, 500 ml, 1000 ml	*L. bulgaricus, S. thermophilus*		1
Milmil	Yakult	100 ml	*B. breve, B. longum, L. acidophilus*	b	1
Joie	Yakult	125 ml	*L. casei*		1
Interbalance L-92	Calpis Ajinomoto Danone	150 ml	*L. acidophilus*		–
Milk Products LAB Drinks					
Yakult	Yakult	65 ml	*L. casei*		1
Yakult 400	Yakult	80 ml	*L. casei*	c	1
LC1 Yogurt Drink	Nestlé Snow	120 ml	*L. johnsonii, S. thermophilus*		–
Milk Products LAB Drinks (sterile)					
Calpis	Calpis	500 ml ***	*L. helveticus, S. cerevisiae*		–
Sour Milk Ameal S	Calpis	160 ml	*L. helveticus*	d	2
Pretio	Yakult	100 ml	*L. casei, L. lactis*	e	2
LAB Drinks					
Rolly Ace	Kagome Rabio	65 ml	*L. casei*		–
Calpis Kids	Calpis	100 ml	*L. helveticus, L. acidophilus*		1

* Health benefit through additives : a) lactoferrin, b) lactoferrin, DHA, vitamin D, c) galactooligosaccharides, calcium, and vitamins.
  Health benefit through fermented products : d) lact-tripeptides, e) γ-amino butyric acid.

** Health claim : 1) Improve gastrointestinal conditions; 2) For people with mild hypertension.

*** need to dilute before drinking

"a coagulated milk product obtained by lactic acid fermentation through the action of *Lactobacillus bulgaricus* and *Streptococcus thermophilus* from milk and milk products and with or without optional additives; and the microorganisms in the final product must be viable and abundant" is now classified in the fermented milk category. Table 19.2 shows the comparison and requirements for fermented milks and yogurt as defined by the previous and current Codex standards and Japanese regulations. In Japan, there is still no legal term for "Yogurt," which is defined as "Fermented Milks" in the Ministerial Ordinance for Milk Products in Japan (4). In the revised Codex standards, a new product category termed "Alternate Culture Yogurt" using LAB, except for *L. delbrueckii* subsp. *bulgaricus* and *S. thermophilus,* was created, allowing classification of fermented milk with probiotic strains. The standards became closer to the Japanese standards in which LAB strains and counts in the products are listed in the label as a requirement for fermented milks. There are still gaps between Japanese and Codex standards in some instances; however, the new Codex standards state that other fermented milks and concentrated fermented milks may be designated with various other names as specified in the national legislation of the country in which the products are sold, or names associated with common usage, provided that such names do not create an erroneous impression in the county of retail sale regarding the character identity of the food.

19.2.3   Food Labeling

The Fair Competition Rules for Fermented Milks and LAB Drinks (5) is a regulation independently provided by an industry association, namely the Association for Fair Trade of Fermented Milks and LAB Drinks. The rules approved by the Fair Trade Commission lay down compulsory requirements on product commercialization and labeling. The Fair Competition Rules updated in 1978 states that the label of fermented milks and LAB drinks should include items such as category name, milk fat and solid nonfat (SNF) contents, ingredient list, shelf life, product volume, name and place of manufacture, and storage procedure on product package. The rules are based on related laws defining concrete and compulsory rules that all manufacturers should follow. It is allowed to place the more familiar word "yogurt" on the package of fermented milks. There are no rules requiring inclusion of the name of bacteria on the package, even in instances where strains other than *L. delbrueckii* subsp. *bulgaricus* and *S. thermophilus* are used for fermentation. Indication of the number of bacteria on the package is not regulated except for FOSHU products but the shelf life of fermented milk and LAB drinks should be indicated. The Fair Competition Rules specify the shelf life as 2 weeks in a refrigerated storage at or below 10°C which is shorter than the European norm of 4 weeks. Probiotics, live microorganisms that when orally ingested, exert health benefit by positively influencing intestinal flora and physiological functions of the host. Because of the importance of the health-

TABLE 19.2

Codex and Japanese Standards of Lactic Acid Bacteria and Ingredients for Fermented Milks

		Japanese regulations				Previous Codex Standard (1975)		Current Codex Standard for Fermented Milks (2003) [2]						
	Conditions	Fermented Milks	Milk Products LAB Drinks	LAB Drinks	Milk Products LAB Drinks (sterile)	Yogurt	Flavored Yogurt	Yogurt	Alternate Culture Yogurt [3]	Acidophilus Milk	Kefir	Kumys	Fermented Milk	Flavored Fermented Milks [4]
Description	*Streptococcus thermophilus*	A	A	A	A	E	E	E	E			A		A
	L. delbrueckii subsp. *bulgaricus*					A	A	A	A	A		E	A	A
	Other *Lactobacillus*								E				A	A
	L. acidophilus									E			A	A
	L. kefiri, Leuconostoc, Lactococcus, Acetobacter										E		A	A
	Yeast	A	A	A	A	A	A	A	A	A	E	E	A	A
Incorporation [1]	Pasteurized milk products / Milk and/or products obtained from milk, Water	A	A	A	A	E	E	A	A	A	A	A	A	A
	Dehydrated milk products					A	A	A	A	A	A	A	A	A
	Non-dairy ingredients [a] and Flavors	A	A	A	A	N	A	N	N	N	N	N	N	A
	Food additives [b]	A	A	A	A	N	A	A	A	A	A	A	A	A
Composition	LAB viable count [7] — Total count (cfu/g)	min 10^7 LAB or Yeast (cfu/ml)		min 10^6 LAB or Yeast (cfu/ml)	no bacteria	viable and abundant		min 10^7	min 10^7	min 10^7	min 10^7	min 10^7	min 10^7	min 10^7
	Labeled microorganisms (cfu/g)								min 10^6	min 10^6			min 10^6	min 10^6
	Yeasts (cfu/g)										min 10^4	min 10^4		
	Solid not-fat (SNF) / Milk protein (%w/w)	min 8%	min 3%	less than 3%	min 3%	min 8.2%		min 2.7%	min 2.7%	min 2.7%	min 2.7%		min 2.7%	min 2.7%
	Milk fat (MF) (%)					Yogurt: min 3%; Partially skimmed yogurt: 0.5–3%; Skimmed yogurt: max 0.5%		less than 15%	less than 15%	less than 10%	less than 10%	less than 10%	less than 10%	less than 10%
	Acidity (w/w lactic acid)							min 0.6%	min 0.6%	min 0.6%	min 0.6%	min 0.7%	min 0.3%	min 0.3%
	Ethanol (% vol/w)											min 0.5%		

TABLE 19.2

Codex and Japanese Standards of Lactic Acid Bacteria and Ingredients for Fermented Milks (continued)

*1	E = essential, A = allowed, N = not allowed.
*2	Sweeteners can be added onto all the fermented milk categories and the label may be described as "sweetened" + designation of Fermented Milks (e.g. sweetened yogurt).
	Concentrated Fermented Milk is described as a Fermented Milk the protein of which has been increased prior to or after fermentation to min 5.6%.
	Whey removal after fermentation is not permitted in the manufacture of fermented milks, except for Concentrated Fermented milk.
*3	Alternate culture yogurt shall be named through the use of an appropriate qualifier in conjunction with the work "yogurt". (e.g. "mild" and "tangy") and the term "alternate culture yogurt" shall not apply as a designation.
*4	"Flavored Fermented Milks" are composite milk products defined at CODEX STAN 206-1999.
*5	Codex defines non-dairy ingredients as nutritive and non nutritive carbohydrates, fruits and vegetables as well as juices, purees, pulps, preparations and preserves derived therefrom, cereals, honey, chocolate, nuts, coffee, spices and other harmless natu
*6	Codex defines food additives as colors, sweeteners, emulsifiers, flavor enhancers, acids, acidity regulators, stabilizers, thickeners, preservatives, and packaging gases. Stabilizers and thickeners can be used in Fermented Milks (Plain) when national leg
*7	"Fermented Milks Heat Treated After Fermentation" described at Codex is not applied to the requirement for viable microorganisms count.

claim expression on the food product label to both industry and consumers, a regulatory system of health claims on food products called Foods for Specified Health Uses (FOSHU) was established in Japan in 1991. The FOSHU system will be described in a subsequent section.

19.2.4   Quality Examination

The Ministerial Ordinance for Milk Products specifies the official method used to count the number of LAB in the product as the Bromo Cresol Purple (BCP) plate count agar (4). The number of *L. delbrueckii* subsp. *bulgaricus* and *S. thermophilus*, *L. acidophilus*, and *Lactococcus* are easily determined by this method. However, LAB with specific nutritional requirements or anaerobes such as *Bifidobacterium* cannot be counted correctly using BCP. On the contrary, some bacteria often found as contaminants in food products, e.g., *Bacillus coagulans*, can grow on this medium since it is not selective. While there are many fermented milk products with bifidobacteria in Japan, there is still no official method for counting Bifidobacteria in such products. The Japanese Association of Fermented Milks and Fermented Milk Drinks is in the process of defining the method for detecting bifidobacteria in order to establish an official method for adoption by the regulatory agencies in the near future.

19.3   Foods for Specific Health Uses (FOSHU)

19.3.1   Health Status of Japanese

In the 20th century, especially after World War II, medical and nutritional sciences made remarkable progress. In Japan the economy has been expanding and nutrition improved in terms of satisfaction of nutrient intake. Infectious diseases drastically decreased because of improvement of hygiene and medical treatment, including the use of antibiotics. Life expectancy of Japanese has increased from 49.63 and 46.92 years in 1935 to 85.23 and 78.32 years in 2002 for females and males, respectively. As a result, the number of elderly increased remarkably with one-fifth of the population now over 65 years of age, and half the population over 50 years of age. Consequently the health insurance system is threatened with bankruptcy because of the sharp increase in medical expenditure. Lifestyle-related diseases including hypertension, obesity, hyper-lipidemia, cardiovascular diseases, diabetes, and cancer are increasing at least partially due to the Westernization of food habits. The National Nutrition Survey of 2001 (7) showed that more than 50% of Japanese adults are at risk for lifestyle-related diseases including mild hypertension, high blood cholesterol, or high blood sugar, but few are aware of their risk status. However, people are becoming more conscious about health and foods are recognized as a means to maintain and promote optimal health.

Foods are not only essential to maintaining the activity of normal life, but also preventing or reducing the risk of certain diseases. While food components can be effective in exerting certain physiological functions, the Pharmaceutical Affairs Law prohibits statements that express or imply that food products exert efficacy like that of drugs registered for the purpose of influencing the structure or function of the body. There is a wide interest in food as a means for health maintenance, especially in developed countries where population aging and medical expenditure are important societal issues. These societies and their governments are starting to examine how and at what level foods should be allowed to carry health benefit claims without confusing consumers. Japan is an ongoing case study with its attempts to allow health claims on food products.

19.3.2   History of Functional Foods and FOSHU

Japanese consumers consider the role of food according to the old concept of isogenicity of food and medicine. A specific study on the "Development and systematic analysis of functional foods" was initiated in Japan in 1984 with support from the Ministry of Education, Science and Culture (MESC). The study report states that food has three functions, (i) the nutrient function to maintain life or growth of the body, (ii) the taste function by some components interacting with the sensory system, and (iii) the body-enhancing function related to body defense or modification of body conditions contributing to health maintenance and prevention of diseases. The study proposed the new concept "functional foods" which focuses on the third function of food and the term functional foods immediately took root internationally (8). According to the proposal, the third function—the body-conditioning function—includes body defense (antiallergy and immune reinforcement), prevention and restoration of health (hypertension, diabetes, and metabolism deficiency in nerve, gut, and internal secretion), and protection from aging (control fat peroxidization). The academic institutions that participated in the project largely contributed to define the functional food concept, and the project generated basic understanding of food functionality and gave rise to the evaluation system for health-claim labeling on functional foods. The Ministry of Health and Welfare, formerly Ministry of Health, Labor, and Welfare (MHLW), subsequently issued an Investigative Commission Report on Functional Foods in 1990.

In 1991, the FOSHU (Foods for Specific Health Uses) system was introduced in Japan as the world's first approval system of health-claim labeling for food products. The first FOSHU product was approved in 1993. Since the Pharmaceutical Affairs Law stipulates that a product aiming at influencing function or body structure should be categorized as a drug, the legal name of functional foods changed to FOSHU, and the concept "foods with functionality" was subsequently modified to "foods with approved label." The term "functional" was reduced to "expectation for the purpose of a Specific Health Use."

The FOSHU system was revised in 1997 to eliminate the 2- to 4-year expiration, strengthen the regulation of quality control, and simplify the documentation for FOSHU application. The type of products was expanded to include tablets and capsule types as FOSHU products, and guidelines (9) for product evaluation and FOSHU application, with concrete examples of expression on Specified Health Uses, were introduced in a 2001 MHLW notice (10). In the same year, the "Food with Health Claims" system was introduced under the Food Sanitation Law (11) and FOSHU was reclassified in the category of Food with Health Claims with "Food with Nutrient Function Claims" specification besides FOSHU (11). In 2004, the MHLW started a discussion to revise the FOSHU system including the development of new rules for disease risk reduction claims and addition of new limited FOSHU category with lighter application documentation.

19.3.3   Current Status of FOSHU

FOSHU is a food containing food-originated component(s) with specific health function(s), contributing to the maintenance or promotion of human health and consumed for the purpose of Specific Health Uses. It is only after the examinations and approval of each product application by the MHLW, based on scientific evidence including the results of human studies, that a food product can indicate the Specific Health Uses such as to help maintain physiological or organic function of the body for health maintenance and promotion. A FOSHU application to the MHLW is made on a voluntary basis. Although a FOSHU application is not required for all functional foods, many food manufacturers made efforts to obtain the approvals for existing products including well-known international brands such as MILO® (Nestlé) (12) and newly developed products such as AMEAL S (CALPIS).

Figure 19.1 shows the position of FOSHU and drugs as it relates to labeling and intended use. As shown in this figure, Foods with Health Claims category are classified between drugs and Foods. Within the category of Foods with health claims FOSHU which requires prior approval for foods and related functional components and Food with Nutrient Function Claims based on a standard type of regulatory approval of some specific nutrients (10). Foods with Nutrient Function Claims are foods containing one or more nutrients at a designated level of 12 vitamins (vitamin A, D, B1, B2, B6, B12, C, E, niacin, folic acid, biotin, and pantothenic acid) and two minerals (calcium and iron), the nutritional and physiological functions of which have already been scientifically proven. Because of this, the manufacturers can indicate functionality of these nutrients on the product package when the specific nutrients are incorporated at a given level without needing to submit documents proving the claims to the MHLW. In contrast, to obtain approval

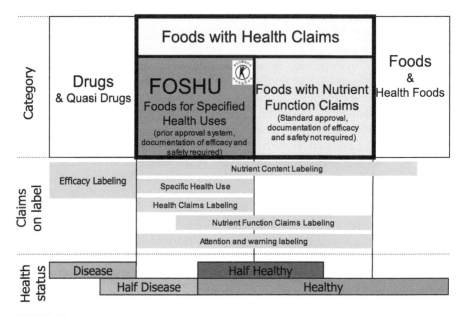

FIGURE 19.1
Food with health claims including FOSHU as of 2004.

TABLE 19.3

The Labeling Standards for Foods with Health Claims

1.	In agreement with the nutritional targets and health policy of the nation.
2.	Expressed to supply nutritional components or to contribute to specific health use (including being helpful to promote or maintain health by influencing the structure or function of the body).
3.	Scientific evidence is adequate and factually described.
4.	Clearly expressed using understandable and correct sentences or terms to convey information to consumers.
5.	Obliged to indicate attention, including appropriate intake manner for prevention of health risk from excess intake or contraindication.
6.	Comply withapplicable Laws including the Food Sanitary Law, the Nutrition Improvement Law, the former Health Promotion Law, and the Pharmaceutical Affairs Law.
7.	Clearly indicateFoods with Health Function (FOSHU or Food with Nutrient Function Claims) as to distinguish products from drugs, and from indicating on the label any referenceto diagnosis, cure, or prevention of diseases.

of For FOSHU health claims, the manufacturers should submit documentation to the MHLW containing the scientific evidence of the efficacy and
safety of the food itself and the related functional components. Once the
FOSHU application is approved, the MHLW allows inclusion of the FOSHU
logo on the product package. Table 19.3 shows the labeling standards of
Foods with Health Claims for both FOSHU and Food with Nutrient Function
Claims. While, the physiological effects of probiotics have been well-documented, probiotics are not nutrients in the classical sense. Thus, when man-

ufacturers intend to indicate the health benefit on the package, FOSHU approval should be sought from the MHLW.

As of September 27, 2004, 454 food products have been approved as FOSHU in Japan (Table 19.4). The total annual sale of FOSHU products is approximately $4 billion (US), and more than half of this figure comes from the probiotic categories. FOSHU products have several health-claim categories including "improve gastrointestinal (GI) conditions," "for those with high blood pressure," "for those with high serum cholesterol," "for those with high serum triacylglycerol," "difficult to bind body fat," "for those concerned with blood sugar level," "promote mineral absorption," "for those concerned with bone health," and "noncarcinogenic/reinforce demineralization." Of all the health claim categories, the claim "Improve GI conditions" is the most common health-claim category. Among FOSHU products carrying this claim, 58 products (17% of total FOSHU product items) are probiotics. Many existing "yogurt" brands in Japan obtained FOSHU approval with similar health claims. Probiotics contributed considerably to the development of the Japanese FOSHU market, not only in the emergence of the market itself but also the establishment of health claims and the required scientific evidence for their evaluation (Table 19.4).

19.4 Probiotics and FOSHU

19.4.1 Appearance of Probiotics with FOSHU Health Claims

Probiotics represent one of the most common food categories with health benefits. FOSHU is a voluntary system, and thus it is not necessary to obtain FOSHU approval for every probiotic with health benefits. However, it could be helpful to understand the current regulation of probiotics concerning indications of health claims.

The FOSHU health claim currently allowed for probiotic products is limited to improvement of GI conditions The claim reads "this product is produced by fermenting yogurt with xxx LAB strain, which reaches one's intestines in an active state (or viable) as to help increase intestinal bifidobacteria and lactobacilli (or good bacillus) and reduce bad microflora, and this promotes the maintenance of a good intestinal environment and regulates gastrointestinal conditions." Instead of the claim "promote the maintenance of a good intestinal environment" or "regulates the gastrointestinal conditions," some claims contain statements like "keep the intestines healthy," or "regulate the balance of intestinal microflora," "useful to improve defecation," or "suitable for those concerned about gut health".

Figure 19.2 provides an example of packaging of a probiotic product with FOSHU approval. According to MHLW rules (10), nutritional compo-

TABLE 19.4

List of FOSHU Products

Health Claim and Functional Component	Number of Product	Product Form
Improve Gastrointestinal Conditions		
Lactic acid bacteria	58	Fermented milk, lactic acid bacterium drink
Lactosucrose	25	Soft drink, jelly, cookie, table sugar
Fructo-oligosaccharides	11	Table sugar, soft drink, candy, pudding
Galacto-oligosaccharides (GOS)	8	Table sugar, vinegar
GOS + polydextrose	1	Soft drink
Soybean oligosaccharides	7	Soft drink, table sugar
Xylo-oligosaccharides	5	Soft drink, vinegar, chocolate, candy
Isomalt-oligosaccharides	3	Table sugar
Raffinose	1	Powdered soup
Lactulose	1	Soft drink
Indigestible dextrin	28	Soft drink, dessert, fish meat paste,
Psyllium husks	20	Noodle, soft drink, cereal, soft drink
Na-Arg + corn fiber	1	Soup
Hydrolyzed guar gum	4	Porridge
Wheat bran	4	Cereal food
Agar	3	Jelly
Beer yeast	1	Fermented milk
Polydextrose	2	Soft drink
Whey fermented products	1	Tablet
For those with high blood pressure		
Sardine peptides	11	Soft drink
Dried bonito oligopeptides	6	Soup, tea, tablet
Lactotripeptides	3	Lactic acid bacteria drink, tablet
Casein dodecapeptides	3	Soft drink
Eucommia leaf glycoside	2	Soft drink
For those with high serum cholesterol (CHO)		
Soy protein	18	Soft drink, hamburger, sausage, soya milk, dessert, soup
Low molecular weight sodium alginate*	9	Soft drink, soup
Chitosan	4	Biscuit
Plant phytosterol	3	Margarine, cooking oil
Soy peptide	2	Soft drink
Psyllium husks*	2	Noodle, soft drink
EPA, DHA	1	Soft drink
Difficult to cling body fat/ For those conscious of body fat / For those high serum tryacylglycerol (TG)		
Diacylglycerol (DG)	5	Cooking oil
DG + phytosterol **	4	Cooking oil
Tea cathekine	2	Tea drink
Globin protein digest	3	Soft drink
MCT	1	Cooking oil

TABLE 19.4

List of FOSHU Products (continued)

Health Claim and Functional Component	Number of Product	Product Form
For those concerned about blood sugar level		
Indigestible dextrin	30	Soft drink, tofu, soup, steamed rice,
Wheat albumin	3	Soup
Guava polyphenols	1	Soft drink
Touchi extract	1	Soft drink
L-alabinose	1	Table sugar
For those concerned about bone health / Promote mineral absorption		
Fractooligosaccharides	5	Soft drink, table sugar
Isoflavone	4	Tea drink
Vitamin K2	3	Natto (fermented soybean)
Milk basic protein	1	Soft drink
Casein phosophopeptide	3	Soft drink, chewing gum, tofu
Calcium citrate malate	2	Soft drink
Heme iron	3	Soft drink, jelly
Non-cariogenic / reinforce demineralization		
CPP-ACP	7	Chewing gum, tablet
Xylitol, Calcium phosphate, fnoran	6	Chewing gum, candy
Sugar alcohols + tea polyphenols etc	3	Chocolate, chewing gum
Maltitol	2	Candy
POs-Ca	1	Chewing gum
Total	339	

As of March 31st, 2003.

* Some products with the health claim "for those who high in blood cholesterol" are also approved the health claim "improvement of GI conditions" in the same product.

** The approved health claim is "for those high in TG and CHO, and mild obesity."

sition with the effective level of the related functional component, approved health claim, daily standard dosage, manner of intake, and caution of intake should be indicated on the packages of FOSHU products (Square B in Figure 19.2). General product information regulated by the Fair Trade Rules for Fermented Milks and LAB Drinks products must also mandatory to be indicated as an en bloc label (Square A in Figure 19.2).

19.4.2   Conditions for FOSHU Approval

An evaluation committee from the MHLW composed of specialists from medical, nutritional, food hygiene, and pharmaceutical fields examines the

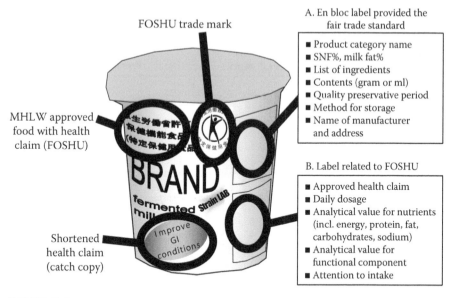

A. En bloc label provided the fair trade standard

- Product category name
- SNF%, milk fat%
- List of ingredients
- Contents (gram or ml)
- Quality preservative period
- Method for storage
- Name of manufacturer and address

B. Label related to FOSHU

- Approved health claim
- Daily dosage
- Analytical value for nutrients (incl. energy, protein, fat, carbohydrates, sodium)
- Analytical value for functional component
- Attention to intake

FOSHU trade mark

MHLW approved food with health claim (FOSHU)

Shortened health claim (catch copy)

FIGURE 19.2
Package example for FOSHU product.

FOSHU application. Conditions for FOSHU approval stipulate that the petitioner show that (i) the food and the incorporated functional component(s) are expected to improve nutrition and contribute to maintaining or promoting health, (ii) indicate the medical and nutritional facts for the health use, and (iii) if possible establish the appropriate daily dose. The final product and the "related functional components" in the food should be confirmed as safe upon normal dietary consumption (e.g., not consumed rarely but rather on a daily basis). To differentiate FOSHU products from drugs, the food and the component claims should not correspond to those for "medical use" listed in the *Pharmacopoeia*. The active component should be described in terms of the physical, chemical, and biological characteristics, as well as the methods of qualitative and quantitative tests for evaluation of the efficacy and quality assurance. The manufacturing process related to quality assurance should also be disclosed to the MHLW.

Generally speaking, a human study conducted according to good clinical practice (GCP), in collaboration with a third party, should be carried out to evaluate health benefits of final products. The results of the study should be published in a peer-reviewed scientific journal to ensure the validity of the results from an objective and scientific points of view. The subjects of the human study should be Japanese. However, human study results in non-Japanese subjects can be utilized for a FOSHU application when the results are applicable to Japanese based on similarities to dietary lifestyle of Japanese (e.g., oral care).

19.4.3   Scientific Evidence of Probiotics in FOSHU Health Claims

Research on intestinal microflora undertaken in the 1980s in Japan generated important findings suggesting that the intestinal microflora is strongly related to the health of the host and that bifidobacteria play a key role as beneficial bacteria. Earlier, Mitsuoka (13) established methods to culture anaerobic intestinal bacteria including bifidobacteria and investigated the ecology of the intestinal microflora. The hypothesis on the relationship between intestinal microflora and health of the host (13) (Figure 19.3) was introduced in 1969 and since then, the influence of microflora on human health has been investigated according to this hypothesis. The composition of intestinal microflora has been shown to change according to age (14) as in Figure 19.4. Based on these pioneer studies, many other studies on the efficacy of probiotics were conducted (15–23) significantly contributing to better understanding of the functionality of probiotic foods and providing the scientific basis for evaluation of the function and health benefits of FOSHU products. Before the introduction of term "probiotics," the concept of health benefit of oral intake of LAB had already been widely known to Japanese researchers and the food industry. Today, the health claim "improve GI conditions" is the most popular claim in the FOSHU product category. This claim is not only used for probiotics but also for prebiotics which are defined as "nondigestible food ingredients that beneficially affect

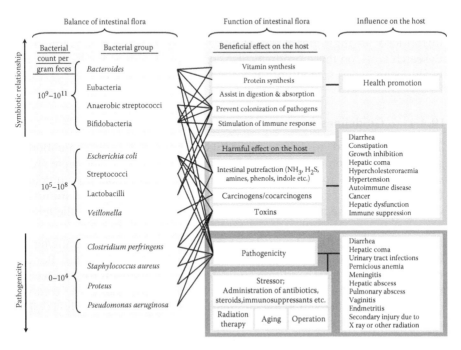

FIGURE 19.3
Interrelationship between intestinal flora and the human body (Reference 13).

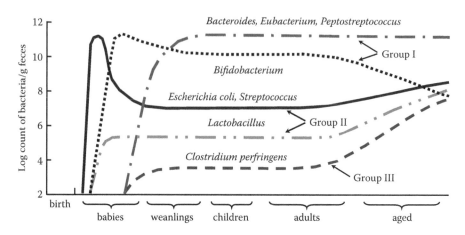

FIGURE 19.4
Changes in the fecal flora with increased age (Schematic model from Reference 14.)

the host by selectively stimulating the growth and/or activity of one or a limited number of bacterial species already resident in the colon and thus attempt to improve host health (24)." Indigestible oligosaccharides (OSs) and dietary fibers are examples of prebiotics. Numerous kinds of the commercial OS were developed by Japanese industries (e.g., fructooligosaccharides, galactooligosaccharides, xylooligosaccharides, soybean oligosaccharides, lactosucrose, isomaltooligosaccharides, and raffinose) and introduced as ingredients for FOSHU products shown in Table 19.4. Research work on these products also contributed knowledge based on the health benefits associated with these FOSHU products (25–30).

19.4.4   Guidelines for FOSHU Application and Evaluation

The MHLW issued guidelines for FOSHU applications and evaluation of some functional components including OSs in 1998 (10). While there are no written guidelines for probiotics in the FOSHU application, the guidelines for an OS provide a good guide on how to establish evidence on "improving GI conditions." The health benefit of OSs is basically exerted by the following mechanism: an orally fed OS that reaches the colon, where beneficial microorganisms in the colon such as bifidobacteria multiply by utilizing the OS, harmful bacteria are suppressed, the balance of intestinal flora is improved and the fecal condition and defecation are then improved. Overall, these effects can be expressed as GI condition improvement. The contents of the guidelines for an OS are summarized in Table 19.5. Briefly, FOSHU applicants should show that the orally fed OS is stable under digestive conditions in an animal model and selectively utilized by bifidobacteria or LAB *in vitro*. Based on this data , human intervention studies using the final product should be conducted to evaluate the efficacy showing improvement

TABLE 19.5

Guideline for Evaluation of FOSHU Application for Products with Oligosaccharides (OS)

Test Items	Requirements
Efficacy Confirmation	
In Vitro and Animal Study on the OS	
1) *In vitro* digestive study	Indigestiblity in the intestinal conditions with digestive juice
2) *In vitro* utilization test by intestinal bacteria	Selective utilization by bifidobacteria or LAB
3) Animal study	Capability to reach into the colon without digestion
Human Study Using Final Product	
1) Intestinal flora analysis	Improvement of intestinal flora to show with increase in beneficial bacterial (%) like bifidobacteria and decrease in harmful bacteria (%) like *Clostridium perfringens*
2) Defecation frequency	Increase in subject with mild constipation
3) Fecal conditions or intestinal environment	Improvement of the quantity, hardness, color or shape of fecal sample, or decrease in pH or putrefactive compounds like ammonia
Safety Confirmation	
1) Acute, sub-acute, sub-chronic toxicity tests*	No side effect in rodent
2) *In vitro* mutagenesis test*	Negative
3) Maximum ineffective dose test in human	No temporary diarrhea
4) Excess dose study on final product in human	No side effect on general health status

* Toxicity test can be excluded when the related functional component has adequate dietary history in humans.

in intestinal flora, intestinal environment, fecal conditions, and defecation frequency and proof of safety.

Scientific evidence for the health claim "improve GI conditions" by probiotic foods should be demonstrated through data indicating improvement of intestinal flora balance, increase in fecal bifidobacteria and, ideally, decrease in *Clostridium perfringens*, lower bacterial metabolite production, and increased defecation frequency in humans. According to a recent discussion of the FOSHU evaluation committee at the MHLW, the capability of probiotic bacterial strains to reach the human intestines as viable bacteria is an important evidence to be proven. Basic knowledge of efficacy in improving GI conditions is obtained from living probiotic bacteria. The probiotic strain is then considered functional component in FOSHU product and as such should be qualified by viable count. If all the strains die in the stomach, the functional component is reduced to dead bacteria or bacterial residue without the capacity to "improve GI conditions" as stated in the FOSHU claims. Thus, such claims should be backed by proof of probiotic viability in the GI tract. Evaluation of FOSHU applications by the MHLW is conducted on a product-by-product basis, and there has been inconsistency of approval criteria in the past. As a result, there are some FOSHU products that were previously approved with bacterial strains incapable to reach the intestines as living bacteria (e.g., *L. delbrueckii* subsp. *bulgaricus* in Meiji Bulgaria Yogurt approved in 1998). To address this issue, it is now recommended that a human trial be conducted in the form of a randomized, double-blind placebo-controlled study, normally performed as a crossover design with 3- by 3-week intake periods. However, some FOSHU products were previously approved based on open studies of efficacy in humans (19) with or without placebo (15, 16, 20–22). The FOSHU regulation itself has been revised several times according to experiences with the FOSHU system and is being updated incrementally. The evaluation criteria are expected to be more standardized in the future to ensure fair and consistent product evaluation.

19.4.5   Dietary Exposure and Safety

The safety of foods is very important, especially for FOSHU products since the MHLW carefully evaluates them for safety. Drugs are normally taken by patients under the control of physicians who know the precise dosage and their side effects based on which the patient takes the drug. Foods are typically consumed by people *ad libitum* without any control. Therefore, the MHLW requires that applicants show safety data including normal food intakes of the FOSHU food or its components at the standard daily dosage level based on sound scientific evidence. Safety of FOSHU food or the food components is indicated by evidence such a long history of safe dietary intake, data showing similarity of dynamics in the body of previously approved food components, or *in vivo* and *in vitro* study results on acute, subacute, and subchronic toxicity and mutagenicity. The dietary history may be typically that 10,000 people have consumed the food for 30 to 100 years. When the food does not have enough history of intake in the diet, the food safety should be documented with the same safety data as those of food additives which are separately regulated by the Food Sanitation Law. According to the MHLW guidelines for FOSHU evaluation on OSs, the excessive-intake test at 3- to 5-times the daily standard dosage should be applied for safety confirmation of the final product. Most LAB species used as probiotics are safe, classified as GRAS (generally recognized as safe) with a long history of consumption (31, 32). The FOSHU applicants for probiotic products generally submit safety evidence including a human study with threefold excessive intake for 2 weeks using the final product and an acute toxicity study in rodents. From 2004, the Food Safety Commmission was

given and access to regulate the safety of the FOSHN products independently from the MHLW evaluation committee, when the FOSHU product
contains new functional component. Safety issue of food becomes more and
more inportant.

19.4.6   Approval Process

FOSHU application documents include the items shown in Table 19.6. Copies
of articles on the scientific research results demonstrating efficacy and safety
published in scientific journals should be attached to the application. A copy
of the unfolded printed package should also be attached to the documents
for evaluation. The applicants must submit the documents to the MHLW
and the local health center. After a hearing involving the applicants and the
MHLW staff, two evaluation committees composed of specialists hold closed
meetings without attendance of the applicants. Before completing the
approval process, the National Institute of Health and Nutrition confirms
the amount of functional components in the final product using the evaluation method specified by the applicant in the FOSHU application documents. In the case of probiotics, the related functional components are the
bacterial strains, whose specific methods of quantification should be developed before the FOSHU application. The methods for probiotic strain evaluation include the identification procedure using a specific selective media
or a polymerase chain reaction (PCR) method using specific DNA probes. It
takes at least 6 months for the MHLW to approve a product as FOSHU
through this evaluation process. In addition, assesment process by the newly
estabished Food Safety Commission makes the approval process longer.

19.5   Efficacy of Probiotics and Future Health Claims

19.5.1   Efficacy of Probiotics

Probiotics not only have the health benefit of "improvement of GI conditions" but also have other health benefits such as those shown in Figure 19.5.
Probiotic strains used in fermented milks and LAB drinks in Japan and their
documented efficacy in healthy humans are listed in Table 19.7. Recently, the
positive effects of probiotics on the immune systems were highlighted by
some research groups for various probiotic strains. The GI tract which has
a large surface area (approximately $400 \ m^2$ in humans) is composed of thin
and soft layers that not only absorb nutrients efficiently but also create humid
and warm environment that promote bacterial growth. Pathogens attack the
mucus membranes of the GI tract, where about 80% of infections occur. As

TABLE 19.6

Items for FOSHU Application Documents

1	Name of the applicant (representative) and the address
2	Name and address of head office and factory
3	Product name
4	Shelf life
5	Content amount
6	Reason for seeking approval and how the intake contributes to the improvement of one's diet and the maintenance/enhancement of health of the entire population
7	Health claims labeling seeking approval
8	List of ingredients and composition
9	Manufacturing process
10	Profile for nutrients and energy
11	Daily amount of intake
12	Considerations and precautions at intake
13	Instructions for preparation, storage, or intake of the product
14	Others
15	Attachments

      a   The articles on an association with corporate seal

      b   Package sample

      c   Explanation how the product how the intake contributes to the improvement of one's diet and the maintenance/enhancement of health of the entire population, daily amount of intake, and considerations and precautions at intake

      d   Summary of scientific literatures filed for each item including Specific Health Uses

      e   List of scientific literatures

      f   Certificate of nutrient analysis and energy

      g   Record of qualification and quantification for the related functional component in the food, and the method for the measurement

      h   Documents for quality control

      i   Reason why some documents are not attached (if necessary)

      j   Scientific literatures for each item including specific health use and safety

a result, the GI tract is one of the largest body organs with immune functions. An enormous amount of food and bacterial antigens are present at the epithelium daily and immunological stimulation of the GI tract plays a large role in maintaining the immune functions of the host. Probiotics are bacteria which can naturally work as stimulators of the immune system of the host while having symbiotic characteristics to the host. Some lactobacilli and bifidobacteria were shown to activate natural immunity including phagocytotic cells (35) and natural killer (NK) cells (41) and acquired immunity to

TABLE 19.7

Documented Efficacies in Healthy Humans on Probiotic Strains in Food Sold in Japan

Probiotic Strains	Product Name in Japan	Improve GI conditions Microflora and Defecation	Immune Reinforcement	Regulate Helicobacter pylori and Gastrititis	Diarrhea Prevention	Improve / Prevent Atopic Dermatitis	Ref.
Lactobacillus delbrueckii ssp. bulgaricus 2038 (LB81 strain)	Meiji Blugaria Yogurt Lb81 *	▲▲					15
Bifidobacterium longum BB536	Bifidus Plain Yogurt *	▲▲					16
Lactobacillus gasseri OLL2716 (LG21 LAB)	Meiji Probio Yogurt Lg21			▲▲			33
Lactobacillus johnsonii La1 (LC1 LAB)	Nestlé LC1 Yogurt	▲▲†	▲▲†	▲▲†			17,34-37
Bifidobacterium animalis DN-173 010	Danoe Bio	▲▲‡					18
Lactobacillus gasseri SBT2055, Bifidobacterium longum SBT2928	Nature Pro GB *	▲▲					19
Lactobacillus rhamnosus	Yogurt E-GG! *	▲▲			▲▲†	▲▲†	20,38-40
Lactobacillus casei Shirota YIT9029	Yakult *	▲▲	▲▲				21,41
Bifidobacterium breve YIT4006 (?)	Milmil *	▲▲					22
Bifidobacterium lactis Bb12	Nan F **	▲▲	▲▲		▲▲†	▲▲†	23,40,42,43

The results in healthy human subjects have been published in scientific journal(s).

† Efficacy confirmation study was performed with a randomized double blind placebo controlled design.

‡ A double blind study was conducted on colonic transient time, which is not corresponded to the Japanese standard to evaluate GI conditions.

* FOSHU approved with health claim of improvement of gastrointestinal conditions.

** The same type of follow-up formula are currently sold outside Japan.

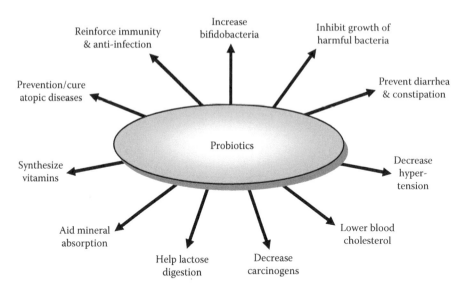

FIGURE 19.5
Beneficial effect of probiotics.

enhance systemic (34) and intestinal IgA (42) in healthy humans. An increase in IgA in breast milk is also observed in mice fed probiotics (44, 45), There is also evidence that probiotic strains do not induce inflammatory reactions, even when the strain reinforced the host immunity (31, 46).

Some probiotic strains may decrease the incidence of infectious diseases, at least partially through modulating or reinforcing immune functions of the host. For instance, prevention against rotavirus diarrhea was observed in infants who were fed a formula with bifidobacteria (43) and therapeutic intake of *Lactobacillus* was found to improve diarrheal conditions in infants (38). Influenza virus was reduced by *lactobacillus* feeding in infected mice (47). Improvement of gastric inflammation with decreases in the number of *Helicobacter pylori* by probiotics was also observed in humans (33, 36, 37). The capability to adhere to the epithelium or intestinal mucus was found in some probiotic strains *in vitro*. These characteristics may contribute to prevent infectious bacterial invasion through competitive exclusion at the epithelium and/or stimulation of the immune functions of the host (48, 49).

The possibility of using probiotics against disease risk or conditions has been reported not only for infectious diseases but also for other diseases. For instance, atopic dermatitis in infants was improved by preventive or therapeutic intake of probiotics (39, 40). In an animal model, serum IgE level decreased following probiotic intake (50). Probiotics could cause reduction in carcinogen levels in the intestines (51), and a decrease in recurrence of superficial bladder cancer in humans (52). There is also preliminary evidence that probiotics may reduce the level of blood sugar and cholesterol, and hypertension (53–55).

The FAO/WHO Expert Consultation group reported that there is emerging evidence indicating that probiotics can be taken by healthy people as a means to prevent certain diseases and modulate host immunity. The report recommended that probiotic products shown to confer defined health benefits on the host should be permitted to describe these specific health benefits (56).

19.5.2   Future Health Claims

Japan has already established a FOSHU regulatory framework to address health claims of foods and their related functional components. However, the approved health claims are currently limited to "improvement of GI conditions" in the case of probiotics. The establishment of this health claim has led to many developments: (i) A large body of scientific research on intestinal microflora and its role in health promotion is now available and widely accepted in the academic field, (ii) The criteria of the health claims were established based on well-accepted scientific evidence, (iii) Many probiotic and prebiotic products were introduced by the industry, thereby enhancing the understanding and usage of these products by consumers, (iv) Probiotic benefits to the GI condition is very understandable and obvious to consumers, (v) The health claim is acceptable to government regulators since the probiotic efficacy is not directly exerted on the body structure or function, as is the case in the category of drugs, but on intestinal flora, an external system of the body. It is evident that probiotics have functions beyond the improvement of GI conditions as suggested by new scientific evidence in humans where emerging evidence include protection from pathogens, promotion of natural defense function, such as boosting of the host immune system. Advanced health claims of probiotics are expected to be approved in the future once the efficacy is properly documented based on the scientific evidence. Regulations could then be revised to include new probiotic health claims beyond the general claim of "improvement of GI condition."

19.5.3   International Harmonization

The international harmonization program led by a Joint FAO/WHO Codex Alimentarius Commission is ongoing, not only to define the categories of fermented milks, but also to create the Guidelines for Use of Nutrition and Health Claims (57). Legislation of health claims and nutrient function claims in Japan leads this worldwide activity. According to the recent draft guideline, three types of claims are listed, including "Nutrient Function Claims," "(Other) Function Claims," and "Reduction of Disease Risk Claims" Of these, the first two types of claims correspond to FOSHU and "Foods with Nutrient Function" claims already approved in Japan. An expert committee on drugs and foods convened by the MHLW recently reported that the "Reduction of Disease Risk Claims" can be allowed in some specific areas, such as osteoporosis. New rules on this claim expansion under FOSHU system are in preparation. This would make the third type of international health claims allowable in Japan.

19.5.4   Regulation on Drugs and Foods

Distinction between food and drugs is always a major issue in food regulations and health claims. The Food Sanitation Law defines foods as all substances to be eaten or drunk excluding drugs and quasidrugsas described in the Pharmaceutical Affairs Law. The latter defines a drug as (i) a substance listed in the *Japanese Pharmacopoeia*, (ii) a substance meant for use in diagnosis, cure, or prevention of diseases in humans or animals, excluding implants and instruments, and (iii) a subtance aimed at influencing the structure or function of the body in humans or animals, excluding implants and instruments. The interpretation of whether the substance corresponds to a drug is totally dependent on the active component, shape (shape of drug, container, package, and design), and the indication including the objectives of usage, efficacy, manner of intake, and sale procedure. A substance with expression of efficacy can only be sold in the market as a drug with prior official approval.

Aging and medical expenditure increases are common social issues facing industrialized societies, which are starting to look to food as means to overcoming/reducing the impact of these issues. Japan has established FOSHU as an advanced system allowing health claims on food products. While FOSHU is gradually becoming popular with Japanese consumers, the awareness of the FOSHU labeling system is still below 50%. Even though functional food is a more understandable and popular term among consumers than FOSHU, the MHLW decided not to use the term "functional" but "Specific Health Uses" as the official term in the FOSHU system. The rationale for such a decision is that the term "functional" may imply a medical effect which is not acceptable to the MHLW that also regulates pharmaceutical products. MHLW is more prone to use "efficacy" or "effect" than "function," which is generally reserved to drugs. It is unfortunate that legal consistency of terms could sometimes cause disservice to consumers, in this case disappearance of the familiar category name "functional food" as an official name.

The health-claim system is continuously being improved in Japan. In 2001, MHLW has slightly relaxed the regulation on product shape as to allow tablet or capsule types of FOSHU products. Although FOSHU refers to food, and thus is not allowed to express "diagnosis," "prevention," and "cure" on the product package, some new expressions/terms are being considered under a FOSHU regulation recently communicated by MHLW (10). These terms are about measurable biomarkers, maintenance of physiological function, or short-term changes in physical conditions such as fatigue (Table 19.8). A Health Promotion Law was introduced in 2003 under which the project "Healthy Japan 21," aimed at promoting a healthy life of

TABLE 19.8

Possible Expression on FOSHU Product

	\<Possible Expression\>	\<Impossible Expression\>
A. Biomarkers on body conditions, which are easily measurable, to be maintained or improved. (Biomarkers measurable by self-check or annual medical examination)	This product contains component (or as main component), thereby	This product improves high blood pressure (hypertension).
	This product helps to maintain blood pressure (or blood sugar level, neutral fat, cholesterol) normal.	The marker should not be directly related to improvement of symptoms or diseases.
B. Physiological functions and organic functions of the body to be satisfactory maintained or improved.	This product satisfactorily maintains (or helps to improve) defecation.	This product is effective on detoxication or promotion of fat metabolism.
	This product heightens (or promotes) absorption (or deposit) of calcium	Apparently related to improvement of diseases.
C. Changes in physical conditions whose body status is subjective and momentary but not continues nor chronic be improved.	This product is suitable (helpful) for people those who feel physical fatigue.	This product is helpful for prevention of aging.
		Scientific evidence is inarticulate.

* MHLW notice in 2001.

Japanese. FOSHU is now regulated under the Food Sanitary Law instead of the Nutrition Improvement Law providing a basis for the new Health Promotion Law. It is anticipated that the MHLW will utilize the FOSHU system to contribute to "Healthy Japan 21" project and address legal hurdles pertaining to the differentiation between drugs and foods.

The regulations for drugs and the other products were originally established to avoid misleading consumers and ensure accurate product labeling. Unfortunately, there are many food products using labels with unapproved expressions/claims regarding efficacy and without adequate scientific evidence. On the other hand, foods like probiotics "Functional yogurts" were introduced by the food industry as health promoting products. Functional yogurts marketed to control gastric *Helicobacter pylori*, prevent food poisoning, control atopic diseases, provide immune reinforcement, or help in prevention of cancer, became very popular in Japan since 2002. Many of these products are sold without FOSHU approval of their health claim because of limitations of the content of the official claim permission. However, TV programs and news media provide prominent coverage of the efficacy of probiotic strains and their potential in health promotion. Thus, consumers who have great interest are likely to know the health benefit of probiotics through media outlets and opt to consume them even without approved health claims. We expect that the regulations will evolve in the future to catch up with consumer trends and take the opportunity to promote consumer health through consumption of popular but beneficial foods such as probiotics.

19.6 Conclusion

Although the Japanese have a shorter history of consumption of dairy products such as yogurt than Western countries, Japan classified LAB and fermented milk as beneficial foods relatively quickly leading to the development of a large probiotics market. The legal definition and standards of products with LAB were established over time in Japan. The legal category, especially for LAB drinks, will change according to ongoing international harmonization process of the Codex. The latter will likely be influenced by the Japanese FOSHU concept given the fact that Japanese LAB drink products and similar types of drinks developed outside Japan are now available in the European and Asian markets.

Many pioneering scientific studies on intestinal microflora have been conducted in Japan. As a result, many beneficial LAB and indigestible oligosaccharides were introduced by the Japanese food industry to improve intestinal flora before the scientific term "probiotics" or "prebiotics" were introduced to consumers. The concept of "functional food" started in Japan as a result of a governmental project and with participation of specialists from academic fields. The concept of functional food sub sequentially gained international acceptance, especially among Japanese consumers who are strong believers in the concept that "food and medicine are isogenic." FOSHU, the world's first health-claim approval system for food products was adopted in Japan and many probiotic products have been approved for "improved GI conditions." Probiotics use has been growing in Japan due to the unique history of Japanese with these products and a sustained research and development effort by both the government and the food industry in Japan. An advanced regulation system like FOSHU was only made possible by the collaboration between industry, government, and academia.

Probiotics, defined as bacteria that when orally ingested exert health benefits to the host, are expected to contribute greatly to health promotion in humans. Recently, scientific evidence on probiotics, including reinforcement of immunity and prevention of atopic diseases or infectious diseases has emerged. Advancement of food regulations by progress in research, international harmonization, and education on the function of probiotics would increase the understanding of the health benefits of probiotics to consumers. This would enhance the contribution of probiotic foods as a means for enhancing human health.

Acknowledgments

We would like to express our deep gratitude to Prof. Emeritus T. Mitsuoka (University of Tokyo), Prof. S. Kaminogawa (Nihon University), and the Applied Science Board for Lactic Acid Bacteria for providing much information and suggestions on research works pertaining to probiotics. We also thank Prof. A. Hosono (Shinshu University), Dr. D. Barclay (Nestec Ltd.), and Dr. J. J. Chen (Nestlé Japan Ltd.) for giving us kind advices during the preparation of this chapter.

Abbreviations

CAC: Codex Alimentarius Commission

FOSHU: Food for Specified Health Uses

GI: Gastrointestinal

JAS: Japan Agricultural Standards

LAB: Lactic acid bacteria

MAFF: the Ministry of Agriculture, Forestry and Fisheries

MESC: the Ministry of Education, Science and Culture

MHLW: the Ministry of Health, Labor and Welfare

OS: oligosaccharide

PCR: polymerase chain reaction

SNF: solid nonfat

References

1. R Fuller. Probiotics in Human Medicine. *Gut*, 32:439–442, 1991.
2. Joint FAO/WHO Working Group. Guidelines for the Evaluation of Probiotics in Food. May 2002. . http://www.who.int/foodsafety/publications/fs_managementprobiotics2/en
3. E Metchnikoff. *The Prolongation of Life*, Heinemann, London, 1907.
4. Ministerial Ordinance concerning Compositional Standards, etc. for Milk and Milk Products. Ministry of Health and Welfare Ordinance. No: 52, 1951; Last amendment: Ministry of Health, Labor and Welfare Ordinance. No: 5, October 2001 (In Japanese).
5. Fair Competition Rules on the Label of Fermented Milk and Lactic Acid Bacteria Drinks. Revised and approved by the Fair Trade Commision, 2001 (In Japanese).
6. Joint FAO/WHO Food Standards Program, CODEX Alimentarius Commission, 26th Session (Rome), 30 June–7 July, 2003. Discussion on Draft Revised Standard for Fermented Milks (ALINORM 03/11–Appendix III p. 40–44).
7. The National Nutrition Survey in Japan. Ministry of Health, Labor and Welfare. 2001 (In Japanese).
8. Arai, T Osawa, H Ohigashi, M Yoshikawa, S Kaminogawa, M Watanabe, T Ogawa, K Okubo, S Watanabe, H Nishino, K Shinohara, T Esaki, T Hirahara. A mainstay of functional food science in Japan – history, present status, and future outlook. *Biosci. Biotechnol. Biochem.*, 65:1–13, 2001.
9. Ministry of Health and Welfare. Notice: Guideline for Application and Evaluation on FOSHU. April 1998.
10. Ministry of Health, Labor, Welfare, Notice on Treatment of "Nutritional Supplement" May 2001.
11. Food Sanitation Law (Law No. 233, 1947, latest amdenment Law No. 55, May 2003, In Japanese).
12. K Uenishi, A Ohta, Y Fukushima, Y Kagawa. Effect of a malt drink containing fructooligosaccharides on calcium absorption and safety of long-term administration. *Jpn. J. Nutr. Diet*, 60:11–18, 2002.
13. T Mitsuoka. Intestinal flora and the host. *Pharmacia*, 5:608–609, 1969.
14. T Mitsuoka, K Hayakawa. Die Faekalflora bei Menchen. I. Mitteilung: Die Zusammensetsung der Faekalflora der verschiedenen Altersgruppen. *Zentralbl. Bakteriol. Parasitenk. Infektionskr. Hyg. I Abt. Orig. Reihe A*, 223:333–342, 1972.
15. H Hara, A Terada, M Takahashi, T Kaneko, T Mitsuoka. Effects of yoghurt-administration on fecal flora and putrefactive metabolites of normal adults (in Japanese with English abstr.). *Bifidus*, 6:169–175, 1993.

16. T Ogata. Effect of *Bifidobacterium longum* BB536 yogurt administration on the intestinal environment of healthy adults. *Microb. Ecol. Health Dis.*, 11:41–44, 1999.

17. T Yamano, M Takada, Y Fukushima, H Iino. Effect of fermented milk containing *Lactobacillus jonsonii* La1 on fecal flora and fecal conditions in healthy young women (in Japanese). *J. Intest. Microbiol.*, 18:15-23, 2003.

18. M Bouvier, S Meance, C Boulev, JL Berta, JC Grimaud. Effects of consumption of a milk fermented by the probiotic *Bifidobaterium animalis* DN-173 010 on colonic transit time in healthy humans. *Biosci. Microflora*, 20:43–48,2001.

19. A Kinumaki, Y Obara, O Kasuga, M Onosawa, H Misawa, K Takagi, E Chiba, K Tani, Y Suzuki, Y Benno. Effects of a new fermented milk prepared using *Lactobacillus gasseri* SBT2055 and *Bifidobacterium longum* SBT2928, both of human origin, on the defecation frequency and fecal microflora of healthy human adults (in Japanese with English abstr.). *Jpn. J. Lactic Acid Bacteria*, 12:92–101, 2001.

20. M Hosoda, F He, M Hiramatsu, H Hasimoto, Y Benno. Effects of *Lactobacillus* GG strains intake on fecal microflora and defecation in healthy volunteers. *Bifidus*, 8:21–28, 1994.

21. R Tanaka, M Ohwaki. A controlled study of the effects of the ingestion of *Lactobacillus casei* fermented milk on the intestinal microflora, its microbial metabolism and the immune system of healthy humans. In: *Intestinal Flora and Diet* (in Japanese with English summary), Proc. of 12th ICPR Symposium on Intestinal Flora, T. Mitsuoka, Ed., Japan Scientific Societies Press, Tokyo, 1991, pp. 85–142.

22. R Tanaka, T Kan, H Tejima, T Kuroshima, S Kodaira, S Suzuki, T Terashima, M Mutai. Bifidobacterium Interplantation –intake for *B. bifidum* 4006 or *B. breve* 4007 (in Japanese). *Jpn. J. Clin. Pediatr.*, 33:2483–2492, 1980.

23. Y Fukushima, S Li, H Hara, A Terada, T Mitsuoka. Effect of follow-up formula containing bifidobacteria (NAN BF) on fecal flora and fecal metabolites in healthy children. *Biosci. Microflora*, 16:65–72, 1997.

24. GR Gibson, MB Roberfroid. Dietary Modulation of the human colonic microbiota: Introducing the concept of prebiotics. *J. Nutr.*, 125:1401–1412, 1995.

25. H Hidaka, T Eida, T Takizawa, T Tokunaga, Y Tashiro. Effects of fructooligosaccharides on intestinal flora and human health. *Bifidobacteria Microflora*, 5:37–50, 1986.

26. Y Bennno, K Endo, N Shiragami, K Sayama, T Mitsuoka. Effects of raffinose intake on human fecal microflora. *Bifidobacteria Microflora*, 6:59–63, 1987.

27. M Ito, Y Yamaguchi, A Miyamori, K Matsumoto, H Kikuchi, K Matsumoto, Y Kobayashi, T Yajima, T Kan. Effect of administration of galactooligosaccharides on the human faecal microflora, stool weight and abdominal sensation. *Microb. Ecol. Health Dis.*, 3:285–292, 1990.

28. M Okazaki, S Fujikawa, N Matsumoto. Effect of xylooligosaccharide on the growth of bifidobacteria. *Bifidobacteria Microflora*, 9:77–86, 1990.

29. H Hara, S Li, M Sasaki, T Maruyama, A Terada, Y Ogata, K Fujita, H Ishigami, K Hara, I Fujimori, T Mitsuoka. Effective dose of lactosucrose on fecal flora and fecal metabolites of humans. *Bifidobacteria Microflora*, 13:51–63, 1994.

30. K Hojo, N Yoda, H Tsuchita, T Ohtsu, K Seki, N Taketomo, T Murayama, H Iino. Effect of ingested culture of *Propionibacterium freudenreichii* ET-3 on fecal microflora and stool frequency in healthy females. *Biosci. Microflora*, 21:115–120, 2002.

31. S Salminen, A von Wright, L Morelli, P Marteau, D Brassart, WM de Vos, R Fonden, M Saxelin, K Collins, G Mogensen, S-E Birkelan, T Mattila-Sandholm. Demonstration of safety of probiotics - a review. *Int. J. Food Microbiol.*, 44:93–106, 1998.

32. G Mogensen, S Salminen, J O'Brien, A Ouwehand, W Holzapfel, C Shortt, R Fonden, GD Miller, D Donohue, M Playne, R Crittenden, B Bianchi-Salvadori, R Zink. Food microorganisms - health benefits, safety evaluation and strains with documented history of use in foods. *IDF Bull.*, 377:4–19, 2002.

33. I Sakamoto, M Igarashi, K Kimura, A Takagi, T Miwa, Y Koga. Suppressive effect of *Lactobacillus gasseri* OLL2716 (LG21) on *Helicobacter pylori* infection in humans. *J. Antimicrob. Chemother.*, 27:709–710, 2001.

34. H Link-Amster, F Rochat, KY Saudan, O Mignot, JM Aeschlimann. Modulation of a specific humoral immune response and changes in intestinal flora mediated through fermented milk intake. *FEMS Immunol. Med. Microbiol.*, 10:55–63, 1994.

35. EJ Schiffrin, F Rochat, H Link-Amster, JM Aeschlimann, A Donnet-Hughes. Immunomodulation of human blood cells following the ingestion of lactic acid bacteria. *J. Dairy Sci.*, 78:491–497, 1995.

36. P Michetti, G Dorta, PH Wiesel, D Brassart, E Verdu, M Herranz, C Felley, N Porta, M Rouvet, AL Blum, I Corthesy-Theulaz. Effect of whey-based culture supernatant of *Lactobacillus acidophilus* (*johnsonii*) La1 on *Helicobacter pylori* infection in humans. *Digestion*, 60:203–209, 1999.

37. CP Felley, I Corthesy-Theulaz, JL Rivero, P Sipponen, M Kaufmann, P Bauer-feind, PH Wiesel, D Brassart, A Pfeifer, AL Blum, P Michetti. Favorable effect of an acidified milk (LC-1) on *Helicobacter pylori* gastritis in man. Lippincott Williams & Wilkins, New York, NY, 13:25–29, 2001.

38. H Szajewska, JZ Mrukowicz. Probiotics in the treatment and prevention of acute infectious diarrhea in infants and children: a systematic review of published randomized, double-blind, placebo-controlled trials. *J. Pediatr. Gastroenterol. Nutr.*, 33:S17–25, 2001.

39. M Kalliomäki, S Salminen, H Arvilommi, P Kero, P Koskinen, E Isolauri. Probiotics in primary prevention of atopic disease: a randomized placebo-controlled trial. *Lancet*, 357:1076–1079, 2001.

40. E Isolauri, T Arvola, Y Sutas, E Moilanen, S Salminen. Probiotics in the management of atopic eczema. *Clin. Exp. Allergy*, 30:1604–1610, 2000.

41. F Nagao, M Nakayama, T Muto, K Okumura. Effects of a fermented milk drink containing *Lactobacillus casei* strain Shirota on the immune system in healthy human subjects. *Biosci. Biotechnol. Biochem.*, 64:2706–2708, 2000.

42. Y Fukushima, Y Kawata, H Hara, A Terada, T Mitsuoka. Probiotic formula enhances intestinal immunoglobulin A production in healthy children. *Int. J. Food Microbiol.*, 42:39–44, 1998.

43. JM Saavedra, NA Bauman, I Oung, JA Perman, RH Yolken. Feeding of *Bifidobacterium bifidum* and *Streptococcus thermophilus* to infants in hospital for prevention of diarrhoea and shedding of rotavirus. *Lancet*, 344:1046–1049, 1994.

44. Y Yasui, J Kiyoshima, H Ushijima. Passive protection against rotavirus-induced diarrhea of mouse pups born to and nursed by dams fed *Bifidobacterium breve* YIT4064. *J. Infect. Dis.*, 172:403–409, 1995.

45. Y Fukushima, Y Kawata, K Mizumachi, J Kurisaki, T Mitsuoka. Effect of bifidobacteria feeding on fecal flora and production of immunoglobulins in lactating mouse. *Int. J. Food Microbiol.*, 46:193–197, 1999.

46. D Haller, C Bode, WP Hammes, EJ Shiffrin. Non-pathogenic bacteria elicit a differential cytokine response by intestinnal epithelial cell/leucocyte co-cultures. *Gut*, 47:79–87, 2000.
47. T Hori, J Kiyoshima, K Shida, H Yasui. Effect of intranasal administration of *Lactobacillus casei* Shirota on influenza virus infection of upper respiratory tract in mice. *Clin. Diagn. Lab. Immunol.*, 8:593–597, 2001.
48. MF Bernet, D Brassart, JR Neeser, AL Servin. Lactobacillus acidophilus La1 binds to cultured human intestinal cell lines and inhibits cell attachment and cell invasion by enterovirulent bacteria. *Gut*, 35:483–489, 1994.
49. AC Ouwehand, S Tolkko, S Salminen. The effect of digestive enzymes on the adhesion of probiotic bacteria *in vitro. J. Food Sci.*, 66:856–859, 2001.
50. K Shida, R Takahashi, E Iwadate, K Takamizawa, H Yasui, T Sato, S Habu, S Hachimura, S Kaminogawa. *Lactobacillus casei* strain Shirota suppresses serum immunoglobulin E and immunoglobulin G1 responses and systemic anaphylaxis in a food allergy model. *Clin. Exp. Allergy*, 32:563–570, 2000.
51. M Hosoda, H Hashimoto, F He, H Morita, A Hosono. Effect of administration of milk fermented with *Lactobacillus acidophilus* LA-2 on fecal mutagenicity and microflora in the human intestine. *J. Dairy Sci.*, 79:745–749, 1996.
52. Y Aso, H Akaza, T Kotake, T Tsukamoto, K Imai, S Naito. Preventive effect of a *Lactobacillus casei* preparation on the recurrence of superficial bladder cancer in double-blind trial. *Eur. Urol.*, 27:104–109, 1995.
53. M Tabuchi, M Ozaki, A Tamura, N Yamada, T Ishida, M Hosoda, A Hosono. Antidiabetic effect of *Lactobacillus* GG in Streptozotocin-induced diabetic rats. *Biosci. Biotechnol. Biochem.*, 67:1421–1424, 2003.
54. JZ Xiao, S Kondo, N Takahashi, K Miyaji, K Oshida, A Hiramatsu, K Iwatsuki, S Kokubo, A Hosono. Effects of milk products fermented by *Bifidobacterium longum* on blood lipids in rats and healthy adult male volunteers. *J. Dairy Sci.*, 86:2452–2461, 2003.
55. M Kawase, H Hashimoto, M Hosoda, H Morita, A Hosono. Effect of administration of fermented milk containing whey protein concentrate to rats and healthy men on serum lipids and blood pressure. *J. Dairy Sci.*, 83:255–263, 2000.
56. Joint FAO/WHO Expert Consultation. Health and Nutritional Properties of Probiotics in Food including Powder Milk with Live Lactic Acid Bacteria. October 2001. http://www.who.int/foodsafety/publications/fs_management/probiotics/en/
57. Codex Committee on Food Labeling. (Proposed) draft guidelines for use of nutrition and health claims. CCFL ALINORM 03/22 APPENDIX VII, 2002.

Index